Methods in Physiology

Published on behalf of The American Physiological Society by Springer

This series describes experimental techniques in cellular, molecular and general physiology. Each book is edited by experts in their field and covers theory and history behind the method, critical commentary, major applications with examples, limitations and extensions of each technique, and vital future directions.

Harry J. Witchel • Michael W. Lee
Editors

Technologies in Biomedical and Life Sciences Education

Approaches and Evidence of Efficacy for Learning

Springer

american physiological society

Editors

Harry J. Witchel
Department of Neuroscience
Brighton and Sussex Medical School
Brighton, UK

Michael W. Lee
Department of Medical Education
Geisel School of Medicine, Dartmouth College
Hanover, NH, USA

ISSN 2628-7471 ISSN 2628-748X (electronic)
Methods in Physiology
ISBN 978-3-030-95632-5 ISBN 978-3-030-95633-2 (eBook)
https://doi.org/10.1007/978-3-030-95633-2

Published on behalf of the American Physiological Society.

This Springer imprint is published by the registered company Springer Nature Switzerland AG
The registered company address is: Gewerbestrasse 11, 6330 Cham, Switzerland

Foreword

The idea for this book was born at Physiology 2019, The Physiological Society's annual meeting in Aberdeen, Scotland, when I heard Harry Witchel speak on "A video on the pathophysiology of heart failure for reinforcing first year medical students' learning: do video techniques that break from medical education tradition undermine the video?" The meeting included many other talks and posters on technology for teaching, ranging from virtual and paperless laboratory practical teaching to gamification, patient simulation, 3D printing, and augmented reality. As I talked with presenters, a question came to mind: what do we know about how well these technologies work?

It has taken years for the higher education community to adopt active learning strategies, with many teachers asking "where is the evidence that active learning is better than what we have always done?" It was not until Scott Freeman and collaborators published their meta-analysis (Freeman et al. 2014) showing learning gains with student-centered teaching that active learning began to infiltrate classrooms more widely. But at the same time, educators everywhere were rapidly adopting new technologies for teaching, without questioning their role in the curriculum and whether these technologies improve student learning and outcomes. How do we know that using technology is not doing our students a disservice by making them too dependent on the technology and unable to function adequately without it? What do we know about how these technologies foster or undermine desired skills and behaviors in addition to teaching content? How can we use technology best to improve learning and enhance student outcomes?

When Harry Witchel and Michael Lee agreed to take on the challenge of finding experts to address some of these questions, it was 2019, and we had no idea that a global pandemic of SARS-CoV-2 was about to disrupt everything we do in higher education. Technology that had been a novelty in education suddenly became essential—the only way we could teach. As a result, the evidence for advantages and disadvantages of educational technology in chapters that follow is more timely than ever. We hope that you enjoy exploring these essays on *Technologies in Biomedical and Life Sciences Education*.

University of Texas at Austin, Austin, TX, USA Dee U. Silverthorn

American Physiological Society eBook Committee, Rockville, MD, USA

America Physiological Society (2022), Rockville, MD, USA

Reference

Freeman, S., Eddy, S. L., McDonough, M., Smith, M. K., Okoroafor, N., Jordt, H., & Wenderoth, M. P. (2014). Active learning increases student performance in science, engineering, and mathematics. *Proceedings of the National Academy of Sciences*, *111*(23), 8410–8415. https://doi.org/10.1073/pnas.1319030111.

Acknowledgments

I gratefully acknowledge Michael Lee for all his hard work, intellectual rigor, and sense of fun on this project. We both thank Dee Silverthorn because it was her inspiration and enthusiasm, based on her unfailing insight and curiosity, that launched this international book project with the American Physiological Society (APS), and so we thank APS. Much of the early inspiration and ideas for this project came from a series of events sponsored by the Physiological Society in London, starting with a journal club meeting including Blair Grubb, Judy Harris, Prem Kumar, and Sarah Hall, plus Kenneth Langlands and Hilary MacQueen. As with all my work with the PhySoc, these events were mediated, administered, and kept on track by Christabel Stokes. I thank and apologize to all of our authors, who have genuinely stimulated me as well as graciously accepted and responded to our extensive but well-meant proddings. My collaborator Carina Westling read and test drove many of the ideas and chapters herewith. My own team of secretaries is the only reason that anything I do works at all, including Terri Desmonds, Chiara Cannata, and Paula Burton. The team at Springer has been most helpful, including Srinivasan Manavalan and Markus Späth. At BSMS, a lot of people have forbearance for my external activities including Hugo Critchley, Juliet Wright, Malcolm Reed, Andrew Dilley, and Lisa Mullen. More forbearance at a personal level came from family including JBS, PLW, and LAW. Most importantly, I thank my many wonderful students, who have taught me everything I really know about teaching.

—Harry J. Witchel, Ph.D.

To my wife Ernestine, my daughter Sofia, and my son Benjamin, thanks for enduring countless nights of me writing, in a never-ending stream, on my laptop alternating between exhilaration and depression over the course of these past 2 years. In words of Gene Fowler, "writing is easy, all you do is sit staring at a blank sheet of paper until drops of blood form on your forehead." A deep thanks to my friend and colleague Andy Butler (may another ACL be in our future); to my best friend JKZ for constantly listening to me talk about this project, incessantly and without context at times; to the APS Book Committee and to Springer, specifically Srinivasan Manavalan and Markus Späth; and to fate (specifically DUS) for providing me the opportunity of meeting and working to Dr. Harry Witchel. Not only has he been an outstanding co-editor, who has been consistently inspiring and intellectually challenging, he is the only person on the planet I can truly talk to about fonts (he genuinely laughed when I bid him "font regards"). Those conversations are only beginning. I am privileged to emerge from this project with both a colleague and a friend in Harry.

—Michael W. Lee, Ph.D.

Contents

Editors and Contributors

About the Editors

Harry J. Witchel is a discipline leader in physiology on the University of Sussex site of Brighton and Sussex Medical School (UK). His team's interdisciplinary research within the Department of Neuroscience resulted in his students winning the Best Research Paper Award at the international ECCE conference (2019); he also researches and innovates on educational topics revolving around human computer interaction and assessment. He has received over 15 teaching awards and recognitions, including three at the national level: from the British Science Association, the Physiological Society, and most recently a National Teaching Fellowship from Advance HE (2021). He is a long-standing member of the Physiological Society, where he currently serves on the Education, Public Engagement, and Policy Committee. His teaching qualification is as a senior fellow (Higher Education Academy); he received his A.B. (college) in biophysics from Columbia University (New York) and his Ph.D. from the Department of Physiology-Anatomy at the University of California at Berkeley.

Michael W. Lee received his masters' and doctoral degrees in medical science, with a focus on medical pharmacology, from the Morsani College of Medicine at the University of South Florida. This was followed by a postdoctoral fellowship at the University of Florida College of Medicine at the Shands Cancer Center. He has served as a founding faculty member at several pharmacy and medical schools where he has trained students both in the classroom and in the laboratory. He is currently an associate professor at the University of Texas at Austin Dell Medical School in departments of medical education and oncology. He is also an associate member at the Livestrong Cancer Institutes. His research interests center around delineating molecular mechanisms of therapeutic agents for cancer and on development of novel educational technology tools for enhancing basic science learning. He has received numerous awards for teaching including a Golden Apple (2011) and a Recognition of Innovation award (2011), and he has been awarded excellence in teaching awards 3 consecutive years (2019, 2020, and 2021). He received a Teaching Career Enhancement Award from the American Physiological Society in 2018. In 2020, he was awarded the designation of Distinguished Teaching Professor, following admittance into the University of Texas at Austin Dell Medical School Academy of Distinguished Educators. Most recently he was admitted to the University of Texas System Kenneth I. Shine Academy of Health Science Education in recognition for teaching and educational research excellence. He is an active member of the International Association of Medical Science Educators (IAMSE) and the American Physiological Society (APS), and he serves on the editorial board of the journals *Pharmacology Research and Perspectives, Kidney and Blood Pressure Research, and Frontiers in Pharmacology.*

Contributors

Peter Alston BPP Education Group, Manchester, UK

Sandra Arango-Caro Donald Danforth Plant Science Center, Saint Louis, MO, USA

John Barrow University of Aberdeen, Aberdeen, Scotland, UK

Cynthia Brame Vanderbilt University, Nashville, TN, USA

Andrew C. Butler Washington University in St. Louis, Saint Louis, MO, USA

Kristine Callis-Duehl Donald Danforth Plant Science Center and Saint Louis University, Saint Louis, MO, USA

Duncan Cole School of Medicine, Cardiff University, Cardiff, Wales, UK

Melanie M. Cooper Michigan State University, East Lansing, Mi, USA

Bryan Dewsbury Florida International University, Miami, FL, USA

Sheila A. Doggrell Queensland University of Technology, Brisbane, Australia

Renée M. Filius Utrecht University, Utrecht, The Netherlands

Daniel Franceschini College of Medicine, University of Central Florida, Orlando, FL, USA

Terry Gleave University of Liverpool, Liverpool, UK

Richard Guy Victoria University, Melbourne, Australia

Heeyoung Han School of Medicine, Southern Illinois University, Springfield, IL, USA

David M. Harris College of Medicine, University of Central Florida, Orlando, FL, USA

Mark Hirst The Open University, Milton Keynes, UK

Ihuoma (Yvette) Igbokwe School of Medicine, Southern Illinois University, Springfield, IL, USA

Victoria Jacoby UCLA David Geffen School of Medicine, University of California, Los Angeles, CA, USA

Christine A. Kauffman College of Medicine, University of Central Florida, Orlando, FL, USA

Philip J. Kellman UCLA David Geffen School of Medicine, University of California, Los Angeles, CA, USA

Rudi Klein Victoria University, Melbourne, Australia

Michael W. Klymkowsky University of Colorado Boulder, Boulder, CO, USA

Sally Krasne UCLA David Geffen School of Medicine, University of California, Los Angeles, CA, USA

Kenneth Langlands University of Wolverhampton, Wolverhampton, UK

Michael W. Lee Department of Medical Education, Geisel School of Medicine, Dartmouth College, Hanover, NH, USA

Kristina Lisk University of Toronto, Toronto, Ontario, Canada

Hilary MacQueen The Open University, Milton Keynes, UK

Christine Massey UCLA David Geffen School of Medicine, University of California, Los Angeles, CA, USA

John C. McLachlan University of Central Lancashire, Preston, UK

Maria Mosley School of Medicine, Southern Illinois University, Springfield, IL, USA

Maria Mylopoulos University of Toronto, Toronto, Ontario, Canada

Izabella Penier University of Central Lancashire, Preston, UK

Joseph Rathner University of Melbourne, Melbourne, Australia

Emma Rengasamy Department of Public Health and Primary Care, University of Cambridge, Cambridge, UK

Louise Robson University of Sheffield, Sheffield, UK

Puspha Sinnayah Victoria University, Melbourne, Australia

Cynthia Taylor University of Rhode Island, Kingston, RI, USA

Shelley Tischkau School of Medicine, Southern Illinois University, Springfield, IL, USA

Christopher Torrens Royal College of Surgeons, Dublin, Ireland

Sabine G. Uijl Utrecht University, Utrecht, The Netherlands

Lisa L. Walsh Donald Danforth Plant Science Center and Saint Louis University, Saint Louis, MO, USA

Emma R. Wester Donald Danforth Plant Science Center and Saint Louis University, Saint Louis, MO, USA

Harry J. Witchel Department of Neuroscience, Brighton and Sussex Medical School, Brighton, UK

Nicole N. Woods University of Toronto, Toronto, Ontario, Canada

Part I
Introduction and Educational Context

Chapter 1
Introduction: Intentional Innovation in Educational Technology and Media to Promote Students' Holistic Development

Michael W. Lee and Harry J. Witchel

> *The future is already here—it's just not very evenly distributed.*
>
> *– William Gibson (Neuromancer, 1984)*

Abstract This chapter introduces the book *Technologies in Biomedical and Life Sciences Education.* In addition to providing the précis for the other chapters of the book, it contextualizes those chapters by addressing how far technology has taken bioscience education and also speculating on where technology needs to go. Technology, particularly in conjunction with innovations in pedagogy, has profoundly influenced the experience of teaching and learning science, and examples are provided, some familiar (e.g., video lectures in flipped classroom) and some emerging. Compared to the experience, the striking outcomes of past innovations in technology and pedagogy have been more weighted toward inclusivity, opportunities, and participation rates. For the future, we propose an intentional development of educational technology and media that supports holistic development of students. This goes beyond design-centered or market-driven research and focuses on student competencies and skills known to play a role in learning and professional skills attainment. These include social, metacognitive, and noncognitive competencies, such as teamwork skills, social behaviors, empathy, and emotional regulation. The challenge is to use intentional innovation to achieve student learning of competencies, by leveraging what computers are particularly good at: timing, repetition, and vigilance. The expected result will probably involve learning with repetition, deeper learning with models, collaborative and cooperative learning, and self-regulated learning. We engage in serious discussion

---

M. W. Lee (✉)
Department of Medical Education, Geisel School of Medicine, Dartmouth College, Hanover, NH, USA
e-mail: Michael.W.Lee@dartmouth.edu

H. J. Witchel (✉)
Department of Neuroscience, Brighton and Sussex Medical School, Brighton, UK
e-mail: h.witchel@bsms.ac.uk

H. J. Witchel, M. W. Lee (eds.), *Technologies in Biomedical and Life Sciences Education*, Methods in Physiology, https://doi.org/10.1007/978-3-030-95633-2_1

on the challenges and obstacles inherent in refocusing media creation and educational technology innovation. We close with an overarching summary of the different chapters and how they fit together into this cohesive narrative.

1.1 Education Technology in the Biosciences: Practical Issues and a Research Problem

When considering a pedagogical method designed to convey knowledge to a learner, both instructors and students should ask two questions: (1) Will the students use it, and (2) does it actually promote learning and development? This book focuses on both, and neither question is trivial. The question of whether students engage with a tool or pedagogy segues immediately to another question: Which students? This leads to serious inquiries about inclusivity and equity. The question of efficacy and efficiency raises even more issues. How and why does it work? We need to understand mechanisms to know: Are all the steps in the pedagogical process necessary for it to be successful? Can it survive being stripped down? These are the kinds of practical questions that serious instructors, faculty leaders, and educational researchers need to consider. The chapters in this book critically examine teaching methods, innovative technologies, and controversial issues in bioscience education that have reshaped and reimagined a student's higher-education experience over the past 10 years.

Although this book is about the plethora of new educational technologies now available for teaching and learning, this is not a book of recipes, nor is it an advertisement for some innovative methods. This book was commissioned to (1) prompt transformative research by (2) engendering extensive critical discussion. This critical approach is needed because some new teaching processes are fads that may not have significant effect on learning (Natividad et al., 2018; Spector, 2020). Indeed, the long-term impact of using educational technologies to enhance student learning remains an ongoing subject of research (Spector, 2020). As educators we wanted to read critically about the evidence for the range of technology-supported techniques in a synoptic way. As editors, we commissioned our authors to produce such an overview, so that, for your own students, you can best decide which techniques are reliable and worth exploring. The chapters in this book address the latest research on efficacy of the most relevant and practical educational technologies and advances. Some controversial issues that we wanted to see explained are also included, and ultimately the book raises serious questions about our values. If you skim through the various chapter abstracts, you will see that the different chapter authors definitely do not trivially agree with one another. Tough questions are asked in a chapter on cheating and in a debate on whether students still need to attend classroom lectures in an era of recorded lectures. The assembled chapters provide a nuanced consideration of what researchers should be researching by asking experts what they are currently speculating about. We believe that this approach will provide an important point of departure for future research, and we are sure that it will be an excellent addition to the discourse.

This chapter will contextualize the rest of the book as a call to action for how education researchers are critical for designing the future of education, and consequently the future of society. In addition to outlining the other chapters of the book, this chapter will situate those chapters as part of the educational movement toward valuing and including intention as well as innovation, competencies as well as knowledge, and noncognitive as well as cognitive skills. Our goal is not only to have smarter students but also to have more inclusive and well-developed students that we will want to work among in the future.

1.2 The Intentional Development of Technology in Teaching and Learning of Bioscience

1.2.1 The Impact of Computers and Educational Technology on Cognitive Processes

When thinking about the approach of developing and deploying educational technology to improve student learning, a quote attributed to Steve Jobs comes to mind:

> Some people say, give the customers what they want. But that's not my approach. Our job is to figure out what they're going to want before they do. People don't know what they want until you show it to them. That's why I never rely on market research. Our task is to read things that are not yet on the page. (Isaacson, 2011)

This design-centered approach that consumer technology companies employ seems to be the pattern that education companies and academic institutions follow in regard to learning and educational technology. Conceive of a tool or technology you think students and instructors will want or may find useful, produce a product, and sell it to schools and students as something they will want, even though it has not been proven to be beneficial. This approach may work well as a business model for consumer products, but a new educational technology may not necessarily benefit the students' learning when developed in this manner (see Chap. 14 on social media). Although it screams innovation, this may not be the best approach for advancing learning and skills attainment. Alternatively, many technology companies conduct market research and use customer (i.e., students and faculty) input to guide innovation and product creation. This approach is helpful so long as instructors and students think long and hard about what the real barriers are to learning and whether or not these issues can even be solved with technology. Yet, in the case of learning, as opposed to producing a new phone or smart car, we are talking about functional benefits here rather than simple aesthetics or convenience. Learning and educational technology should be supported by evidence and grounded educational research that drives improvements in knowledge attainment, retainment, or application. The goal should be student development.

1.2.1.1 Technology Is Not Solely the Answer

In 1966, the architect Cedric Price made the provocation, "Technology is the answer, but what was the question?" The belief that science and technology is always the best way to solve everything—and that it *will* solve everything—is sometimes called technological solutionism, while critics in the humanities and social sciences have labeled it scientism; Chap. 2 (inclusivity) pleads a strong case that educators, and their financial managers should not blindly follow technological innovations without question or nuance; those who do may find themselves in the kind of ethical cul-de-sac in which science found itself during the first half of the twentieth century. Ethical contemporary scientists must learn the lessons from the enabling roles played by early scientists in eugenics, weapons of mass destruction, and fascist regimes. For educators, this means that cultural progress in our institutions is at least as important as technology. Chapter 9 on qualitative feedback also strongly argues that the relational elements of education (e.g., student-faculty research projects) are better at teaching than the transactional elements; there are more meaningful ways of teaching than "I give you a test, you give me your answers, I give you a mark." These kinds of status-based hierarchies are openly questioned in Chap. 3 (empowering students as partners), which suggests that we as academics can empower students in their own learning, so long as we are prepared to trust our students enough to relinquish some of our control. Whereas authoritarians have long used technology to control others, Chap. 3 suggests ways that we can use educational technology to provide students with a greater degree of self-determination.

As we see in the chapters of this book, in many cases (with some notable exceptions) the barriers to learning and mastery are more about the receptiveness of the learner to engaging with the material, rather than the media via which the information or skills are conveyed (Clark, 1983; Kozma, 1994). Thus, technology is not always the answer, particularly when the right question may not have even been properly formulated. Indeed, learners learn best when they are struggling and questioning their own knowledge (Bjork et al., 2013). The question may be the following: Are there technologies that can increase student engagement with the process of struggling to learn? We need tools and activities that enhance engagement and are satisfying for students (e.g., gamification or personalization) while also driving learning gains. Thus, the driving vision of educational technology development should not *only* be market research or some prescient vision of what people want. Rather, educational research that reveals how each student learns best, so learning can be individualized and inclusive, should be a driving force behind educational technology innovation. This way, the decision can be made as to whether or not a high-tech solution needs to be leveraged (or created), especially if a low-tech option exists and can be used instead. This is not to say that technological improvements for the sake of change should *not* be undertaken. Oftentimes, the "design-thinking" technological solution can in fact make tools that drive knowledge or skills attainment more accessible, more visually appealing, sometimes in a more cost-effective manner. However, from an educator's perspective, rather than prioritizing product development for profit's sake and user satisfaction, innovation in an educational space needs to be carefully, deliberately, and intentionally thought out in a way that is backed by classroom research. Thus, tools that actually advance learning and address classroom needs can be prioritized over simple technological flashiness that may not align with research findings.

1.2.1.2 How Students Learn with Technology Is the Answer

At the center of the issue of using technology to drive learning is recognizing what computers are actually good at and matching it with the educational need. According to the 2018 consensus study report by the US National Academy of Sciences "How People Learn II" (National Academies of Sciences, Engineering, and Medicine, 2018), the advantages information conveyed by learning technology include:

- **Learning through repetition**, which can be easily automated and tracked with learning technology
- **Deeper learning with models**, both via use of intelligent tutor systems and construction of model systems
- **Collaborative and cooperative learning**, driven by the ability to work on assignments and programs together and simultaneously
- **Self-regulated learning**, as a consequence of having the ability to test one's thinking and mental models

Educators need to go beyond learning through repetition. Yes, computers are very good at reproducibly eliciting and tracking user actions and interactions (i.e., they are better than human beings at timing, repetition, and vigilance). However, we need new ways of applying educational technologies that can harness both what we know about human learning while also incorporating the strengths of computers. These new methods should do more than simply encourage factual recall; they should lead to the holistic evolution our students need to undergo in order to personify their professional identities and competencies in the future. These include improved information transfer, the ability to work in teams, development of empathy, and skills acquisition.

As we will see in the chapters that lay ahead, there are a number of existing and emerging technological approaches that rely on learning theory and empirically derived data intentionally designed to improve thinking skills, as well as drive attainment and retention of knowledge. For example, educational technology that incorporates adaptive learning and games (see Chap. 5) allows students to enter into learning at an individual level, which permits personalization of learning. Much of this technology incorporates retrieval practice to improve long-term retention of knowledge (see Chap. 4); psychology has shown that the timing of learning reinforcement determines the biology of neuronal synapse formation, yet traditional education is mass produced with students being taught and tested in lockstep (Ebbinghaus, 1913; Kelley & Whatson, 2013; Korte & Schmitz, 2016). Fortunately, there are testing platforms that take advantage of this well-established psychology

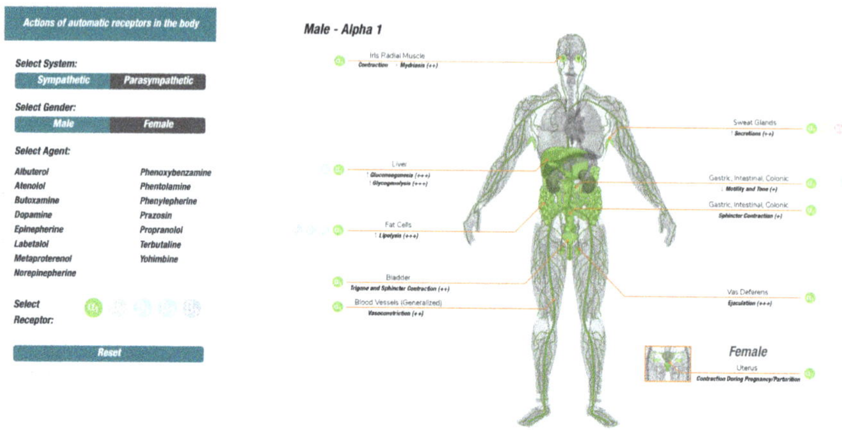

Fig. 1.1 The Integrated Pharmacology Atlas©. An electronic tool to help students visualize and interact with the autonomic nervous system anatomy, physiology, and pharmacology. Image courtesy of Michael W. Lee, Ph.D

phenomenon. Other approaches are also being developed to elicit students' deeper thinking and elaborate on their true understanding of a concept. As discussed in Chap. 8, beSocratic is an example of an assessment program that requires students to articulate their thinking rather than simply picking a single best answer choice in a multiple-choice question (MCQ).

Another example of using technology designed to improve cognition is leveraging the superior visualization and animation that computers offer. This allows developers and instructors to reimagine how complex information can be visualized and learned. The Integrated Pharmacology Atlas (Lee & Zahedivash, 2021) is an example of one such tool. Here information about autonomic nervous system anatomy, physiology, and pharmacology is linked together in the context of a virtual human (Fig. 1.1) (Zahedivash & Lee, 2018).

This format allows students to interact with the different autonomic receptors to learn their location, function, and how drugs can alter function both therapeutically and adversely at offsite receptors. Because the autonomic receptors are presented in their natural organ ecosystems in the human body and students can interact with the receptors and see their function, the barrier for entry into this complex domain of biology is reduced. This is an application where digital technology drives a student's initial learning by providing them with a more intuitive organizational model that they can interact with and manipulate.

In a related example, educational technology can also be designed to enhance engagement when paired with non-technology-based pedagogies like problem-based learning (PBL) by reframing how case information is presented to students. Creation and implementation of educational electronic health records (EHR) programs early in a curriculum to present clinical cases helps students contextualize knowledge and get acclimated to the manner in which these cases

Fig. 1.2 Educational electronic health records (EHR) paired with small group activities. Media and educational technology can be used to organize, focus, and augment basic science learning in a clinical, patient-centered context. Image courtesy of Michael W. Lee, Ph.D

are encountered in clinical practice. Pairing educational EHR programs with PBL, for example, can lead to rich, structured learning events, where students discuss basic science content in the context of a patient (Fig. 1.2) (Steinel et al., 2019).

The common thread linking these tools is that they encourage students to interact with knowledge rather than simply being passive receptacles receiving knowledge from a resource.

1.2.2 Impact of Computers and Educational Technology on Noncognitive Processes

There is a documented need for strong noncognitive attributes, such teamwork skills, social behaviors, empathy, emotional regulation, and professionalism (among others), due to the professional nature of health- and science-related careers (Bennett & Gadlin, 2012; Chakraborti et al., 2008; Martin & Ochsner, 2016; McNair et al., 2005; O'Connell & Pascoe, 2004; Ritter et al., 2017). Furthermore, there appears to be a relationship between noncognitive attributes and assessment performance, such that performance can be enhanced with teamwork and active learning (Adam et al., 2012; Beckerson et al., 2020). Therefore, finding ways to cultivate noncognitive attributes and skills is a priority as students' progress in their professional training. In contrast to cognitive ability, these attributes are more difficult to teach and measure. While it remains unclear how well computers and technology can be used to teach or improve these noncognitive skills in a temporally durable manner, there are some promising approaches that merit further investigation, which are discussed in the following sections.

1.2.2.1 Teamwork Skills and Social Behaviors: The Influence of Communication Using Social Media Technology

The capability of computers and technology to facilitate communication provides an avenue to influence students' noncognitive skills. Good communication between students can improve engagement and enhance socialization, which may lead to learning gains (Goldman & Wong, 2020; Luiz Adrian et al., 2015). Communication is also a product of teamwork training and experiences (Bennett & Gadlin, 2012; Chakraborti et al., 2008; O'Connell & Pascoe, 2004). While technology has the potential to improve communication in ways that can enhance teaching and learning, these tools also come with some notable liabilities that need to be carefully considered ahead of time. The challenge then is to leverage technology and media in ways that favor group activities, teamwork, collaboration, and disincentivize isolation and disengagement that can accompany the use of technology, so that students' noncognitive growth flourishes.

As discussed in later chapters of this text, there are a number of Web-based tools that can be employed to facilitate communication between students to enhance learning. Social media stands out among available forms of technology-based communication tools because of its widespread use and the native familiarity students have with it. As Rengasamy et al. discuss in Chap. 14, social media, if used appropriately, is a powerful means to link students together and foster communication, so long as students have access to a Wi-Fi-enabled device (Chap. 2). However, they stress that it is critical to set expectations on the use of social media ahead of time to ensure professional and respectful communication. This is because, like older methods of technology-based communication, such as email, messages can accumulate so quickly that they can be missed (or even ignored), leading to professionalism issues related to timeliness, participation, and completion of assignments, for example. Likewise, electronic "real-time" social media tools like Facebook or Twitter are often rife with incorrect, or even harmful, information and messaging that can lead to confusion, de-motivating students. This in part stems from the crowd sourced aspects of social media platforms and the disconnected nature inherent in virtual communication. Consequently, rapid and more readily accessible routes of communication may actually elicit socially aberrant behavior and unprofessional actions.

Similar maladaptive noncognitive phenomena are also seen with Web-based media sharing sites used to communicate information to students. Video-based Web sites, like YouTube and Vimeo, that are often used in structured science curricula contain social media communication abilities, and students are able to comment on videos, sometimes unprofessionally. In addition, these venues are also host to erroneous and unprofessional content, which is often not vetted or supervised by content experts or instructors. To address these issues, whether with social media or media-based Web sites, providing guidance and expectations of student behavior and outcomes around their use and acceptable communication is clearly key (see Chaps. 7 and 14). So too is role modeling professional behavior through explicit instruction and feedback. This can also be achieved with instructor-created or instructor-curated videos or media content.

Examples of technology that bring together the positive aspects of social media-like communication with sanctioned content and faculty supervision are now beginning to emerge. For example, Perusall® is a simple but apt example of how technology can be used to enhance communication among students in the form of textbook annotations (Miller et al., 2018). In the Perusall® tool, students access reading material in digital form (e.g., a textbook, instructor developed material, PDFs, etc.) and are able to engage in a social media discussion with other students as they progress through the text, leaving comments and responses for each other. In this way, students are able to learn new material and resolve misunderstandings in real time or near real time from their peers (www.Perusall.com). This facilitates communication and teamwork among students. It may also be a means to teach professionalism as well, since it has the features of a social media tool in an academic environment and the comments can be viewed by others including the instructor. In theory, a tool like this that puts social media components in a professional context with accountability (more on this below) within a course has the potential to improve noncognitive attributes and interactions, but more research is needed to determine whether this is actually occurring. Tools like this can also be paired with active learning pedagogies such as flipped classroom.

1.2.2.2 Emotional Regulation, Empathy, and Professionalism: A Role for Instructor-Guided Technology and Activities

Technology such as simulation has been used to teach both basic science knowledge as well as procedural skills in a variety of science and health science disciplines (Klein, 2020). It may also facilitate improvements in noncognitive skills such as empathy and emotional connection of individuals in certain contexts (Harris et al., 2015; Reinemann et al., 2015). Using simulation to mimic real-life exercises, where the actions of a learner can impact the health of a simulated patient, has been shown to influence students' emotional connections to the patient (see Chap. 16) (Bearman et al., 2015). In many programs, the simulation exercise is paired with a debrief, where students gather together after the simulation to discuss the experience (Fanning & Gaba, 2007). This can provide a venue for students to form normed behaviors as they interact with other groups of students and faculty and discuss the patient, their responses, and team members actions during the simulation. Using simulation together with debrief sessions can also be used to improve professionalism that is backed up by learning theory (Corliss & Lee, 2021; Lee, 2016).

Instructors and teachers can also help students develop noncognitive skills through role modeling and direct instruction. In Chap. 9, Uijl et al. weigh the merits and challenges of delivering feedback to students via audio or video formats. In this case the feedback is offered in a one-way manner for the student to consume. However, students often respond very positively to video feedback because it is personalized and relational. For it to be successful though, one-way communication requires a culture of feedback. If paired thoughtfully with active learning exercises or self-regulated learning (SRL) approaches, this technology-mediated feedback could lead to more efficient learning.

Finally, technologies and media should also be used to reassure students on the purpose of a chosen pedagogy or activity. The dramatic surge in the use of active learning and SRL approaches in the classroom has enhanced student responsibility, which can cause stress, and it has also led students to question the purposes of learning activities. Frequently, instructors think that they have explained the purpose of using a pedagogy to deliver a topic clearly, when in fact to some students, it may not be clear at all. For example, Owens et al. found that getting students to approach discussion in a scientific manner using an active learning approach was often met with resistance because students did not see the purpose (Owens et al., 2020). Silverthorn (2020) recommends that instructors explain the rationale for active learning *repeatedly* to enhance student acceptance of active teaching approaches (Silverthorn, 2020). Frequently, our own perceptions about how we learn best and what constitutes good learning are skewed from reality (Bjork et al., 2013; Deslauriers et al., 2019). Clearly communicating with students and briefly describing the underlying learning theory and sharing data on the effectiveness of the pedagogy (technology or otherwise) may help to increase buy-in from the students and may increase compliance and engagement.

1.2.2.3 Impact of Computers on Metacognition and Self-Regulated Learning

Finally, given how important learner characteristics (e.g., engagement, motivation, and self-efficacy) are in the mastery of material, it bears consideration as to whether technology can help students develop these characteristics by influencing self-regulation and metacognition. Metacognition is defined as "the ability to monitor and regulate one's own cognitive processes and to consciously regulate behavior, including affective behavior" (National Academies of Sciences, Engineering, and Medicine, 2018). A student's emotional state as well as their belief that they can accomplish a task (self-efficacy) can both influence metacognition and, in turn, academic performance (Artino et al., 2010). Thus, in many ways educational technology and computers are well positioned to help students build or strengthen metacognition by promoting self-regulation in learning (National Academies of Sciences, Engineering, and Medicine, 2018). However, designing educational technology with the intention to drive attainment of these attributes may be the critical key to success. Educational technology, media, or Web-based tools that intentionally prompt students to think about how they are learning, such as having them articulate how and why they come up with an answer, can *directly* improve metacognition and self-regulated learning (National Academies of Sciences, Engineering, and

Medicine, 2018). Properly structured media and educational technology can help students do this (Badger et al., 2019). However, students still need to believe that the endpoints of studying are valuable and justify the personal (as well as financial) investment. Simply getting students to think and verbalize their thinking is not enough; their emotions associated with achievement are also important in determining outcomes as well (Pekrun et al., 2007). Artino et al. show that task value (how interesting, important, or useful a task is) is positively associated with enjoyment and negatively associated with emotions like boredom, whereas self-efficacy is negatively associated with anxiety (Artino et al., 2010). Thus, the impact of a tool on the emotional state of the learner needs to be carefully considered. Indeed, assessing the impact of such tools is difficult, because the very act of measuring it may influence the learner and alter their emotional state and interaction with the learning modality or technology.

In contrast to direct effects of technology on metacognition and SRL, pairing active learning with technology (e.g., flipped classroom teaching) may hold the keys to *indirectly* helping students improve critical thinking skills, because teachers (and fellow students) have more classroom time to invest in offering feedback and helping to correct thinking when the background facts are presented to students by machines (see Chap. 6) (van Vliet et al., 2015). However, the ultimate outcome of combining technology that supports self-regulated learning, together with in-class active learning sessions and small group work, on student metacognitive development is unclear and merits more exploration (see Chap. 6). For this reason, current efforts aimed at using technology to enhance metacognition and promote development of SRL integrate multiple forms of technology with pedagogical methods (Azevedo et al., 2010). That said, there are many open questions surrounding how achievable this is in a classroom with time constraints, (potentially) more than one instructor with different priorities, and a heterogeneous class population.

1.3 Educational Research Should Develop Intentionally

1.3.1 Past: Significant Outcomes That Have Arisen from Innovations in Educational Technology

Although technology has caused many changes in the process of education, and although it has made great strides in accessibility for a subset of students, the results of all these technologies, in terms of learning, have been disappointing for the average student (Spector, 2020), despite the influence of visionaries like Steve Jobs. Educational technology does not advance in a vacuum. Since the millennium, there have been high-level demands for an educational revolution that includes research (Cech & Kennedy, 2005; DeHaan, 2005; Handelsman et al., 2004), and demands for educational reform have been mooted since the 1980s (AAAS, 2011; AAMC, 1984). Much of this has been driven by the belief that a learning society is propelled by technological knowledge and that the creation of a scientifically literate workforce has not kept pace with demand or international competition (National Academy of Sciences, 2007). Reformers have suggested educational technologies may be able to solve the two major problems that prevent many students from embracing and learning science, especially at the introductory university level:

1. **Students do not like engaging with lectures**. Students complain about the lectures, especially that they are bored by memorizing facts rather than exploring concepts on their own (Cech & Kennedy, 2005). Furthermore, there are high levels of attrition not only in science, both in early science courses, but also in public engagement with science, where potential scientists are turned off by the rote learning (Udovic et al., 2002).
2. **Students do not learn efficiently from lectures**. Students taught by lectures have poor retention of information after the course ends (Arons, 1983). Also, students taught by lectures have minimal or no understanding or conceptual knowledge of what biology really is, especially of three elements: problem-solving processes, epistemology of science, and subtle/precise definitional distinctions that appear on concept inventories (AAAS, 2011).

These two problems have been made much more pressing by the fact that the amount of bioscience information that there is to learn has exploded, while the durations of degree programs have remained constant (or in some cases have decreased due to shifting emphasis on experiential learning). Thought leaders have suggested that the solutions to these problems should come from innovations in: (A) pedagogies, (B) educational technologies, and (C) a combination of the two. The issue is that technology alone seems like an unlikely savior for education—its advances usually require a concurrent change in pedagogy and also in large-scale institutional and social organization. For example, we now have platforms that pick out plagiarism, and so they find it constantly, which creates problems for both faculty and students. The innovations that reformers are looking for are probably a combination of educational tech and pedagogy.

Massive Open Online Courses (MOOCs), which were the obvious combination of educational tech and new pedagogy, mooted the question of whether it is possible to mass produce formal university education at a scale that was previously unimaginable (Waldrop, 2013). The answer was "MOOCs do not work as hoped, but there are still new opportunities" (Jordan, 2015; Konnikova, 2014), especially for informal learning, continuing professional development, computer programming, and training teachers. Obviously, other educational technologies have changed how students access and learn some things, but are they also mostly hype? Are we really revolutionizing education, or are we just making incremental changes around the edges?

We are not the first to ask this (Elrod, 2010; Spector, 2020). Robert Kozma made an eerily prescient prediction in 1994, before streaming TV, online banking, or Amazon were let loose on the public:

> In the not-too-distant future, we will be faced with a situation where telephone, cable television, and digital computer technologies will merge [...] If by then [...] we have not forged a relationship between media and learning—this capability may be used primarily for interactive soap operas and online purchasing of merchandise with automatic funds transfer.

Kozma's point was that technology could provide an extraordinary opportunity for improving education but that this opportunity could easily be squandered on consumerism and entertainment, especially if led by technologists for profit. We propose that major educational advances will require intention. There is a visionary example of intentional use of technology and new media that delivered a clearly beneficial educational outcome: the television show Sesame Street. It was developed and researched by Children's Television Workshop, so it has been one of the most carefully researched educational interventions, both for individuals and at scale (Ball & Bogatz, 1970; Cook et al., 1975; Fisch et al., 1999; Mares & Pan, 2013). No doubt Sesame Street has made children more literate, but did it really create the learning revolution via technology as intended? Yes, Sesame Street delivered literacy and social norms to millions of children all over the world, also providing guidance to teams developing local language versions within their cultural norms. Admittedly, the learning that Sesame Street activated was not instantaneous, but it was almost effortless to the children, cost-free to the parents, accessible to all, and extraordinarily inclusive to the societies of its time.

There is no Sesame Street counterpart for university bioscience that has made young people better biologists, although efforts at adding video to biology teaching have been made by many organizations (e.g., Khan Academy and Crash Course), but these have not revolutionized the net learning of our students (Petrilli, 2018). Perhaps the clearest example of the revolution in bioscience education can be seen in the photographs of undergraduate medical students in Fig. 1.3. The upper photo shows learning—and learners—from the 1950s at a well-known medical school in London, and the lower photo shows contemporary medical students at the same university. Note first the simulation suite, informal clothes, and the fact women were previously excluded and are now a majority of the students at that medical school; these pictures tell an undeniable story of previously unimaginable change, with vastly expanded technological knowledge, access by students to that technological knowledge, and inclusivity at center stage. What is missing from the photos is the fact that the contemporary medical school trains four times as many students per yearly cohort. Education thought leaders may have taken for granted the historically unprecedented change in opportunities and participation rates in higher education (Our World in Data, 2014). The educational revolution has already happened, and is still happening, but the revolution is more a story of inclusivity and social learning rather than of "machines that teach." Technology has played a supporting role in the educational revolution all around us, but advances in culture and finance have been the stars of the show for driving genuine leaps in education.

Fig. 1.3 A historical and contemporary photograph of undergraduate medical students from the same London medical school. Courtesy of Barts Health Archives and QMUL

1.3.2 *Present: Intentional Research and How Technology Can Smooth Educational Barriers*

To address the above problems of (1) engagement and (2) proper learning, educational committees and commentators made three major recommendations, proposing that all bioscience educators should use:

1. Educational processes that are evidenced by research for efficacy (Handelsman et al., 2004)
2. Active learning pedagogies (Freeman et al., 2014; Theobald et al., 2020)
3. Educational curricula that reflect and incorporate scientific research (National Research Council, 2003)

Chapter 18 in this volume focuses on the future of educational research in the biosciences, and they conclude that researchers need to focus on outcomes in the learner rather than educational processes. In some ways, evidence-based education (recommendation 1) has been slow in making progress into the classroom. Nearly two decades ago, Handelsman et al. (2004) bemoaned the problem in *Science*:

> … reform has been initiated by a few pioneers, while many other scientists have actively resisted changing their teaching. So why do outstanding scientists who demand rigorous proof for scientific assertions in their research continue to use and, indeed defend on the basis of the intuition alone, teaching methods that are not the most effective? Many scientists are still unaware of the data and analyses that demonstrate the effectiveness of active learning techniques. Others may distrust the data because they see scientists who have flourished in the current educational system. Still others feel intimidated by the challenge of learning new teaching methods or may fear that identification as teachers will reduce their credibility as researchers. (Handelsman et al., 2004)

The most evidenced form of education is the broad category of active learning pedagogies (see Chaps. 4 and 6) (Michael, 2006); as opposed to passively listening to lectures, these active pedagogies benefit from the cognitive process of active learning by formally incorporating special activities in the classroom (or other locales) that force the student to confront and think about the concepts to be learned, often with a group of peers (Bonwell & Eison, 1991). Surprisingly, although the evidence in favor of active learning pedagogies is overwhelming (Freeman et al., 2014; Theobald et al., 2020), a slow but steady stream of literature suggests (1) that a reasonable percentage of students resist or dislike active learning (Deslauriers et al., 2019; Owens et al., 2020; Seidel & Tanner, 2013; Silverthorn et al., 2006), (2) that students do not believe it is particularly useful for their learning (Deslauriers et al., 2019; Kauffman et al., 2019), (3) that educational institutions can make rules that prevent its proper implementation (Rop, 2002; Silverthorn, 2020), and (4) that it does not result in superior learning when performed by ordinary bioscience instructors (Andrews et al., 2011).

The research that is plainly needed regards the role that educational technology can play in smoothing those barriers. This is especially important in light of the proliferation of commercial study preparation and self-quizzing platforms that seemingly fill the gap and accelerate learning. We need to know, in what ways can technology help to reduce student resistance to active learning pedagogies. For example, the basis of flipped classroom pedagogies requires that students prepare before their classroom session by reading or watching a video on background information, so that the student can apply the information or make conceptual connections during the supervised classroom activities; yet, many students arrive for flipped classrooms without preparing (Rahman et al., 2015), hoping to glean background information from their classmates during group work (see Chap. 6).

Researchers need to develop technologies that address this issue, by facilitating or encouraging students to be prepared for classroom sessions. These technological interventions may include gamification or avatars that are increasingly found in automated tutoring systems. The technologies could also seek to minimize student resistance to active learning in the classroom; such technologies would be even more useful if they nest with the active learning exercise in such a way that the students actually experience the benefits of preparing through success in the active learning sessions.

Another open door for future innovations in educational technology concerns assessment, both formative and summative. The benefits of using computer platforms for providing formative assessments is that they would allow students to practice retrieval and test their knowledge both repeatedly and at their own pace. Although technologies might provide formative assessment experiences, they will not hasten the profusion of high-quality education if the best formative learning providers are commercially limited and not inclusive. The other limitations for computerized formative testing are that currently the test formats (e.g., multiple-choice tests, see Chap. 8) and the feedback formats (nondiagnostic, canned explanations) are insufficient for students to learn from them unless extensive investment is made to detail that scientific content. Technology could contribute to organizing this as an open resource project, where curation, categorization, and prevention of duplication could be used by the teaching community.

The development of technologies for summative testing has succeeded in many of its goals (Yan et al., 2016) but still has a long way to go. Currently, preventing misconduct (e.g., impersonation, collaboration, and access to prohibited resources) requires in-person attendance or labor-intensive monitoring; the current computerized proctoring technologies also raise a variety of privacy issues (Hylton et al., 2016). Effortless verification of summative testing will become increasingly important as summative tests become shorter, lower stakes, and more frequent (see Chap. 4) (Schuwirth & Van der Vleuten, 2011). However, even the most innovative technologies for detecting misconduct will also require concurrent developments in our institutions and expectations, because accuracy and attrition are no longer primary educational goals. As stated by the American Physiological Society, "The future of learning in physiology should provide all students a learning environment that maximizes their academic success" (APS, 2020). "All students" includes cheaters (see Chap. 10 on cheating in biosciences), which means that our rules and regulations must still support students who are bad faith actors. Educators would really benefit from these changes in detecting cheating, but the implementation of such technologies will only be of benefit if our institutions (and how we address cheating) change with it.

1.3.3 Future: Intentionally Developing Teaching Research by Mechanistic Analysis

To design the education of the future, many education researchers have embraced the call for using the scientific method when investigating pedagogy (Freeman et al., 2014). So why don't we have a clear answer as to which active learning pedagogy is best? Why are some instructors using flipped classroom and others team-based learning? Plainly because there is no single best pedagogy for all situations, the best pedagogy depends on the particular student and the specific topic: context determines the efficiency and efficacy of each pedagogy. The two editions (2000 and 2018) of the USA National Academy of Science's *How People Learn* demanded that educational researchers consider context, although in slightly different ways. In *How People Learn: Brain, Mind, Experience, and School* (*HPL1*), the context was based on the situation:

> Asking which teaching technique is best is analogous to asking which tool is best—a hammer, a screwdriver, a knife, or pliers. In teaching as in carpentry, the selection of tools depends on the task at hand and the materials one is working with. Books and lectures **can** be wonderfully efficient modes of transmitting new information for learning, exciting the imagination, and honing students' critical faculties. (National Research Council, 2000)

Similarly, 18 years later the National Academy of Science's *How People Learn II* reinforced the idea of context, but in a sociological framework:

> it is now possible to move beyond the idea of an "average" learner to embrace and explain variation among individuals. It will be valuable to have more interdisciplinary research that examines how individual variation and developmental and contextual factors [...] influence the lifelong learning process and learning outcomes. (National Academies of Sciences, Engineering, and Medicine, 2018)

Educators are beginning to explicitly investigate differences between different cohorts, and ultimately different students (Theobald et al., 2020). As part of the inclusivity program to understand these applied learning questions, we need to pick apart the detailed mechanisms that drive learning simultaneously and in sequence over the longer term. Contextual understanding requires careful analysis, in which the workings of a longer-term pedagogy are examined piece by piece. For example, team-based learning (TBL) has several elements including the following: (A) an approach to factual presentation based on student preparation before class, (B) an assessment approach using iRATs and tRATs, (C) a social approach (persistent teams), and (D) a learning activity approach (active learning). In order to understand what the effects of the assessment approach are, we need to see experiments running TBL without iRATs or tRATs. It is not enough to determine simply that students learn less without these assessments; we need to understand specifically and mechanistically how they learn less. Each pedagogical element potentially has effects on six broad areas (motivation, engagement, cognitive strategies, metacognitive ability, knowledge, and competencies). Thus, each pedagogical element, such as assessments, may affect one or more momentary mechanism(s) for learning (e.g., internalization of motivation, cognitive elaboration, increased self-efficacy, etc.). Will educational researchers be willing to analyze the effects of the individual elements and how they synergize for different cohorts? Is this even ethically possible?

It would certainly clarify the pedagogy if interested investigators measured one-by-one the effectiveness of each element of the pedagogical method or technology, such that each mechanistic element (and its contribution to learning) was proven. The evidence gathered by these granular experiments may not be perfect proofs, but they would certainly make the teaching and learning process clearer. Such a program of research will not be the result of a single investigator, and the mechanistic links may not be proven in the order of the effects on learning. Nevertheless, the resulting accumulated information would help predict the impact of a range of pedagogies on student learning. This would provide valuable guidance for when instructors consider making changes (e.g., as a result of course time constraints) to a teaching method or tool. For example, when instructors chop and change teaching approaches, such chopping and changing, however well-intentioned, could ultimately reduce the effectiveness of the pedagogy, so decisions of this nature should be guided by mechanistic data and studies. Future research needs to go beyond investigating broad effects on the average student, because calculating amalgamated statistics alone, without mechanistic analysis, is more like market research on consumer preference (so-called A-B tests by Google) than scientific research (Gallo, 2017; Google, 2021).

1.4 Emotional, Social, and Employment Competencies in the Future Role of Bioscience Education

1.4.1 The Next Purpose of Higher Education and Its Enhancement by Technology: Emotional Education and Social Competencies

Computers are useful tools to transmit and reinforce information, but they rarely teach or reinforce noncognitive elements that we consider central to a student's development: emotional, social, and metacognitive. As a result, using computers and technology in education could end up reinforcing isolated facts and may prevent students from developing what bioscience teachers hope to convey: (A) competencies such as problem-solving, (B) employability skills such as teamwork, and (C) a sense of awe for the complexity and beauty of biological systems. When taken together with the many competing roles that vie for students' time as they progress through an educational system (e.g., future goals, family, leisure), it could be argued that computers encourage a strategic or surface learning approach (see Chap. 4 on engagement), where they learn none of the above. Thus, we need to consider how technology aligns and reinforces the purposes of education. These purposes of higher education were eloquently articulated by Australia's Commonwealth Department of Education, Science, and Training (in the Higher Education

Report for the 2003–2005 Triennium, p. 1) (Australia Department of Education, Skills, and Training, 2003):

1. Inspire and enable individuals to develop their capabilities to the highest potential throughout their lives (for personal growth and fulfillment, for effective participation in the workforce, and for the benefit of the economy and society).
2. Advance knowledge and understanding.
3. Aid the application of knowledge and understanding to the benefit of the economy and society.
4. Enable individuals to adapt and learn, consistent with the needs of an adaptable knowledge-based economy at local, regional, and national levels.
5. Enable individuals to contribute to a democratic, civilized society and promote the tolerance and debate that underpins it.

Science and biomedical education can address many of the purposes of education laid out here by virtue of the inquiry and uncertainty that is inherent in science itself. The key question that needs to be considered, however, is what role do *computers* and *technology* play in helping us further these purposes of education? Arguably, computers and technology can facilitate 2, 3, and 4, but what about 1 and 5?

1.4.2 Educational Technology Researchers Should Consider Competencies and Employability Skills

The aforementioned purposes of education are more than just aspirational goals of education. They also embody skills and attributes that are desired by employers. Thus, it should come as no surprise that many accrediting bodies for health profession and health science programs have crafted competencies that address these purposes. This includes not only knowledge but also "soft skills" that are alluded to in the fifth purpose listed above. For example, the AAMC (American Association of Medical Colleges) has generated and published a list of core competencies for *entering* medical students that seek to address these types of skills (AAMC, 2021). The first two categories of competencies listed relate to inter- and intrapersonal skills, including ethical responsibility for self and others, cultural competency, social skills, teamwork, capacity for improvement, and communication, among others. Again, these are competencies that the AAMC expects students *entering* medical school to have, and thus, it would be expected that they acquire them in college or prior to admittance to medical school. This argues for a role in teaching and monitoring these skills in undergraduate science curricula (in countries where a college degree is needed to secure entrance into medical school) as part of a continuum of effort on development of "soft skills." Importantly, however, these competencies relate to key workplace skills, which would be predicted to reinforce democratic equality, civility, and tolerance embodied in the fifth purpose of education in the Australian framework. Putting aside the issue of how to actually assess or measure these competencies in an entering medical student, we need to consider that these skills, in contrast to knowledge, are among the most difficult to teach.

At the core of these AAMC competencies are ideas like sensitivity, responsibility, and awareness. Considering the observation that diminished social interaction with peers, driven in part by increased use of technology and time spent on devices and the Internet, can lead to emotional dissonance, increased levels of loneliness, and greater isolation in teenagers, one might question how increased technology use would help students meet these competencies (Twenge et al., 2019). So, the question arises: does the dosage or frequency of technology use in a curriculum matter? Is it possible to undermine the learning of "soft skills" if too much educational technology or media is employed in a curriculum? It may instead depend on how these technologies are used, in what context, how frequently, and whether they are paired with traditional pedagogical methods, and whether they foster collaboration or competition among student peers. While competencies are useful targets, without guidance on how to achieve them or the role technology plays in mediating their attainment, schools may be inconsistently addressing said competencies or worse, failing to achieve the long-term outcomes they hope for in their graduates.

1.4.3 Fostering Emotional and Social Development: Competition or Collaboration and the Role of Technology

Taking into consideration societal expectations and accrediting body competencies, it should now be apparent that the emotional, noncognitive, and metacognitive development of students needs to be cultivated during a student's education. As a consequence, there has been a shift in health science education toward more collaboration rather than competition, because competition focuses on the individual, whereas collaboration develops interpersonal communication and soft skills. This has manifested itself in many ways, including increased use of group work and active learning in the classroom (e.g., PBL, TBL, etc.) and a transition from numerical grades (and licensing exams) to pass/fail. There is also a role for technology, educational tools, and media to facilitate collaborative student learning (Johnson et al., 2013). However, there is a tendency for technology to lead to polarization of opinion, formation of factions, and unprofessional behavior (see Chap. 14 on social media in bioscience education). Careful monitoring of student interactions with collaboration-based technology obviously helps, so too does clear messaging ahead of time on how to use these tools, the goals, and the expectations for behavior and outcomes. We believe that this is an area of future research that needs to be prioritized and advanced. In particular, the short- and long-term impacts of different types of educational technology and media tools on all students' emotional, noncognitive, and metacognitive development need to be more carefully evaluated, not just on their ability to collaborate.

Although there is an important role for collaboration, there is also a role for competition in learning too (Corell et al., 2018). For example, self-competition can serve as a source of internal motivation, where one seeks to outperform their own prior performance. This involves processes, such as goal setting, motivation, and self-monitoring. Devices and technology can be used to monitor self-progress, performance, timing, and accuracy, which can be compared to historical performance. A widely known example is the Fitbit, which promotes self-competition by allowing you to set goals and encouraging you to exceed them, and this may alter a wearer's behavior as well (Sullivan & Lachman, 2016). Yet, like so many areas of human psychology related to learning, there are conflicting results in the literature. Vandercruysse et al. (2013) reported that competition may only be weakly associated with motivation and is not significantly related to learning gains, albeit in a nonscience discipline-based computer game (Vandercruysse et al., 2013). In contrast, Lei et al. found that introducing a competitive element into PBL in their medical school program boosted performance and diagnostic accuracy slightly compared to a noncompetitive form of PBL (Lei et al., 2016). Many competitive paradigms (including meritocracy) are explicitly designed to be exclusive, so the research community needs to delineate how and when to employ technology that promotes collaboration versus competition in a context-specific manner to shape emotional and social development. These are tough questions that require both more information and genuine reflection in the community. We need reliable evidence addressing whether collaboration is better for formation and cultivation of emotional, social, and professional skills in teams, whereas competition may have a role in motivating engaging for knowledge acquisition. The argument starts with leader boards and class rankings, but it points us toward much bigger questions about what are we doing to our students, and what values are we actually teaching.

This brings the argument full-circle back to the quote from Steve Jobs, cited at the beginning of the chapter. What is really missing in educational technology development is proper consideration of the critical importance of emotional, noncognitive, and metacognitive skills development in our students. If we are going to intentionally develop technologies that support what we actually want students to learn, then designer-driven development focused on easily justified profitability, easily measurable factual learning, and easily seen exciting user interfaces will not get us there. This leaves technologists with a development strategy based on market research. As educators we think this is not good enough. We need technologies focused on measuring student development in a way that is relevant for educational research, not market research, because market research is unlikely to tell us what we need to find out. Although surveyed students may not say that they want these soft skills and surveyed faculty may not think that educational technology should focus on helping students developing soft skills, as higher education becomes more inclusive and focuses more on employability, these are definitely the skills that students will need. Thus, perhaps we should replace the words "want" in the Steve Jobs quote with the word "need" to better reflect the true priorities of educational technology and media creation.

1.5 How the Book Is Structured to Address These Issues

The focus of each chapter is to survey and understand the strengths and weaknesses of mediated content (including face-to-face learning supplemented by media) in comparison to traditional face-to-face learning. Each chapter includes a general summary of technological approaches to an aspect of life sciences education. The field of educational technologies and multimedia is too wide to be covered comprehensively, but we think we have captured the most common and well-known activities that instructors in the biomedical and life sciences are familiar with. Each chapter also addresses the evidence—or lack thereof—supporting each element of digitally mediated learning. The typical chapter will start with education at present, consider the technologies that are changing education both inside and outside the classroom, and end with a meditation on the future. Notably, we consider contextual factors and key procedural details that have a clear effect on efficacy and efficiency: What are the differences that make a difference? The final part of each chapter proposes future research to answer uncertainties about efficacy, performance and unanticipated effects; these suggestions also consider the limitations of the current research and how life science education research procedure might be improved, particularly for quantitative research. A theme that emerges from these chapters is that we need to better understand human learning and psychology and how this impacts, or is impacted by, technology. This needs to be done intentionally, because it will not emerge spontaneously from researchers promoting their pet projects.

1.5.1 The Four Parts of the Book

The book's 18 chapters are organized into four broad parts, each reflecting a key question related to the consequences of incorporating media into education. The first part is called "Introduction and Educational Context," and it will contextualize the role that technology and social learning play in both institutional and societal development as well as in the psychological development of the student. Chapter 2 is focused on inclusivity, Chap. 3 on institutional cultures, and Chap. 4 on student engagement. Collectively, these chapters weigh how educational technology is both a bridge and barrier to student inclusivity, engagement, and empowerment.

The second part is called "How Educational Technology Shapes the Classroom Experience." It considers how digital technologies enable students to be active and have agency in their own learning and how this improves both retention and conceptual understanding. Chapter 5 concerns adaptive learning and gamification, Chap. 6 concerns flipped classroom, Chap. 7 considers supplementary videos, and Chap. 8 computer-based assessment. Chapters 9 and 10 make more cultural statements about education, with a chapter on video feedback and the importance of a culture for feedback and another chapter on cheating in a world of technology-enhanced learning.

The third part is called "How Educational Technologies Transcend the Classroom," and it starts with a debate in Chaps. 11 and 12, interrogating the value of face-to-face attendance in a classroom in our era of lecture recording and whether students need to be present in a live situation. How successfully does digitally mediated education in STEM subjects cope with changes in the social interactions as they become more mediated and less face-to-face? Chapter 13 addresses various out-of-classroom approaches, such as MOOCs, distance learning, and science education at scale. Chapter 14 tackles the benefits and concerns of using social media tools and co-creation in teaching and learning. This part is particularly timely given the COVID-19 pandemic and the massive shift toward online delivery of lectures and small group activity. As such, we commissioned Chap. 15 on the role of educational technologies in mitigating the impact of the COVID-19 pandemic on teaching, learning, and the higher education environment—including mental health.

The final part, which is entitled "The Future and Research", covers some of the most cutting-edge technologies used in biomedical and health science education. From virtual reality and augmented reality in Chap. 16 to digital anatomy tables and specialist computationally enhanced high-fidelity simulations in Chap. 17, these tools provide interactivity with knowledge and skills that goes beyond other frequently employed methods of learning (i.e., standardized patients, books, visual and audio media). Chapter 18 sums up the future of research in the area of bioscience education, taking a more synoptic view.

1.5.2 Individual Chapter Summaries

Chapter 2: Inclusivity

In addition to building educational technology around learning efficiency, attributes, such as accessibility, inclusivity, and equal access, should also be considered early on in the design and/or selection process. Taylor, Dewsbury, and Brame set forth a compelling argument that many educational technology advancements, driven by for-profit ventures, lack these attributes and thus marginalize the very students who need it the most. In addition to cost, the issues of access, control, and data privacy are also addressed. Accessing these learning tools via the Internet is not generally free, and the bandwidth available to users is also split down socioeconomic lines as well. These and other issues, including their etiologies and potential solutions, are presented in Chap. 2.

Chapter 3: Empowering Students as Partners

In an effort to engage medical students who had previously been motivated by extrinsic rewards, universities have been experimenting with empowering students as partners. Han et al. show how this partnership involves, including students in the process of curriculum design, as well as having students represented on many faculty committees that both run and strategically design polices for the school. To give students more empowerment in deciding about their own education, students are able to interact with their own data via dashboards. As students are given more agency in their own education, they engage more in the educational processes they have a say in.

Chapter 4: Engagement
This chapter considers how student engagement relates to improving learning. Although most practicing educators feel that they can recognize engagement when they see it, this chapter considers whether it is even possible to research engagement given its definitional ambiguities. Whereas momentary engagement, such as in a psychology lab, has been greatly delineated by learning science, longer-term engagement—such as with a school or a course—remains controversial and difficult to define. Witchel et al. review the proven methods for improving engagement and motivation, but they conclude that mechanisms and the relationship between momentary engagement and longer-term learning need to be more carefully researched.

Chapter 5: Adaptive Learning
The influence that cognitive and educational psychology play in learning and educational technology is stressed throughout this book but is most notable in regard to adaptive learning and retrieval practice. In Chap. 5, Kellman et al. discuss how adaptive learning and gamification can be used to improve engagement and long-term retention of information. Indeed, tools that incorporate adaptive learning or gamification perhaps come the closest to tailoring learning to the learner, reinforcing areas where the learner is weak or deficient. Learners may be more likely to engage with such a tool or system because of the tailored approach. The chapter presents several existing examples of programs that serve as proof of concept for the use of adaptive learning in the biomedical sciences along with data and rich references for readers interested in implementing such an approach at their home institution.

Chapter 6: Flipped Classroom
Of all of the pedagogical techniques currently in use, few have as intimate a reliance on technology as flipped classroom teaching. In this pedagogical approach, students receive basic content knowledge prior to class via video, audio, or, in some cases, digital pre-reading, followed by in-class application of this content. Presently, data supporting the short- and long-term effectiveness of this teaching approach are thin, and more work is needed to understand how to optimally deliver flipped sessions. In Chap. 6, Lee and Butler explore the data, as it stands currently, behind the different components of flipped classroom learning and identify key areas where more research is needed.

Chapter 7: Supplementary Videos
Supplementary videos, as opposed to videos derived from lecture capture of live sessions, are becoming an important learning resource for students. Indeed, both "sanctioned" videos (those created by faculty and staff for use by their institution's students) and "unsanctioned" videos (those created by companies, other students,

medical residents, or other outside entities) are incredibly popular with students. This may in part be due to the perception that learning can be expedited and optimized when in video form, and students routinely watch videos at 2× speed to acquire the information faster. Only a few authors have tested whether or not the information internalization and encoding from videos is different (superior or inferior) compared to other methods such as reading. After having watched an educational video, many students are overconfident about their knowledge. Guy and Witchel review the body of literature on best practices on the design of videos; the effectiveness of learning from videos is not as strong as one might expect, suggesting that there may be an opportunity for improvement and innovation.

Chapter 8: Assessment

The role of assessment in education has matured over time as we have learned more about the impact assessment has on learning and memory retention. Traditionally, assessment has been thought of as simply a means to motivate students to learn and to evaluate their learning using no-stakes, low-stakes, or high-stakes tests. However, a number of technology-associated assessment forms have emerged that can facilitate learning in the classroom. Often in the biomedical and life sciences, these tests are made up of unelaborative types of items such as MCQs. Deeper learning and knowledge construction can be better determined by grading intensive short answer and essay questions. In Chap. 8, Klymkowsky and Cooper provide an in-depth discussion of a program they have developed for assessing cognitive processes and knowledge construction, known as beSocratic.

Chapter 9: Video and Audio Feedback

The role for feedback in learning is well established. What is less certain is the impact of *how* feedback is delivered on student learning. In Chap. 9, Uijl and Filius present research and approaches that support the role of feedback in fostering deep learning, as opposed to surface learning. They outline different factors that can improve feedback to students including the quality of the feedback, feedback literacy, and feedback culture. This is followed by discussions on the different ways feedback can be delivered and ideas for future research and the future.

Chapter 10: Unprofessional Behavior and Cheating

Considering that many of the learning and educational tools we are discussing in this book can be used online, the potential for students to cheat or adversely take advantage of the system for illicit gain exists. Research on cheating is rare due to concerns about both privacy and institutional reputation. Yet, it remains relevant and controversial, because our goals for including all students means including cheaters as well. In Chap. 10, McLachlan et al. discuss the various ways that students cheat on assessments and writing assignments. They also outline characteristics that seem enriched among cheaters, commercial programs designed to detect cheating, and important areas for future research.

Chapters 11 and 12: Lecture Attendance and Performance in the Era of Online Learning
An enduring debate that has persisted in academia at all levels, and in particular in biomedical and life sciences, is whether or not students need to attend lectures to derive the most benefit for learning. In Chaps. 11 and 12, authors present counter-arguments *for* and *against* the need to attend lectures and in-class sessions in an era of online learning and lecture capture videos. Both sets of authors admit that there are many past studies that have linked exam performance with attendance but also that there are a number of exceptions where exam performance was not associated with attendance. The role of lecture capture recordings and how the students view them may determine whether the link is found; assuredly, lecture capture improves accessibility. As discussed in these chapters and others, these nuances play a large role in shaping whether or not students take anything away from the in-person sessions. There is a distinct possibility that we are measuring the wrong thing. In-person attendance may have more impact on a student's socialization, teamwork skills, and professionalism as opposed to knowledge mastery. It may very well be time to rethink what we are trying to research and match those expectations with the proper measurement instrument.

Chapter 13: Distance Learning and MOOCs
Given the impact of the COVID-19 on higher education in 2020 and 2021, it has become increasingly clear that distance learning, social media, remote learning, and even simulation can fill the gap to some degree and make up for a lack of in-person interactions. As laid out in Chap. 13 by Alston et al., there are many compelling successes achieved by distance learning (besides education during the pandemic) that support its continued use and study. The authors discuss definitions and the different types of distance learning, including MOOCs (massive open online course), xMOOCs (extended massive open online course), cMOOCs (connectivist massive open online course), and SPOCs (small private online course). Rich examples and best practices for optimizing these distance learning modalities are provided together with studies and data supporting the effectiveness of distance education.

Chapter 14: Social Media
Despite a number of promising attributes, a wholesale adoption and use of social media-based tools and technology in higher education has not occurred. In many ways, this is surprising, considering that social media has the potential to support online and distance learning. However, issues such as trolling and unprofessional behavior, which are commonplace in nonacademic settings, along with the expectation that responses to social media posts be immediate, have dampened enthusiasm. Rengasamy et al. discuss social media in the context of biomedical and life science education using numerous case studies and examples. They highlight how targeted use of social media can enhance the connectivity of students and offer best practices for engaging students in the academic use of social media. Getting out in front of the issue and managing expectations for professional behavior on social media are paramount. Co-creation using online social media tools is another potential course of action, one that rank and file faculty at teaching institutions may be less

familiar with. These strategies are outlined in more depth, and tips on how to implement co-creation-based projects are explored with case study examples.

Chapter 15: Emergency Remote Learning and COVID
Many of the technologies discussed in this book have figured prominently in our collective response in academia to the COVID-19 pandemic. Like distance learning (Chap. 13), many of the tools and approaches have a track record of use, with some best practices and data to support their effectiveness and guide their implementation. However, a key issue during the pandemic is that most faculty do not have experience with distance learning teaching (either the experience of actually delivering lessons remotely or the technology used to deliver lessons). Arango-Caro et al. deftly navigate the tools used to deliver synchronous and asynchronous learning online, including video conferencing programs and course management software. In addition to covering the history of these tools they offer advice on how to use of these tools effectively with students. They also cut to the heart of one of the most difficult issues facing instructors who have had to go all remote, the online delivery of labs. Finally, they provide a timely and rich discussion of virtual assessments covering tools and the potential for cheating which intersects with Chap. 10.

Chapter 16: Simulations
Simulation has been around in various incarnations for many years. In the health sciences, this has included standardized patients and high-fidelity simulations. However, the promise of simulation and the anticipated learning gains have yet to be fully achieved. Harris et al. take this discrepancy head on in Chap. 16 as they review the history and research supporting the development and implementation of simulation in health science education.

Chapter 17: AR/VR
Similar to simulation technology, augmented reality (AR)- and virtual reality (VR)-based technologies have great promise but are still in the early stages of their use, and thus their effectiveness is poorly understood. John Barrow outlines some of the more successful and promising uses of AR and VR including virtual microscopy for learning histology and cell structure and imaging molecules, which can allow the user to better appreciate molecular interactions and enzyme action. These techniques go further than simply visualizing though, because they afford the user the ability to "interact" with the medium and even offer the ability to learn procedural skills. What is clearly lacking is rigorous mechanistic assessment of the impact these technologies have on learning and retention.

Chapter 18: Future of Biomedical Education Research
Lastly, in Chap. 18, Lisk et al. emphasize a tighter focus on learning theory and learner outcomes in guiding life science and biomedical teaching. They focus on supporting students to develop adaptive expertise. The structure and contextual background of how content is delivered matters. Using integrated instructional methods including causal biomedical information illuminates the clinical signs and symptoms. They assert that it is this causal information that links the clinical features of the disease together. Learning in this way can alter how future knowledge is

learned and retained. This can also be incorporated into nascent learning and educational technology, as can productive failure and self-explanation, another approach that has been found to help learners develop content knowledge and problem-solving skills. They go on to discuss these approaches as well as retrieval practice and the role of motivation, environment, and context in mediating student performance. They conclude their chapter with a list of preconditions that should be met before introducing the learning principles and practices they discuss to ensure optimal effect.

References

AAAS. (2011). Vision and change in undergraduate biology education: A call to action. American Association for the Advancement of Science. Retrieved from http://visionandchange.org/finalreport

AAMC. (1984). *Physicians for the twenty-first century: Report of the project Panel on the General Professional Education of the Physician and College Preparation for Medicine*. Association of American Medical Colleges.

AAMC. (2021). *Core competencies for entering medical students*. American Association of Medical Colleges. Retrieved from https://www.aamc.org/services/admissions-lifecycle/competencies-entering-medical-students

Adam, J., Bore, M., McKendree, J., Munro, D., & Powis, D. (2012). Can personal qualities of medical students predict in-course examination success and professional behaviour? An exploratory prospective cohort study. *BMC Medical Education, 12*, 69. https://doi.org/10.1186/1472-6920-12-69

Andrews, T. M., Leonard, M. J., Colgrove, C. A., & Kalinowski, S. T. (2011). Active learning not associated with student learning in a random sample of college biology courses. *CBE Life Sciences Education, 10*(4), 394–405. https://doi.org/10.1187/cbe.11-07-0061

APS. (2020). American Physiological Society Institute on Teaching and Learning Virtual Week "The Future of Learning in Physiology" [Press release].

Arons, A. B. (1983). Achieving wider scientific literacy. *Daedalus, 112*(2), 91–122. Retrieved from http://www.jstor.org/stable/20024855

Artino, A. R., La Rochelle, J. S., & Durning, S. J. (2010). Second-year medical students' motivational beliefs, emotions, and achievement. *Medical Education, 44*(12), 1203–1212. https://doi.org/10.1111/j.1365-2923.2010.03712.x

Australia Department of Education, Skills, & Training. (2003). *Higher education report for 2003 to 2005 triennium* (1442-7680 (print)). Retrieved from Department of Education, Skills, & Training.

Azevedo, R., Johnson, A., Chauncey, A., & Burkett, C. (2010). Self-regulated Learning with MetaTutor: Advancing the science of learning with MetaCognitive tools. In M. S. Khine & I. M. Saleh (Eds.), *New science of learning: Cognition, computers and collaboration in education* (pp. 225–247). Springer.

Badger, C., Horrocks, S., Turton, C., & Lewis, H. (2019). Using technology to promote metacognition. *Impact: Journal of the Chartered College of Teaching*. Retrieved from https://impact.chartered.college/article/using-technology-promote-metacognition/

Ball, S., & Bogatz, G. (1970). *The first year of Sesame Street : An evaluation: A report to the children's television workshop*. Educational Testing Service.

Bearman, M., Palermo, C., Allen, L. M., & Williams, B. (2015). Learning empathy through simulation: A systematic literature review. *Simulation in Healthcare, 10*(5), 308–319. https://doi.org/10.1097/sih.0000000000000113

Beckerson, W., Anderson, J., Perpich, J., & Yoder-Himes, D. (2020). Research and teaching: An introvert's perspective: Analyzing the impact of active learning on multiple levels of class social personalities in an upper level biology course. *Journal of College Science Teaching, 49*(3), 47–57. https://doi.org/10.2505/4/jcst20_049_03_47

Bennett, L. M., & Gadlin, H. (2012). Collaboration and team science: From theory to practice. *Journal of Investigative Medicine, 60*(5), 768–775. https://doi.org/10.2310/JIM.0b013e318250871d

Bjork, R. A., Dunlosky, J., & Kornell, N. (2013). Self-regulated learning: Beliefs, techniques, and illusions. *Annual Review of Psychology, 64*, 417–444. https://doi.org/10.1146/annurev-psych-113011-143823

Bonwell, C., & Eison, J. (1991). *Active learning: Creating excitement in the classroom*. ERIC Clearinghouse on Higher Education.

Cech, T., & Kennedy, D. (2005). Doing more for Kate. *Science, 310*(5755), 1741. https://doi.org/10.1126/science.1123580

Chakraborti, C., Boonyasai, R. T., Wright, S. M., & Kern, D. E. (2008). A systematic review of teamwork training interventions in medical student and resident education. *Journal of General Internal Medicine, 23*(6), 846–853. https://doi.org/10.1007/s11606-008-0600-6

Clark, R. E. (1983). Reconsidering research on learning from media. *Review of Educational Research, 53*(4), 445–459. https://doi.org/10.3102/00346543053004445

Cook, T. D., Appleton, H., Conner, R. F., Shaffer, A., Tamkin, G., & Weber, S. J. (1975). *Sesame street revisited*. Russell Sage.

Corell, A., Regueras, L. M., Verdú, E., Verdú, M. J., & de Castro, J. P. (2018). Effects of competitive learning tools on medical students: A case study. *PLoS One, 13*(3), e0194096. https://doi.org/10.1371/journal.pone.0194096

Corliss, S. B., & Lee, M. W. (2021). Assessment and teaching of professional attitudes and behaviors. In W. B. Jeffries & K. Huggett (Eds.), *An introduction to medical teaching* (3rd ed.). Springer.

DeHaan, R. L. (2005). The impending revolution in undergraduate science education. *Journal of Science Education and Technology, 14*(2), 253–269. https://doi.org/10.1007/s10956-005-4425-3

Deslauriers, L., McCarty, L. S., Miller, K., Callaghan, K., & Kestin, G. (2019). Measuring actual learning versus feeling of learning in response to being actively engaged in the classroom. *Proceedings of the National Academy of Sciences of the United States of America, 116*(39), 19251–19257. https://doi.org/10.1073/pnas.1821936116

Ebbinghaus, H. (1913). *Memory: A contribution to experimental psychology*. Teachers College Press.

Elrod, S. (2010). Project Kaleidoscope 2.0: Leadership for twenty-first century STEM education. *Liberal Education, 96*, 24–33.

Fanning, R. M., & Gaba, D. M. (2007). The role of debriefing in simulation-based learning. *Simulation in Healthcare, 2*(2), 115–125. https://doi.org/10.1097/SIH.0b013e3180315539

Fisch, S. M., Truglio, R. T., & Cole, C. F. (1999). The impact of sesame street on preschool children: A review and synthesis of 30 years' research. *Media Psychology, 1*(2), 165–190. https://doi.org/10.1207/s1532785xmep0102_5

Freeman, S., Eddy, S. L., McDonough, M., Smith, M. K., Okoroafor, N., Jordt, H., & Wenderoth, M. P. (2014). Active learning increases student performance in science, engineering, and mathematics. *Proceedings of the National Academy of Sciences of the United States of America, 111*(23), 8410–8415. https://doi.org/10.1073/pnas.1319030111

Gallo, A. (2017). A refresher on A/B testing. Retrieved from https://hbr.org/2017/06/a-refresher-on-ab-testing

Goldman, J., & Wong, B. (2020). Nothing soft about 'soft skills': Core competencies in quality improvement and patient safety education and practice. *BMJ Quality and Safety, 29*, 619–622.

Google. (2021). Create an A/B test. Retrieved from https://support.google.com/optimize/answer/6211930?hl=en

Handelsman, J., Ebert-May, D., Beichner, R., Bruns, P., Chang, A., DeHaan, R., Gentile, J., Lauffer, S., Stewart, J., Tilghman, S. M., & Wood, W. B. (2004). Education. Scientific teaching. *Science, 304*(5670), 521–522. https://doi.org/10.1126/science.1096022

Harris, D., Bellew, C., Asmar, A., Cendan, J., & Johnson, T. (2015). High fidelity patient simulations (HFPS) in medical physiology improve non-cognitive skills in first year medical students. *The FASEB Journal, 29*(S1), 687.616. https://doi.org/10.1096/fasebj.29.1_supplement.687.16

Hylton, K., Levy, Y., & Dringus, L. (2016). Utilizing webcam-based proctoring to deter misconduct in online exams. *Computers in Education, 92-93*, 53–63.

Isaacson, W. (2011). *Steve Jobs*. Simon & Schuster.

Johnson, T. R., Lyons, R., Chuah, J. H., Kopper, R., Lok, B. C., & Cendan, J. C. (2013). Optimal learning in a virtual patient simulation of cranial nerve palsies: The interaction between social learning context and student aptitude. *Medical Teacher, 35*(1), e899–e907. https://doi.org/10.3109/0142159X.2012.714884

Jordan, K. (2015). *Massive open online course completion rates revisited: Assessment, length and attrition. The International Review of Research in Open and Distributed Learning, 16*(3) https://doi.org/10.19173/irrodl.v16i3.2112

Kauffman, C. A., Derazin, M., Asmar, A., & Kibble, J. D. (2019). Patterns of medical student engagement in a second-year pathophysiology course: Relationship to USMLE Step 1 performance. *Advances in Physiology Education, 43*(4), 512–518. https://doi.org/10.1152/advan.00082.2019

Kelley, P., & Whatson, T. (2013). Making long-term memories in minutes: A spaced learning pattern from memory research in education. *Frontiers in Human Neuroscience, 7*, 589. https://doi.org/10.3389/fnhum.2013.00589

Klein, B. A. (2020). The evolution of high-fidelity patient simulation in medical education: A literature review. *The FASEB Journal, 34*(S1), 1–1. https://doi.org/10.1096/fasebj.2020.34.s1.07438

Konnikova, M. (2014, November 7). Will MOOCs be flukes? *The New Yorker*. Retrieved from https://www.newyorker.com/science/maria-konnikova/moocs-failure-solutions

Korte, M., & Schmitz, D. (2016). Cellular and system biology of memory: Timing, molecules, and beyond. *Physiological Reviews, 96*(2), 647–693. https://doi.org/10.1152/physrev.00010.2015

Kozma, R. B. (1994). Will media influence learning? Reframing the debate. *Educational Technology Research and Development, 42*(2), 7–19. https://doi.org/10.1007/BF02299087

Lee, M. W. (2016). Using retrieval practice to teach professionalism to medical students: A proposal. *Medical Science Educator, 26*(4), 719–722. https://doi.org/10.1007/s40670-016-0304-1

Lee, M. W., & Zahedivash, A. (2021). Integrated pharmacology atlas. Retrieved from https://pharmatlas.dellmed.utexas.edu

Lei, J.-H., Guo, Y.-J., Chen, Z., Qiu, Y.-Y., Gong, G.-Z., & He, Y. (2016). Problem/case-based learning with competition introduced in severe infection education: An exploratory study. *Springerplus, 5*(1), 1821. https://doi.org/10.1186/s40064-016-3532-3

Luiz Adrian, J. A., Zeszotarski, P., & Ma, C. (2015). Developing pharmacy student communication skills through role-playing and active learning. *American Journal of Pharmaceutical Education, 79*(3), 44–44. https://doi.org/10.5688/ajpe79344

Mares, M.-L., & Pan, Z. (2013). Effects of Sesame Street: A meta-analysis of children's learning in 15 countries. *Journal of Applied Developmental Psychology, 34*, 140–151.

Martin, R. E., & Ochsner, K. N. (2016). The neuroscience of emotion regulation development: Implications for education. *Current Opinion in Behavioral Sciences, 10*, 142–148. https://doi.org/10.1016/j.cobeha.2016.06.006

McNair, R., Stone, N., Sims, J., & Curtis, C. (2005). Australian evidence for interprofessional education contributing to effective teamwork preparation and interest in rural practice. *Journal of Interprofessional Care, 19*(6), 579–594. https://doi.org/10.1080/13561820500412452

Michael, J. (2006). Where's the evidence that active learning works? *Advances in Physiology Education, 30*(4), 159–167. https://doi.org/10.1152/advan.00053.2006

Miller, K., Lukoff, B., King, G., & Mazur, E. (2018). Use of a social annotation platform for pre-class reading assignments in a flipped introductory physics class. *Frontiers in Education, 3*. https://doi.org/10.3389/feduc.2018.00008

National Academies of Sciences, Engineering, & Medicine. (2018). *How people learn II: Learners, contexts, and cultures*. The National Academies Press.

National Academy of Sciences. (2007). *Rising above the gathering storm: Energizing and employing america for a brighter economic future*. The National Academies Press.

National Research Council. (2000). *How people learn: Brain, mind, experience, and school: Expanded edition*. The National Academies Press.

National Research Council. (2003). *BIO2010: Transforming undergraduate education for future research biologists*. The National Academies Press.

Natividad, G., Spector, J. M., & Evangelopoulos, N. (2018). Research topics and trends in educational technology. In G. Natividad, J. M. Spector, & N. Evangelopoulos (Eds.), *An analysis of two decades of educational technology publications: Who, what and where* (pp. 85–94). Springer.

O'Connell, M. T., & Pascoe, J. M. (2004). Undergraduate medical education for the 21st century: Leadership and teamwork. *Family Medicine, 36(Suppl)*, S51–S56.

Our World in Data. (2014). Gross enrollment ration in tertiary education, 1970-2014. Retrieved from https://ourworldindata.org/grapher/gross-enrollment-ratio-in-tertiary-education?country=East+Asia+%26+Pacific~Europe+%26+Central+Asia~Latin+America+%26+Caribbean~Middle+East+%26+North+Africa~North+America~South+Asia~Sub-Saharan+Africa

Owens, D. C., Sadler, T. D., Barlow, A. T., & Smith-Walters, C. (2020). Student motivation from and resistance to active learning rooted in essential science practices. *Research in Science Education, 50*(1), 253–277. https://doi.org/10.1007/s11165-017-9688-1

Pekrun, R., Frenzel, A. C., Goetz, T., & Perry, R. P. (2007). The control-value theory of achievement emotions: An integrative approach to emotions in education. In P. A. Schutz & R. Pekrun (Eds.), Emotion in education (pp. 13–36). Academic Press.

Petrilli, M. J. (2018). From cat videos and cooking tips to the history of the punic wars. Retrieved from https://www.educationnext.org/from-cat-videos-cooking-tips-to-history-of-punic-wars-educational-content-youtube/

Rahman, A. A., Aris, B., Rosli, M. S., Mohamed, H., Abdullah, Z., & Zaid, N. (2015). Significance of preparedness in flipped classroom. *Advanced Science Letters, 21*(10), 3388–3390. https://doi.org/10.1166/asl.2015.6514

Reinemann, J., LaBell, K., Bellew, C., Asmar, A., Cendán, J., & Harris, D. M. (2015). Integration and improvement of teamwork skills in first year medical students by using physiology-based high fidelity patient simulations. *Medical Science Educator, 25*(4), 503–512. https://doi.org/10.1007/s40670-015-0161-3

Ritter, B. A., Small, E. E., Mortimer, J. W., & Doll, J. L. (2017). Designing management curriculum for workplace readiness: Developing students' soft skills. *Journal of Management Education, 42*(1), 80–103. https://doi.org/10.1177/1052562917703679

Rop, C. J. (2002). The meaning of student inquiry questions: A teacher's beliefs and responses. *International Journal of Science Education, 24*(7), 717–736. https://doi.org/10.1080/09500690110095294

Schuwirth, L. W., & Van der Vleuten, C. P. (2011). Programmatic assessment: From assessment of learning to assessment for learning. *Medical Teacher, 33*(6), 478–485. https://doi.org/10.3109/0142159x.2011.565828

Seidel, S. B., & Tanner, K. D. (2013). "What if students revolt?"—Considering student resistance: Origins, options, and opportunities for investigation. *CBE—Life Sciences Education, 12*(4), 586–595. https://doi.org/10.1187/cbe-13-09-0190

Silverthorn, D. U. (2020). When active learning fails… and what to do about it. In J. J. Mintzes & E. M. Walter (Eds.), *Active learning in college science: The case for evidence-based practice* (pp. 985–1001). Springer.

Silverthorn, D. U., Thorn, P. M., & Svinicki, M. D. (2006). It's difficult to change the way we teach: Lessons from the Integrative Themes in Physiology curriculum module project. *Advances in Physiology Education, 30*(4), 204–214. https://doi.org/10.1152/advan.00064.2006

Spector, J. M. (2020). Remarks on progress in educational technology. *Educational Technology Research and Development, 68*(3), 833–836. https://doi.org/10.1007/s11423-020-09736-x

Steinel, N., Palmer, G. C., Nowicki, E., Lee, E., Nelson, E., Whiteley, M., & Lee, M. W. (2019). Integration of microbiology, pharmacology, immunology, and infectious disease using active teaching and self-directed learning. *Medical Science Educator, 29*(1), 315–324. https://doi.org/10.1007/s40670-018-00689-8

Sullivan, A. N., & Lachman, M. E. (2016). Behavior change with fitness technology in sedentary adults: A review of the evidence for increasing physical activity. *Frontiers in Public Health, 4*, 289. https://doi.org/10.3389/fpubh.2016.00289

Theobald, E. J., Hill, M. J., Tran, E., Agrawal, S., Arroyo, E. N., Behling, S., Chambwe, N., Cintrón, D. L., Cooper, J. D., Dunster, G., Grummer, J. A., Hennessey, K., Hsiao, J., Iranon, N., Jones, L., 2nd, Jordt, H., Keller, M., Lacey, M. E., Littlefield, C. E., … Freeman, S. (2020). Active learning narrows achievement gaps for underrepresented students in undergraduate science, technology, engineering, and math. *Proceedings of the National Academy of Sciences of the United States of America, 117*(12), 6476–6483. https://doi.org/10.1073/pnas.1916903117

Twenge, J. M., Spitzberg, B. H., & Campbell, W. K. (2019). Less in-person social interaction with peers among U.S. adolescents in the 21st century and links to loneliness. *Journal of Social and Personal Relationships, 36*(6), 1892–1913. https://doi.org/10.1177/0265407519836170

Udovic, D., Morris, D., Dickman, A., Postlethwait, J., & Wetherwax, P. (2002). Workshop biology: Demonstrating the effectiveness of active learning in an introductory biology course. *BioScience, 52*(3), 272–281. https://doi.org/10.1641/0006-3568(2002)052[0272:Wbdteo]2.0.Co;2

Vandercruysse, S., Vandewaetere, M., Cornillie, F., & Clarebout, G. (2013). Competition and students' perceptions in a game-based language learning environment. *Educational Technology Research and Development, 61*(6), 927–950. https://doi.org/10.1007/s11423-013-9314-5

van Vliet, E. A., Winnips, J. C., & Brouwer, N. (2015). Flipped-class pedagogy enhances student metacognition and collaborative-learning strategies in higher education but effect does not persist. *CBE Life Science Education, 14*(3). https://doi.org/10.1187/cbe.14-09-0141

Waldrop, M. M. (2013). Online learning: Campus 2.0. *Nature, 495*(7440), 160–163. https://doi.org/10.1038/495160a

Yan, D., von Davier, A., & Lewis, C. (2016). *Computerized multistage testing: Theory and applications*. CRC Press/Taylor & Francis.

Zahedivash, A., & Lee, M. W. (2018). Development of an open-access, web-based interactive tool to learn autonomic nervous system physiology and pharmacology. *Advances in Physiology Education, 42*(1), 64–67. https://doi.org/10.1152/advan.00125.2017

Chapter 2
Technology, Equity, and Inclusion in the Virtual Education Space

Cynthia Taylor, Bryan Dewsbury, and Cynthia Brame

Abstract Advances in technology have been occurring at an exponential rate, and the increased sophistication of these advances can be seen in the proliferation of new educational tools. As educational technology continues to immerse itself in the higher education space, it behooves administrators and practitioners to consider the ways in which this technology, either by design or by its alignment with polarizing capitalist framework models, may exclude students traditionally marginalized in education. When these systems are cultivated in environments with adversarial relationships between instructors and students, then requisite solutions are necessarily crafted in ways that will be fundamentally exclusive. We argue in this essay that inclusion in the virtual education space only occurs when key stakeholders are mindful and intentionally attentive to the structures that undergird surveillance capitalism and educational "solutionism" assumptions of technological innovation, and the factors that govern how and why people engage with technology writ large. This requires examining the written assumptions companies identify in their policies and applying a critical lens to the problems technology is purported to solve. A focus on recentering humanity as the driving philosophy of inclusive pedagogy can also improve how the academic community engages, uses, and operationalizes technological advancement in the virtual education space.

C. Taylor
Department of Sociology and Anthropology, University of Rhode Island, Kingston, RI, USA
e-mail: cynthia_taylor@uri.edu

B. Dewsbury (✉)
Florida International University, Miami, FL, USA
e-mail: bdewsbur@fiu.edu

C. Brame
Center for Teaching, Vanderbilt University, Nashville, TN, USA

Department of Biological Sciences, Vanderbilt University, Nashville, TN, USA
e-mail: cynthia.brame@Vanderbilt.edu

H. J. Witchel, M. W. Lee (eds.), *Technologies in Biomedical and Life Sciences Education*, Methods in Physiology, https://doi.org/10.1007/978-3-030-95633-2_2

2.1 Introduction

Today, the term "educational technology" is commonly associated with the idea of computer-assisted instruction and/or learning. However, educational technologies encompass an array of tools, techniques, and processes that are of perceived pedagogical value. As Veletsianos and Moe (2017) explain, educational technology, or ed tech, may refer to any "efforts to design, develop, and use technology to achieve a never-ending array of desirable educational outcomes, including improving learning, increasing retention rates, enhancing teaching effectiveness, reducing costs, and increasing access." Systematic investigations of the impact of educational technologies on instruction and learning outcomes may be a modern phenomenon, but the use of technology in attempts to enhance educational experiences is as old as formal education itself (for a more comprehensive history of educational technology, see Spector, 2015).

With the introduction of personal computers in both classrooms and households beginning in the 1980s, recent decades have witnessed an increasing optimism that machine-based educational technologies and online learning will revolutionize STEM education and institutionalized learning at large (Neumann & Waight, 2020). Proponents of the use of cutting-edge educational technologies (such as new digital, telecommunications, and multimedia tools) have long heralded their potential benefits in "helping to scale up standardized instruction, facilitate differentiated instruction, expand opportunities for practice, and increase student engagement" (Ganimian et al., 2020, p. 6). Additionally, the widespread adoption of such tools has been promoted as a promising solution for addressing persistent issues surrounding diversity, equity, and inclusion in the STEM classroom and beyond.

Hollenbeck and Hollenbeck (2009) note a variety of ways that computer-based educational technologies can be used to enhance inclusion in multicultural classrooms. For example:

- The use of online translation services can help educators tailor the delivery of content in ways that are more inclusive to cultural and linguistic diversity.
- Multimedia content can be utilized to deliver more personalized and individualized content and provide multisensory experiences that accommodate different learning styles and preferences.
- Various software can be used to assess learning outcomes, identify individual learning challenges, and provide more expedited feedback to learners.
- Educational and informational technologies can be used to enhance communication and collaboration between students and facilitate the creation of cooperative learning environments that contribute to students' self-esteem and sense of belonging.

These potential benefits of educational technology integration in the classroom, however, are often offset by organizational norms, institutional constraints, competing educational priorities, and lack of professional development among educators that limit the impact of digital technologies on student learning (Peck et al., 2002).

Additionally, inequities surrounding students' access to learning technologies and inclusion in digital environments can result in an expansion, rather than narrowing, of opportunity gaps—especially for the most vulnerable populations (Gonzales et al., 2020; Hawkins & Oblinger, 2006; Moore et al., 2018). Unfortunately, when it comes to educational technology adoption and integration, there is often a mismatch between "techno-promoter dreams" and the "realities" of student experiences (Peck et al., 2002). As Veletsianos and Moe (2017) observe "throughout its history, educational technology proponents have assumed positive impacts, promoting an optimistic rhetoric despite little empirical evidence of results—and ample documentation of failures."

The ongoing Covid-19 pandemic and the remote learning environments it has necessitated have shed light on widespread social, economic, and institutional barriers to effective and equitable educational technology integration (see Chap. 15 in this volume). At the same time, the pandemic has also been a boon to educational technology developers and companies and has accelerated the pace at which new educational technologies are being integrated into higher education curricula. Many of these technologies are of questionable pedagogical value and perpetuate untested, outdated, and/or inadvisable practices surrounding teaching and learning, as discussed in the third section of this chapter. Equally problematic are the business models underlying popular educational technologies (such as learning management systems and plagiarism/cheating detection software), which demand that data privacy and ownership be surrendered through widespread tracking and surveillance of students.

Existing scholarship has focused disproportionately on the promise of educational technology and neglected its perils, particularly pertaining to inclusion. This chapter therefore aims to highlight pressing tech-equity issues that educators must contend with as they seek technological solutions to current challenges. In what follows, we discuss how the unreflective adoption of educational technologies, without consideration of the ecosystems in which they emerge and are implemented, is more likely to reinscribe and exacerbate existing inequalities in STEM education, rather than reduce them. We focus specifically on how current trends in educational technology perpetuate the dehumanizing and exclusionary logics of surveillance capitalism (Zuboff, 2015: p.75). Surveillance capitalism is defined as "A rogue mutation of capitalism marked by concentrations of wealth, knowledge, and power unprecedented in human history [...] that claims human experience as free raw material for hidden commercial practices of extraction, prediction, and sales" (Zuboff, 2019). We examine how the design and implementation of certain educational technologies may facilitate the "extraction, commodification and control" of student data (in service of surveillance capitalism and behavioral modifications) and undermine efforts toward inclusion, equity, and justice in digital environments. We argue that efforts to design more inclusive digital learning environments must include critical evaluations of both the problems educational technology is enlisted to solve and the business models underlying technological "solutions" to these problems.

A keen reader will notice that many of the sources we use to construct arguments in this chapter come from educational technology blogs and news sites and, as such, are not peer-reviewed. This is, in part, a reflection of the fact that research in the field of educational technology cannot keep up with the pace at which new technologies are currently being developed and adopted. Given that blogs and journalistic sources are not constrained by the lengthy peer-review process, they can provide more up-to-date information on current trends and issues in educational technology. In the absence of evidence-based research on novel technologies and their impacts, such sources are of value to both researchers and educators. We recognize that there are, however, potential problems with relying on non-peer-reviewed sources, and so we were careful to focus our attention only on publications that were authored by credible and respected scholars in the educational technology arena.

In contrast to other chapters in this volume, our evaluations and recommendations regarding educational technologies and their associated pedagogies are informed primarily by critical theory rather than positivism and reductionism. While we do draw from empirical research to support our arguments, we are of the firm belief that discussions on the impacts of educational technologies on inclusion and equity must be deliberately anti-reductionist in their approach. This chapter is not intended to provide a comprehensive review of technologies, pedagogies, and best practices as they pertain to inclusion. Rather, the goal of this chapter is to encourage readers to question the underlying assumptions, ideologies, and agendas that drive current trends in educational technology design, implementation, and research and cultivate greater awareness of how these issues intersect with issues of inclusion and equity in the educational arena.

The arguments put forth here have been shaped primarily by observations surrounding educational technology use in the United States, as this is the context the authors are working in and most familiar with. The US-centric lens through which we examine educational technology leaves out many important trends, issues, and discussions specific to other regions of the world. At the same time, many of the problems we identify and describe here, specifically as they relate to surveillance capitalism and the exclusionary/discriminatory potential of technology implementation, are of global relevance. We hope readers will find a discussion of these issues to be of value, regardless of the geographic and cultural context in which they are working.

2.2 Barriers to Inclusion in STEM and Educational Technology

Understanding the barriers marginalized students in STEM face in general is critical to understanding the barriers that might present themselves in educational technology, because the underlying social assumptions associated with those barriers may determine who thrives or who is left behind or manipulated in the virtual environment. In the United States, disparities still exist in STEM academic outcomes. Twenty-nine percent of Black students beginning their Bachelor's degree in STEM left the institution entirely, and 36% changed to a non-STEM major. Comparatively, 20% of White students left their institution entirely, and 28% switched to a non-STEM major (Chen, 2013). Black and Hispanic professors make up only 4% of the USA STEM professoriate, a paltry proportion considering that they make up 16% of the general population (Lynch & Lyle-Gonga, 2019). Women are less likely to drop out of postsecondary education in the United States, but they are much more likely to switch to a non-STEM major than male students (Chen, 2013). This switching of disciplines without leaving higher education entirely may partly explain why the percentage of women decreases as one goes deeper into the STEM career pathway. The "chilly climate" studies from the 1980s are often credited as the beginning of higher education's focus on the role of belonging for women on campus (Heller et al., 1985). However, even more recent evidence points to the unwelcoming culture of STEM driving women away. Hunter (2019) showed that for their sample of students, 56% of the women switchers cited "competitive, unsupportive STEM culture" as a reason for switching compared to 46% of men. Students with disabilities, an oft ignored population in STEM disciplines, also leave STEM at higher rates than their counterparts (Vogel et al., 2008). Additionally, a lack of epistemological diversity (ways of knowing) in STEM curricula (Garcia-Olp et al., 2020) may leave some students feeling unwelcome or completely excluded from the STEM experience based on the perception that science is incompatible with certain cultural backgrounds and worldviews, such as Indigenous communities that transmit knowledge by storytelling. Studies have identified several reasons why a culture of exclusion both exists and persists in science, and a full review of these reasons is outside the scope of this essay. In brief, a culture of exclusion occurs when there is direct knowledge dissemination: the teacher states information, provides examples, and presents questions or problems for the students to answer, and in return the student listens, follows directions expeditiously, and answers the questions (Louie, 2017). Students who do not learn in this way are excluded as "bad students," despite the fact that this teaching method largely omits the sensemaking, experimentation, communication, and creativity that are central to professional scientific practice.

It is noteworthy to understand that institutions of higher education by-and-large have always generally been subservient to the social order of their time. With rare exceptions, college classroom pedagogies and structures were consequences of social hierarchies and not drivers of a different social order. In this light, STEM instruction was no different. However, unique properties of science and its underlying philosophy resulted in this exclusion, taking on a unique flavor. Many of the reasons for the development of the exclusive culture in science harken back to the late sixteenth century, when falsifiability became the bedrock of the scientific method and took social precedence on how the world should be interpreted (Voegelin, 1948). This philosophy also absolved scientific practice and pedagogy of responsibility to individuals, whose access to or impact by science were subsumed by the focus on a larger, "scientific" truth (Voegelin, 1948). This means the psychosocial contexts that impact the learning process in particular, though well-documented in the psychology and sociology literature, are routinely considered fringe elements of excellent science classroom teaching (Gasparatou, 2017).

Oblivious to this scholarship, curriculum designers, pedagogues, and STEM instructional policy often continue to privilege content distribution over the cultivation of inclusive environments (Petersen et al., 2020). In the absence of this, the psychological contingencies that result in things like stereotype threat (Steele & Aronson, 1995), reduced sense of belonging (Master & Meltzoff, 2020), and low sense of STEM identity (Kim et al., 2018) preclude many capable students from full entry into the discipline. This is in addition to contingencies associated with lack of direct access to resources. Black and Hispanic students in many colleges and universities are more likely to come from under-resourced school districts, be low-income, and be a first-generation college student (Crosnoe, 2009). Collectively, this means that the psychosocial factors are not only unfortunately predictive of academic outcomes but are also simultaneously divorced from any actual measure of students' ability to do STEM. Instructors are able to mitigate the potential negative impacts of psychosocial factors by being explicit with their curricula intentions (Brame, 2019; Dewsbury & Brame, 2019). Pedagogies that focus on malleable characteristics and are not reinforcing of fixed traits are key in this endeavor (Yeager & Dweck, 2012). Theobald et al. (2020) propose the "heads and hearts" hypothesis for achieving active learning in an inclusive classroom to reduce the achievement gap:

- Extensive and highly focused efforts geared toward improving performance—meaning that students work hard on relevant tasks
- Scaffolded exercises designed to address specific deficits in understanding or skills
- Immediate feedback
- Repetition
- Treating students with dignity and respect
- Communicating confidence in students' ability to meet high standards
- Demonstrating a genuine interest in students' intellectual and personal growth and success

While barriers to inclusion in STEM are not new, they may be compounded when educational technologies introduce additional barriers. Inclusion in the educational environment, regardless of medium, assumes three things: (1) equitable access to all of its features, (2) an environment constructed so that all stakeholders are able to thrive regardless of background, and (3) a design that fosters stakeholders' agency in shaping the construction of the environment. Inclusivity is meant to shift the onus for making equity succeed from the marginalized students to actors within the institution; put another way, the institution should "meet students halfway" (Naylor & Mifsud, 2019). By allowing a few mostly for-profit actors to shape and determine the parameters of the virtual learning environment, opportunities abound for the violation of any or all of these principles.

Despite possessing one of the largest economies in the world, the United States lags behind most Westernized countries in terms of the ubiquity of Internet access (Howard et al., 2010). Many rural and low-income communities still struggle to access high-speed Internet or in some cases any broadband access at all (Drake et al., 2019; Ford et al., 2020; Whitacre & Mills, 2007). The economic structures and disparities in physical infrastructures between more urban and highly remote areas partly explain this. Internet access is still essentially viewed as a private good (Crock Bauerly et al., 2019; Eng et al., 1998), which means that acquiring it is stratified along economic means, except in situations where it is accessed as a by-product of another environment like employment or attendance at school. In the case of school, however, this means that students must remain tethered to the physical environment in order to reap the benefits of this service. There are consequences to this umbilical relationship for educational environments that go virtual. For students from low economic means, leaving the physical campus removes access to this service, thus stratifying, among classroom members, who might continue to enjoy the benefits and who may not (Asanov et al., 2021; Jordan et al., 2021).

The reality of this stratification was laid bare in the spring of 2020, when a pandemic caused by a lethal strain of the coronavirus forced all institutions of higher education to transition their instruction to virtual environments. What quickly followed was the recognition (a first for many) that many of their students did not have the option to quickly transition home and engage their academics at the same level. Institutions of higher education recognized, lamentably anew, that campuses were not only simply a place for the pursuit of credentials but also an opportunity for marginalized students (assuming they have sufficient financial support for on campus study) to feel safe and to access resources they otherwise may not be able to. In cases where institutions were not able to continue to house marginalized students, Internet access was derived by congregating in parking lots, restaurants, and public spaces. There are some examples of institutions that rose to meet this challenge (Nuere & de Miguel, 2020), but these examples are disappointingly few. It is not entirely clear what the physical infrastructure of the future university will be, but many studies have documented trends in the past decade that suggest many classrooms will migrate to a virtual space (Martin et al., 2011). This may have ramifications for those without full private access to the Internet.

Access to virtual environments does not necessarily guarantee that all participants thrive. Educational technology, particularly at institutions of higher education, has assumed a certain comfort level with the ways in which naive users can engage its products. This is a problematic notion when one considers the rate at which technology has increased in recent years. Some technologies have been both cutting edge and rendered completely obsolete within two decades. This means that demographics whose technological prowess have not kept pace with this advancement can be shut out of this environment if their needs are not considered. Three groups in particular are likely to experience barriers in this scenario.

First, students with limited access to technology may not navigate virtual environments as easily as those for whom access was assumed prior to matriculating to higher education. The proliferation of smartphones, and the reduction in price of accessing this technology, means that a greater percentage of individuals are embedded in a technological space, but this does not necessarily mean that prowess in this space translates to nonmobile environments. While Internet access is ubiquitous enough such that even low-resourced secondary schools are able to engender a decent amount of exposure, the level of sophistication associated with their tech curricula is likely to vary. This presents not so much a barrier to comfort in the virtual environment as a consumer but the degree to which students from marginalized backgrounds are motivated to pursue majors that place them in technology-based professions. This perpetuates the aforementioned problem of representation in technology disciplines, which in turn reduces the likelihood that these technologies will be built to be inclusive of this population in the future (Binkerd & Moore, 2002).

Second, educational technology and much of higher education in general is designed with a traditionally aged individual entering college as an emerging adult (Roberts, 2011). While recent reports have encouraged institutions to pay closer attention to adult learners, the image of the modal college student is still a largely 4-year degree seeking individual in the 18–22 demographic. Even the term nontraditional is slowly becoming an inaccurate description of adult learners as recent reports suggest an increasing percentage of college students are nontraditional. Barriers present themselves if educational technology is not cognizant of learners who, in pursuing opportunities later in life, might be new to this environment and may thus need a different type of onboarding process (Park & Choi, 2009).

Third, virtual learning environments are not uniform in terms of their alignment with UDL (Universal Design for Learning) principles, because using them well subjectively benefits from teacher training (Scott et al., 2015). Advocacy from this demographic has resulted in a marked improvement in the design and accessibility of these platforms for students with disabilities, but these improvements have fallen short of an accepted and guaranteed standard around which this design takes place. Inclusion in the design of virtual technology occurs when its purveyors consider a broader perspective of who users might be. If that perspective is based on a narrow stereotype of the user, barriers are likely to persist.

Lastly, inclusion does not simply refer to the mere presence of individuals from different backgrounds (diversity). Rather, it is an active process, where the institution deliberately seeks out the voices of the students, so they know who and what they are including. It is only authentic when students have access to the power structures of the environment [see Chap. 3 (student empowerment) in this volume]. A student can access virtual environments and perhaps be successful within it academically but may still have no agency with respect to how the virtual environment is constructed and enacted for particular purposes. We use the term agency here as described by the Freirean tradition of dialogue (Freire, 1972), where the cultivation of relationships between practitioner and student determines the shape, structure, and potential of the educational experience. Dialogue in this context not only determines the process but also is ongoing throughout the academic experience. In this way, educational technology is used in service of a dialogic philosophy and not the reverse. It is in its misunderstanding of the central role of dialogue that educational technology approaches a dangerous line. For STEM teaching, this may be a challenging philosophy to fully embrace. Conventional graduate training models in the United States privilege the acquisition of content expertise and treat pedagogical skill as at best a happenstance suite of skills cultivated through the process of doctoral study. STEM classrooms then envision education solely through the lens of the content that is disbursed and not the individual that is cultivated. Dialogue recognizes that the psychosocial context of the student matters if they are to fully engage in the learning experience, and thus pedagogical mastery must necessarily speak to that. Once student voices become central to the design of the classroom, all of the classroom's elements are appropriated with their needs in mind. In this way, dialogue becomes the mechanism through which true inclusion is achieved. Students, as passive users of virtual environments, remain oblivious both to their exclusion from the shaping process and also more potentially nefarious uses of this technology. Dewsbury and Brame (2019) remind us that no educational technology is fundamentally inclusive; for inclusion to occur with an educational technology, a dynamic humanism must undergird the pedagogy, regardless of environment. Specific to educational technology, this will necessarily involve a greater degree of transparency than what is currently offered informing users about the data the platforms collect, ways in which that data is analyzed, and the full suite of decisions that these analyses are used to inform. At its best, educational technology can facilitate this type of dialogue—reducing the barriers that time and space and power differentials create. Rather than being posited as an imagined solution for educational problems, it can be better construed as a mechanism to enhance equity.

2.3 The Rhetoric Versus Reality of Educational Technology Solutionism

As is so often the case with new technologies, more consideration has historically been paid to the question of what problems can educational technology solve, than to the question of what problems educational technology (and its associated pedagogies) can produce—and reproduce (Teräs et al., 2020). This imbalanced social imagination surrounding the impact of new technologies is part of a broader ideological trend referred to as "technological solutionism" (Morozov, 2013; Teräs et al., 2020), whereby technological interventions are assumed to offer the quickest and most efficient means for addressing social problems. As an ideological project, solutionism is sustained by investments in "technological fixes" at the expense of alternatives (Weinberg, 1978). Of course, using technologies to solve problems is nothing new—that has always been an integral part of the "human strategy" (Johnston, 2018). However, widespread enchantment with technological innovation as a social "cure-all" is a distinctive feature of modernity. Johnston (2018: 48) explains that the modern "allure" of innovation is accompanied by "a constellation of other beliefs and values, such as confidence in reliable innovation and progress, trust in the impact and effectiveness of new technologies, and reliance on technical experts as general problem-solvers." At the same time, recent years have witnessed a growing public skepticism surrounding governmental and corporate promotion of "technological fixes" (Baum & McPherson, 2019). Societal acceptance of new technologies necessitates a particular framing of the issues at stake and the "problems" that technology is being asked to solve.

Public buy-in to educational technology's solutionism is contingent largely on a perpetuation of the narrative that education is in some way "broken" (Teräs et al., 2020; Watters, 2014b). Undoubtedly, there are many persistent problems in education that warrant attention and intervention. Yet, as educational technology critics remind us, it behooves us to be aware of how broken education narratives are leveraged by Silicon Valley-esque corporations and start-ups to rhetorically position education in a perpetual state of crisis and to encourage the adoption of untested "solutions" (Teräs et al., 2020; Watters, 2020a). *What* exactly is broken, and according to *whom*, is rarely defined explicitly in edupreneurs' broken education claims (Teräs et al., 2020; Weller, 2015). Rhetorical silences surrounding how problems are "set" matter greatly, since *how* exactly problems are defined is revealing of what imagined futures are deemed desirable and, hence, what interventions are prescribed. Teräs et al. (2020: 874) explain:

> It should be understood that 'education is broken' can mean different things when looked at from different worldviews. If education is seen as a conveyor belt to create graduates for the capitalist economy, then education might look like something to be (Weller, 2015) fixed with technology in order for it to be more efficient. If education is seen as something that should promote holistic human growth for a more democratic and just society, leveraging digital technologies might look somewhat different.

Under the broken education narrative, technological change is often depicted as an opportunity for "revolution" in teaching and learning (Weller, 2015). Paradoxically, however, pedagogy seems to have been peripheral (if it was considered at all) in the development of some of the most heavily marketed and most widely implemented technologies in education today. Take Zoom, for example, which was developed primarily for "enterprise customers," rather than educators and students (Zoom Video Communications, 2020), or learning management systems (LMS)—such as Blackboard, Canvas, and Moodle—which were initially designed for the purpose of facilitating and integrating administrative tasks and, as such, were not created with the needs of learners, or user experience, in mind (Watters, 2014a). Other technologies, currently being marketed under the guise of educational technology (such as exam proctoring and plagiarism detection software), have little to no pedagogical value at all and primarily function to facilitate surveillance and data tracking to further for-profit agendas (Morris & Stommel, 2017; Watters, 2017, 2020b). Rarely are the technological "solutions" promoted by "Big Ed Tech" backed by any actual research on educational impact. In the absence of such data, we must ask what the "effectiveness" of technological interventions means to different stakeholders. From the vantage point of the companies themselves, learning outcomes may be less important measures of effectiveness than are things like ease of integration and user lock-in.

Drawing from the work of Christensen (1997), Watters (2014b) observes that most purported "disruptive innovations" in educational technology are not, in fact, actually accompanied by radical changes in technology, or pedagogy, at all. In other words, for prevailing Silicon Valley models of educational technology development, it is not the technologies themselves that are disruptive, or innovative, but rather the business models behind them. Silicon Valley makes most of its money selling data that users provide for free (McElroy, 2019; Williamson, 2017b). It is no coincidence that learning management systems, such as Canvas, Blackboard, and Moodle (which have become the backbone of e-course delivery in the United States and Europe), are becoming similar to Facebook in terms of their layout, their systems of push notifications, and their constant tracking and datafication of user activity—which function to personalize content and increase user engagement. Much like Facebook (and other social media platforms), the algorithms underlying big ed tech's "personalization" of content and modes of behavioral conditioning are driven by big data analytics, dependent on the widespread tracking of user engagement, and designed primarily with profit generation and user lock-in in mind (Friesen, 2018).

Unfortunately, as is the case with social media, buy-in to disruptive business models in educational technology is often also accompanied by a loss of privacy, ownership, and agency on the part of users. For examples, Watters (2014a) notes that the submission of student work on the LMS is accompanied by a surrendering of intellectual property (in the same way that user-created content on social media becomes property of the platform), and that this work is then used as "data" by the company to improve its algorithms. Students are often not aware that their work is being used in this way and have little ability to opt out, since both classroom activities and administrative operations are becoming increasingly centralized on the LMS. Moreover, they are denied the necessary administrative privileges to access to learning materials, personal communications, and the content they created once the semester ends.

Just as alarming are the ways in which the educational technologies are currently being enlisted in systems of behavior management and modification. Unsurprisingly, behaviorist approaches are often baked into learning software, and the impact of the behaviorist imagination is readily apparent in their design, development, and implementation (Amory, 2012; Watters, 2018a, 2018b). Extending from the work of B.F. Skinner and E.L. Thorndike, behaviorism treats learning as a product of operant conditioning and relies on rewards, punishments, and reinforcements to shape and modify student behaviors (Watters, 2015). For Skinner, the impacts of operant conditioning on behavior could be better enhanced through the use of machines and programmed instruction, which allowed students greater flexibility to move through educational materials at their own pace, and could provide more immediate feedback on their performance (Casas, 2002). Skinner's ideas on "teaching machines" serve as the foundation for what has become known as "personalized" learning in educational technology, where interactive content is strategically tailored and delivered through datafication and automation, based on the preferences and past performance of users (Friesen, 2018). "Schedules of reinforcement," à la Skinnerian conditioning, are employed through stimulus-response-reward systems that combine sounds and images with addition or subtraction of points, notifications, level advancements, rewards, and so on (Flippen, 2012; Friesen, 2018; Reddy et al., 2016).

We do not deny that games, simulations, incentives, and personalized content delivery have educational value. However, we do believe that an overreliance on behaviorist-inspired educational technologies can be problematic on both pedagogical and ethical grounds. First, they tend to reduce measures of learning to changes in observable and measurable changes in behaviors, at the expense of the development of more complex intellectual, affective, and social skills (Weegar & Pacis, 2012). By ignoring the more humanistic aspects and experiences of learning, behaviorist approaches often treat learning as a behavioral product, rather than a social process (Vygotsky, 1978), and favor educational outcomes that are more easily measurable (albeit not always good indicators of deep learning or student growth). Second, behaviorist approaches favor the cultivation of compliance and standardized task performance over agency and empowerment and run the risk of dehumanizing learners. In many ways, behaviorism is antithetical to the core values of education, especially in its denial of free will and its focus on manipulation of the behavior toward uniformity. Third, many of the behaviorist-inspired schedules of reinforcement offered by educational technologies are not, in fact, necessarily designed to facilitate learning (although this may be an outcome). Rather, they are designed primarily to keep users interacting with the platform itself and to collect data on frequency of use, time of use, purpose of use, etc. to better facilitate user lock-in and profit generation in the future (Watters, 2019; Zuboff, 2019).

Morris (2017b) is especially critical of the ways in which LMS usage gets conflated with student engagement and learning [see Chap. 4 (Engagement) in this volume] and how student usage data can be used in untoward ways in educational settings. Using the example of the popular Hero K12 platform, he examines how the company marketed its learning management system to educators and administrators, specifically as a tool for behavioral monitoring and modification, and he warns of the dangers in using LMS usage data to create profiles of learners for the purpose of identifying "problematic" students and enabling "corrective" modifications. Regardless of whether or not student usage data is *actually* used in this way, the fact that LMS users have no power to resist constant surveillance and tracking of behaviors raises important questions surrounding privacy, power, and the dehumanizing effects of reducing learners to data points. Morris (2017b) writes:

> The LMS threatens the same reduction of human complexity to simple data. I say "simple" because even when data is nuanced and complex, it fails to be an accurate representation of a human being. This is not to say data cannot indicate certain behaviors, nor that it is useless, only that it has limitations. But it is not those limitations that are advertised, not those limitations that we're trained to observe; instead, we are encouraged to see data as descriptive, not just indicative. And when that happens, a surfeit of data erects a barrier between students, teachers, and administrators. But most importantly, and least spoken about, data as a determinant erects a barrier between a student and themselves.

We would be wise to consider the ways in educational technology's behaviorist emphasis on user persuasion, behavior management, and surveillance undermine students' agency, empowerment, and sense of belonging, as well as the ways in which these underlying design principles may work to perpetuate existing inequalities. Despite widespread social assumptions, technologies (educational or otherwise) are never neutral (Benjamin, 2019b). Rather, they are always imbued with ideologies and imagined futures in both their design and implementation. Existing societal logics, norms, and hierarchies are both inscribed in and reinscribed by new technologies. In other words, technology and social arrangements are mutually constructed and legitimized through a process known as coproduction (Jasanoff, 2004). Issues of equity and inclusion in educational technology then extend much deeper than visible divides in accessibility—down to the very architecture of devices and software themselves and to the pedagogies, policies, and imaginations that drive their creation, adoption, and integration in learning environments in the first place. Responsible and ethical educational technology use requires that we look beyond seemingly inevitable and innocuous technologies and critically reflect on the ways in which bias and discrimination gets "coded" into their design and implementation (Benjamin, 2019a, 2019b).

Recent criticisms of AI-assisted exam proctoring software underscore this point. Such software relies on algorithms to categorize bodies, behaviors, eye movements, gestures, etc. as "normal" or "suspicious" in efforts to detect cheating remotely. As Swauger (2020) explains "Algorithmic test proctoring encodes ideal student bodies and behaviors and penalizes deviations from that ideal by marking them as suspicious, which threatens students with academic misconduct investigations and exclusion from the educational community." Human decisions on *what* behaviors and bodies should be coded as "normal" and "ideal" by algorithms are shaped by social biases and perpetuate existing patterns of discrimination, especially as they pertain to race, gender, class, and ability status. Swauger cites several examples to support this claim—from proctoring algorithms disproportionately flagging students with disabilities as suspicious to software being unable to detect the faces of students of color (to name a few).

On May 9th, 2021, the *New York Times* published a deeply unsettling report on how the learning management system Canvas was used by Dartmouth Medical School to "retroactively track student activity during remote exams without their knowledge" (Singer & Krolik, 2021). Following a review of "data snapshots" obtained from Canvas, Dartmouth's faculty-led Committee on Student Performance and Conduct accused 17 students of cheating. These accusations (and the subsequent disciplinary actions imposed on students) were based on Canvas activity logs that the Committee interpreted as evidence that students were accessing course materials during remote examinations. However, as the *Times* report indicates, Canvas automatically generates activity data, even when students are not actively using the platform—raising further questions about the accuracy, legitimacy, and consequences of equating data with student behavior, intent, and integrity. Space limitations prevent us from exploring the nuances of this case (including how Dartmouth's actions are both rooted in and work to perpetuate a culture of mistrust on campus) and so we encourage readers to review the original article. As the authors observe, "Dartmouth's drive to root out cheating provides a sobering case study of how the coronavirus has accelerated colleges' reliance on technology, normalizing student tracking in ways that are likely to endure after the pandemic" (Singer & Krolik, 2021).

In a critical review of the current state of educational technology in STEM, Woodard (2019) argues that "(w)hat's increasingly clear is that the real digital divide of the future will be between students who can afford protection from the bad effects of technology for as long as possible, and those who can't." We argue that this inequitable future is already here and elaborate on this idea through a further examination of current trends surrounding the use of surveillance technologies in STEM education.

2.4 Surveillance Capitalism and Higher Education

Zuboff (2019) describes surveillance capitalism as "an economic system centered around the commodification of personal data with the core purpose of profit-making" (e.g., Pokemon Go harvests players location and movement data, see Teachout, 2016). The term has been sharply brought into focus from the meteoric rise of the tech industry and the psychological tools it employs to keep users engaged, even when it is to their detriment and to society's disadvantage (Bakshy et al., 2015; Tufekci, 2015). Surveillance capitalism arises from the value system of Silicon Valley, which focuses on society as "a network," where problems can be solved by innovation and combining data, which requires freedom, deregulation, and libertarianism (Levina & Hasinoff, 2017). As an economic system, libertarian capitalism operates with a flawed morality, or without one at all (Reiman, 1981, 1987; Wilson, 1995). In libertarian capitalism, most interactions are limited contracts of exchange with clear endpoints (Lewchuk & Clarke, 2011: pp. 38–40), as opposed to ongoing relationships, such as between teacher and student. Therefore, when technology is applied to higher education systems, it is worth pausing to explore how a system built on modularity and short-term profit is overlaid on a process that is supposed to centralize humanism and relationships in service of knowledge construction. When the bill of goods regarding educational technology is sold, it is certainly marketed as having noble intentions, but clear examples from the field point to, at best, some thoughtlessness with respect to its implementation.

Concerns with how student data is used did not begin with the advent of surveillance capitalism. Partly in response to a climate of student data abuse, the Family Education Rights Protection Act (FERPA) was enacted (with nine subsequent amendments) to place limits on who is able to access student data within colleges and universities (Penrose, 2011). Some of the most conservative interpretations of this law prevent students from viewing even the email addresses of their peers without explicit permission. However, this law mostly governs the access that individuals external to the university operations (including parents of the students) have to student records. It fails to address (a) how the university itself might approach their own voluminous data sets and (b) how educational technology companies might use even anonymized data to further their own aims. The current academic environment is significantly different from 1974, when FERPA was first enacted, or perhaps from even more recent times, when compared to today. On a higher education campus, student data exists in ubiquity. Magnetic strips on card swipes means that every building, parking lot, and institutional event entry is potentially recorded. The merging of educational technology into the higher education experience means that many academic behaviors as minute as simply logging onto an online system is captured with a unique timestamp and IP address. What is particularly crucial in this situation is that by registering to be part of the academic community, the student has no way to opt out or even manage any aspect of this technological relationship. This relationship is governed by the rules of capitalism and neoliberalism, paradigms that likely generated the issues educational technology positioned themselves to solve. This airtight, unbreakable connection means that the power structures within the ivory tower and the ways in which that power structure is brought to bear on inclusive practices is particularly crucial. Conventional models of education in colleges and universities do not message to students that they have any agency in questioning how educational technology is constructed and used for pedagogical and other purposes. Three considerations are important if educational technology is to be leveraged for inclusion and not simply used in a way that enables students to be unwilling, unconscious participants in a surveillance system.

2.4.1 Leveraging Technology for Inclusion Rather Than Surveillance

First, administrators and practitioners need to be mindful of how technology can drive administrator and instructor narratives of who students are. This approach has been seen in several prominent social examples. O'Neill (2016) provides several specific examples of how overreliance on opaque AI algorithms can reinforce discrimination against people from historically marginalized backgrounds, particularly for computer-assisted parole decisions and "predictive policing" (Angwin et al., 2016; Howard & Borenstein, 2018; Robinson & Koepke, 2016). Algorithms in these contexts were marketed as benign assessments. Essentially, they were equations that compiled a cadre of humanly determined salient factors to assess what the future behavior of an individual might be or the ways in which a consequence involving said humans represents an outcome and not the outcome itself. Therefore, an overfocus on the proxy can lead to erroneous conclusions, if the proxy becomes divorced from that which it is purported to assess. The unfortunate ramifications of this have been seen in the ways in which K12 teaching has been evaluated (Rubel et al., 2020), the ranking of colleges and universities (Shin & Toutkoushian, 2011), and how potential for committing a crime is determined (Joh, 2017). In even darker areas, it has been used to target demographics perceived to be vulnerable or more liable to be coerced in accepting a particular product (O'Neill, 2016). In the United

States, for-profit universities used carefully tailored algorithms to target low-income communities to aggressively market “opportunities” for college degrees mostly in the business and health sectors. This was then followed by duplicitous marketing strategies that promised returns on investments for the students in terms of successful career placement. Many students did not complete, and those who did were not accessing the professional options they were promised, leaving both groups with mountains of debt (Chen, 2013; Goodell, 2016). Marketing strategies relied on the political powerlessness of this demographic to not hold the institutions accountable.

Educational technology versions of this exist within the walls of individual campuses, and if this technology is engaged without appropriate dialogic structures, it can lead to similar outcomes that will continue to exclude students. For example, secondary school academic performance data that is posited as predictive of success in higher education can be interpreted in a couple of different ways. A deficit approach might suggest a deterministic model, where prior academic struggles are indicative, if not a guarantor, of the student also struggling in the college. Another framing might be to consider the pedagogical approaches that provide the support structures necessary for that same student to have a different experience in that subject area. Without an inclusive mindset, technology simply reinforces and fine-tunes an existing deficit mindset and provides more quantitative tools to further that opinion. Where this is especially problematic is the reality that those default, deficit mindsets are rife with sexist, ableist, and racist assumptions about student ability. Thoughtlessly used technology and the uncritical gathering of data simply provide a mechanism to codify existing bias structures. This bias is evident in many aspects of data collection, analysis, and interpretation. A different way to think about this process might be to ask some pertinent questions with respect to data collection. For example, why collect certain types of data? Because data is collected, does it necessarily need to be incorporated into a predictive or analytical model? These were the types of questions that arose from the bioinformatics community and led to a paradigm called no-boundary thinking (NBT) (Huang et al., 2013), which sought to challenge how bioinformaticians and cell and molecular scientists frame their collection of biological data. In NBT, the framing of the research question is as important as the collection itself and is necessary to determine what data should be collected and in which context. A similar thought process can apply here. Instead of outsourcing our understanding of student behavior to technological snap shots, dialogic pedagogical models should drive how curricula approaches are developed.

Second, for inclusive practices to occur in the presence of technology, institutions of higher education should be mindful of how dialogic their relationships with students are. Dialogic education models do not simply refer to principles of shared governance or soliciting student feedback. It recognizes at its core that the student-professor relationship need not be adversarial, such that any interaction between these two key actors must be one, where there are clear winners and losers. A more collaborative approach builds all elements of the course structure including assessments as an opportunity for accountability for everyone and not just the students. Conventional pedagogies tend to posit that this relationship is adversarial and in so doing set up metrics within curricula that might more accurately be seen from the student perspective as roadblocks rather than milestones. Grades are like prizes to be won and are symbols of professional value; recommendation letters are luxury goods granted by the professoriate only to those who have demonstrated value in conventional ways. Through this lens, when students are deemed to have fallen short, the reasons proffered are more likely to refer to a behavioral deficit of the students' own making, which can be corrected by "fixing them" (the students) or recruiting different types of students entirely.

Taking this mindset into virtual environments simply results in a different manifestation of the same culture of exclusion. The consequences of this mentality have been seen in the development of tools that supposedly limit cheating by freezing Web browsers (Kainz et al., 2015), virtually monitoring the physical environment of the online test taker (Winneg & Soske, 2007), and even invasive tools like monitoring facial cues (Gleim et al., 2017). Companies such as Proctorio, a learning integrity platform that provides ID verification and online exam proctoring services to schools and universities, posit these technologies as necessary to solve adversarial relationships that are well within the pedagogue's wheelhouse to reconstruct without resorting to policing. In essence, monitoring tools made by companies like Proctorio represent the actualization of an adversarial instructor-student relationship (Swauger, 2020). As a core marketing strategy, it presents itself as the solution to students whose first inclination is to be dishonest (Lang, 2013). These technologies remain in demand even in light of data that document its potential for bias (Huff & Cooper, 1987), misidentification (White et al., 2015), and the harmful consequences that have come from these errors. More importantly, the use of these products does not address some of the more fundamental issues with high stakes summative exams and how they are structured in STEM courses. Instead of assuming that students will cheat, the inclusively minded practitioner would instead ask why would students (or anyone for that matter) want to cheat (Simkin & McLeod, 2010; see Chap. 10 (cheating) in this volume). In the process of addressing this question, the practitioner might find that the ways in which stakes are structured impact whether or not the assessment would actually reflect authentic learning. For the student, overcoming those stakes becomes a barrier to be surmounted and not an opportunity to demonstrate intellectual understanding. In this lens, any opportunity (including cheating) to overcome the stakes will be taken. One can argue that there will be individuals who, regardless of circumstance, will seek to take advantage of opportunities to gain a free ride. The question to be asked however is if the role of the practitioner is to seek out this individual. In other words, the collaborative approach sets up a culture that appeals to the students' better angels and discourages their demons. There may need to be a willingness to assume that in that process, an extreme minority of bad actors might take advantage of that atmosphere over time. Inclusion therefore is communicating and even codesigning course goals and assessment structures that engage students as full participants who are pursuing meaningful ends rather than overcoming assessments designed by and for others. This re-visioning could also result in assessments that are more meaningful to both instructors and students than single-answer multiple-choice questions while still respecting the realities of large class sizes and the ways that grades are often important for students' future endeavors.

Third, administrators must be mindful of the blurry lines between protection, efficiency, and unconscionable surveillance. Protection and efficiency are central to why a mindless public volunteers intimate details about their social and financial lives on media platforms or how an unchecked government justifies its "need" to harvest limitless data from its citizens. Both philosophies have been used in the academic space to make a similar case for campus and classroom operations, but in the interest of inclusive practices, it behooves us to read the fine print more carefully. From the student perspective, most data collection is pitched in terms of efficiency. The student ID card swipe allows ease of access to everything from the fitness center to college games to specific rooms where they may work or take classes. This system not only negates carrying around physical keys but also creates a digital and physical fence around the campus that "secures" the student within it. It also creates, however, a digital footprint that allows administrators to determine who was where and for how long. Said footprint allows schools to tie physical behavior with things like cafeteria use or attendance to a specific course. Similarly, courseware typically associated with textbook companies promise magical pedagogical properties of their own. Predictive algorithms claim to measure "performance" based on several dubious variables, like "time on a question" or "number of attempts used." The studies on many of these variables are mixed, but because of proprietary considerations, the user is never privy to what goes into the algorithm (Junco & Clem, 2015; Williamson, 2017a). The instructor may simply see a color-coded bar next to a student's name that indicates whether or not they are "in trouble" in the course or doing well. What makes this veil of secrecy particularly problematic is the fact that students have no agency in this entire process.

By paying tuition and registering for a course, students are assuming that the best and most equitable decisions are being brought to bear on their academic experience. If instructors are not doing due diligence on the technology behind these platforms, then they may simply be providing conduits for the ever-expanding harvesting of student data. This is a real concern since, in the course of their undergraduate career, students are likely to use dozens of different coursewares, learning management systems, early intervention systems, and financial record platforms. A key consideration here is the recognition that because data *can* be collected does not mean it should. For example, Bowen (2012) argues that instead of investing in technology that tracks (Jones et al., 2020), rewards, and punishes student attendance, we should be seeking to design courses where students clearly recognize the added value of being physically present for the course, whether that attendance be a physical or a virtual one. If the task value is not apparent, or worse, a student can be equally successful by not attending, attention should be paid to improving the course design and not fixate on student attendance. Inclusive practice means that at the very least, there needs to be greater transparency in this process. Dialogic relationships mean that students need to be aware of how their data is used, when they register for academic experiences, and have some consenting power with respect to the additional processes that information might inform, particularly if that use is for purposes extraneous to the course.

2.5 Conclusions and Recommendations

The challenges educators and students in higher education face are numerous and daunting. The Covid-19 pandemic has cast a brighter light on, and intensified, persistent issues surrounding accessibility, equity, and inclusion—and has introduced new and unexpected obstacles to contend with. Amidst all of the struggles and uncertainties of our current day, it is no surprise that we look to technology for quick solutions. Yet, there is a danger in assuming that technologies offer the most efficient and effective "fixes" to complex social problems. For all of the problems that technologies "solve," they also create new ones. The problems *produced* by technologies are often not readily apparent until *after* these technologies have become deeply entrenched in, and irremovable from, society (Macgilchrist et al., 2020). There is also a danger in assuming that, by default, these technologies are inherently neutral and equitable in their design. Until we lessen the glaring inequities surrounding *who* exactly has the power to imagine and shape the future through innovation, we should assume that the "default design" of new technologies may in fact be biased and/or discriminatory (Benjamin, 2019b).

Current trends in educational technology (especially pertaining to data tracking and surveillance) are worrisome and warrant more critical evaluations of technologies before they are haphazardly adopted by institutions and educators. Such evaluations must look beyond the rhetoric surrounding innovation and focus on "understanding the 'here-and-now' realities rather than future possibilities and potentials of educational technology"(Selwyn, 2010, p. 69). They must also be closely attuned to the social, political, and economic contexts in which educational technology is designed, marketed, implemented, and engaged with. Responsible and ethical decisions surrounding educational technology adoption require careful consideration of the interests, and input, of diverse stakeholders and must factor in outcomes that extend "beyond learning" (Biesta, 2006; Selwyn, 2010).

In evaluations of educational technology, Morris (2017b) urges us to consider:

> How does this tool represent a politics of oppression—the surrender of privacy, data, authorship, authority, agency, as well as issues of representation, equity, access? Who owns the tool and what are their goals? How is the production of this tool funded? What influence does the maker of this tool have on culture more broadly writ? What labor is rewarded and what labor is erased? What is the relationship between this tool and the administration of the institution? Who must use this tool and who is trained to use this tool, and is that labor compensated? These are all important questions to ask, and the answers may play a role in the adoption of any given tool in a classroom or learning environment.

While we wholeheartedly agree with Morris that these questions are crucial, we also acknowledge that the burden of asking these questions should not fall on instructors alone. Both educators and students are frequently left out of discussions on the potential benefits and consequences of educational technology integration at the institutional level (Morris, 2017b) and often have little say in which classroom technologies are adopted, available, and/or required for use. To be sure, issues of institutional accountability, transparency, and openness (or lack thereof) surrounding educational technology decisions are crucial issues that warrant greater attention but are outside the scope of this chapter.

Still, educators are not completely powerless in enacting changes to the current climate surrounding educational technology. Though they may not be able to resist institutional requirements surrounding the implementation of learning management systems and other technologies, they can make better decisions about how these technologies are used and to what end. They can, and should, involve students in these decisions when possible (see Chap. 3 in this volume) and create opportunities to discuss how the use of classroom technologies intersect with larger social issues surrounding data privacy, intellectual property, surveillance, consent, and systemic inequities. They can also enhance students' sense of empowerment and agency surrounding technology use, by providing opportunities for students to voice specific technology-related concerns and allowing students to opt in, or out, of using certain technology features.

Educational technology researchers can also enact change, by challenging their own assumptions surrounding the applicability and efficacy of educational technology integration and adopting more humanistic research methods. We recommend that future research on educational technologies not be reduced simply to measures of effectiveness surrounding content-specific instruction and learning. Rather, they should be broadened to include questions related to how educational technologies can facilitate, or undermine, a student's sense of agency and belonging in learning environments and how they function to perpetuate, widen, or reduce opportunity gaps and other social inequalities. It is unlikely that insights to these questions can be gained through quantitative-based assessments of educational technology alone. In alignment with Morris and Stommel's (2018) observation that "far too much work in educational technology starts with tools, when what we need to start with is humans," we recommend that that more attention be paid to qualitative methods in educational technology research moving forward. More qualitative-based inquiries can help center human experiences in evaluations of educational technology, thus contributing to a more nuanced understanding of the actual ways in which educational technologies are engaged and the myriad social spaces in which this engagement occurs (Morris, 2017a).

The viewpoints put forth in this chapter are not intended as a recommendation for educational technology abstinence, nor should they be taken as such. There are undoubtedly benefits to, and good reasons for, educational technology incorporation in both physical and virtual environments. We recognize that educational technologies can be, and in reality *are*, used in various ways by educators to promote dialogue, inclusion, agency, and access among students. Though certain educational technologies may be more, or less, conducive to these goals over others, we also acknowledge their actual impacts are largely determined by how they are employed and for what purpose. We suspect that many of our readers are already finding ways to creatively offset and/or subvert the more problematic aspects of educational technology design and implementation through pedagogical approach.

References

Amory, A. (2012). Instructivist ideology: Education technology embracing the past? *Interactive Learning Environments, 20*(1), 41–55.

Angwin, J., Larson, J., Mattu, S., & Kirchner, L. (2016, May 23). Machine bias: There's software used across the country to predict future criminals. And it's biased against blacks. *ProPublica*. https://www.propublica.org/article/machine-bias-risk-assessments-in-criminal-sentencing. Accessed 28 Feb 2021.

Asanov, I., Flores, F., McKenzie, D., Mensmann, M., & Schulte, M. (2021). Remote-learning, time-use, and mental health of Ecuadorian high-school students during the COVID-19 quarantine. *World Development, 138*, article 105225.

Bakshy, E., Messing, S., & Adamic, L. A. (2015). Exposure to ideologically diverse news and opinion on Facebook. *Science, 348*(6239), 1130–1132.

Baum, S., & McPherson, M. (2019). The human factor: The promise & limits of online education. *Daedalus, 148*(4), 235–254.

Benjamin, R. (2019a). Introduction: Discriminatory design, liberating imagination. In R. Benjamin (Ed.), *Captivating technology: Race, carceral technoscience, and liberatory imagination in everyday life*. Duke University Press. https://read.dukeupress.edu/books/book/2588/chapter/1386029/IntroductionDiscriminatory-Design-Liberating

Benjamin, R. (2019b). *Race after technology: Abolitionist tools for the New Jim Code* (1st ed.). Polity.

Biesta, G. J. J. (2006). *Beyond learning: Democratic education for a human future* (1st ed.). Routledge.

Binkerd, C. L., & Moore, M. D. (2002). Women/minorities in computer science: Where are they? no attention no retention. *Journal of Computing Sciences in Colleges, 17*(5), 8–12.

Bowen, J. A. (2012). *Teaching naked: How moving technology out of your college classroom will improve student learning*. Wiley.

Brame, C. J. (2019). *Science teaching essentials: Short guides to good practice*. Academic.

Casas, M. (2002). *The use of Skinnerian teaching machines and programmed instruction in the United States, 1960-1970*. https://eric.ed.gov/?q=The+Use+of+Skinnerian+Teaching+Machines+and+Programmed+Instruction+in+the&id=ED469942

Chen, X. (2013). *STEM attrition: College students' paths into and out of STEM Fields*. Statistical Analysis Report. NCES 2014-001. National Center for Education Statistics.

Christensen, C. M. (1997). *The innovator's dilemma: When new technologies cause great firms to fail*. Harvard Business Review Press.

Crock Bauerly, B., McCord, R. F., Hulkower, R., & Pepin, D. (2019). Broadband access as a public health issue: The role of law in expanding broadband access and connecting underserved communities for better health outcomes. *The Journal of Law, Medicine & Ethics, 47*(2_suppl), 39–42. https://doi.org/10.1177/1073110519857314

Crosnoe, R. (2009). Low-income students and the socioeconomic composition of public high schools. *American Sociological Review, 74*(5), 709–730.

Dewsbury, B., & Brame, C. J. (2019). Inclusive teaching. *CBE—Life Sciences Education, 18*(2), fe2. https://doi.org/10.1187/cbe.19-01-0021

Drake, C., Zhang, Y., Chaiyachati, K. H., & Polsky, D. (2019). The limitations of poor broadband internet access for telemedicine use in rural America: An observational study. *Annals of Internal Medicine, 171*(5), 382–384.

Eng, T. R., Maxfield, A., Patrick, K., Deering, M. J., Ratzan, S. C., & Gustafson, D. H. (1998). Access to health information and support: A public highway or a private road? *Journal of the American Medical Association, 280*(15), 1371–1375.

Flippen, C. H. (2012). Educational technologies and learning theories: Behaviorism. *Educational Technology & Learning Theories*. http://edtechtheory.weebly.com/behaviorism.html

Ford, S., Buscemi, J., Hirko, K., Laitner, M., Newton, R. L., Jr., Jonassaint, C., Fitzgibbon, M., & Klesges, L. M. (2020). Society of behavioral medicine (SBM) urges congress to ensure efforts to increase and enhance broadband internet access in rural areas. *Translational Behavioral Medicine, 10*(2), 489–491.

Freire, P. (1972). *Pedagogy of the oppressed*. 1968 (Myra Bergman Ramos, Trans.). Herder.

Friesen, N. (2018). *Personalized learning technology and the new behaviorism: Beyond freedom and dignity*. https://www.academia.edu/36111979/Personal_learning_new_behaviorism2_docx

Ganimian, A., Vegas, E., & Hess, F. M. (2020, September 10). *Realizing the promise: How can education technology improve learning for all?* Brookings Institution. https://www.brookings.edu/essay/realizing-the-promise-how-can-education-technology-improve-learning-for-all/

Garcia-Olp, M., Nelson, C., Hinzo, A., & Young, D. A. (2020). Indigenous epistemologies: Implementing indigenous practices and perceptions to the area of STEM. In C. Haeussler Bohan (Ed.), *Curriculum & teaching dialogue* (Vol. 22, pp. 197–215). Information Age.

Gasparatou, R. (2017). Scientism and scientific thinking. *Science & Education, 26*(7–9), 799–812.

Gleim, G. W., Caldwell, S. R., Mentz, J., & Yu, S., Gleim Conferencing LLC. (2017). *System and method for validating honest test taking*. U.S. Patent Application 15/226,757.

Gonzales, A. L., McCrory Calarco, J., & Lynch, T. (2020). Technology problems and student achievement gaps: A validation and extension of the technology maintenance construct. *Communication Research, 47*(5), 750–770. https://doi.org/10.1177/0093650218796366

Goodell, J. W. (2016). Do for-profit universities induce bad student loans? *The Quarterly Review of Economics and Finance, 61*, 173–184.

Hawkins, B., & Oblinger, D. (2006). The myth about the digital divide. *Educause Review, 41*(4), 12–13. https://er.educause.edu/articles/2006/1/the-myth-about-the-digital-divide

Heller, J. F., Puff, C. R., & Mills, C. J. (1985). Assessment of the chilly college climate for women. *The Journal of Higher Education, 56*(4), 446–461.

Hollenbeck, J., & Hollenbeck, D. (2009). *Using technology to bridge the cultures together in the multicultural classroom*. https://eric.ed.gov/?id=ED519991

Howard, A., & Borenstein, J. (2018). The ugly truth about ourselves and our robot creations: The problem of bias and social inequity. *Science and Engineering Ethics, 24*(5), 1521–1536.

Howard, P., Busch, L., & Sheets, P. (2010). Comparing digital divides: Internet access and social inequality in Canada and the United States. *Canadian Journal of Communication, 35*(1), 109–128.

Huang, X., Bruce, B., Buchan, A., Congdon, C. B., Cramer, C. L., Jennings, S. F., Jiang, H., Li, Z., McClure, G., McMullen, R., & Moore, J. H. (2013). No-boundary thinking in bioinformatics research. *BioData Mining, 6*(1), 1–6.

Huff, C., & Cooper, J. (1987). Sex bias in educational software: The effect of designers' stereotypes on the software they design 1. *Journal of Applied Social Psychology, 17*(6), 519–532.

Hunter, A. B. (2019). Why undergraduates leave STEM majors: Changes over the last two decades. In E. Seymour & A. B. Hunter (Eds.), *Talking about leaving revisited*. Springer. https://doi.org/10.1007/978-3-030-25304-2_3

Jasanoff, S. (Ed.). (2004). *States of knowledge: The co-production of science and the social order* (1st ed.). Routledge.

Joh, E. E. (2017). Feeding the machine: Policing, crime data, & algorithms. *William & Mary Bill of Rights Journal, 26*, 287–302.

Johnston, S. F. (2018). The technological fix as social cure-all: Origins and implications. *IEEE Technology and Society Magazine, 37*(1), 47–54. https://doi.org/10.1109/MTS.2018.2795118

Jones, K. M., Asher, A., Goben, A., Perry, M. R., Salo, D., Briney, K. A., & Robertshaw, M. B. (2020). "We're being tracked at all times": Student perspectives of their privacy in relation to learning analytics in higher education. *Journal of the Association for Information Science and Technology, 71*(9), 1044–1059.

Jordan, K., David, R., Phillips, T., & Pellini, A. (2021). Education during the COVID-19 crisis opportunities and constraints of using EdTech in low-income countries. *Revista de Educación a Distancia (RED), 21*(65).

Junco, R., & Clem, C. (2015). Predicting course outcomes with digital textbook usage data. *The Internet and Higher Education, 27*, 54–63.

Kainz, O., Cymbalák, D., & Jakab, F. (2015). Adaptive web-based system for examination with cheating prevention mechanism. *Lecture Notes on Software Engineering, 3*(2), 90–94.

Kim, A. Y., Sinatra, G. M., & Seyranian, V. (2018). Developing a STEM identity among young women: A social identity perspective. *Review of Educational Research, 88*(4), 589–625.

Lang, J. M. (2013). *Cheating lessons*. Harvard University Press.

Levina, M., & Hasinoff, A. A. (2017). The Silicon Valley ethos: Tech industry products, discourses, and practices. *Television & New Media, 18*(6), 489–495.

Lewchuk, W., & Clarke, M. (2011). *Working without commitments: The health effects of precarious employment*. McGill-Queen's Press-MQUP.

Louie, N. L. (2017). The culture of exclusion in mathematics education and its persistence in equity-oriented teaching. *Journal for Research in Mathematics Education, 48*(5), 488–519.

Lynch, E., & Lyle-Gonga, D. M. (2019). Where have all the black academicians gone? A case study mixed methods evaluation of shortages in black faculty in higher education. *National Social Science, 49*(2), 54–64.

Macgilchrist, F., Allert, H., & Bruch, A. (2020). Students and society in the 2020s. Three future 'histories' of education and technology. *Learning, Media and Technology, 45*(1), 76–89. https://doi.org/10.1080/17439884.2019.1656235

Martin, S., Diaz, G., Sancristobal, E., Gil, R., Castro, M., & Peire, J. (2011). New technology trends in education: Seven years of forecasts and convergence. *Computers & Education, 57*(3), 1893–1906.

Master, A., & Meltzoff, A. N. (2020). Cultural stereotypes and sense of belonging contribute to gender gaps in STEM. *International Journal of Gender, Science and Technology, 12*(1), 152–198.

McElroy, E. (2019). Data, dispossession, and Facebook: Techno-imperialism and toponymy in gentrifying San Francisco. *Urban Geography, 40*(6), 826–845.

Moore, R., Vitale, D., & Stawinoga, N. (2018). *The digital divide and educational equity: A look at students with very limited access to electronic devices at home* (Insights in education and work) (p. 14). Center for Equity in Learning, ACT.

Morozov, E. (2013). *To save everything, click here: The folly of technological solutionism, and the urge to fix problems that don't exist*. Allen Lane.

Morris, S. M. (2017a, May 1). *Critical digital pedagogy and design*. Sean Michael Morris. https://www.seanmichaelmorris.com/critical-digital-pedagogy-and-design/

Morris, S. M. (2017b, September 8). *Reading the LMS against the backdrop of critical pedagogy*. Sean Michael Morris. https://www.seanmichaelmorris.com/reading-the-lms-against-the-backdrop-of-critical-pedagogy/

Morris, S. M., & Stommel, J. (2017, June 15). *A guide for resisting Edtech: The case against Turnitin*. Hybrid Pedagogy. https://hybridpedagogy.org/resisting-edtech/

Morris, S. M., & Stommel, J. (2018). Critical digital pedagogy: A definition. In *An urgency of teachers: The work of critical digital pedagogy*. Hybrid Pedagogy. https://criticaldigitalpedagogy.pressbooks.com/chapter/chapter-1/

Naylor, R., & Mifsud, N. (2019). *Structural inequality in higher education: Creating institutional cultures that enable all students*. National Centre for Student Equity in Higher Education. www.ncsehe.edu.au/publications/structural-inequality-retention-equity-students/

Neumann, K., & Waight, N. (2020). The digitalization of science education: Déjà vu all over again? *Journal of Research in Science Teaching, 57*, 1519–1528. https://doi.org/10.1002/tea.21668

Nuere, S., & de Miguel, L. (2020). The digital/technological connection with Covid-19: An unprecedented challenge in university teaching. *Technology, Knowledge and Learning, 1–13*. https://doi.org/10.1007/s10758-020-09454-6

O'Neil, C. (2016). *Weapons of math destruction: How big data increases inequality and threatens democracy*. Penguin.

Park, J. H., & Choi, H. J. (2009). Factors influencing adult learners' decision to drop out or persist in online learning. *Journal of Educational Technology & Society, 12*(4), 207–217.

Peck, C., Cuban, L., & Kirkpatrick, H. (2002). Techno-promoter dreams, student realities. *Phi Delta Kappan, 83*(6), 472–480. https://doi.org/10.1177/003172170208300614

Penrose, M. M. (2011). In the name of Watergate: Returning FERPA to its original design. *NYU Journal of Legislation & Public Policy, 14*, 75–114.

Petersen, C. I., Baepler, P., Beitz, A., Ching, P., Gorman, K. S., Neudauer, C. L., Rozaitis, W., Walker, J. D., & Wingert, D. (2020). The tyranny of content: "Content coverage" as a barrier to evidence-based teaching approaches and ways to overcome it. *CBE—Life Sciences Education, 19*(2), article 17.

Reddy, S., Labutov, I., Banerjee, S., & Joachims, T. (2016). Unbounded human learning: Optimal scheduling for spaced repetition. In *Proceedings of the 22nd ACM SIGKDD international conference on knowledge discovery and data mining* (pp. 1815–1824).

Reiman, J. (1987). Exploitation, force, and the moral assessment of capitalism: Thoughts on Roemer and Cohen. *Philosophy & Public Affairs, 16*(1), 3–41.

Reiman, J. H. (1981). The fallacy of libertarian capitalism. *Ethics, 92*(1), 85–95.

Roberts, S. (2011). Traditional practice for non-traditional students? Examining the role of pedagogy in higher education retention. *Journal of Further and Higher Education, 35*(2), 183–199.

Robinson, D., & Koepke, L. (2016). *Stuck in a pattern: Early evidence on "predictive policing" and civil rights*. A report from Upturn. https://www.teamupturn.com/reports/2016/stuck-in-a-pattern. Accessed 28 Mar 2021.

Rubel, A., Castro, C., & Pham, A. (2020). Algorithms, agency, and respect for persons. *Social Theory and Practice, 46*(3), 547–572.

Scott, L. A., Temple, P., & Marshall, D. (2015). UDL in online college coursework: Insights of infusion and educator preparedness. *Online Learning, 19*(5), 99–119.

Selwyn, N. (2010). Looking beyond learning: Notes towards the critical study of educational technology. *Journal of Computer Assisted Learning, 26*(1), 65–73. https://doi.org/10.1111/j.1365-2729.2009.00338.x

Shin, J. C., & Toutkoushian, R. K. (2011). The past, present, and future of university rankings. In J. C. Shin, R. K. Toutkoushian, & U. Teichler (Eds.), *University rankings: Theoretical basis, methodology and impacts on global higher education* (pp. 1–16). Springer.

Simkin, M. G., & McLeod, A. (2010). Why do college students cheat? *Journal of Business Ethics, 94*(3), 441–453.

Singer, N., & Krolik, A. (2021, May 9). Online cheating charges upend Dartmouth Medical School. *The New York Times*. https://www.nytimes.com/2021/05/09/technology/dartmouth-geisel-medical-cheating.html

Spector, J. M. (2015). *The SAGE encyclopedia of educational technology*. SAGE.

Steele, C. M., & Aronson, J. (1995). Stereotype threat and the intellectual test performance of African Americans. *Journal of Personality and Social Psychology, 69*(5), 797–811.

Swauger, S. (2020). Our bodies encoded: Algorithmic test proctoring in higher education. Chapter 6. In J. Stommel, C. Friend, & S. M. Morris (Eds.), *Critical digital pedagogy*. Hybrid Pedagogy. ISBN: 978-0-578-72591-8 https://cdpcollection.pressbooks.com/chapter/our-bodies-encoded-algorithmic-test-proctoring-in-higher-education/

Teachout, B. R. (2016). Gotta collect it all: Surveillance law lessons of Pokemon Go. *Stanford Law Review Online, 69*, 83–91.

Teräs, M., Suoranta, J., Teräs, H., & Curcher, M. (2020). Post-Covid-19 education and education technology 'solutionism': A seller's market. *Postdigital Science and Education, 2*(3), 863–878. https://doi.org/10.1007/s42438-020-00164-x

Theobald, E. J., Hill, M. J., Tran, E., Agrawal, S., Arroyo, E. N., Behling, S., Chambwe, N., Cintrón, D. L., Cooper, J. D., Dunster, G., Grummer, J. A., Hennessey, K., Hsiao, J., Iranon, N., Jones, L., 2nd, Jordt, H., Keller, M., Lacey, M. E., Littlefield, C. E., … Freeman, S. (2020). Active learning narrows achievement gaps for underrepresented students in undergraduate science, technology, engineering, and math. *Proceedings of the National Academy of Sciences of the United States of America, 117*(12), 6476–6483.

Tufekci, Z. (2015). Facebook said its algorithms do help form echo chambers, and the tech press missed it. *New Perspectives Quarterly, 32*(3), 9–12.

Veletsianos, G., & Moe, R. (2017). The rise of educational technology as a sociocultural and ideological phenomenon. *Educause Review*. https://er.educause.edu/articles/2017/4/the-rise-of-educational-technology-as-a-sociocultural-and-ideological-phenomenon

Voegelin, E. (1948). The origins of scientism. *Social Research, 15*(4), 462–494.

Vogel, S. A., Holt, J. K., Sligar, S., & Leake, E. (2008). Assessment of campus climate to enhance student success. *Journal of Postsecondary Education and Disability, 21*(1), 15–31.

Vygotsky, L. S. (1978). In M. Cole, V. John-Steiner, S. Scribner, & E. Souberman (Eds.), *Mind in society: The development of higher psychological processes* (Revised ed.). Harvard University Press.

Watters, A. (2014a, September 5). *Beyond the LMS*. Hack Education. http://hackeducation.com/2014/09/05/beyond-the-lms-newcastle-university

Watters, A. (2014b). *The monsters of education technology*. CreateSpace Independent Publishing Platform.

Watters, A. (2015, February 10). *Education technology and Skinner's box*. Hack Education. http://hackeducation.com/2015/02/10/skinners-box

Watters, A. (2017, April 7). *Education technology's completely over*. Hack Education. http://hackeducation.com/2017/04/07/prince

Watters, A. (2018a, April 26). *Teaching machines, or how the automation of education became "personalized learning."* Hack Education. http://hackeducation.com/2018/04/26/cuny-gc

Watters, A. (2018b, October 18). *B. F. Skinner: The most important theorist of the 21st century*. Hack Education. http://hackeducation.com/2018/10/18/skinner

Watters, A. (2019, August 28). *Education technology and the age of surveillance capitalism*. Hack Education. http://hackeducation.com/2019/08/28/surveillance-ed-tech

Watters, A. (2020a, May 6). *Educational crises and ed-tech: A history*. Hack Education. http://hackeducation.com/2020/05/06/crisis

Watters, A. (2020b, July 20). *Building anti-surveillance ed-tech*. Hack Education. http://hackeducation.com/2020/07/20/surveillance

Weegar, M. A., & Pacis, D. (2012). A comparison of two theories of learning—Behaviorism and constructivism as applied to face-to-face and online learning. In *Proceedings E-leader conference, Manila*. https://www.gcasa.com/conferences/manila/ppt/Weegar.pdf

Weinberg, A. M. (1978). *Beyond the technological fix [Detrimental and unforeseen side effects]* (ORAU/IEA-78-5(O); CONF-780222-16). Institute for Energy Analysis. https://www.osti.gov/biblio/7036775

Weller, M. (2015). MOOCs and the Silicon Valley NARRATIVE. *Journal of Interactive Media in Education, 2015*(1), Art. 5. https://doi.org/10.5334/jime.am

Whitacre, B. E., & Mills, B. F. (2007). Infrastructure and the rural—urban divide in high-speed residential Internet access. *International Regional Science Review, 30*(3), 249–273.

White, D., Dunn, J. D., Schmid, A. C., & Kemp, R. I. (2015). Error rates in users of automatic face recognition software. *PLoS One, 10*(10), e0139827.

Williamson, B. (2017a). Learning in the 'platform society': Disassembling an educational data assemblage. *Research in Education, 98*(1), 59–82.

Williamson, B. (2017b). Who owns educational theory? Big data, algorithms and the expert power of education data science. *E-Learning and Digital Media, 14*(3), 105–122.

Wilson, J. Q. (1995). Capitalism and morality. *Public Interest, 121*, 42–60.

Winneg, D., & Soske, J. (2007). Systems, methods and apparatus for monitoring exams. U.S. Patent Application 11/603,259.

Woodard, J. (2019, August 20). Rotten STEM: How technology corrupts education. *American Affairs Journal, 3*(3). https://americanaffairsjournal.org/2019/08/rotten-stem-how-technology-corrupts-education/

Yeager, D. S., & Dweck, C. S. (2012). Mindsets that promote resilience: When students believe that personal characteristics can be developed. *Educational Psychologist, 47*(4), 302–314.

Zoom Video Communications. (2020, April 2). A message to our users. *Zoom Blog*. https://blog.zoom.us/a-message-to-our-users/

Zuboff, S. (2015). Big other: Surveillance capitalism and the prospects of an information civilization. *Journal of Information Technology, 30*(1), 75–89.

Zuboff, S. (2019). Surveillance capitalism. *Esprit, 5*, 63–77.

Chapter 3
Institutional Culture of Student Empowerment: Redefining the Roles of Students and Technology

Heeyoung Han, Maria Mosley, Ihuoma (Yvette) Igbokwe, and Shelley Tischkau

Abstract Student engagement in medical education is progressing beyond passive participation in instructional activities toward embracing student empowerment that will develop independent and engaged healthcare professionals. This book chapter reorients the current understanding of student engagement by reviewing the importance of an institutional culture of empowerment and students as partners. We will provide concrete examples of groundbreaking leadership in student autonomy and empowerment undertaken at Southern Illinois University School of Medicine (SIU SOM), a recipient of the ASPIRE-to-Excellence Award in Student Engagement and the ASPIRE Stella Award Winner (The ASPIRE Stella Award is to recognize commitment to excellence in various areas. SIUSOM has received ASPIRE-to-Excellence Award in curriculum development, assessment, social accountability, simulation, as well as student engagement.) from the Association for Medical Education in Europe (AMEE). The coachability curriculum is a student-developed curriculum addition of intrapersonal and interpersonal skills. Some year 4 students choose to teach our year 1 students as tutors in problem-based learning (PBL). A reformed clerkship curriculum has abandoned shelf exams and instead focuses on clinical immersion and each student's personalized education plans based on their educational needs. Programmatic assessment has been instituted that includes a progression committee that values and integrates student-adviser discussions and remediation plans. A recent initiative has developed an educational data dashboard system to embrace the institutional culture of student empowerment through transparency. The chapter concludes with ideas to facilitate an institutional culture of student empowerment.

H. Han (✉) · M. Mosley · I. (Yvette) Igbokwe · S. Tischkau
School of Medicine, Southern Illinois University, Springfield, IL, USA
e-mail: hhan@siumed.edu

H. J. Witchel, M. W. Lee (eds.), *Technologies in Biomedical and Life Sciences Education*, Methods in Physiology, https://doi.org/10.1007/978-3-030-95633-2_3

3.1 Introduction

Students' active involvement and participation in instructional tasks and school activities for academic success is the essential element in the current understanding of student engagement (see Chap. 4 on engagement in this volume). A recent integrated and holistic view of student engagement expands the object of student engagement from instructional activities toward a more sustainable, individual, and institutional goal, embracing a learner's holistic development and a lifelong reflective process as competent professionals (Kahu, 2013).

Considering the institutional goal to develop independent and competent healthcare professionals in the community, a rationale for student engagement should be extended to consider the development of autonomy and empowerment of students in medical education (Allen et al., 2019; Fredricks et al., 2004). As medical learners progressively develop their knowledge and clinical skills for patient care during undergraduate and graduate medical education training, they need to strive toward professional independence, patient ownership, and ultimately professional identity. Students need to learn and develop skills that will sustain and help them succeed in their profession after their medical school degree is completed. This requires them to grow with significant autonomy and empowerment, which will become critical to cement their ownership of their own professional development (Sawatsky et al., 2020). With progressively increased autonomy and empowerment, students can truly engage in their own educational process to develop knowledge, skills, a sense of responsibility, and a professional identity (Allen et al., 2019). Without autonomy and empowerment, medical learner engagement can be superficial and incomplete (Sawatsky et al., 2020).

The meaning and importance of students' empowerment are explained by self-determination theory (Deci & Ryan, 2012; Reeve, 2012). According to self-determination theory, an individual has natural tendencies to develop the wholeness of self, which can be facilitated or discouraged by environmental factors. In this developmental process, a student's intrinsic motivation is essential and developed in an environment supporting three psychological needs: autonomy, competence, and relatedness (Deci & Ryan, 2012). Autonomy is a student's feeling of themselves having control of their decisions and actions. Competence does not indicate knowledge and skills but rather a student's ability to navigate in and interact with learning environments, where they have autonomous opportunities to exercise and express their capacities. Relatedness is a student's sense of belonging and feeling of being cared for by others. A student's feeling of belonging and bonding with people and the school would create a virtuous cycle to continuously motivate them to autonomously participate in school activities (Finn, 1989). These psychological needs are interconnected, and a student's autonomy becomes the essential driver as the student is the ultimate change agent in their own learning. A student's autonomy means neither independence nor entitlement. Rather, it is the process of the student's representing the self, *being empowered*.

Given the importance of learner autonomy and empowerment, the reconceptualization of the power in educational program processes is pivotal in health professions education. Without this shift, the practice of student engagement and any related application of technology could potentially be misleading. This should be examined from the perspective of identifying the needs of healthcare education and its specific students, rather than simply looking at what has already been done in other arenas and then applying those techniques. Merely adopting student engagement practices from other educational settings without critical reflection on student empowerment in the specific medical professional context will not accurately reflect the needs of the students and has the potential to turn students off, leading to a disengagement, rather than engagement (Lamb et al., 2020).

This book chapter explores the nature of an institutional culture of student empowerment as revisiting the concept of curriculum writer and reader. Concrete examples of student empowerment practices across the curriculum at Southern Illinois University School of Medicine (SIU SOM), a recipient of the ASPIRE-to-Excellence Award in Student Engagement from the Association for Medical Education in Europe (AMEE), are detailed. Finally, current initiatives at SIU SOM that aim to redirect the institutional culture of student engagement using a technological process to ultimately empower students and increase student ownership of their own learning are discussed.

3.1.1 *Reconceptualizing the Actors of Teaching and Learning: Writers vs. Readers*

In exploring the institutional culture of student empowerment, we need to revisit some common assumptions surrounding the traditional role of the actors in the teaching and learning process. From the traditional perspective, the faculty and school provide high-quality curricula and instruction, and students are the recipients of the program. For this reason, many educators have focused on improving instructors' teaching strategies and technological interventions at medical schools (Guze, 2015; Kay & Pasarica, 2019; Moran et al., 2018). If one follows this mindset, then well-designed curricula and instructional activities would bring about student engagement. The assumption here is that students are the recipients of the curriculum created by faculty and the institution and that students' engagement is a reaction to what they have been given. The school and faculty are the arbiters of the learning process, and students are merely compliant with what has been enacted. The school and teachers are the writers and creators of engagement, while students are the readers and followers who accomplish engagement in a regulated way. As the school provides activities that students are expected to engage in, the school also possesses an institutional vantage point to monitor and assess students' actions and engagement.

This provider-based structure of student engagement has critical limitations. White et al. (2014) observed that their medical students were "checked out" and disengaged from preclinical instructional activities, even when the curriculum was designed based on active learning principles for student participation. The authors discussed how the students' intellectual maturity may not have yet reached the development required for active and adult learning. While this reflection may provide a valid point, the study also questioned the fragile premise that an active learning curriculum without true student empowerment in their learning processes may not necessarily result in student engagement. The underlying reason why students are "checked out" from instructional activities should be explored as part of the curricular design and development process. A critical question that must be asked is who is to engage whom and in what, which is blurred in the curriculum co-construction process.

This structure of power around educational processes can be redistributed and redefined to promote students' authentic engagement experiences (Mann, 2001). The quality of the instruction and learning experiences can be based on a true partnership between faculty and students (Cook-Sather et al., 2014). In this approach, the dichotomy of writer and reader is being blurred as student engagement is co-constructed through collaborative curriculum development by both parties. Students give feedback to the governing bodies, so that the curriculum is a living entity, where the "experts" (faculty) can refine what is provided in response to real "needs" expressed by students. This iterative feedback process allows for a living curriculum that grows with students as they grow in their profession. Noting the importance of power-sharing among faculty and students, Lizzio and Wilson (2009) argued that an effective instructional environment is not created by the control of student experiences but by constant responsiveness to students' needs in complex and challenging learning environments. In this situation, both students and faculty should share a common goal to prioritize (a) students' professional growth and (b) constant curriculum improvement throughout the responsive feedback process. Students also should not misuse the shared power for selfish reasons by advancing personal gripes. This effort to create true collective ownership through shared power is essential for student engagement experiences. Blurring the dichotomy of curriculum providers (traditionally faculty) and curriculum consumers (traditionally students) and building the true partnership are keys to truly successful student engagement.

The assumptions discussed above indicate the need for a cultural transformation to embrace student empowerment at an institution. In the following section, we will further discuss the institutional culture of student empowerment.

3.2 Institutional Culture of Student Empowerment

Student engagement in health profession education is achieved through an institutional culture of *student empowerment* (Peters et al., 2019). The institutional culture of student empowerment is a shared set of values, assumptions, and beliefs about how students gain and exercise authority to manage their learning and professional development processes.

The existing literature has reported various ways that schools accommodate students' empowerment (Behrend et al., 2019; Carey, 2013; Geraghty et al., 2020; Hsih et al., 2015; Milles et al., 2019). One of these efforts involves participatory institutional representation structures of student representations (Carey, 2013). Student representatives occupy seats on curriculum committees to speak for their peers on educational issues. Student ownership of learning can be demonstrated by active feedback and the resultant curricular change. For example, Hsih et al. (2015) implemented a comprehensive process, where students played a significant role in directing a curriculum change by reviewing course evaluation data, gathering information through town hall meetings and surveys, generating solutions to identified issues, and discussing these changes with course directors. Geraghty et al. (2020) also reported on a student curricular board model. In their model, approximately 11% of medical students served on the curricular board and have created a sustainable approach through which senior student members train new member students about program evaluation, curriculum development, and conflict resolution. Milles et al. (2019) presented an innovative approach to student engagement, in which students occupied a codirector role and worked with faculty to change the existing curriculum from a discipline-based model to one that integrated material across disciplines. These initiatives highlight the importance of an institutional culture that legitimizes medical students as an empowered and capable body to create and change their curriculum at the institutional level.

While this literature clearly contains excellent examples of instances where students can contribute to educational practices, it remains unclear what student empowerment truly means and how they are empowered. One may worry that students would feel entitled to control their learning environments and curriculum in favor of their self-focused preferences. Liao (2020) noted the concern about the culture of entitlement in medical education. Learners tend to demand that faculty and the school meet their desire for an easy and efficient way of learning, although this runs counter to what is psychologically established about reliable ways to learn and retain knowledge (Bjork & Bjork, 2020). Noting the issue in medical education, Liao (2020) proposed promoting a culture of empowerment as a way to discontinue the detrimental value of entitlement. His notion of empowerment is to empower the students to deal with complexity and difficulties in order to develop as competent doctors rather than "learned helplessness (p. 586)." Student empowerment cannot occur unless students have "a legitimate sense of ownership over the management of their education" (Carey, 2013, p. 1301), which often entails navigating complexity and learning from hardship. This is consistent with the self-determination theory that underlines the importance of students' competency to navigate the complexity and autonomy for their decisions, promoting their motivation in their development.

Faculty and institutional efforts to delegate power and ownership to students are critical for student empowerment. Likewise, students' willingness and ability to embrace this legitimate transfer of power is imperative, because the institutional culture of student empowerment must be a socially co-constructed practice among students, faculty, and institutions for the democracy of education (Lizzio & Wilson, 2009). While students may think of themselves as legitimate stakeholders, students nonetheless often perceive their participation as optional and voluntary, assuming that it is ultimately the job of faculty and their institution to satisfy students as consumers and improve their instructional service (Carey, 2013; Lizzio & Wilson, 2009; Milles et al., 2019). Because students make a financial transaction for their education, resulting in long-term student debt, they may think of themselves as consumers, especially given the discourse of markets in education (Molesworth et al., 2009). If students are accustomed to the view of "students as consumers," they would prioritize their own emotions being satisfied. Once students embrace the view of "students as partners," they must readjust their priorities toward long-term educational values (Liao, 2020).

Students' repudiation of this mindset and reorientation toward true ownership of their learning is critical. From a faculty perspective, there is a delicate balance in delegating power. Faculty may view student empowerment with true ownership as a real risk to their authority and jobs, which may result in difficulty in sharing power. Faculty would perceive their power to be generated and maintained through owning instructional processes they provide to the consumers. An institutional culture of student empowerment does not mean faculty giving up power but redefining their hierarchical relationship to a partnership with students for the sake of sustainable effects of student engagement and professional success. The student-faculty-institution relationship should move past the "service recipient" and "service provider" mindset and embrace a transformative culture toward student ownership and empowerment.

Student empowerment for meaningful engagement is a sensitive topic, as it involves restructuring the power and epistemic status among faculty and students in their teaching and learning practice. Generally, a curriculum is perceived to be *owned* by faculty, which creates an institutional expectation of a curriculum *governed* only by faculty. An institutional culture of student empowerment crosses this boundary, requiring that traditional instructional practices be dismantled and transformed for co-ownership of education. In the following section, we will describe examples of the institutional culture of student empowerment at SIU SOM that have been developed and evolved for decades.

3.3 Student Engagement and Empowerment as a Central Value in Curriculum

3.3.1 Case at Southern Illinois University School of Medicine

Southern Illinois University School of Medicine (SIU SOM) has a long history of innovation in medical education (Fair et al., 2011). In 1976, SIU SOM became the first medical school to publish a complete set of objectives for their medical degree.

In 1985, SIU SOM boldly added the criteria that all students must pass a final clinical assessment utilizing standardized patients in order to graduate, a groundbreaking move at the time. SIU SOM was also an early adopter of problem-based learning, a student-centered model of education that is now commonplace in most medical schools. More recently, in 2016, SIU SOM ambitiously revamped its traditional 3rd-year clerkship schedule in order to grant students increased autonomy and flexibility (Klamen, 2015; Klamen et al., 2019). Since 2019, SIU SOM has implemented the programmatic assessment that promotes students' formative professional development processes. These innovations share a common attribute of student engagement based on the institutional endeavor to cultivate student empowerment. In this section, we will further characterize innovations that have encouraged student engagement and empowerment at SIU SOM and will describe ongoing projects that utilize new technologies to further accomplish these goals.

3.3.2 *Problem-Based Learning*

Clerkship students' inability to identify patient problems despite high achievement in classroom learning prompted Dr. Howard Barrow's invention of problem-based learning (PBL) at McMaster in the 1970s (Barrows & Tamblyn, 1980). PBL is defined as learning derived from working toward the understanding or resolution of unsettled, puzzling problems (Barrows & Tamblyn, 1980). In medical education, a PBL curriculum allows students to learn medical concepts and skills, as well as the underlying science, using real patient problems in an integrative inductive manner, thereby enhancing interdisciplinary knowledge. PBL resets the traditional power structure of teaching and learning. Learning becomes active and centered around the students' inquiry process toward solving problems, rather than a teacher's distillation and telling of the issues in a didactic form. In PBL, students' struggles with messy patient problems allow integration of cross-disciplinary information and understanding of difficult biomedical concepts in the context of real-life patient scenarios, rather than exclusively emphasizing academic or test-taking achievement.

PBL was first implemented as a formal curriculum with the novel approach of standardized patients for performance-based assessment at SIU SOM in the 1980s (Barrows, 1986, 1988; Barrows et al., 1987). PBL forms the foundation of the hybrid, "preclinical" curriculum during years 1 and 2 at SIU SOM. The curriculum consists of a set of multiple patient cases provided in a virtual interactive patient information database system. The system has an inquiry-based interface, where students have to input questions in the database to acquire patient information. Each case is designed to direct inquiry into basic sciences that underlie the presented clinical scenario. The students manage the learning process and define group roles (Barrows, 1988). During the group discussion, students generate a comprehensive list of differential diagnoses from the chief complaint and initial history. Subsequently, students conduct a virtual physical exam, noting pertinent findings. During this process, students constantly evaluate new patient information to narrow the list of differential diagnoses. Importantly, as they work, students identify potential learning issues (LI's), which are topics that require additional investigation or "learning." LI's issues serve as the framework for the self-directed, open-ended learning that occurs between PBL group sessions. LI's may include topics from both the basic and clinical sciences, spanning the areas of physiology, pathology, differential diagnosis, patient evaluation testing, management, and social considerations.

During self-directed learning, students utilize various resources, such as textbooks, lectures, websites, primary literature articles, etc., as they see fit. Resource sessions (i.e., lectures provided by faculty) are made available for use as a resource. Frequently, these resource sessions take place after the LI has already been identified. The goal is to have the students solidify and refine the knowledge obtained on their own with the lectures if they choose to do so. During the subsequent small group session, students share what they have learned about the LI's and discuss how the newly attained knowledge bears upon the current patient case. Through this process of independent learning and group discussion, PBL groups not only arrive at a final diagnosis but also, more importantly, use the scenario to collaboratively identify learning issues, learn the relevant material, think through relationships between basic and the clinical picture, and teach each another along the way.

An astute reader may have noticed that, so far, we have not yet mentioned any involvement of the facilitator. This is not an accidental omission but rather key to an accurate characterization of tutor group responsibilities. The role of the tutor group leader is to facilitate, rather than truly lead, the group's discussion and learning (Barrows, 1988). When questions naturally arise during the group's investigation and discussion of the case, the facilitator guides the students and ensures that they generate relevant LI's to cover the material to sufficient depth. The facilitator engages students using a metacognitive approach rather than simply supplying students with a pre-generated list of learning issues. By engaging with students and asking questions to challenge students' critical thinking, facilitators monitor the breadth and depth of topic coverage.

Assessment and evaluation in PBL occur through active participation in feedback sessions that take place midway through and at the end of each unit. Students and facilitators meet to assess themselves and each other. Areas of evaluation include active participation, knowledge, responsibility, leadership, and teamwork. Importantly, this feedback is discussed, and constructive criticism is addressed within the group before it is submitted online. Students and facilitators develop skills of both giving and receiving actionable feedback and are able to utilize this feedback toward their personal development, leading to not only an increase in knowledge as the year progresses but also a significant amount of personal and professional growth.

As the foundational method of content delivery, PBL requires an institutional culture that empowers students to engage with the problem-solving process. PBL is based on the core belief that students are inherently competent in navigating complex problems, which leads to their autonomous decision-making processes regarding what to study to understand and solve the problem (Deci & Ryan, 2012). SIU SOM continuously reminds both students and faculty of its core guiding principle that it is the students that are ultimately responsible for their own learning.

It is the students that identify what they need to know, how to acquire information, reason through the significance of each piece of data, evaluate and expand their own knowledge, identify areas for studying, decide on their preferred resources, diagnose and manage patients' chief complaint, and evaluate their own learning processes overall.

Far from being passive recipients of a lecture, students in PBL are active participants in their own learning as well as the creation and advancement of their group study. Although the facilitator ensures, through engagement and guiding questions, that key subjects are explored at sufficient breadth and depth, and to encourage shared participation among group members, this curriculum allows for a great deal of flexibility and room for individualization of each student's learning process. The collaborative, adaptive nature of PBL learning in small groups organically fosters better teamwork, student initiative and investment, and personal development (Neville, 2008; Pluta et al., 2013; Schmidt, 2010). It also provides opportunities for facilitators to develop as coaches who guide students' learning on the side and nurtures a constructive faculty-student relationship (Barrows, 1988).

We often observe tension in curriculum committee meetings between advocates for traditional lecture-based learning and proponents of PBL's active learning ethos. There has been constant organizational change at SIU SOM as part of the process of facing internal and external environment challenges. For example, the established active culture can be challenged by various contexts, such as students' anxiety of board exam (USMLE STEP 1) preparation that would be in favor of lecture-based instruction, where students are told what to study. Some new faculty may not fully understand how to be a facilitator. Variation and idiosyncrasy in tutoring and divergence from core PBL principles are challenging (Cianciolo et al., 2016). The culture of active learning in PBL might go off the rails if not for continuous and explicit efforts to embed it in the established culture of the institution. Such efforts include continuing faculty development, new tutor coaching, and the presence of medical education faculty on the committees who are specialized in education. Due to the organized efforts, the core value and culture of student empowerment and autonomy in PBL have been secured, while variation in technique still occurs (Cianciolo & Regehr, 2019).

3.3.3 Student as Educators: PBL Tutors and Pathology Instructors

In the previous section, we discussed the importance of problem-based learning at SIU SOM and described how this curriculum naturally fosters student empowerment for tutor group members. Though we mentioned briefly that the tutor group leader ("tutor") role may be filled by either a faculty member or, alternatively, a qualified 4th-year medical student, we would like to further discuss here this important opportunity for student empowerment. By leading a tutor group, senior students can occupy a unique educational role that helps develop their teaching skills while facilitating the learning of younger students. Additionally, supervising these group meetings and guiding student discussions can also serve as a knowledge "refresher" for senior students on the basic science and clinical medicine topics covered in PBL scenarios. In this way, senior students benefit from a review of the 1st- and 2nd-year material as they near the end of medical school and the beginning of residency training in their chosen field.

Positioning students as the cocreators of their own curriculum can certainly seem imprudent or untested to educators who are accustomed to a more traditional approach. Nevertheless, research in higher education has consistently shown that situating students as partners in, rather than mere recipients of, their education can be mutually beneficial for both students and staff. One such study by Deeley and Bovill (2017) found that a "partnership approach" to student assessment led to increased motivation, a sense of engagement, and a feeling of belonging to a community of learning. Pereira et al. (2020) found that relationships between students and staff improved, following the implementation of a partnership initiative in a medical curriculum. Staff members reported a greater understanding of the students' points of view, and students were able to view instructors as mentors, role models, and colleagues. Students who take an active role in education also experience such benefits as improved leadership skills, managerial skills, acquisition of new knowledge, familiarity with pedagogy, confidence in skill demonstration, and advancement of metacognitive skills to aid in their own learning (Murphy et al., 2017).

Naturally, such an adjustment in the structure of curricula is not without its downsides or disadvantages. As Pereira et al. (2020) reported, such changes require a significant time and effort commitment from both students and staff involved with the project. Such a significant responsibility may act as an obstacle toward recruiting new participants or retaining current ones, although students in both the Pereira et al. study and our own experiences generally report satisfaction with their involvement and feel that their time commitment reaped even greater rewards.

Our experiences at SIU SOM echo what the research has shown regarding the mutual benefits of a partnership approach to medical student education. Both 2nd-year tutor group members and 4th-year student group leaders ("tutors") report a positive experience with the curriculum. Interestingly, many 2nd-year students actually prefer having year 4 (Y4) students as their group leaders over faculty. This is largely because senior students tend to be concise and efficient and have a good memory for which concepts are most essential to master during the early years of medical school. Having just recently been 1st or 2nd year students themselves, Y4 group leaders have the advantage of this proximity and often clearly recall what was most essential to learn as a junior student. They have a unique perspective and are often quite effective at ensuring that students learn what they need to without becoming unnecessarily bogged down in irrelevant details or information that is too advanced. This can be a difficult balance to achieve, especially for faculty group leaders who have perhaps never experienced a PBL-style curriculum themselves or have significantly less proximity to the role of the student. In this way, Y4 students serve as a bridge between the roles of the novice junior students and the group leaders (Cianciolo et al., 2016). They are in the ideal position to help guide younger students and facilitate key educational discussions. Furthermore, serving as a tutor for junior students is also beneficial to the Y4 group leaders. It allows the student group leader to solidify their own knowledge and gives them insight into just how much they have truly learned over their time in medical school, building confidence in their role as both a healthcare provider and a teacher.

Recently, the involvement of students as educators in the medical curriculum has created another avenue to facilitate student engagement at SIU SOM. In the summer of 2020, SIU introduced a new opportunity for student engagement in Y2 education. Six 4th-year students, selected for their demonstrated aptitude in pathology and teaching, were offered the chance to provide a portion of the pathology education to 2nd-year medical students. All six students eagerly accepted and were supplied with lecture slides created by the previous head of pathology education but were given the autonomy to utilize other presentation materials or modalities of education. Student educators were given the freedom to present the material in any form they wished (be it via traditional lectures with interactive components, question and answer sessions, flipped classroom sessions, etc.), and each received ~1 credit hour for their involvement.

So far, based on our internal survey data, the student-led pathology curriculum has been received positively by both 4th-year presenters and 2nd-year students. While the curriculum is presently still in the early stages of implementation, junior students seem to be comfortable having 4th years as presenters. They report that the 4th-year students seem to have an excellent understanding of the content that they are presenting and are well equipped to deliver the lectures and answer questions. One student commented that the 4th-year presenters were actually more engaging and interactive than many faculty lecturers. Most significantly, 2nd-year students report that they appreciate older students' unique insight about which topics are "high-yield" for not only examinations but also real clinical practice. Fourth-year students serving as pathology educators have the advantage of relative proximity to both examinations (e.g., medical school tests and USMLE board exams) and to recently completed clinical rotations. They often use this insight to approach the pathology lectures with a unique perspective.

The student-led pathology curriculum has served as an opportunity for students to develop teaching skills with a very high level of autonomy. Like in the PBL curriculum, students who serve as pathology educators have truly moved from mere recipients of the curriculum to co-constructors of medical education. While Y4 students reported being initially intimidated by this aspect, they expressed gratitude for the practice, as they will soon be moving into the role of a resident physician with increased teaching responsibilities. One student commented that his experience as a pathology educator has changed how he will approach his own learning in the future. He noted that in order to effectively teach, a deep and comfortable understanding of the knowledge is imperative. As a result, he will continue to ask himself, "How would I teach this?" when learning new material during residency. In early feedback, SIU Y4 students report that they truly feel empowered to enact their own vision for the pathology curriculum and feel supported by the institutional culture to exercise their own autonomy, rather than ask a faculty member to guide or micromanage their work. One student suggested that this opportunity should be made permanently available as an elective to 4th-year students interested in medical education. The experience serves as an excellent learning opportunity and aligns perfectly with SIU's philosophy around student engagement, autonomy, and hands-on learning. Overall, 4th-year students have thus far responded positively to the opportunity to serve as pathology educators and appreciate the autonomy of operating mostly independently in a teaching role.

3.3.4 Curricular Changes in Clerkships: Toward Clinical Immersion and Socialization

An institutional culture of student empowerment is imperative for the clerkship curriculum of any medical school. Most contemporary medical school curricula in the USA are based on the 2 × 2 model, where students spend two preclinical years that are similar to traditional classroom-based learning and the other 2 years on clinical learning that is mainly based on the apprenticeship model of situated learning (Lave & Wenger, 1991). The impact of clerkship learning is most meaningful when students can fully engage and focus on the patient care roles and clinical tasks of the medical workplace (Chen & Teherani, 2015). This experience would provide holistic developmental experiences through reflective practices in the workplace.

However, our longitudinal qualitative study found that the traditional clerkship curriculum did not work for students' engagement and immersion in clinical learning as much as was previously expected (Han et al., 2015). The traditional 3rd-year clerkship curriculum presented several structural problems for student engagement and immersive clinical experiences (Han et al., 2015). First, students' clinical reasoning was not progressing at a pace that met learning expectations (Williams et al., 2011). Students could not participate in forming differential diagnoses and plans in clerkships, because thinking in clinical practices moves very quickly toward patient care in cases where the diagnoses have already been established. As a result, students could not observe expert physicians' thought processes (Han et al., 2015). Second, students' educational relationships with attending physicians were too brief to be meaningful, as students were assigned to as many different subspecialty attending physicians as possible. Because expectations for their work were seldom established, students rarely felt they were valuable members of the clinical team. Third, due to the lack of any long-term learning relationships, attending physicians' evaluations of students tended to be superficial, which led to students primarily engaging in impression management rather than in meaningful feedback-seeking behaviors (Han et al., 2015). Fourth, students felt an inherent tension between their time spent working clinically with patients and the burden of studying for shelf and board exams. Because shelf exams were utilized so heavily for clerkship evaluation, there was a notable pressure for students to use as much time as possible to study for written tests, rather than to prioritize spending as much time as possible on patient care or other clinical responsibilities. Last, the 3rd-year schedule was completely "one size fits all," with no opportunity for students to personalize or individualize their own experiences. The apprenticeship model commonly implemented in medical schools did not work, as it failed to provide students with legitimate peripheral participation given the structural issues (Han et al., 2015; Klamen, 2015).

To face these challenges and limitations, in 2016, SIU SOM implemented a new type of 3rd-year clerkship curriculum that focuses on clinical immersion and socialization into medicine (Klamen, 2015). The final result of this effort is a reimagined and revitalized clerkship schedule, with eight foundational rotations in different specialties taking place over the first 8 months of the 3rd year. This new clerkship curriculum allows students to focus on socialization into medicine and finding their specialty. Quite remarkably, these rotations include no classroom lectures and no required shelf exams to facilitate students' immersion in clinical work. One strength of this clerkship model is that students are not pressured to minimize their clinical time in order to study for shelf exams. Students are often paired with the lowest possible number of physicians (ideally a single preceptor for a 4-week rotation) in order to maximize continuity, build trust, and allow for more meaningful longitudinal feedback based on a long-term learning relationship.

The assessment model has also been reimagined. Rather than one final summative performance assessment at the end of the clerkship, student performance is evaluated and reflected continually throughout the rotation using a customized "on-the-fly" feedback form focusing on narrative comments. Therefore, students are able to consider and implement preceptor feedback in real time, using it to improve and grow into more proficient clinicians and team members.

Furthermore, after completing 8 months of core clinical clerkships, students spend the remaining 15 weeks of their 3rd-year clinical rotations completing a personalized schedule of clerkships titled the Personalized Education Plan (PEP) period. The PEP period is designed so that students may further immerse themselves in any specialty of their choosing and address their own individual learning needs. Such a curriculum affords each student the opportunity to further practice skills that they feel call for improvement, or to spend additional time in fields that will aid them in professional development and specialty choice. In this way, students are the active designers of their own clinical curriculum. The PEP period gives students autonomy over their own education and empowers them to set their own goals and make their own plans.

The results of this reimagined clerkship curriculum have, so far, been positive (Klamen et al., 2019). Students reported a higher number of procedures and surgeries per week of rotations, likely due to both decreased pressure to spend time on nonclinical study activities and increased trust between the preceptors and students. Reassuringly, this increased clinical engagement has not been associated with any negative trade-off regarding exam scores. Students have scored at least as well as they did under the old clerkship model on their summative clinical competency exams, despite the exam being administered several months earlier under the new schedule. Board exams (Step 2 Clinical Knowledge and Clinical Skills) scores also remained stable compared with previous years, demonstrating that the lack of mandated shelf exams did not negatively impact students' testing performance (Klamen et al., 2019). Student engagement in clinical settings has settled nicely as an institutional culture through the collective efforts of faculty and students during the clerkship curriculum change process.

3.3.5 Coachability Curriculum: A Student-Driven Education in the "Soft" Nonacademic Attributes of Medicine

One excellent example of student engagement in which SIU SOM students felt empowered to blur the lines between curriculum writers and readers is the coachability curriculum. As described in the earlier section, year 3 clerkships have changed to embrace students' clinical immersion and individualized learning paths. This new environment requires students to be proactive in their learning processes rather than remain passive recipients of clerkship activities. Given this, a group of students realized that they needed to develop their intrapersonal and interpersonal skills (IIS) in order to enhance their clinical reasoning, skills, and professionalism. The development of these skills, augmented by clinical knowledge, allows students to continuously identify and improve areas of weakness in the social and academic aspects of their clinical experience. With faculty support, students designed, developed, and implemented the *coachability curriculum*. This longitudinal curriculum consists of several small group sessions during the 1st and 2nd years of medical school, in which junior students meet with senior students to discuss a general approach to learning in medical school. The content of these sessions emphasizes the importance of having a "growth mindset" (Dweck, 2016) (as opposed to a "fixed mindset") through interactive activities and group discussions. Lombardo et al. (2019) noted that the defensive attitude seen in medical students and physicians concerning mistakes and learning was a great hindrance in implementing IIS. As a result, emphasizing a growth mindset early allows one to create an environment that is receptive to the improvement of IIS (Lombardo et al., 2019). Topics include intrapersonal skills like acquiring and implementing feedback, setting SMART (specific, measurable, achievable, relevant, and time-bound) goals, utilizing feedback, maximizing clerkship advising resources, mitigating burnout, as well as interpersonal skills like self-reflection, conflict resolution, and emotional intelligence. Each session asks the students to reflect on their own similar situations and practice scenarios as well as possible applications in the future. Student leaders are able to drive home the topics by providing their own similar experiences during their clerkship. This format is particularly important, because these skills are solidified through reflection and feedback (Lombardo et al., 2019). Furthermore, the reflection on situations allows the student to utilize and integrate multiple concepts of IIS (Lombardo et al., 2019). By using early coaching during the first 2 years of medical school, the coachability curriculum aims to foster important attitudes and nonacademic attributes in students during the "preclinical" years. This approach seeks to allow students to be better positioned for the rigorous and sometimes challenging clinical experiences they will face during their clinical rotations (Klamen et al., 2017).

It is important to remember that the creation of the coachability curriculum came from the students' autonomous decisions. This suggests that many of our students are already motivated and invested in their education. Regardless of whether the most salient factors are intrinsic or extrinsic, this motivation based on their autonomy drives education and student engagement at SIU SOM. As a result, the implementation and success of the curriculum at other schools will vary, depending on their students' motivation, autonomy, and empowerment levels. Even at SIU SOM, students' participation levels during coachability sessions vary depending on the topic and timing. Barriers like time and our split campus with the year 1 students make this difficult. With the more frequent use of online meetings, we should be able to overcome the split campus. Proficiency in IIS is imperative to the treatment of patients. Clinical knowledge and IIS are a different side of the same coin (Lombardo et al., 2019). Medicine is always evolving and changing, and thus physicians must too. PBL teaches the student how to acquire information once the formal learning structure of the medical school is gone. Concurrently, the coachability session augments core studies and provides a mechanism to become a better physician to patients and a better colleague to staff.

3.3.6 *Programmatic Assessment for Collective Reflective Practices*

Assessment practices define the characteristics of a learning culture, as performance assessment inherently imposes power through decision-making regarding success and failure. The recent movement toward the concept of programmatic assessment invites a reflection on how assessment practices can constructively shape learning culture rather than existing as a fixed judgment about learning and performance (van der Vleuten et al., 2012, 2015). Programmatic assessment is an approach, where the assessment is *for learning* rather than *of learning*, by using multiple data points that are fundamentally necessary to guide students on their developmental processes. Moreover, "One of its prime virtues is that it enables assessment to move beyond the dominant psychometric discourse with its focus on individual instruments, towards a systems approach to assessment design underpinned by empirically grounded theory (van der Vleuten et al., 2012, p. 205)." This indicates that it is ultimately the learner who ought to benefit from assessment; therefore, assessment practices should be informative to learners (van der Vleuten et al., 2012). Programmatic assessment requires an institutional culture change in order to empower students to own education and utilize assessment itself as a developmental, rather than judgmental, process. The student empowerment inherent in programmatic assessment aims to bring about students' meaningful engagement in their learning and professional development processes.

Adopting this innovative concept of assessment, SIU SOM has conducted an organizational change process in which traditional assessment practices have been reworked into a framework of programmatic assessment. Previously, in the traditional assessment practices, the Year One Student Progress Committee (YOSPC), composed of a group of faculty, decided whether each student passed or failed each unit and notified students of their decision. In the newly adopted model of programmatic assessment, the Year One Student Competency Committee (YOSCC) has replaced the YOSPC, and the committee work focuses on providing intermediate evaluation and formative feedback to every student during and after each unit.

The advisor system that we have put in place plays a pivotal role in this formative feedback practice of programmatic assessment. The positive outcome of utilizing academic advisors in programmatic assessment has been reported in recent literature (Rich et al., 2020). In a qualitative study on programmatic assessment in a residency program, Rich et al. (2020) reported that based on a longitudinal relationship with a learner, an academic advisor plays a crucial role in the learner's sense-making and interpretation of their performance (i.e., application of knowledge) in the workplace. At SIU SOM, students meet regularly with their faculty advisors, who facilitate students' reflections on their learning progress, areas for improvement, learning strategies, and outcomes for coaching rather than providing solutions. Students' learning in all four SIU SOM graduation competencies—clinical skills, knowledge, self-directed learning, and professionalism—is reviewed. Faculty and students collaboratively create personalized remediation plans if needed. At the end of the curricular year, the YOSCC reviews all assessment data points in all units, as well as the reflective discussions by students and their advisors, to create a summative evaluation and recommendation to the Student Progress Committee. This advisor system brings together the coaching aspect and student empowerment, as students are active and can drive discussions with their advisers. This provides for a faculty-student partnership that focuses on students' professional growth.

The organizational change from a traditional assessment model to programmatic assessment at SIUSOM was based on several guiding principles of student empowerment: (1) students should be able to access their activity data and information on their performance and interpret the meaning of each data point; (2) students should be aware of their patterns of learning progress toward all graduation objectives longitudinally; (3) it is the student's responsibility to identify problems and create remediation plans; and (4) the role of advisor and the committee is as academic coaches to foster the students' individualized developmental process until the summative decision at the end of the Year 2 curriculum.

This culture change requires that SIUSOM transforms its assessment practice to be inclusive and educational data management processes to be more transparent and accessible for all stakeholders, especially students. This institutional transformation also necessitates the transformation of the medical school information technology (IT) practice (Spickard et al., 2016). In the next section, we will discuss more details of how changing IT practice could facilitate an organizational change toward student engagement and empowerment.

3.4 Technological Transformation Processes for Student Empowerment

Educational technology in health professions education is a pedagogical process that has evolved from a narrow techno-centric view to a broader systems-based approach (Han et al., 2013). A technological process that shares the vision of student empowerment within a medical education program can catalyze growth toward the achievement of the institutional culture of student engagement.

At SIU SOM, following the model at Vanderbilt University (Spickard et al., 2016), the school's IT practices are being transformed to facilitate institutional change. The goal of the IT transformation is to improve *educational data transparency and empowerment by* developing *an educational data dashboard system*. One of the problems we encountered at the medical school was that our existing educational data (regarding student performance, activities, feedback, reflections, etc.) unsystematically resided in isolated places within each department, making the task of overseeing students' longitudinal and cross-sectional performance difficult. Whereas our school's curriculum is innovative and strongly aligned with student empowerment and engagement, as discussed in the earlier section, our IT system did not fully support this ethos of student empowerment. Instead, the ownership of each educational data set was at the control of the faculty and educational staff. Data on student educational performance and feedback on their learning process were typically infrequent and in paper format. This created a barrier to systematically understanding students' longitudinal and cross-sectional progress. There was a clear need for an overarching strategy and governance structure for educational data capture and usage. The Educational Informatics Committee was formed to address this curriculum need.

3.4.1 Development of a Dashboard for Empowering Different Stakeholders

The Educational Informatics Committee's work has focused on two strategic plans: (1) digitalization and centralized management of educational data for easy access and use and (2) development of a dashboard to empower multiple stakeholders. A dashboard is a visualized report system based on learning analytics of an individual, group, and organization (Chan et al., 2018). The committee advocated for discontinuing the process of capturing performance data on paper documents that are often stored in isolated locations and moved to an online centralized data platform across the medical curriculum. Different data capturing systems, such as exam software, evaluation forms, and a student information system, were connected to a coordinated, centralized database that is securely stored and managed at the school. With the benefit of the digitized and centrally located educational data, the committee was able to develop a dashboard system in which individual and programmatic progress is stored, monitored, and analyzed efficiently and effectively for convenient access by both learners and authorized faculty.

The dashboard system presents all educational data and analytic information for stakeholders at various levels: macro vs. micro, individual vs. program, and cross-sectional vs. longitudinal. The dashboard has four major competency areas: knowledge, clinical skills, self-directed learning, and professionalism. Assessment data and activity data, such as progress exams without a stake and clinical logbooks, are presented and assigned one of three performance indicators: green (satisfactory), yellow (concern), or red (unsatisfactory). With data analysis in the dashboard at a group or curriculum level, curriculum administration can compare historical performance trends and identify areas for improvement and strengths of the curriculum. This dashboard, combined with the programmatic assessment initiative discussed in the earlier section, functions as a tool to assist reflections for all stakeholders, especially students.

The role of the dashboard is to provide stakeholders, including students, with a clear picture of their learning performance and progress. With access to this comprehensive information, students have up-to-date performance information in each assessment activity at all times. Thus, the dashboard functions as formative feedback for students to construct follow-up learning goals and action plans. In this way, students can be empowered to exercise informed decision-making proactively in each step of their learning processes rather than waiting for a decision on the summative pass or fail (remediation) decision decreed by faculty. The process of deciding whether remediation is necessary and, through what means, can be co-created with students, as they continuously reflect on their learning and make progress empowered by the dashboard process.

3.4.2 Implementation of a Dashboard

Implementation of a prototype of the dashboard began with the Year 1 curriculum in 2019 and expanded to include the Year 2 curriculum in 2020. This gradual implementation was intentionally designed so that the institutional change toward a centralized educational data repository and analysis can be adopted smoothly (Kotter, 1996).

The dashboard technology has different user groups. The first target group includes each unit director, curriculum director, nurse educator, and curriculum coordinator. The student progress committee also uses the dashboard for their decision-making at the end of the curricular year. These groups have access to data patterns and trends at the program level, as well as at the individual student level. The Educational Informatics Committee had to ensure that this dashboard process was functioning at the program level. With positive feedback and acceptance by the aforementioned groups, the Educational Informatics Committee plans to invite the next group, consisting of faculty advisors, to utilize the dashboard. These advisors have access to their individual student advisees' data. Additionally, a focus group of students will also have access to their own data as a pilot project in which any technical glitches and student requests can be addressed. After ensuring that all stakeholders' needs are met at a program level and an individual level, the committee plans to fully implement the dashboard system for all students in 2021.

This planned change process aims to ensure each stakeholder's acceptance of the new technological process of the dashboard in the joint effort to increase meaningful student engagement. By improving transparency of their educational data and decision-making, education informatics is able to further facilitate an institutional culture for student empowerment.

3.4.3 *Future Plan: Embracing Student Individualized Learning*

The institutional culture of student empowerment implemented through the dashboard system process will necessarily evolve. As mentioned earlier, the dashboard is not a tool to decide students' success or failure but rather a reflection tool that helps students monitor and manage their learning process with informed data analysis.

By enabling the students' reflections and professional growth, the dashboard function allows students to embrace their individualized learning paths. Each student has different developmental needs and, therefore, a different learning path. This idea becomes apparent at the time of clinical learning, when our students have the ability to choose rotations for their personalized educational plan (PEP) period in the late 3rd year and early 4th year as electives. Each student has different specialty interests and goals, and their learning and assessment cannot be standardized. This variability in their learning experiences and assessment is inherent in authentic workplace learning. Rather than being forced to fit into a standardized practice, the assessment system and the dashboard process should evolve to accommodate the diverse students' learning paths (van der Vleuten et al., 2012). This flexibility in learning, assessment including remediation, and the dashboard system should work together to facilitate learners' personalized developmental pathways toward independent practice in their future workplace (Heeneman & Schut, 2020).

3.5 Conclusion

The institutional culture of student empowerment can be facilitated by a curriculum process based on active learning, relationship-based workplace learning experiences, and assessment practices focusing on formative development. In this culture of empowerment, students are able to be engaged in holistic and reflective development processes as legitimate agents of change in their own learning and professional growth. The IT process focusing on transparency of educational data analysis and management can support this institutional culture of student empowerment. Technology should not remain as a mere tool to support teaching and learning activities but evolve as a system reflecting all stakeholders, who will collectively work together to achieve the goal of student learning and development into independent, competent health-care providers.

References

Allen, M., Gawad, N., Park, L., & Raîche, I. (2019). The educational role of autonomy in medical training: A scoping review. *Journal of Surgical Research, 240*, 1–16. https://doi.org/10.1016/j.jss.2019.02.034

Barrows, H. S. (1986). A taxonomy of problem-based learning-methods. *Medical Education, 20*(6), 481–486.

Barrows, H. S. (1988). *The tutorial process*. Southern Illinois University School of Medicine.

Barrows, H. S., & Tamblyn, R. (1980). *Problem-based learning: An approach to medical education*. Springer.

Barrows, H. S., Williams, R. G., & Moy, R. H. (1987). A comprehensive performance-based assessment of fourth-year students' clinical skills. *Journal of Medical Education, 62*(10), 805–809. https://doi.org/10.1097/00001888-198710000-00003

Behrend, R., Franz, A., Czeskleba, A., Maaz, A., & Peters, H. (2019). Student participation in the development of interprofessional education courses: Perceptions and experiences of faculty members and the students. *Medical Teacher, 41*(12), 1366–1371. https://doi.org/10.1080/0142159x.2019.1638895

Bjork, R. A., & Bjork, E. L. (2020). Desirable difficulties in theory and practice. *Journal of Applied Research in Memory and Cognition, 9*(4), 475–479.

Carey, P. (2013). Student engagement: Stakeholder perspectives on course representation in university governance. *Studies in Higher Education, 38*(9), 1290–1304. https://doi.org/10.1080/03075079.2011.621022

Chan, T., Sebok-Syer, S., Thoma, B., Wise, A., Sherbino, J., & Pusic, M. (2018). Learning analytics in medical education assessment: The past, the present, and the future. *AEM Education and Training, 2*(2), 178–187. https://doi.org/10.1002/aet2.10087

Chen, H. C., & Teherani, A. (2015). Workplace affordances to increase learner engagement in the clinical workplace. *Medical Education, 49*(12), 1184–1186. https://doi.org/10.1111/medu.12888

Cianciolo, A. T., Kidd, B., & Murray, S. (2016). Observational analysis of near-peer and faculty tutoring in problem-based learning groups. *Medical Education, 50*(7), 757–767. https://doi.org/10.1111/medu.12969

Cianciolo, A. T., & Regehr, G. (2019). Learning theory and educational intervention: Producing meaningful evidence of impact through layered analysis. *Academic Medicine, 94*(6), 789–794. https://doi.org/10.1097/Acm.0000000000002591

Cook-Sather, A., Bovill, C., & Felten, P. (2014). *Engaging students as partners in learning and teaching: A guide for faculty*. Josey-Bass.

Deci, E., & Ryan, R. (2012). Self-determination theory. In P. A. V. Lange, A. W. Kruglanski, & E. T. Higgins (Eds.), *Handbook of theories of social psychology* (Vol. 1, pp. 416–437). SAGE.

Deeley, S. J., & Bovill, C. (2017). Staff student partnership in assessment: Enhancing assessment literacy through democratic practices. *Assessment & Evaluation in Higher Education, 42*(3), 463–477. https://doi.org/10.1080/02602938.2015.1126551

Dweck, C. S. (2016). *Mindset: The new psychology of success*. Ballantine Books.

Fair, M., Diamond, S., Stagnaro-Green, A., & Sondheimer, H. (2011). *Beyond Flexner site visit report: Southern Illinois University School of Medicine*. Retrieved from http://beyondflexner.org/wp-content/uploads/2015/04/Beyond_Flexner_Report-SIU-FINAL1.pdf

Finn, J. D. (1989). Withdrawing from school. *Review of Educational Research, 59*(2), 117–142. https://doi.org/10.2307/1170412

Fredricks, J. A., Blumenfeld, P. C., & Paris, A. H. (2004). School engagement: Potential of the concept, state of the evidence. *Review of Educational Research, 74*(1), 59–109. https://doi.org/10.3102/00346543074001059

Geraghty, J. R., Young, A. N., Berkel, T. D. M., Wallbruch, E., Mann, J., Park, Y. S., Hirshfield, L. E., & Hyderi, A. (2020). Empowering medical students as agents of curricular change: A value-added approach to student engagement in medical education. *Perspectives on Medical Education, 9*(1), 60–65. https://doi.org/10.1007/s40037-019-00547-2

Guze, P. A. (2015). Using technology to meet the challenges of medical education. *Transactions of the American Clinical and Climatological Association, 126*, 260–270.

Han, H., Resch, D. S., & Kovach, R. A. (2013). Educational technology in medical education. *Teaching and Learning in Medicine, 25*(suppl 1), S39–S43. https://doi.org/10.1080/10401334.2013.842914

Han, H., Roberts, N. K., & Korte, R. (2015). Learning in the real place: Medical students' learning and socialization in clerkships at one medical school. *Academic Medicine, 90*(2), 231–239. https://doi.org/10.1097/ACM.0000000000000544

Heeneman, S., & Schut, S. (2020). Meaningful mapping of remediation in longitudinal and developmental assessment models. *Medical Education, 54*(10), 866–868. https://doi.org/10.1111/medu.14258

Hsih, K. W., Iscoe, M. S., Lupton, J. R., Mains, T. E., Nayar, S. K., Orlando, M. S., Parzuchowski, A. S., Sabbagh, M. F., Schulz, J. C., Shenderov, K., Simkin, D. J., Vakili, S., Vick, J. B., Xu, T., Yin, O., & Goldberg, H. R. (2015). The Student Curriculum Review Team: How we catalyze curricular changes through a student-centered approach. *Medical Teacher, 37*(11), 1008–1012. https://doi.org/10.3109/0142159x.2014.990877

Kahu, E. R. (2013). Framing student engagement in higher education. *Studies in Higher Education, 38*(5), 758–773. https://doi.org/10.1080/03075079.2011.598505

Kay, D., & Pasarica, M. (2019). Using technology to increase student (and faculty satisfaction with) engagement in medical education. *Advances in Physiology Education, 43*(3), 408–413. https://doi.org/10.1152/advan.00033.2019

Klamen, D. L. (2015). Getting real: Embracing the conditions of the third-year clerkship and reimagining the curriculum to enable deliberate practice. *Academic Medicine, 90*(10), 1314–1317. https://doi.org/10.1097/ACM.0000000000000733

Klamen, D. L., Ghareeb, A., Crouse, M., & Cianciolo, A. (2017). *Student engagement at its finest—A coachability curriculum for (and by) medical students*. Paper presented at the AMEE 2017, Helsinki, Finland.

Klamen, D. L., Williams, R., & Hingle, S. (2019). Getting real: Aligning the learning needs of clerkship students with the current clinical environment. *Academic Medicine, 94*(1), 53–58. https://doi.org/10.1097/Acm.0000000000002434

Kotter, J. P. (1996). *Leading change*. Harvard Business School Press.
Lamb, S., Chow, C., Lindsley, J., Stevenson, A., Roussel, D., Shaffer, K., & Samuelson, W. (2020). Learning from failure: How eliminating required attendance sparked the beginning of a medical school transformation. *Perspectives on Medical Education, 9*(5), 314–317.
Lave, J., & Wenger, E. (1991). *Situated learning: Legitimate peripheral participation*. Cambridge University Press.
Liao, L. (2020). Entitlement: The big problem in medical education we will not talk about. *Medical Teacher, 42*(5), 585–587. https://doi.org/10.1080/0142159x.2019.1636956
Lizzio, A., & Wilson, K. (2009). Student participation in university governance: The role conceptions and sense of efficacy of student representatives on departmental committees. *Studies in Higher Education, 34*(1), 69–84. https://doi.org/10.1080/03075070802602000
Lombardo, L., Ehlers, J., & Lutz, G. (2019). Is your mind set?—how are intra- and interpersonal competences dealt with in medical education? A multi-professional qualitative study. *BMC Medical Education, 19*(1), article 317. https://doi.org/10.1186/s12909-019-1748-y
Mann, S. J. (2001). Alternative perspectives on the student experience: Alienation and engagement. *Studies in Higher Education, 26*(1), 7–19. https://doi.org/10.1080/03075070020030689
Milles, L. S., Hitzblech, T., Drees, S., Wurl, W., Arends, P., & Peters, H. (2019). Student engagement in medical education: A mixed-method study on medical students as module co-directors in curriculum development. *Medical Teacher, 41*(10), 1143–1150. https://doi.org/10.1080/0142159x.2019.1623385
Molesworth, M., Nixon, E., & Scullion, R. (2009). Having, being and higher education: The marketisation of the university and the transformation of the student into consumer. *Teaching in Higher Education, 14*(3), 277–287.
Moran, J., Briscoe, G., & Peglow, S. (2018). Current technology in advancing medical education: Perspectives for learning and providing care. *Academic Psychiatry, 42*(6), 796–799. https://doi.org/10.1007/s40596-018-0946-y
Murphy, R., Nixon, S., Brooman, S., & Fearon, D. (2017). "I am wary of giving too much power to students:" Addressing the "but" in the Principle of Staff-Student Partnership. *International Journal for Students as Partners, 1*(1). https://doi.org/10.15173/ijsap.v1i1.3055
Neville, A. J. (2008). Problem-based learning and medical education forty years on. *Medical Principles and Practice, 18*, 1–9. https://doi.org/10.1159/000163038
Pereira, J. V.-B., Vassil, J. C., & Thompson, R. E. (2020). Students as partners in an Australian medical program: Impact on student partners and teachers. *International Journal for Students as Partners, 4*(2), 110–121. https://doi.org/10.15173/ijsap.v4i2.4175
Peters, H., Zdravkovic, M., Costa, M. J., Celenza, A., Ghias, K., Klamen, D., Mossop, L., Rieder, M., Nadarajah, V. D., Wangsaturaka, D., Wohlin, M., & Weggemans, M. (2019). Twelve tips for enhancing student engagement. *Medical Teacher, 41*(6), 632–637. https://doi.org/10.1080/0142159x.2018.1459530
Pluta, W. J., Richards, B. F., & Mutnick, A. (2013). PBL and beyond: Trends in collaborative learning. *Teaching and Learning in Medicine, 25*(sup1), S9–S16. https://doi.org/10.1080/10401334.2013.842917
Reeve, J. (2012). A self-determination theory perspective on student engagement. In S. L. Christenson, A. L. Reschly, & C. Wylie (Eds.), *Handbook of research on student engagement* (pp. 149–172). Springer.
Rich, J. V., Fostaty Young, S., Donnelly, C., Hall, A. K., Dagnone, J. D., Weersink, K., Caudle, J., Van Melle, E., & Klinger, D. A. (2020). Competency-based education calls for programmatic assessment: But what does this look like in practice? *Journal of Evaluation in Clinical Practice, 26*(4), 1087–1095. https://doi.org/10.1111/jep.13328
Sawatsky, A. P., Santivasi, W. L., Nordhues, H. C., Vaa, B. E., Ratelle, J. T., Beckman, T. J., & Hafferty, F. W. (2020). Autonomy and professional identity formation in residency training: A qualitative study. *Medical Education, 54*(7), 616–627. https://doi.org/10.1111/medu.14073
Schmidt, H. (2010). A review of the evidence: Effects of problem-based learning on students and graduates of Maastricht medical school. In H. Van Berkel, A. Scherpbier, H. Hillen, & C. Van

Der Vleuten (Eds.), *Lessons from problem-based learning* (pp. 227–240). Oxford University Press.

Spickard, A., Ahmed, T., Lomis, K., Johnson, K., & Miller, B. (2016). Changing medical school IT to support medical education transformation. *Teaching and Learning in Medicine, 28*(1), 80–87. https://doi.org/10.1080/10401334.2015.1107488

van der Vleuten, C. P. M., Schuwirth, L. W. T., Driessen, E. W., Dijkstra, J., Tigelaar, D., Baartman, L. K. J., & van Tartwijk, J. (2012). A model for programmatic assessment fit for purpose. *Medical Teacher, 34*(3), 205–214. https://doi.org/10.3109/0142159x.2012.652239

van der Vleuten, C. P. M., Schuwirth, L. W. T., Driessen, E. W., Govaerts, M. J. B., & Heeneman, S. (2015). Twelve tips for programmatic assessment. *Medical Teacher, 37*(7), 641–646. https://doi.org/10.3109/0142159x.2014.973388

White, C., Bradley, E., Martindale, J., Roy, P., Patel, K., Yoon, M., & Worden, M. K. (2014). Why are medical students 'checking out' of active learning in a new curriculum? *Medical Education, 48*(3), 315–324. https://doi.org/10.1111/Medu.12356

Williams, R. G., Klamen, D. L., White, C. B., Petrusa, E., Fincher, R. M. E., Whitfield, C. F., Shatzer, J. H., McCarty, T., & Miller, B. M. (2011). Tracking development of clinical reasoning ability across five medical schools using a progress test. *Academic Medicine, 86*(9), 1148–1154. https://doi.org/10.1097/ACM.0b013e31822631b3

Chapter 4
From Psychology Laboratory to Student Development: Untangling Momentary Engagement from Longer-Term Engagement in Bioscience Education

Harry J. Witchel, Rudi Klein, Puspha Sinnayah, and Joseph Rathner

Abstract Student engagement and its measurement remain contested by different research disciplines that aim to evaluate education. In computer analytics, engagement is momentary, and longer-term engagement arises from accumulated momentary engagement. By contrast, policymakers see engagement as fundamentally longer-term, emerging from the student's identification and sense of belonging. The definition of engagement is a long-debated topic, and this lack of agreement is a major problem for the field. Behavioral engagement is dedicated time and effort, which may be necessary and is certainly contributory to the outcomes of learning and development. Various engagement theories have included behavioral, cognitive, emotional, and social elements. Some engagement paradigms have subsumed other contributors to learning, such as motivation, cognitive strategies, and social communities. Despite much recent progress in our understanding of learning science, the relationship of the benefits of momentary to longer-term engagement is insufficiently researched. We and others have attempted to tease apart engagement's causes and consequences. Some technological approaches (e.g., gamification) boost momentary engagement to kindle longer-term engagement; this may be the mechanism of active

H. J. Witchel (✉)
Department of Neuroscience, Brighton and Sussex Medical School, Brighton, UK
e-mail: h.witchel@bsms.ac.uk

R. Klein
First Year College, Victoria University, Melbourne, VIC, Australia
e-mail: rudi.klein@vu.edu.au

P. Sinnayah
First Year College, Victoria University, Melbourne, VIC, Australia

Institute for Health and Sport, Victoria University, Melbourne, VIC, Australia
e-mail: Puspha.Sinnayah@vu.edu.au

J. Rathner
University of Melbourne, Melbourne, VIC, Australia
e-mail: joseph.rathner@unimelb.edu.au

H. J. Witchel, M. W. Lee (eds.), *Technologies in Biomedical and Life Sciences Education*, Methods in Physiology, https://doi.org/10.1007/978-3-030-95633-2_4

learning and frequent formative assessments. The opposite pedagogical approach uses longer-term, internalized motivation as a substrate that allows consistent triggering of momentary engagement by traditional pedagogies. Whereas metrics of behavioral engagement are sometimes agreed upon, the emotional-cognitive-behavioral meta-construct of engagement may cause more confusion than clarity when evaluating outcomes such as development and learning. Thus, using engagement as an outcome, instead of measuring learning and development directly, solves some policy problems, but it exacerbates many scientific problems regarding mechanism.

4.1 The Problem with Engagement Research: Too Many Definitions

4.1.1 Different Definitions of Engagement

The goal of this chapter is to map out the meaning and relevance, particularly in the context of the biosciences in higher education, of engagement during learning (Azevedo, 2015; Fredricks et al., 2016; Kahu, 2013; Reschly & Christenson, 2012; Sinatra et al., 2015). Practitioners, researchers, and policymakers all agree that engagement is a good thing, although they may be talking about different things (Ashwin & McVitty, 2015; Witchel, 2013). Although engagement is familiar and seemingly obvious to education practitioners ("I know it when I see it"), researchers sometimes define it as at least three separate constructs: (1) momentary engagement (e.g., interest and paying attention, from minutes to hours) is considered by computer analytics and psychology laboratories; (2) longer-term engagement (e.g., attending class and regularly making an effort at studying during a semester) is important to sociologists; and (3) school completion (retention) and dropout (attrition) (Reschly & Christenson, 2012) is important to policymakers because dropout is not an event but a longer-term process of failing emotional engagement and identification with the school (see below). The relationship between these constructs is poorly understood, as is the relationship between momentary engagement and longer-term learning (Schmidt et al., 2018; Shernoff & Hoogstra, 2001). Thus, the value of engagement research is open to interpretation because various researchers have presented engagement as any or all of an activity, a metric, a cause, an input, an outcome, or a value unto itself (Aparicio et al., 2021). This chapter aims to start a discussion among bioscientists on what engagement means in educational practice.

Researchers often define engagement by combining multiple learning parameters into a meta-construct that is difficult to formulate as a composite measure. Engagement as a meta-construct continues to be persistently problematic because its definition is not agreed (Azevedo, 2015; Sinatra et al., 2015), to the point where it has been likened to a blind man describing an elephant (Baron & Corbin, 2012; Eccles, 2016). Educational researchers have proposed different definitions for engagement, which can be seen in a spectrum from narrow to all-inclusive. Kuh, who established the most prevalent measurements of engagement, the USA-based

National Survey of Student Engagement (NSSE, pronounced "Nessie") program (Kuh, 2009a; Olivas, 2011), has one of the most restrictive definitions, limited to activities. The NSSE, which depends mostly on polling student perceptions, aligns engagement with "quality of effort and involvement in productive learning activities." Kuh (2001) claims, "What students **do** during college counts more in terms of desired outcomes than who they are or even where they go to college." Student effort, as in human capital and investment (Dowd et al., 2011), is a core concept for the accountability aspects of engagement (Kuh, 2009a) and its related precursors, involvement (Astin, 1993), and integration (Tinto, 1975). Educational technology focuses on metrics that are supposed to reflect activities and associated cognitive engagement (Jovanović et al., 2021). Some researchers have also included extracurricular activities as part of engagement (Finn, 1989; Pascarella & Terenzini, 1980), as these so-called enhancement activities can enhance learning (as well as social development) (Kuh, 1993; Thomas & Jones, 2017).

By contrast, some authors define engagement very broadly, going far beyond activities. Historically, the broader inputs related to engagement have been captured by dividing the meta-construct into behavioral (activities), cognitive (attitudes and strategies), and emotional components (Fredricks et al., 2004; Kahn, 1990). The one definition we sidestep in this chapter is the idea that student cohort engagement includes having students represented on educational committees and in designing the curriculum (Carey, 2013; Peters et al., 2018); these activities are plainly a very intense form of student engagement, but we describe them as participation, partnership, or part of the wider "students as partners" movement (Cook-Sather et al., 2014; see Chap. 3 on student empowerment in this volume). The spectra of possible definitions of engagement, including student outcome metrics, are reviewed in (Azevedo, 2015; Fredricks et al., 2016).

This rest of this chapter will show how researchers need to grasp the thorny topic of engagement in order to make advances in how students drive their own learning. This first section will outline the central role (literally) that engagement plays in leading to learning and other benefits. There follows a description of the processes by which momentary engagement, particularly elements of cognitive engagement, results in greater learning. Next, there is a section on how longer-term engagement is not simply a random collection of moments of engagement but is often crystallized around internalized motivation ("I want to be a biologist") or intrinsic motivation ("I like studying biology"). After that, a short survey of techniques that have been successfully used to boost student motivation leads into the classical way to motivate students: assessment. The section on assessment shows that innovations are still being made in how assessment is used to increase both learning and motivation for longer-term engagement. A section on active learning follows, in order to demonstrate how innovations in pedagogy depend upon the precise way students find motivation. The final section wraps up by showing that momentary engagement and longer-term engagement need to be discussed separately and that much more research is needed to understand the link between them, particularly to make sense of all the different types of active learning.

4.1.2 Engagement as a Facilitator of Learning Benefits

Engagement (as well as motivation) is proposed to be an upstream facilitator of student learning (and personal development) (Kuh, 2001; Ogbu, 2003, p.20). This suggests you can partially predict a student's outcomes by measuring aspects of the student's input process (engagement); however, simple associations between the NSSE and performance have often been disappointing (Campbell & Cabrera, 2011), and demonstrating this connection often required statistical unmasking (Kuh et al., 2008) and untangling (Pike et al., 2008). By comparison, the relationship between future student performance and previous student academic performance is nearly always statistically apparent and usually dominates most models (Campbell & Cabrera, 2011; Kuh et al., 2008).

Motivation is meant to drive engagement, but some researchers subsume aspects of motivation into engagement (Reschly & Christenson, 2012; Sinatra et al., 2015), which creates mechanistic overlap. Motivation and attitudes are presumed to be facilitators of engagement, and if they are present but behavioral engagement is prevented, comparatively little learning and development will occur (Rotgans & Schmidt, 2011). This presents further definitional ambiguity because emotional engagement may have significant overlap with motivation, and attitudes may have significant overlap with both motivation and cognitive engagement. Behavioral engagement or effort is necessary for most learning to occur—attitudes and feelings are not enough.

The model shown in Fig. 4.1 proposes the main causal pathways, in which motivation tends to lead to engagement, and engagement tends to lead to learning. Motivation and preparedness are substrates—longer-term situations that are

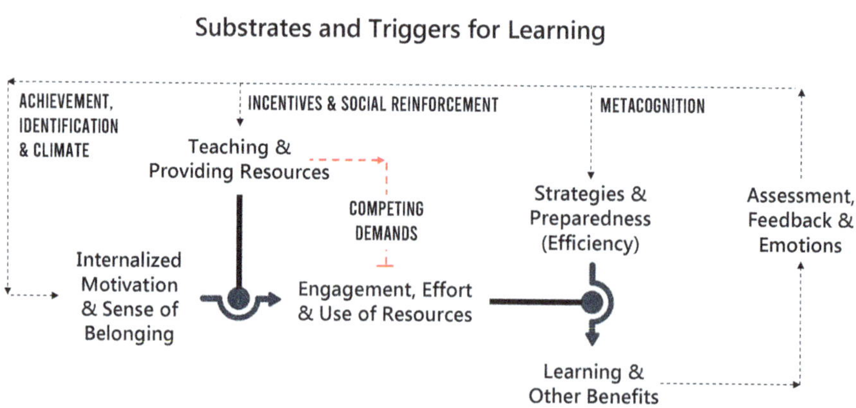

Fig. 4.1 Simplified causal model for learning with clear divisions between constructs. Triggers are shown as matchsticks, and necessary substrates are shown as looping arrows. Engagement is an input and mediator to learning and benefits. The red dashed line shows inhibition

relatively stable—which are (mostly) necessary but not sufficient for learning benefits.[1] They work synergistically with teaching and engagement, which are triggers—a pool of states or opportunities, like a box of matches—which catalyze the substrate. Substrate and trigger together provide a momentary impetus for learning; more of either will drive downstream effects. For example, engagement with the information is a necessary trigger that can lead to learning, but it will fail to lead to learning if the student lacks the substrate of cognitive elaboration strategies (or background knowledge) that are needed to learn the current information. Thus, the efficiency of the engagement (and therefore the correlation coefficient between engagement and learning) depends on the individual student having the required preparedness as a substrate. Because the substrate and the trigger work together, there are two approaches to increase a student's engagement: (1) intensify the substrate by increasing or internalizing motivation, or (2) multiply the triggers by sharpening the teaching and learning resources. This model simplifies the actual situation and allows testing of simple hypotheses with direct mechanisms.

This model implies that engagement's benefits—and its efficiency—reflect more than simply time-on-task and accumulated actions for learning, for example, as measured by computer analytics. In an example of data mining for blended learning with educational technology, Jovanović et al. (2021) analyzed medical students in problem-based learning over a 3-year period; the researchers divided causes of student learning into broadly two categories: (A) issues internal to the student (i.e. substrates) that are undetectable by the online platform versus (B) externally observable parameters such as engagement with the learning management system (LMS, the triggers). The LMS engagement parameters' effects on students' grades were shown to be statistically supported, as in previous studies (Gitinabard et al., 2019). However, by using mixed effect linear regression models, they showed that the total variation due to all their measured analytics of engagement factors was comparatively small, whereas the variation associated with unmeasured factors internal to the students was large. Thus, they found that there was such a thing as self-regulated (internal) learners (Winne, 1996), who consistently performed well, and that measuring engagement (external) explained only a small fraction of their performance (Jovanović et al., 2021). By making the educational institution a constant, they could conclude that the vast majority of this high-performing cohort's variability in final grades was determined by efficiency of their learning (or by unobserved forms of engagement). Thus, not all time spent in a learning environment results in equal learning benefits; for students' efforts at learning to succeed, the best students have something internal (e.g., preparedness or cognitive strategies) that underpins their ability to learn efficiently (see Fig. 4.1). The benefits of engagement depend not only on what you do (e.g., measurable analytics) but also on how well you do it (strategies) and on how well you inhibit extraneous behaviors (efforts at self-regulation) (Witchel et al., 2014).

[1]Note that in this text "necessary" and "sufficient" are not absolute or binary but act in degrees. They could be called "facilitatory," but the distinguishing features in this model would be lost.

The benefits and outcomes that researchers have linked to engagement are wide-ranging, including academic success (e.g., grades, educational completion), post-baccalaureate achievement (e.g., further education and graduate income), learning (some of which is unmeasurable by test scores (Corno & Mandinach, 1983)), and student development (Kuh & Umbach, 2004); other benefits may include expertise, social development, specific competencies, and employability skills. Accurate measures of engagement could be valuable for triaging additional support to underperforming students (Finn, 1989), evaluating educational interventions by researchers, and assessing student development in soft skills that are difficult to formally assess (e.g., process-based learning, such as research projects) (Wilson et al., 2016). The NSSE, which relates directly to satisfaction and persistence (but not learning), as revealed primarily on surveys, assembles five major subconstructs: level of academic challenge, active and collaborative learning, student-faculty interaction, supportive campus environment, and enriching educational experiences (Kuh, 2009a), the last being most strongly supported by data (Kuh et al., 2013). The other subconstructs ("benchmarks") do not statistically arise in data from all institutions (Kuh, 2009b), nor are they reproduced by all investigators (Campbell & Cabrera, 2011).

4.1.3 Disengagement Versus Lack of Engagement

These statistical inconsistencies might relate to the fact that some researchers have proposed that disengagement (also called disaffection) be considered as a separate concept from engagement (Chipchase et al., 2017; Martin et al., 2012; Wang & Degol, 2014); disengagement is universally agreed to be undesirable because of its association with attrition, delinquency, and crime (Li & Lerner, 2011; Wang & Fredricks, 2014). Thus, disengagement may not be the same as lack of engagement (Reschly & Christenson, 2012; Skinner et al., 2008), so that the causes and effects are qualitatively different for disengagement versus engagement, rather than simply reversed. Possible contributors to this difference focus on (A) sense of belonging (Dowd et al., 2011; Hurtado & Carter, 1997), (B) the self-regulating elements of metacognition (e.g., family support versus social media interruptions (Rosen et al., 2011; Wong, 2008)), and (C) how the definitions incorporate the inhibition of disruptive activities (Finn, 1989; Witchel, 2013; Witchel et al., 2014). Students with a low sense of belonging or poor self-regulation may easily make a disruption to an entire class, let alone their own studies. When comparing openly disruptive behavior to physical withdrawal or listless behavior, disengagement can appear qualitatively different from lack of engagement. Blaming students for certain kinds of disengagement can be unfair and may take the form of blaming the victim (Horton, 2005; Ogbu, 2003). For students who have a sense of not belonging and still feel marginalized, their motivation to engage will be rightly diminished (Hurtado & Carter, 1997; Hurtado & Ruiz Alvarado, 2015). Likewise, not all students have a level playing field to engage, particularly nontraditional students with full-time jobs or caring roles.

4.1.4 *The Challenges in Defining Engagement Necessitate New Mechanistic Measurements*

Is it possible that engagement is a meaningless or useless term that we should abandon? In education research, there seem to be so many exceptions to the causal connection between observed engagement and students' academic performance (Kauffman et al., 2018; Jovanović et al., 2019; see Chap. 12 on attendance) that the discussion about engagement needs to be a discussion about mechanisms, and the definitions need to be explicit. Rather than dividing the elements of engagement according to how we detect them (cognitive, emotional, behavioral), we may want to group them by how they function mechanistically in the learning process (see Table 4.1).

As stated above, the most serious problem with research into engagement is that the definitions conflate three different types of outcomes: momentary effects versus longer-term versus large-scale and institution-wide investigations (Sinatra et al., 2015). These forest-versus-the-trees perspectives in measuring engagement have been called "grain size," and they may represent distinct learning or forgetting mechanisms. Assuredly, the nomenclature for discussing these learning perspectives needs to be clarified. Table 4.2 shows the results of a recent literature search showing three different terms that refer to the same idea: short-term engagement, momentary engagement, and moment-to-moment engagement; almost all of the papers referred to one term but not to the others. Note that the term "micro engagement" (Sinatra et al., 2015) is already used for another phenomenon that is related to longer-term engagement (Handelsman et al., 2005; Taylor et al., 2011).

The remainder of this chapter will sketch a picture of how life sciences education relates to the two facets of engagement: momentary versus longer-term. It will attempt to synthesize these into a question of how we might employ technology—a tool—to take us beyond access of learning materials. We are searching for aspects of the educational climate (whether technology or pedagogy) that we can improve to maximize learning for all, given that some aspects of the education climate are elements of our wider society. Our destination should be a holistic view of student development that is fundamentally social, human, and inclusive, connecting momentary engagement to longer-term engagement and its benefits.

Table 4.1 Elements of engagement grouped mechanistically rather than by cognitive, emotional, and behavioral categories

Facilitator/motivation		Indicator/committed effort		Effecting procedure/skills		Result/feedback	
E	Anxiety/anticipation[a]	B	Time on task	C	Cognitive strategies	C	Metacognitions[b]
E/C	Belonging	B	Attendance	C	Elaboration	E	Higher needs (SDT)
C	Self-efficacy	C	Mental effort	C	Visualization	E/C	Competence
C	Beliefs	C	Focus on task	B/C	Repetition/retrieval	E/C	Autonomy
C	Values & attitudes	B	Inhibit irrelevant tasks	C	Misconceptions	E	Relatedness
C	Goal clarity (career)	C	Attention	B	Following instructions	E	Achievement emotions
B	Conflicting demands	C	Listening	B	Taking notes	E	Enjoyment
E/C	Curiosity/interest					E	Confusion

This table is not comprehensive; it provides a few examples of elements of engagement (and disengagement) along with their traditional grouping as cognitive (C), emotional (E), or behavioral (B)

[a]Emotions are not only precursors but also results and may also function as indicator responses

[b]Metacognitions are explicitly part of planning and hence also function as precursors, but metacognitions are less related to motivations than to strategies and procedures

Table 4.2 Database searches of Google Scholar for papers that refer to different names for short-term engagement

Primary search term	Primary search	Percentage of main search that also includes secondary term: 1	2	3
1. Short-term engagement	1290		0%	0%
2. Momentary engagement	293	0.8%		0.8%
3. Moment-to-moment engagement	168	0.6%	1.8%	

4.2 Momentary Engagement and Learning

Momentary student engagement is the set of dedicated learning processes that take place during a short time in a single task or activity (de Vreede et al., 2019) such as during a single classroom session, when watching an educational video, or in private study activity. Momentary engagement while learning can include methods that encourage thinking more deeply about the topic. From a neuroscience perspective, these activities facilitate (A) encoding sensations and thoughts as meaningful information in short-term memory, (B) consolidating short-term memories into long-term memory (which includes elaboration with previous knowledge), and (C) retrieval (Van Hoof & Doyle, 2018). Momentary engagement behaviors including attention and boredom (short-term disengagement) are easily measurable by methods, such as subjective rating (Shernoff & Hoogstra, 2001), eye tracking, fidgeting, and physiological responses (D'Mello et al., 2017; Witchel et al., 2014), but do not necessarily predict long-lasting learning, nor do they predict longer-term behaviors. The link between individual engagement activities that can be measured in the span of a laboratory session or a single class may not translate in a classroom cohort into sustainable learning and development across a semester.

A major *caveat* in attempting to apply discoveries from the psychology laboratory onto longer-term educational contexts concerns motivation. A bored and disengaged research participant in a psychology laboratory can simply choose not to continue to participate in the experiment, and their data is excluded from the study. By contrast, in a school environment, this same individual who does not engage or do any of the work is at risk of failing. In most laboratory research, the data is biased to reflect the motivated research participants and not the unmotivated, who are absent from the data.

Cognitive neuroscience (or *learning science*) has put forward two broad approaches that have empirically been shown to improve learning outcomes (Table 4.3): **connecting memories** and **timing of practice**. Several reviews (Davidesco & Milne, 2019; Gooding et al., 2017; Weinstein et al., 2018) and commentaries (Putnam et al., 2016) have been written recently about the *science of learning*. Five key learning strategies are clearly evidenced as optimizing learning (Davidesco & Milne, 2019; Weinstein et al., 2018). These are:

Table 4.3 Principles of study that improve learning and retention

1. Connecting memories
(a) Elaboration: Connecting the new concepts to other information, either previously known or new
(b) Dual coding: Using images to support text
(c) Concrete examples: Using real-world examples to illustrate abstract concepts (Davidesco & Milne, 2019)
2. Timing of practice
(a) Retrieval practice: Retrieval practice involves strategies that require the learner to actively recall material from memory (Dobson, 2013; Roediger & Butler, 2011), as opposed to passive study strategies, which involve rereading, reviewing, and rewatching/listening to learning material (as an audience member).
(b) Spaced practice: Spaced practice is often described as the opposite to cramming (Rodriguez et al., 2021). Where cramming is the attempt to learn the entire content of a subject in a very short period prior to a test, spaced practice encourages students to delay their retrieval practice by setting aside small amounts of time to regularly engage with subject content over the longer-term.

4.2.1 Connecting Memories

A consistent way to improve retention is to have newly presented information connected to other concepts or mental constructs. The associative links between new learning content and established conceptual schema can be explained by Hebbian learning theory (Nicoll, 2017). Memories are maintained by long-term potentiation, and if a specific memory is connected to other memory circuits, the well-connected memory will be reinforced more often (Van Hoof & Doyle, 2018). There are three well accepted methods from learning science to link new knowledge to other established ideas or concepts: (1) elaboration of the new information, especially by making meaning of it by linking it to "how" or "why" information, or through discussion; (2) linking the new information to a picture, so that new names and abstract concepts can be remembered visually in a mnemonically meaningful way; (3) linking the new information to concrete examples in order to make abstract information meaningful or salient to the student's other knowledge

4.2.1.1 Elaboration

Hirshman (2001) defined elaboration as, "A conscious, intentional process that associates to-be-remembered information with other information in memory," which may be from the student's prior knowledge and experience or with contiguously presented new information. Hamilton (1997) succinctly states, "Essentially, elaboration is encoding the original content in a different but related way," which allows the student to build a more detailed mental schema of their knowledge: for example, when the learner explains concepts or ideas in their own words. This is possibly best exemplified in peer-to-peer learning when the learner tries to answer their peer's content-related questions (Green et al., 2014).

Elaborative interrogation is when students ask—and attempt to answer—*how* and *why* questions about the concepts they are studying (McDaniel & Donnelly, 1996). Initially their answers may be sought from the study materials, but eventually the students should formulate their answers from memory and ideally check that the answers are correct. Elaborative interrogation is associated with robust increases in learning, and it is associated with deep learning because it involves higher order questioning (Roediger & Pyc, 2012).

Asking students to design and draw a concept map is a widespread technique among educators for encouraging visual elaboration. In linking concepts together, the learner relates both new and prior knowledge together to form a graphical representation of how the content is interrelated (Hilbert & Renkl, 2008). When properly trained in how to develop concept maps, students report that the concept maps assist in helping students identify the relationships between core concepts and generally see concept maps as a beneficial learning tool (Fonseca et al., 2020).

4.2.1.2 Dual Coding

A simple illustration can easily and rapidly convey more information than two paragraphs of text (Mayer & Gallini, 1990), which aids memory encoding. Words are ultimately less memorable than pictures; when educators (or textbook authors) provide both an image and text to represent the same information, learning is often enhanced, an instance of dual-coding theory (Paivio & Csapo, 1973). Plainly the images must be chosen wisely, being helpful, clear in meaning, and relevant to the concepts to be learned; otherwise, they may lead to cognitive overload (Mayer & Moreno, 2003).

4.2.1.3 Concrete Examples

Abstract concepts can be particularly difficult to visualize, which makes encoding and learning more difficult. Providing concrete examples is the teaching strategy of translating abstract (but generalizable) ideas and concepts into more relatable (specific) examples (Weinstein et al., 2018). Scientific content can be made concrete for allied health students by contextualizing the content within clinical scenarios; when bioscience content is presented as part of a clinical context, it adds credibility and relevance for nursing students, who feel greater confidence in "visualizing" the concepts (Gordon & Hughes, 2013). Analogies can be used as another tool for developing concrete examples, in which a new abstract concept is described in terms of a familiar concrete thing (Newby & Stepich, 1987). For example, an analogy for action potential propagation might be a *Mexican wave* (Kurbel, 2014). Analogies have been proposed as a valuable tool in teaching human biosciences at many educational levels (Bobulescu, 2020; Gamba, 2003; Krieger, 2017; Sundrud & Hueftle, 2009).

4.2.2 *Timing of Rehearsal*

From a psychological and educational efficiency perspective (Carnine, 2000), forgetting is *the* fundamental problem in learning. Limited information can be stored simultaneously in short-term memory. Only a percentage of such information is encoded into long-term memory. Over time an extensive portion of long-term memory is forgotten if not reinforced (Cowan, 2008, 2014; Murre & Dros, 2015). The most powerful method for long-term memory retention of semantic information, particularly for answering questions on tests, is repeated intentional recall (rather than merely passive revision) (Schmidmaier et al., 2011; Szpunar et al., 2013). Whereas it has long been known that spacing of rehearsal is more effective than immediate repetition (Ebbinghaus, 1885), educators and students have often favored immediate repetition in the form of cramming and writing lines (Hogan, 1985; Sommer, 1990). Only recently has our understanding of memory reinforcement become detailed enough to understand that the most useful reinforcement of memory occurs just before it would otherwise be forgotten (Benjamin & Tullis, 2010). The newfound clarity of these psychological principles has resulted in a burgeoning popularity of spaced repetition software (e.g., Duolingo (Munday, 2016), Anki (Deng et al., 2015) and Supermemo (Wozniak, 2021)) that explicitly employ these timing principles with computerized flashcard learning. This digital flashcard learning is based on combining three practices: retrieval practice based on cues, spaced repetition, and interleaved learning.

4.2.2.1 Retrieval Practice

Retrieval practice, also called *test-enhanced learning* (Roediger, Agarwal, et al., 2011), is a set of techniques that require the students to recall information that has previously been studied. Common methods of retrieval practice include practice exam questions, flashcard study and many other methods of cued recall (Wissman et al., 2012). Traditional paper flashcards have long been used for foreign language self-study, allowing for the students to test themselves using the cue on one side and to receive instant feedback by simply turning over the card. Flashcards can be sorted and shuffled to allow a student to regularly self-test; this has made them commonplace as a study method (Wissman et al., 2012). However, the rigidity and repetitiveness of drill-and-practice activities can make flashcard study feel monotonous (Kuiper & de Pater-Sneep, 2014; Seibert Hanson & Brown, 2020). The use of flashcards has proliferated, as other subjects can be learned by flashcards if the cues are carefully designed (e.g., as questions). This growth has been further fueled by the advent of digital flashcard programs and websites that incorporate elements of gamification (e.g., Quizlet, StudyBlue, Cram, Flashcard Machine, etc.) (Waluyo & Bucol, 2021).

Not only does retrieval practice improve memory retention (Deng et al., 2015; Dobson, 2013; Dunlosky et al., 2013; Roediger & Karpicke, 2006b), but when combined with other learning methods, the retrieval practice can also increase behavioral engagement. Audience response systems (ARS, "clickers") are a commonly cited tool for making traditional face-to-face lectures more engaging and active (Cain et al., 2009; Latessa & Mouw, 2005; Miller et al., 2003). When an ARS was used to present questions to students during in-person lectures, student engagement and ratings of motivation were increased, as well as exam scores for students who used the "clickers" (Gauci et al., 2009). Instructors also found that the use of pop-up quizzes in the lecture allowed them to rapidly diagnose misconceptions by students on learning content and address them immediately during in the lecture. Szpunar et al. (2013) ran experiments showing that students exposed to retrieval practice intermittently during a video lecture retained more of the presented information. In a carefully designed controlled study, they showed that it was the retrieval practice that was improving the learning and not simply reexposure. In a classroom setting, students willingly spend more time watching videos of learning content when there were pop-up quiz items embedded in the video lecture (Haagsman et al., 2020). Frequent quizzes can also have a diagnostic value for identifying at risk students and allow targeted interventions (Roediger, Putnam, et al., 2011) as well as providing timely formative/diagnostic feedback to students as they progress through the subject (Page et al., 2017; Rathner et al., 2013; Sinnayah et al., 2019).

4.2.2.2 Spaced Practice

Spaced practice describes a study technique where there is an optimal delay of time between learning content for the first time and revising the same content (Carpenter, 2014; Smolen et al., 2016). It seems counterintuitive that, for longer-term memory, drilling material immediately after learning is less effective than drilling after a delay, yet this is one of the longest standing results of psychology (Bjork & Allen, 1970; Ebbinghaus, 1885; Melton, 1970). Neuroscience may provide a mechanism to explain this remarkable finding; spaced practice may be aligned with additional protein synthesis needed to maintain synapses that form during long-term potentiation (Kelley & Whatson, 2013). Reviewing learning content at intervals after the initial learning experience presumably reinforces those synapses retaining the encoded memories.

There is a wealth of laboratory data that supports spaced practice for memory retention (especially of decontextualized word pairs), but there seems to be less certainty about its efficacy in real learning environments (Carpenter, 2014; Davidesco & Milne, 2019). Spaced practice appears to be effective in language vocabulary learning and underlies language learning packages like Duolingo (Tabibian et al., 2019); note that this learning is conscious learning rather than subconscious language acquisition (Munday, 2016).

The effectiveness of spaced practice appears to be dependent on the gap in time between episodes of content revision. With increasing mastery, the optimal delay between revision sessions greatly increases (Cepeda et al., 2008). For example, Dobson (2011) asked students to read textbook extracts on human biology and then to *recall* the information originally learned (retrieval practice) at spaced intervals between the initial learning and the final assessment 10 days after the initial learning activity. Students who practiced the learning content at expanding intervals (immediately, day 1, and then day 3) did better on the final assessment (day 10) than their peers who reviewed the learning material at uniform intervals (day 1, day 2, and day 3 after learning the content). Computer programs and online platforms are well suited to keeping track of these delays for individual facts to-be-learned.

The timing of homework tasks relative to in-class learning activity can also alter student performance on a summative final exam (Cadaret & Yates, 2018). In a physiology laboratory class, students were randomized to receive homework assignments that were released online at either a 1-day delay or 5-day delay after the completed the class; students in both cohorts were required to complete the assignment within 36 h of the release of the homework. After an entire semester, students who completed their homework 5 days after each class typically did 8% better on the final summative exam (week 15) than students in the same class who completed the same homework 1 day after class. Although the students' short-term assessment outcomes were unaffected by the delay between learning and practice, long-term retention (final exam performance on relevant items) was improved by the prolonged interval between learning task and homework task, consistent with *spaced practice* learning approach. Mixing learning tasks, especially between learning topics, is known as interleaved learning; the benefits of interleaved learning remain controversial (Brunmair & Richter, 2019; Carvalho & Goldstone, 2019; Linderholm et al., 2016; Pani et al., 2013).

4.3 Longer-Term Engagement and Student Motivation: Internalizing Extrinsic Toward Intrinsic

4.3.1 Engagement Depends on Motivation

Longer-term engagement is assumed to result in positive educational benefits (Kuh et al., 2008), and it is presumed to be dependent on maintaining student motivation (Allen, 1999; Kahu, 2013). Motivation is defined as the psychological link that allows thoughts (for cognitive scientists (Wittrock, 1978)) or instincts and drives (for behaviorists and ethologists (Eibl-Eibesfeldt, 1979; Mowrer, 1939; Tolman, 1938)) to influence behaviors. Motivation is crucial to understanding engagement because of its potential role in student attrition. Attrition is the clearest example, where engagement is related to learning and development; when students drop out of higher education before completion, their development and learning are worse off (Ishitani, 2006). In the 1950s, when higher education participation rates were much lower (Schofer & Meyer, 2005), attrition was presumed to be caused by academic failure and lack of intellectual potential. Research on student attrition in the 1970s responded to the recent student protests ("tune in, turn on, drop out") by adding a psychological and motivational perspective, which acknowledged that some attrition was due to voluntary withdrawal of academically capable students (Heist, 1968; Tinto, 1975). Researchers have now revisited these two approaches (low motivation or intellectual ability), which both focused on the mental situation of the student (thus blaming the victim), to a more sociological perspective accounting for interaction between the student and the institution. Motivation and academic promise are not completely internal qualities of the student that are fixed on arrival but are receptive to feedback, emotions, and a sense of belonging (Fig. 4.1).

4.3.2 Motivation Is Categorized Between Internal and External

The theoretical backdrop for what is motivation and how motivation works remain controversial. There is a long history of dividing motivations into intrinsic motivation (McGregor, 1960), where the behavior is inherently rewarding or enjoyable unto itself (e.g., satisfying instinctual needs like food or drives like status), or extrinsic motivation. Extrinsic motivations are impelled by incentives or expectations of rewards that follow but are separable from the activity itself. Gamification is a technological approach for adding extrinsic motivation to the learning process. It prompts retrieval practice by driving momentary engagement repeatedly using many small extrinsic (emotional) rewards (see Chap. 5 in this volume). Mobile phone applications have been developed that are intended to extrinsically motivate and engage the student in learning (e.g., Quitch (Söbke, 2018)). These games use competition (in the form of a leader board) to encourage participation (Sterling et al., 2020). Gamification improves learning by increasing student engagement and by making it compulsive (Kim & Werbach, 2016). The use of gamified learning applications on mobile devices (apps) has been shown to decrease attrition and improve academic performance in undergraduate business students (Pechenkina et al., 2017). App engagement is positively correlated to higher student grades. A systematic review of the literature has shown that gamification apps can improve student learning in science classes (Loganathan et al., 2019). This makes retrieval practice incidental to the enjoyment (or thrill) of the competition. These games also extrinsically motivate metacognitive development by providing instant feedback to students, revealing the gaps in the competitors' own understanding of the learning content. However, caution with gamification is needed for different students (Tomé Klock et al., 2020), because gamifying a classroom can undermine motivation and reduce student learning (Hanus & Fox, 2015), despite the fact that gamified elements of a course can be compulsive (Evans et al., 2014), and on average aid knowledge acquisition (Gentry et al., 2019; Majuri et al., 2018).

In order for intrinsic motivation to explain educational persistence, which includes forestalling rewards, motivation theories have embraced the concept of "higher needs," such as in Maslow's Hierarchy of Needs (Maslow, 1943), for example, when a person pursues a higher education degree because they have an almost instinctual need for esteem. It is important to distinguish needs, which are necessary for a person to live a healthy life, from wants, which are optional; needs have threshold values above which the person thrives, whereas wants can potentially be far in excess of what the person requires or can use (e.g., wealth). Although needs have been mostly abandoned by psychologists studying motivation (Wahba & Bridwell, 1976), they remain a key feature of sociology (Dover, 2013) and are the foundation of the most studied theory of motivation in education, self-determination theory (SDT) (Ryan & Deci, 2000).

Deci and Ryan's self-determination theory is based around three fundamental independent higher needs: relatedness, competence, and autonomy, autonomy being the most researched and evidenced by the literature (Ryan & Deci, 2020). These higher needs are meant to be intrinsically motivating and, thus, rewarding unto themselves; by contrast, extrinsic motivations based on incentives (such as money, grades, gamification, and various academic punishments) can conceivably be used to spur engagement and effort, but SDT focuses on the idea that ultimately these behaviorist interventions lead paradoxically to worse motivation and performance after the incentives are extinguished or even before (Chib et al., 2012; Deci, 1971; Deci et al., 2001). The supposed near-worthlessness of external incentives is a core principle of SDT (and much of positive psychology), but this position remains controversial and may only apply in limited situations (Cameron 2001; Cameron et al. 2001; Cameron & Pierce, 1994). A key feature of SDT is that incentives provide weaker or more detrimental motivation the more external they are.

According to SDT, extrinsic motivations can be described as four stages (see Fig. 4.2), which are hierarchically organized into increasingly internalized forms of motivation. External regulators are traditional incentives that can be physical (e.g., money, after-school detention), whereas integrated motivations are fundamental to the student's goals. For example, with the external incentive of money offered for learning, the student may search for a way to obtain the money without doing any studying. By contrast, if the student is motivated by making a scientific discovery in the future (internalized), although current study may not feel intrinsically enjoyable, there is no way for the student to rationalize engineering a situation where they make a scientific discovery in the future without doing the learning now. This internalized motivation is proposed to elicit persistent motivation, deep rather than surface learning, and persistence rather than attrition during a program of study. Internalization of motivation forms the basis of intention and self-regulated activities for longer-term work activities such as educational completion (Deci et al., 1996). However, internalization of career motivations is considered a less reliable drive for education (and its completion) than intrinsic motivation for learning, and they do not have identical educational results, as in the cases of tech billionaires Mark Zuckerberg and Bill Gates, who both dropped out of Harvard University. The endpoints of a baccalaureate degree may seem arbitrarily set by the institution, so arguments about attrition often focus on lack of motivation, whereas the student's motivation for learning may simply be starting a business rather than becoming an academic.

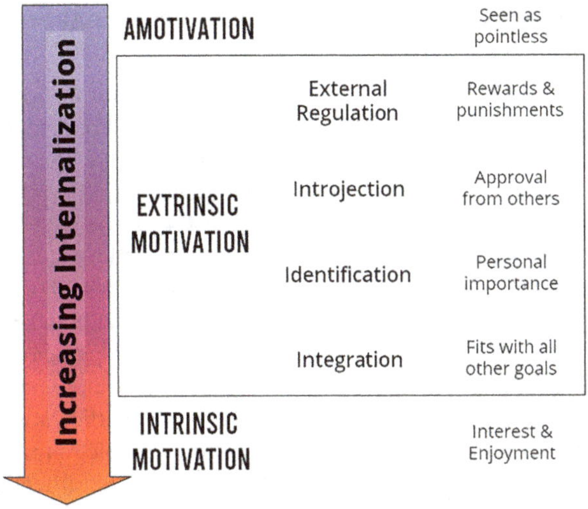

Fig. 4.2 Extrinsic vs. intrinsic motivation from a self-determination theory perspective. Forms of motivation closer to the bottom of the image are considered more internalized. With permission, adapted from Ryan and Deci (2020) Contemporary Educational Psychology 61: 101860

4.3.3 Intrinsic Motivation vs. Internalized Instrumental Motivation

Whereas professional educators and SDT theorists presume that intrinsic motivation and a love of knowledge should be the bedrock of liberal education, it has long been recognized that universities play an instrumental role in society (i.e., preparing students for the workforce and economy) and that students usually prioritize their instrumental goals (e.g., improving their career prospects) for being at university (Kerr, 2001). It is well-established that the motivations of most university students are instrumental, rather than a fundamental love of knowledge (Silverthorn, 2020). For some time in the USA, over 75% of entering university students have rated "to get a better job" and "to make more money" as "very important" for attending university, and these reasons are becoming more important (Astin, 1993; Wyer, 2013). To achieve the full benefit of internalized motivation, students need to have fully integrated the values and goals of their course of study with their instrumental goals for attending university. The mismatch between students' internal motivations and the university courses' values lead to disagreement about whether students or institutions should be held accountable for attrition and disengagement from science (Astin, 1993; Cech & Kennedy, 2005; Jeno et al., 2018).

4.4 Interventions to Improve Longer-Term Engagement

Kuh et al. (2013), working with the American Association of Colleges and Universities, in association with Liberal Education and America's Promise (LEAP) and the NSSE, have highlighted a set of instructional practices that are familiar to most academics that have been shown to be effective for increasing longer-term student engagement (i.e., retention and attrition, see Table 4.4); these have been labeled by Kuh et al. as high-impact practices (HIPs). As with most longer-term engagement data, these are based primarily on observational data, much of which is survey-driven. Table 4.4 shows the percentage of students in Kuh et al.'s large USA-based data set for 2012 who stated that they currently or planned to participate in one of the stated HIPs; according to their data, participating in one of these practices substantially increased the students' 6-year graduation rates, and participating in more than one such practice additionally substantially improved completion rates (up to three HIPs). Importantly, these practices have a particularly powerful retention effect on historically excluded minorities (their data considered Latino/a students at California State University at Northridge), groups which typically have drastically lower completion rates. In a fascinating critique of the NSSE, Campbell and Cabrera (2011) examined data from one university in detail and found that the only elements of the five NSSE data benchmarks that predicted GPA (student grade point average) were those questions relating to enriching experiences, including most of the high-impact practices.

The mechanisms by which the high-impact practices work are proposed to relate to key features that Kuh et al. (2013) have identified (see Table 4.5). In a broad sense, these demonstrate a substrate and trigger relationship, leading to motivation and engagement, where the substrate is a sense of belonging to a community (even a future community or a profession), whereas the triggers are based on goal clarity (Kluger & DeNisi, 1996); goal clarity is when the momentary learning activity/teaching relates clearly to the student's goals, motivating students to apply effort or engage with the task. For many students, who may be quite obliging, belonging to the university plus their parents' expectations provide enough motivation to drive all their activities (Zhang et al., 2011). However, for first-in-family scholars or students from historically excluded groups, the substrate and trigger together are necessary for motivating them to dedicate time and effort to each learning activity.

Table 4.4 High-impact practices for first or final year students (see Kuh et al., 2013)

High-impact practice	Percentage of student participation
Learning communities (year 1 only)	18%
Service learning	48%
Study abroad	14%
Student-faculty research	20%
Internships	49%
Senior Culminating Experience	33%

Table 4.5 Key features that make up high-impact practices that fuel longer-term engagement

Focus	Key feature	Example activity[a]
G	Performance expectations set at appropriately high levels	First-year seminar
[b]	Significant investment of time and effort by students over an extended period of time	Dissertation
B	Interactions with faculty and peers about substantive matters	External lecturer series
B	Experiences with diversity, wherein students are exposed to and must contend with people and circumstances that differ from those with which students are familiar	Service learning field assignment
G	Frequent, timely, and constructive feedback	Student-faculty Research
G	Periodic, structured opportunities to reflect and integrate learning	Portfolio
G/B	Opportunities to discover relevance of learning through real-world applications	Internship
B	Public demonstration of competence	Graded oral presentation

G, goal clarity; B, sense of belonging to a community
[a]Successful activities require due care and supervision. For complete description of key requirements see Kuh et al. (2013)
[b]This is an indicator of engagement/effort

4.4.1 Short-Term Interventions to Increase Student Motivation

There are a wealth of successful brief (<8 h) interventions for increasing student motivation (Lazowski & Hulleman, 2016) that have been tested experimentally, such as *transformational experiences*, which are meant to help students see the relevance of biology in their own lives (Heddy & Sinatra, 2013). Transformational experiences occur when the student learning changes the way they see the world, such as in informal experiences and discussion, field trips, or explicit examples where taught content is related to newspapers and other daily experiences. Transformative experiences are theoretically based on three elements of learning:

1. Motivated use
2. Expansion of perception
3. Experiential value

Motivated use is application of school content to everyday life when out of school. *Expansion of perception* is when students perceive deeper layers of meaning by "reseeing" the world (events, objects, issues, or themselves) through the lens of taught content. *Experiential value* is the appreciation of greater value for everyday life when reseen from the perspective of the scientific principles they were taught. Incorporating transformational experience into one's teaching design has been shown to improve learning, reduce misconceptions (Heddy & Sinatra, 2013), and increase motivation to learn (Pugh et al., 2017).

Improving student motivation can also be achieved by improving students' clarity of their own learning goals—particularly if the students start with poor goal clarity. Morisano et al. (2010) demonstrated that poorly performing students significantly improve their end of term grade point average (from 2.25 to 2.9 out of 4) by performing an intensive (2.5 h in a single session) online goal setting intervention at the beginning of the term.

Reeve et al. (2004) showed that a brief teacher training intervention led to a clear increase on student engagement as measured by ratings of classroom behavior, which was due to teachers increasing their autonomy-supporting behaviors. Based on Kuh's (Kuh et al., 2011, p. ix) data he proposes, "When students [...] take responsibility for activities that require daily decisions and tasks, they become invested in the activity and more committed to the college and their studies." In the classroom, teachers can support autonomy by creating well-organized learning environments and activities that catalyze students' willingness to engage in learning. Teachers can scaffold autonomy with three clear behaviors: bolstering meaningful choices, providing rationales, and answering students' questions (Ryan & Deci, 2020). According to self-determination theory, helping to satisfy the student's need for autonomy in this way should help to internalize the student's motivation away from extrinsic and more toward intrinsic motivation (see Fig. 4.2).

4.4.2 Communities Fuel Social Engagement

Although some students have such internalized motivations that they almost do not need to meet with other students (Kauffman et al., 2018), some students identified as at-risk for attrition (such as children in the Check and Connect program (IES WWC, 2021; Sinclair et al., 2003)) need additional support based on social interventions, often geared toward creating a sense of belonging (Juvonen et al., 2012). Thus, another collection of methods to motivate students to engage with the topic focus on creating communities of learners (Engle & Conant, 2002). A variety of different learning communities can be aimed for the following: online (Garrison et al., 2010; Rourke & Kanuka, 2009) and in person, including communities of practice (Wenger, 1999, p. 125) and communities of inquiry (Pardales & Girod, 2006), which are quite different from each other (see Table 4.6). Perhaps the most basic of these is the "learning community" movement (Kuh et al., 2013; Tinto, 2003) in the USA, in which first-year students register together and share some small classes regularly with the same group (e.g., freshman seminar or intensive writing); this is particularly important at USA-style universities that allow students massive choices in what modules they enroll in and thus who they see (and how big the classes are). At their heart, a community for learners creates a persistent peer group that see each other regularly and then makes the learners accountable to these peers, which means that the individual students have a more powerful external authority than just the teacher.

Table 4.6 Communities of practice vs. communities of inquiry

Community of practice	Community of inquiry
Central vs. peripheral members	Mutual respect
Power structures	Focus on equality, democratic
In jokes, knowing laughter	Take one another's ideas seriously, attentive
Conversational shorthand, jargon, and shortcuts to communication	Detects and unpicks underlying assumptions
Discussions start in the middle	Deliberative
Used in workplaces where the same people reappear	Ideal for schools and universities

In addition, the learning community leads eventually to internalization because each student must act as an authority to assess their peers.

4.4.3 *Sense of Belonging Is a Substrate to Internalization of Motivation*

Much recent research on attrition among diverse student cohorts have focused on the sense of belonging (Hurtado & Carter, 1997), a concept with a strong overlap to SDT's relatedness (Furrer & Skinner, 2003). In a first-year university cohort, a diminished sense of belonging is an independent predictor of both students' consideration of leaving and actual attrition (Hausmann et al., 2009). Transition pedagogies have been developed to address first-year students' potential lack of sense of belonging, which a number of universities in Australia and elsewhere have adopted to increase early engagement (Kift et al., 2010; Sinnayah et al., 2019). Disengagement, as quantified by attrition, may be caused by environmental triggers that lead to a sense of not belonging or not wanting to belong; drastic disengagement is triggered by demands that seem hopeless, worthless, or insulting. This pointlessness triggers the substrate of not fitting in, leading directly to apathy. By contrast, identification with the degree program is a substrate (i.e., moderately necessary) for longer-term engagement and internalization.

Identification with the degree program is made up of two motivational elements: sense of belonging and valuing the institution (Voelkl, 2012). Identification and goal clarity come together task-by-task to internalize motivation; thus, a sense of belonging fairly directly facilitates internalization of motivation. Using structural equation modeling, Hausmann et al. (2009) found a strong relationship between sense of belonging and institutional commitment, which in turn affected intention to persist. Hausmann et al. (2009) also attempted to increase belonging by sending the students in the intervention group supportive letters with small gifts branded with the university's name, logo, or colors (e.g., baseball caps, ID holders, magnets, decals, etc.). The intervention significantly improved the maintenance of sense of belonging measures in white students (as well as improving their actual persistence), but not in

African American students. The two groups also differed in that college GPA had much larger effect on actual attrition than did intention to persist in the African American students, whereas the roles were reversed for white students. This partial success of such a simple and inexpensive intervention highlights the role of the institution in motivation, because sense of belonging is a precursor to internalization of academic motivation (Ryan & Stiller, 1991; Won et al., 2018).

4.5 Assessment Drives Engagement and More

Testing has three potential effects on engagement and learning: (1) motivating behavioral engagement and effort via rewards or punishments (Wormald et al., 2009), (2) triggering feedback that clarifies goals and strengthens elaboration (Butler & Roediger, 2008), and (3) reinforcing cognitive engagement directly by retrieval practice (Roediger, Agarwal, et al., 2011). Technologies such as self-quizzing platforms can strengthen learning of facts by the latter two effects, but not all students feel motivated to engage with these technologies. Innovations in pedagogical assessment approaches are helping to reconnect the motivating effect of assessment with feedback and cognitive reinforcement. These changes also link longer-term engagement with momentary engagement.

4.5.1 The Tradition of Using Assessment to Drive Behavioral Engagement

Most educators have heard the adage that "assessment drives learning" (Wormald et al., 2009), because traditional high-stakes assessments incentivize study by determining students' opportunities in later life. However, many commentators have critiqued the problems with this system (Boud & Molloy, 2013; Henderson et al., 2019). High-stakes assessments usually occur at the end of a course, so they provide feedback to the student when the information is too late to be of use. A systematic review (the EPPI review, Harlen et al., 2002) concluded that high-stakes exams are disliked by students, they create anxiety among girls, they create (rather than reinforce) low self-esteem among underperformers, they focus academic efforts on one-way information transmission, and they convince students and teachers that these old methods of teaching are the basis of learning.

An education system that puts such great emphasis on evaluation produces students with strong extrinsic orientations toward grades and social status (Harlen et al., 2002). If such students have good metacognitive skills in this evaluation-focused context, they may develop a strategic approach to engaging with learning. Students with a strategic approach aim toward efficiency, by directing their study toward their perceptions of the assessments rather than toward the full content of the course (Entwistle, 2000). Snyder (1971) observed that the final exam reflected "a hidden curriculum" (unintentionally emphasized by the instructor) that allowed many savvy students to figure out what elements of the publicized curriculum could be safely ignored and to study only those parts of the course that will likely be on the test. The main driver for much student effort, especially for strategic students, can be summarized by "Will this be on the test?" (Marchant, 2002; Silverthorn, 2020). Strategic approaches to engagement are commonly observed in medical education (Norman, 1987), which often has a competitive focus (e.g., class ranking), a strong orientation toward examinations, and students with established academic skills.

Many students, particularly those with insufficient metacognitive skills, will only engage with the material based on a surface approach to learning. In surface learning, the student views the material as isolated facts, and their approach to the learning tasks is based on coping (Entwistle, 2000; Marton & Säljö, 1976). Students with a surface approach often believe they have a strategic approach and follow similar strategies, but less effectively and with less effort. According to educational theorists, the best students use a deep approach to learning in which the student attempts to connect new facts to their previous knowledge using meaning, evidence, principles, and logic (Entwistle, 2000). Figure 4.3 illustrates how these approaches to learning might result from the student's motivation (down-to-up on the graph), metacognition (away-toward), and knowledge (left-right). Apathetic students have

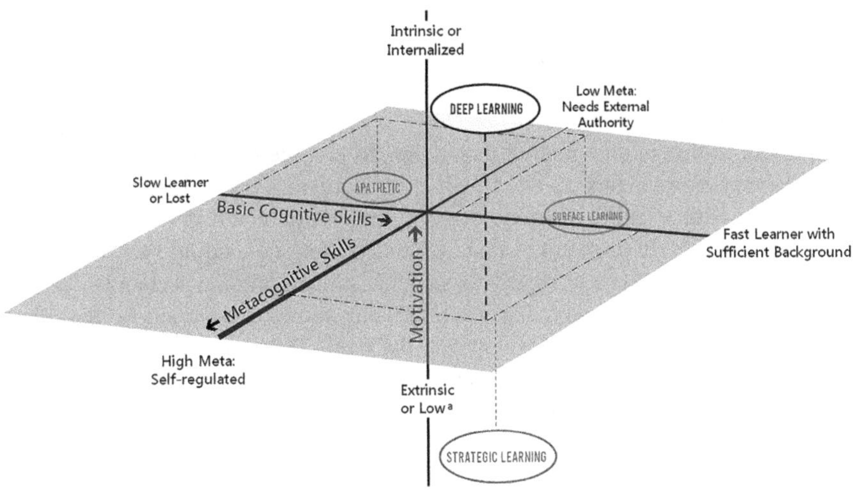

[a]Very low motivation, especially when combined with modest academic ability, may result in withdrawal from the course, failure, or other types of incomplete results

Fig. 4.3 Schematic of how each student's study approach (in ovals) is determined in part by their traits in three dimensions: basic cognitive skills (horizontal axis) versus motivation (vertical axis) versus metacognitive self-regulation (depth axis). Different student approaches to learning appear at different points of the 3-D space. Surface learners who lack metacognitive insight may believe that they are being strategic

low motivation (downward), their basic cognitive skills (such as memorization and discipline-based preparedness) are minimal (leftward), and their self-regulatory, metacognitive activities (skills such as organization and disciplined studying) are low (behind the plane of the image). By contrast, strategic learning and deep learning differ in their motivation, in that strategic learning has more extrinsic motivation (downward), whereas deep learning has internalized motivation (upward); however, both of those student approaches can have high basic cognitive skills (right) and high metacognitive skills (in front of the plane of the image).

4.5.2 Lower-Stakes Assessment Programs to Drive Behavioral Engagement

Whereas assessments have long been used as incentives for extrinsic motivation, educators are researching whether meaningfully designed assessment programs can lead to internalized motivations that persist. This has led to a shift in medical educational from assessment *of* learning to assessment *for* learning (Martinez & Lipson, 1989; Schuwirth et al., 2017). Early modules of study for clinical degree programs, which often focus on the fundamental scientific knowledge of biomedicine rather than clinical skills or practice, were traditionally the least motivating aspect of clinical study, because the student was the passive recipient of potentially irrelevant facts; this learning was considered teacher-centered, rather than student-centered, because the instructor taught what they knew, often without reference to how much the students in front of them might need the information (Barrows & Tamblyn, 1980).

Progress testing is one method that claims to provide early years students with both relevant assessments and a growing sense of mastery in order to motivate engagement with scientific learning that they previously could not connect to their instrumental goals. In progress testing, all students (e.g., including year 1 students) have to take the same test as final year students, although the expectations for year 1 performance are much lower. This kind of testing allows the students to "feel" their clinically relevant knowledge develop over the years (Schuwirth & van der Vleuten, 2012). Thus, early years students can repeatedly see how and when their basic science learning is relevant to clinically relevant tests. These formative experiences fuel a more relational (rather than transactional) and iterative approach to learning, in which the students are entering a community of practice. However, progress testing has the risk of eliciting powerful negative emotions in early years students who are unable to do well on the test (Herrmann et al., 2020; Nouns & Georg, 2010); this frustration can diminish engagement. Progress testing only really works for resilient students such as advanced students in medical school, because they have a proven growth mindset (Yeager & Dweck, 2012). Another obvious limitation of progress testing is that it only works for uniform educational programs where all students end up learning mostly the same material.

As opposed to high-stakes testing, low-stakes assessments can iteratively motivate without crushing the students with anxiety. In a classic study of the use of clickers (audience response devices) to kindle active learning during classroom lectures, Freeman et al. (2007) found that in a gateway biology course with a traditionally high failure rate, addition of small summative weighting of in-class clicker questions resulted in lower failure rates, higher exam scores, and higher attendance rates compared with an ungraded version of the same active learning exercises using paper cards. Likewise, in flipped classroom and team-based learning, where active learning sessions are only meaningful if the students have prepared, on arrival to most classes students are subjected to readiness tests (e.g., iRATs and tRATs (Team Based Learning Collaborative, 2013)), which we will call *accountability testing*, because small summative weighting (extrinsic incentives) are usually linked to these tests to motivate students to engage in a timely fashion.

Programmatic testing is another low-stakes assessment innovation that many medical schools are currently interested in (Ellis & Hogard, 2016; Schuwirth et al., 2017; Wilkinson & Tweed, 2018). Programmatic testing replaces (or supplements) a single high-stakes exam with many low-stakes exams spaced over the year; the lower stakes assessments are more formative and bite-sized. This type of assessment program leads to a more relational approach because students interact much more with their feedback. There are many other advantages to making assessments more piecemeal in this way.

1. Anxiety at each test is reduced because each assessment is a low-stakes event that individually does not have serious consequences if a student fails it.
2. Formative learning experiences are increased for the student.
3. Progress is monitored carefully, so students with problems are identified early for support.
4. The assessment of students is fairer because a single poor performance does not result in significant consequence.
5. The last-minute cramming style of study is de-prioritized, because assessments occur too often; cramming poorly schedules reinforcement, so it is a poor cognitive strategy for learning.

The main query that programmatic assessment raises is how, and how well, replacing high-stakes assessments with a collection of low-stakes assessments can maintain standards (Schut et al., 2020).

4.5.3 Quality Feedback Can Drive Metacognitive Engagement and Reflection

Feedback is a mediator from within the assessment process that can affect all pathways toward the engagement needed to learn (Nicol & Macfarlane-Dick, 2006; Fig. 4.1; see Chap. 9 on feedback in this volume). Not all feedback is equally valuable to learners (Henderson et al., 2019), and there has been much criticism of one-way, performance-related information (Boud & Molloy, 2013). Simply returning a test to a student with a letter grade on top (providing emotional engagement alone) does not give the student sufficient goal clarity (Kluger & DeNisi, 1996) for what actions to take or what concepts are misunderstood. Black and Wiliam (1998) claim, "The way in which test results are reported to pupils so that they can identify their own strengths and weaknesses is critical. Pupils must be given the means and opportunities to work with evidence of their difficulties". With quality feedback, assessment becomes a self-reflective learning tool for students and an opportunity for teachers to present experiences for learning improvement.

Good feedback should be (1) valued by the student (Stiggins, 2005), (2) clarifying, (3) timely (Henderson et al., 2019), (4) actionable, and (5) iterative in order to motivate both the cognitive and behavioral engagement needed for learning. Hattie and Timperley (2007) provide a model for feedback, which focuses on helping the student address three questions: *Where am I going? How am I going? Where to next?* When a test provides feedback about whether or not students know particular items and the students guide their future study efforts accordingly, it promotes learning by making later studying or encoding more effective; thus, testing with feedback can enhance metacognitive learning by means of this mediating process (Roediger & Karpicke, 2006b). One approach to encouraging students to reflect on summative assessment task is referred to as "exam wrappers" (Lovett, 2013). Exam wrappers direct students to engage in metacognitive processes—reflecting on their study approach, test performance, and strategies for improvement in the future. This type of process can lead to improvement in academic performance (Edlund, 2020), although this is not a consistent observation (Soicher & Gurung, 2017; Stephenson et al., 2017).

Assessments and feedback integrate with each other, and an iterative format encourages engagement directed at appropriate learning. Thus, feedback can be in the form of quantitative feedback to inform students about their grades and qualitative feedback that allows students to review their understanding of the content. For example, Voelkel (2013) describes a two-stage approach to online tests, where the first stage is characterized by a formative test in which timely qualitative feedback is provided to students. Students have multiple attempts in the first stage to achieve at least 80%, and upon achieving this, they are given access to the second stage of the test. The second stage is summative in nature and contains variants of the questions used in the first stage. This staged use of qualitative/formative and quantitative/summative tests reportedly improves performance not only of good students but also of weak ones.

4.5.4 *Assessment Leads to Cognitive Engagement with Learning Strategies*

Testing can enhance later retention more than additional exposure to the material, even when tests are administered without feedback (Szpunar et al., 2013). This surprising phenomenon is called the test-enhanced learning (see above section on retrieval practice), and although it has been studied by cognitive psychologists sporadically over the years, there is a renewed effort to learn why testing is effective and to apply testing in educational settings (Roediger & Karpicke, 2006a, 2006b). The act of taking the test has proven to be the actuator in promoting memory retention, rather than just the process of going over the material again in preparation for the test (Larsen et al., 2008; Szpunar et al., 2013). Repeated testing has been shown to promote better retention than taking a single test (Wheeler & Roediger, 1992). The benefits of repeated testing have proven greater still when testing is distributed over time (Karpicke & Roediger, 2007; Larsen et al., 2008). The mechanism for the learning benefits seems to be retrieval practice, which means that formative tests (often with minimal incentives) can also be effective (Freeman et al., 2007). The retrieval functions as a kind of active learning with clear motivation to the student.

4.6 Active Learning Pedagogies

An active learning pedagogy is any form of instructional design that shifts the focus from what the teacher does (students as passive learners) to what the student does (students as active learners). When they work correctly, active learning pedagogies create a link between momentary engagement and longer-term engagement. Active learning pedagogies may help motivate students who have a strategic engagement approach (or a surface approach) to engage in effective learning processes both frequently and regularly. The benefits of active learning pedagogies are that they (1) provide students with real-time dialogic feedback, making it a powerful way to address misconceptions and difficult learning (Olimpo & Esparza, 2020), as well as (2) provide STEM instructors with an orthodoxy for increasing classroom engagement and reducing attrition in STEM (Freeman et al., 2014; Idsardi, 2020; Hodges, 2020; Owens et al., 2020). These pedagogies strengthen both conceptual and factual learning, by compelling students to engage productively using both (a) emotional/motivational elements (social support, social authority/enforcement, sense of belonging, and self-efficacy) and (b) cognitive elements (elaboration, spaced retrieval practice, and frequent feedback).

Active learning *pedagogies* need to be distinguished from active learning, the cognitive learning strategy that underpins constructivist theories of knowledge (National Research Council, 2000, p. 11). Active learning is "students' efforts to actively construct their knowledge, both inside the classroom and in the broader world"(ACER, 2011). Almost all descriptions of active learning distinguish it from

Table 4.7 Brief classroom techniques for adding active learning to a lecture or other session

Classroom intervention	Citation
Clickers/audience response Systems/polling devices	Cheng and Wang (2018)
Socratic questioning	Paul and Elder (2007)
Think-pair-share	Prahl (2017)
Notes exchange between students	Philpott (2009, p. 144)
One-minute writing (evaluation, prediction, synthesis)	Stead (2005)
Most confusing/muddiest concept (at end of class, 1st year)	Angelo and Cross (1993)

passive learning, which is usually associated with just listening (or watching a video podcast (Jham et al., 2008)). Note that "in the broader world" implies that a student can passively listen to a lecture and then actively learn by doing problem sets at home, because "there is a place for lectures and readings in the classroom if students have sufficiently differentiated domain knowledge to use the expository materials in a generative manner" (Schwartz & Bransford, 1998). According to Bonwell and Eison's (1991) classic, *Active Learning: Creating Excitement in the Classroom*, active learning pedagogies are "instructional activities involving students in doing things and thinking about what they are doing." A list of brief active learning interventions is in Table 4.7 (Faust & Paulson, 1998). At their core, active learning pedagogies require students to elaborate their new knowledge, so they problem-solve, discuss, research and answer questions, or engage in other processes to drive the learning *while in the classroom*. Of course, there are alternative planned learning activities that can take place elsewhere (e.g., field trips, community service, final year research projects (Kuh et al., 2013)). But plainly, from their book's title, Bonwell and Eison were very interested in applying activating learning activities to what happens in the classroom. In part their end point and mechanism were to create excitement and combat boredom, but this is not inevitable for active learning pedagogies (Seidel & Tanner, 2013; Shekhar et al., 2015; see section on resistance below).

The theoretical bases of active learning are carefully formulated (Mintzes, 2020). Frequently, active learning pedagogies engage cognitive strategies from learning science (elaboration, retrieval, test enhanced learning, social learning) (Michael, 2006). Furthermore, the act of "doing" associated with active learning pedagogy means students *must* engage with learning content—if they have trouble doing so, they get real-time diagnostic feedback of gaps in their learning experience (McKeachie et al., 1987; Michael, 2006). Thus, active learning in the classroom promotes higher-order thinking processes and forces surface learners to confront their metacognitive abilities (McKeachie et al., 1987), which students can find either exciting or frustrating (Bonwell & Eison, 1991; Seidel & Tanner, 2013; Silverthorn, 2020). Active learning can increase the student's intrinsic motivation to learn the topic and improve their abilities by providing course-relevant opportunities for the student to explore their own attitudes, values, and habits (Bonwell & Eison, 1991; Gleason et al., 2011). In fact, to many instructors in the arts and humanities, active learning is simply called "learning"; that is, seminars, writing essays, long projects and studio critiques are all taken for granted outside STEM (see Table 4.8). By

Table 4.8 Activities tapping into active learning and career relevance that are easily incorporated into most courses

Mid-length active learning pedagogical interventions
Vignettes or case studies
Reflective portfolios
Model construction
Low-fidelity simulations
Role plays (especially for clinical or communication skills)
Team debates (especially for small classes)

Table 4.9 Active learning pedagogies at the curriculum level

Curricular approach	Relevant entry citation
Problem-based Learning	Wood (2003)
Flipped classroom	Talbert (2017)
Peer learning	Crouch and Mazur (2001)
Discovery learning	Balim (2009)
Inquiry-based learning	Pedaste et al. (2015)
Team-based learning	Team Based Learning Collaborative (2013)
Conceptual change strategies	Sinatra and Pintrich (2002)

contrast, in STEM, a desire for objectivity and reproducibility encouraged a tradition of fact-based foundational learning manifesting as multiple-choice tests and passive lectures (Chauncey & Ebel, 1962; Hilgard, 1969).

Several innovative classroom reformulations are based around active learning pedagogies (Table 4.9); each method involves making multiple changes in how the class routinely operates and is assessed. Problem-based learning (PBL) is based around solving open-ended, real-world problems in a tutorial-sized group, such that the students develop their own learning objectives; it was developed at McMaster University in Canada (Davis & Harden, 1999). Usually lectures are not provided, so students must assess the knowledge they will need and find methods (e.g., library research) or resources to solve the problem. Team-based learning (TBL) is a format involving breaking a large class into small teams for solving a problem or analyzing a case; it was developed by Larry Michaelsen (Michaelsen et al., 2008; Team Based Learning Collaborative, 2013). Usually, background information is provided in the form of pre-class reading, videos, or home study. The classroom session, which is usually several hours in duration, commences with two short accountability testing phases (individual Readiness Assurance Test (iRAT) following by a team Readiness Assurance Test (tRAT)), followed by the main segment of the class when the members of each team solve the problem or analyze the case together. The main risk in TBL (as in all group work) is social loafing/free riding, or noncompliant team members whose disruptive behaviors can adversely affect other students in their team (Pieterse & Thompson, 2010); these problems are disincentivized by peer evaluations (Levine, 2008). Flipped classroom is another form of active learning that is similar to TBL but without teams, in which in-class activity sessions involve problem solving and other activities that include group work, whereas most factual information is provided for home study via reading or video, and all students are expected to prepare for class in advance (see Chap. 6 on flipped classroom in this volume).

4.6.1 Student Resistance and Other Barriers to Active Learning

Although extensive evidence supports the idea that active learning pedagogies increase both engagement and learning for many students (Freeman et al., 2007, 2014; Michael, 2006; Theobald et al., 2020), many problems have been documented when instituting these pedagogies (Silverthorn, 2020; Silverthorn et al., 2006; White et al., 2014), so a critical approach is required. Active learning pedagogies have often been sold to fellow educators as a threshold concept, where the idea alone is sufficient to result in immediate benefits to the students and instructors (Freeman et al., 2014). In fact, it is more like a tool kit of interlocking methodologies, and like any method, if the precise procedure does not fit to the context, good teachers can run into serious problems and even disasters (Silverthorn, 2020; see Chap. 1 Introduction in this volume). Silverthorn (2020) teases apart potential barriers when instituting active learning pedagogies into four categories (see Table 4.10).

Some studies claim that students enjoy active learning pedagogies (Bonwell & Eison, 1991; Ebert-May et al., 1997; Freeman et al., 2007), but self-regulated students (e.g., introverted, talented, or strategic learners) can sometimes experience it as ill-paced and limiting to their autonomy for learning using their own method (Deslauriers et al., 2019; White et al., 2014; Wiltbank et al., 2019), making them sullen or hostile (Felder & Brent, 1996); student feedback about the instructor might then become highly negative due to these emotions. Active learning depends upon the buy-in of the students, and if the students do not recognize their additional responsibility and the consequent need for additional cognitive engagement (or extra metacognitive effort, see Fig. 4.4), active learning will paradoxically undermine classroom engagement (Rathner & Schier, 2020). For example, the student's lack of compliance with active learning activities (e.g., failing to do pre-class reading) can undermine the subsequent in-class group activity that depends on this foundational knowledge, thus short-circuiting the active learning approach. Although in theory active learning creates a virtuous cycle toward motivation (because identification and goal clarity in the student create more motivation than is needed to study via active learning (Kluger & DeNisi, 1996)), in fact sometimes active learning

Table 4.10 Categories of active learning barriers according to Silverthorn (2020)

Active learning barriers
Student resistance to change
Teacher reluctance to change
Administrative roadblocks
Physical setting

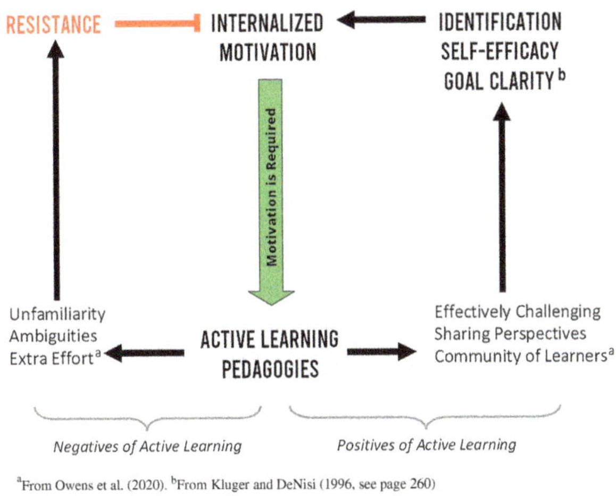

Fig. 4.4 Schematic showing how resistance affects motivation to active learning pedagogies

pedagogies can lead to student resistance (Felder & Brent, 1996; Owens et al., 2020), so that the active learning consumes motivation rather than creating it (see Fig. 4.4). In particular, there is a suggestion that in larger classes students will passively resist active learning as the term draws onward (Freeman et al., 2014; Shekhar et al., 2015), although this may be controversial (Ebert-May et al., 1997).

In conclusion, active learning pedagogies can be used to transform momentary engagement into longer-term engagement, conceptual thinking, and metacognitive abilities. However, wholesale incorporation of active learning curricular formats requires a well-matched set of learning activities, assessments, institutional approval, and appropriate physical space. The student cohort needs to be extensively (and repeatedly (Silverthorn, 2020)) prepared, convinced, and supported to embrace the format, particularly as it demands explicit metacognitive engagement that holds students in lockstep (White et al., 2014; Wiltbank et al., 2019). In biology, this is meant to kindle higher-level learning (through deliberate thinking practice) while reducing attrition (via support and inclusion). When performed successfully, students (1) are supported to become autonomous learners and (2) love it (Ebert-May et al., 1997), which Theobald et al. (2020) call the heads and hearts hypothesis. There may be some subgroups of students that on average benefit most from active learning; Theobald et al. (2020) present data suggesting that underrepresented minority students in STEM are one such group.

4.7 Conclusion

Student engagement will remain a controversial topic because researchers do not agree on what it is; science educators and education funders only agree that it is important because attrition is wasteful for students and bad for the institution. The missing puzzle piece in this controversy is the relationship between the benefits of momentary engagement (as measured by learning science experiments in laboratories) and the benefits of longer-term engagement (as measured by observation—questionnaires and contact metrics—at the end of term). To make longer-term gains, technological approaches such as the television show Sesame Street (Fisch et al., 1999; Mares & Pan, 2013) and gamification (see Chap. 5 in this volume on adaptive learning and gamification) trigger exciting momentary engagement to kindle longer-term engagement and learning (Fig. 4.5a). Interventions to improve momentary engagement often occur at the level of the student interest, which typically requires vast efforts in making the instruction/instructor consistently more exciting.

By contrast, much current research focuses on how internalizing motivation can be the substrate for repeated episodes of momentary engagement (see Fig. 4.5b). For example, high-impact practices (e.g., student-faculty research) almost presume internalized motivation to maintain longer-term engagement throughout the duration of the program. The kindling approach to longer-term engagement is more like play, whereas the motivated approach is more like work, and the motivated approach presumes a more mature approach to learning, or potentially previous enculturation of the student. Active learning as conceived by Bonwell and Eison (1991) was a more kindled approach, although many curricular approaches to active learning pedagogies (e.g., problem-based learning) use a more motivated approach. High-stakes testing does not really fit with the playful aesthetic of the kindling approach. The students as partners approach (see Chap. 3), where students serve on faculty committees and help design the curriculum, is based on having students having, and wanting to express, internalized motivation. Note that the two approaches in Fig. 4.5 are not necessarily mutually exclusive, so a person may experience a kindled approach in an early part of schooling and later have a more motivated approach.

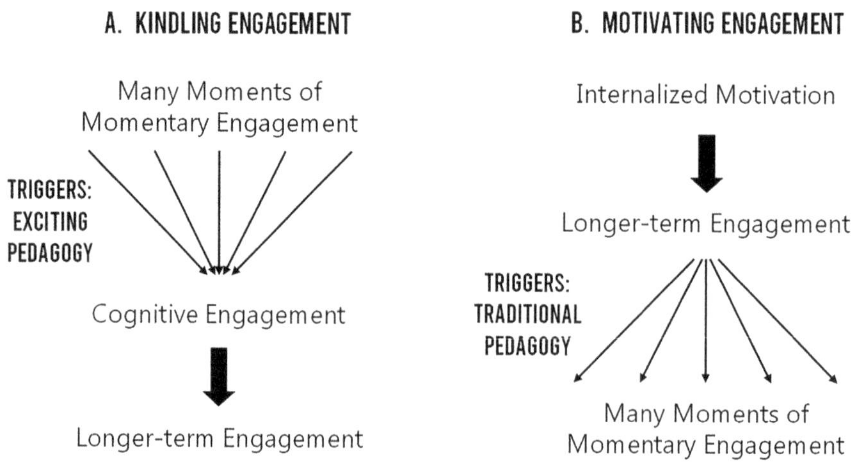

Fig. 4.5 Comparing two approaches to elicit longer-term engagement: (**a**) kindling many moments of momentary engagement (e.g., gamification) versus (**b**) motivating longer-term engagement via internalization and identification

The two approaches may need to be simultaneous for nontraditional students and commuters, because their competing personal demands modulate daily motivation and engagement (see Fig. 4.1); although commuters may have established internalized motivation, they benefit from logistics and social structures that support their momentary engagement (Klein et al., 2019; Thomas & Jones, 2017).

Figure 4.5b shows the process starting with internalized motivation, but plainly such motivation has precursors and does not come out of nowhere, although some instructors assume the students already have it when they arrive. Internalized motivation is relational, which is why inclusivity and belonging are so important (see Chap. 2 on Inclusivity in this volume). By contrast, most educational technologies (such as digital flashcards) are primarily transactional (Garrison, 2000) and based on accumulated short-term interactions. From a research perspective, motivational parameters fuel the disconnection between psychology labs, which assume that experimental participants can maintain their motivation for about an hour, and longer-term educational contexts, because there is no guarantee that a real student can maintain their motivation in a topic for an entire year.

Technology can play a role in active learning by using adaptive learning programs, but getting students to study using stripped down educational technologies that drive efficient learning (e.g., simple drilling programs) almost presupposes internalized motivation. Most learning technologists do not see the irony that longer-term motivation is required as a substrate in order to drive the efficient learning strategies identified by learning science (e.g., technologies such as digital flashcards). Many kinds of active learning (e.g., drilling) require substantial intrinsic motivation from the student because the social element is virtual or imagined. Thus, much technological innovation has focused on mediating or simulating these social and enjoyable moments, such as by gamification and leader boards. Students who feel resistance to active learning need a positive feedback cycle between active learning and internalized motivation, which would fuel the active learning (Fig. 4.4). However, only some forms of active learning (e.g., social learning) sustain internalized motivation, while all active learning requires it.

Our main critique is that much discussion about engagement conflates the two approaches in Fig. 4.5. In particular, often discussions about how educational technology can kindle learning and engagement in a university context fail to acknowledge that technology mostly adds extrinsic triggers to the educational climate. Technology rarely directly addresses the substrates (Fig. 4.1), such as intrinsic or fully internalized motivation, which depend on personality (Komarraju et al., 2009), values, and career goals. Identifying motivational substrates as separate from engagement is important because the majority of students—even among self-regulated students (e.g., service teaching for medicine)—are generally uninterested in learning for learning's sake and are instead focused on their grade/mark and instrumental outcomes (Chan, 2016; Kerr, 2001; Silverthorn, 2020). Consequently, many of these students are behaviorally engaged, but their emotional engagement (e.g., motivation) and cognitive engagement (e.g., attitudes) will diverge from their behavior. This makes the cognitive-emotional-behavioral meta-construct of engagement a problem for research on the mechanism between the individual elements. For engagement to be a meaningful research concept, it should be clearly distinguished from motivation and from learning; motivation should be seen as a precursor to drive engagement, and engagement should be a process that enables learning (Fig. 4.1).

Thus, we strongly recommend that future research publications on engagement focus on both the mechanisms and benefits of engagement and that:

- The different engagement activities of the student (e.g., attendance, online contact metrics, home study with textbooks) need to be explicitly described and recognized as interchangeable.
- The self-regulation and internalization of motivation of the student cohort (on average, see Fig. 4.3) must be highlighted, because teaching postgraduate medical students is different from first-generation scholars.
- The different "grain sizes" of engagement (e.g., longer-term and momentary) need to be explicitly labeled in papers and be recognized terminologically by databases.
- The links between momentary and longer-term engagement and their outcomes (i.e., lab learning versus school performance) need to be discussed mechanistically.
- Both momentary and longer-term engagement research need to be valued—especially by engaging faculty with inclusivity, competencies, and employability.

References

ACER. (2011). *Student engagement briefing, 2011 AUSSE survey*. Australian Council for Educational Research. https://www.acer.org/files/AUSSE_2011-Australasia-University-Engagement-Briefing.pdf

Allen, D. (1999). Desire to finish college: An empirical link between motivation and persistence. *Research in Higher Education, 40*(4), 461–485.

Angelo, T. A., & Cross, K. P. (1993). *Classroom assessment techniques* (2nd ed., pp. 119–158). Jossey Bass.

Aparicio, G., Iturralde, T., & Maseda, A. (2021). A holistic bibliometric overview of the student engagement research field. *Journal of Further and Higher Education, 45*(4), 540–557. https://doi.org/10.1080/0309877X.2020.1795092

Ashwin, P., & McVitty, D. (2015). The meanings of student engagement: Implications for policies and practices. In A. Curaj, L. Matei, R. Pricopie, J. Salmi, & P. Scott (Eds.), *The European higher education area*. Springer. https://doi.org/10.1007/978-3-319-20877-0_23

Astin, A. (1993). *What matters in college? Four critical years revisited*. Jossey-Bass.

Azevedo, R. (2015). Defining and measuring engagement and learning in science: Conceptual, theoretical, methodological, and analytical issues. *Educational Psychologist, 50*(1), 84–94. https://doi.org/10.1080/00461520.2015.1004069

Balim, A. G. (2009). The effects of discovery learning on students' success and inquiry learning skills. *Eurasian Journal of Educational Research, 35*, 1–20. https://www.academia.edu/2167480/The_Effects_of_Discovery_Learning_on_Students_Success_and_Inquiry_Learning_Skills?from=cover_page

Baron, P., & Corbin, L. (2012). Student engagement: Rhetoric and reality. *Higher Education Research and Development, 31*(6), 759–772. https://doi.org/10.1080/07294360.2012.655711

Barrows, H. S., & Tamblyn, R. M. (1980). *Problem-based learning: An approach to medical education* (Vol. 1). Springer.

Benjamin, A. S., & Tullis, J. (2010). What makes distributed practice effective? *Cognitive Psychology, 61*(3), 228–247.

Bjork, R. A., & Allen, T. W. (1970). The spacing effect: Consolidation or differential encoding? *Journal of Verbal Learning and Verbal Behavior, 9*, 567–572.

Black, P., & Wiliam, D. (1998). Assessment and classroom learning. *Assessment in Education: Principles, Policy & Practice, 5*(1), 7–74. https://doi.org/10.1080/0969595980050102

Bobulescu, I. A. (2020). The Old West analogy for acid-base buffering. *Advances in Physiology Education, 44*(2), 210–211. https://doi.org/10.1152/advan.00033.2020

Bonwell, C. C., & Eison, J. A. (1991). *Active learning: Creating excitement in the classroom. 1991 ASHE-ERIC higher education reports*. ERIC Clearinghouse on Higher Education, The George Washington University.

Boud, D., & Molloy, E. (2013). Rethinking models of feedback for learning: The challenge of design. *Assessment & Evaluation in Higher Education, 38*(6), 698–712. https://doi.org/10.1080/02602938.2012.691462

Brunmair, M., & Richter, T. (2019). Similarity matters: A meta-analysis of interleaved learning and its moderators. *Psychological Bulletin, 145*(11), 1029–1052.

Butler, A. C., & Roediger, H. L. (2008). Feedback enhances the positive effects and reduces the negative effects of multiple-choice testing. *Memory & Cognition, 36*, 604–616. https://doi.org/10.3758/MC.36.3.604

Cadaret, C. N., & Yates, D. T. (2018). Retrieval practice in the form of online homework improved information retention more when spaced 5 days rather than 1 day after class in two physiology courses. *Advances in Physiology Education, 42*, 305–310.

Cain, J., Black, E. P., & Rohr, J. (2009). An audience response system strategy to improve student motivation, attention, and feedback. *American Journal of Pharmaceutical Education, 73*(2), 21. https://doi.org/10.5688/aj730221

Cameron, J. (2001). Negative effects of reward on intrinsic motivation—A limited phenomenon: Comment on Deci, Koestner, and Ryan (2001). *Review of Educational Research, 71*(1), 29–42.

Cameron, J., Banko, K. M., & Pierce, W. D. (2001). Pervasive negative effects of rewards on intrinsic motivation: The myth continues. *The Behavior Analyst, 24*(1), 1–44.

Cameron, J., & Pierce, W. D. (1994). Reinforcement, reward, and intrinsic motivation: A meta-analysis. *Review of Educational Research, 64*(3), 363–423.

Campbell, C. M., & Cabrera, A. F. (2011). How sound is NSSE?: Investigating the psychometric properties of NSSE at a public, research-extensive institution. *The Review of Higher Education, 35*(1), 77–103.

Carey, P. (2013). Student as co-producer in a marketised higher education system: A case study of students' experience of participation in curriculum design. *Innovations in Education and Teaching International, 50*(3), 250–260.

Carnine, D. (2000). *Why education experts resist effective practices. Report of the Thomas B. Fordham Foundation*. Thomas B. Fordham Foundation. https://files.eric.ed.gov/fulltext/ED442804.pdf

Carpenter, S. K. (2014). Spacing and interleaving of study and practice. In V. A. Benassi, C. E. Overson, & C. M. Hakala (Eds.), *Applying the science of learning in education: Infusing psychological science into the curriculum* (pp. 131–141). Society for the Teaching of Psychology.

Carvalho, P. F., & Goldstone, R. L. (2019). When does interleaving practice improve learning? In J. Dunlosky & K. A. Rawson (Eds.), *The Cambridge handbook of cognition and education* (pp. 411–436). Cambridge University Press.

Cech, T., & Kennedy, D. (2005). Doing more for Kate. *Science, 310*(5755), 1741–1742.

Cepeda, N. J., Vul, E., Rohrer, D., Wixted, J. T., & Pashler, H. (2008). Spacing effects in learning: A temporal ridgeline of optimal retention. *Psychological Science, 19*, 1095–1102.

Chan, R. Y. (2016). Understanding the purpose of higher education: An analysis of the economic and social benefits for completing a college degree. *Journal of Education Policy, Planning and Administration, 6*(5), 1–40.

Chauncey, H., & Ebel, R. L. (1962). Testing, tests, and testing service organizations. *American Behavioral Scientist, 6*(3), 74–77.

Cheng, L. T., & Wang, J. W. (2018). Enhancing learning performance through classroom response systems: The effect of knowledge in a global economic environment. *Journal of Teaching in International Business, 29*(1), 49–61.

Chib, V. S., De Martino, B., Shimojo, S., & O'Doherty, J. P. (2012). Neural mechanisms underlying paradoxical performance for monetary incentives are driven by loss aversion. *Neuron, 74*(3), 582–594.

Chipchase, L., Davidson, M., Blackstock, F., Bye, R., Clothier, P., Klupp, N., Nickson, W., Turner, D., & Williams, M. (2017). Conceptualising and measuring student disengagement in higher education: A synthesis of the literature. *International Journal of Higher Education, 6*(2), 31–42.

Cook-Sather, A., Bovill, C., & Felten, P. (2014). *Engaging students as partners in learning and teaching: A guide for faculty*. Jossey-Bass.

Corno, L., & Mandinach, E. B. (1983). The role of cognitive engagement in classroom learning and motivation. *Educational Psychologist, 18*(2), 88–108.

Cowan, N. (2008). What are the differences between long-term, short-term, and working memory? *Progress in Brain Research, 169*, 323–338.

Cowan, N. (2014). Working memory underpins cognitive development, learning, and education. *Educational Psychology Review, 26*(2), 197–223.

Crouch, C. H., & Mazur, E. (2001). Peer instruction: Ten years of experience and results. *American Journal of Physics, 69*(9), 970–977.

Davidesco, I., & Milne, C. (2019). Implementing cognitive science and discipline-based education research in the undergraduate science classroom. *CBE Life Sciences Education, 18*, es4.

Davis, M. H., & Harden, R. M. (1999). AMEE medical education guide no 15: Problem-based learning: A practical guide. *Medical Teacher, 21*(2), 130–140.

Deci, E. L. (1971). Effects of externally mediated rewards on intrinsic motivation. *Journal of Personality and Social Psychology, 18*(1), 105–115.

Deci, E. L., Koestner, R., & Ryan, R. M. (2001). Extrinsic rewards and intrinsic motivation in education: Reconsidered once again. *Review of Educational Research, 71*(1), 1–27.

Deci, E. L., Ryan, R. M., & Williams, G. C. (1996). Need satisfaction and the self-regulation of learning. *Learning and Individual Differences, 8*(3), 165–183.

Deng, F., Gluckstein, J. A., & Larsen, D. P. (2015). Student-directed retrieval practice is a predictor of medical licensing examination performance. *Perspectives on Medical Education, 4*(6), 308–313.

Deslauriers, L., McCarty, L. S., Miller, K., Callaghan, K., & Kestin, G. (2019). Measuring actual learning versus feeling of learning in response to being actively engaged in the classroom. *Proceedings of the National Academy of Sciences, 116*(39), 19251–19257.

de Vreede, T., Andel, S., de Vreede, G. J., Spector, P., Singh, V., & Padmanabhan, B. (2019). What is engagement and how do we measure it? Toward a domain independent definition and scale. In *Proceedings of the 52nd Hawaii international conference on system sciences*.

D'Mello, S., Dieterle, E., & Duckworth, A. (2017). Advanced, analytic, automated (AAA) measurement of engagement during learning. *Educational Psychologist, 52*(2), 104–123.

Dobson, J. L. (2011). Effect of selected "desirable difficulty" learning strategies on the retention of physiology information. *Advances in Physiology Education, 35*, 378–383.

Dobson, J. L. (2013). Retrieval practice is an efficient method of enhancing the retention of anatomy and physiology information. *Advances in Physiology Education, 37*, 184–191.

Dover, M. A., (2013). Human needs: Overview. In *Encyclopedia of social work*. Oxford University Press. https://doi.org/10.1093/acrefore/9780199975839.013.554. https://philarchive.org/archive/DOVHNO

Dowd, A. C., Sawatzky, M., & Korn, R. (2011). Theoretical foundations and a research agenda to validate measures of intercultural effort. *The Review of Higher Education, 35*(1), 17–44.

Dunlosky, J., Rawson, K. A., Marsh, E. J., Nathan, M. J., & Willingham, D. T. (2013). Improving students' learning with effective learning techniques: Promising directions from cognitive and educational psychology. *Psychological Science in the Public Interest, 14*(1), 4–58.

Ebbinghaus, H. (1885). *Memory: A contribution to experimental psychology* (H. A. Ruger & C. E. Bussenius, Trans. (1913)). https://doi.org/10.1037/10011-000. https://web.archive.org/web/20050504104838/, http://psy.ed.asu.edu/~classics/Ebbinghaus/index.htm

Ebert-May, D., Brewer, C., & Allred, S. (1997). Innovation in large lectures: Teaching for active learning. *Bioscience, 47*(9), 601–607.

Eccles, J. (2016). Engagement: Where to next? *Learning and Instruction, 43*, 71–75. https://doi.org/10.1016/j.learninstruc.2016.02.003

Edlund, J. E. (2020). Exam wrappers in psychology. *Teaching of Psychology, 47*(2), 156–161.

Eibl-Eibesfeldt, I. (1979). Human ethology: Concepts and implications for the sciences of man. *Behavioral and Brain Sciences, 2*(1), 1–57.

Ellis, R., & Hogard, E. (2016). Programmatic assessment: A paradigm shift in medical education. *All Ireland. Journal of Higher Education, 8*(3), article 2951.

Engle, R. A., & Conant, F. R. (2002). Guiding principles for fostering productive disciplinary engagement: Explaining an emergent argument in a community of learners classroom. *Cognition and Instruction, 20*(4), 399–483.

Entwistle, N. (2000). Promoting deep learning through teaching and assessment: Conceptual frameworks and educational contexts. In *Teaching and learning research programme conference*. ESRC. https://www.arasite.org/RMdatabase/entwistle2000.pdf

Evans, D. J., Zeun, P., & Stanier, R. A. (2014). Motivating student learning using a formative assessment journey. *Journal of Anatomy, 224*(3), 296–303.

Faust, J. L., & Paulson, D. R. (1998). Active learning in the college classroom. *Journal on Excellence in College Teaching, 9*(2), 3–24.

Felder, R. M., & Brent, R. (1996). Navigating the bumpy road to student-centered instruction. *College Teaching, 44*, 43–47.

Finn, J. D. (1989). Withdrawing from school. *Review of Educational Research, 59*, 117–142.

Fisch, S. M., Truglio, R. T., & Cole, C. F. (1999). The impact of Sesame Street on preschool children: A review and synthesis of 30 years' research. *Media Psychology, 1*(2), 165–190.

Fonseca, M., Oliveira, B., Carreiro-Martins, P., Neuparth, N., & Rendas, A. (2020). Revisiting the role of concept mapping in teaching and learning pathophysiology for medical students. *Advances in Physiology Education, 44*(3), 475–481.

Fredricks, J. A., Blumenfeld, P. C., & Paris, A. H. (2004). School engagement: Potential of the concept, state of the evidence. *Review of Educational Research, 74*(1), 59–109.

Fredricks, J. A., Filsecker, M., & Lawson, M. A. (2016). Student engagement, context, and adjustment: Addressing definitional, measurement, and methodological issues. *Learning and Instruction, 43*, 1–4.

Freeman, S., Eddy, S. L., McDonough, M., Smith, M. K., Okoroafor, N., Jordt, H., & Wenderoth, M. P. (2014). Active learning increases student performance in science, engineering, and mathematics. *Proceedings of the National Academy of Sciences of the United States of America, 111*(23), 8410. https://doi.org/10.1073/pnas.1319030111

Freeman, S., O'Connor, E., Parks, J. W., Cunningham, M., Hurley, D., Haak, D., Dirks, C., & Wenderoth, M. P. (2007). Prescribed active learning increases performance in introductory biology. *CBE—Life Sciences Education, 6*(2), 132–139.

Furrer, C., & Skinner, E. (2003). Sense of relatedness as a factor in children's academic engagement and performance. *Journal of Educational Psychology, 95*(1), 148–162.

Gamba, G. (2003). Analogy for explaining intermediate metabolism. *Advances in Physiology Education, 27*(3), 156–157. https://doi.org/10.1152/advan.00015.2003

Garrison, D. R., Anderson, T., & Archer, W. (2010). The first decade of the community of inquiry framework: A retrospective. *The Internet and Higher Education, 13*(1-2), 5–9.

Garrison, R. (2000). Theoretical challenges for distance education in the 21st century: A shift from structural to transactional issues. *International Review of Research in Open and Distributed Learning, 1*(1), 1–17.

Gauci, S. A., Dantas, A. M., Williams, D. A., & Kemm, R. E. (2009). Promoting student-centered active learning in lectures with a personal response system. *Advances Physiology Education, 33*, 60–71.

Gentry, S. V., Gauthier, A., Ehrstrom, B. L. E., Wortley, D., Lilienthal, A., Car, L. T., Dauwels-Okutsu, S., Nikolaou, C. K., Zary, N., Campbell, J., & Car, J. (2019). Serious gaming and gamification education in health professions: Systematic review. *Journal of Medical Internet Research, 21*(3), e12994.

Gitinabard, N., Xu, Y., Heckman, S., Barnes, T., & Lynch, C. F. (2019). How widely can prediction models be generalized? Performance prediction in blended courses. *IEEE Transactions on Learning Technologies, 12*(2), 184–197.

Gleason, B. L., Peeters, M. J., Resman-Targoff, B. H., Karr, S., McBane, S., Kelley, K., Thomas, T., & Denetclaw, T. H. (2011). An active-learning strategies primer for achieving ability-based educational outcomes. *American Journal of Pharmaceutical Education, 75*(9), article 186.

Gooding, H. C., Mann, K., & Armstrong, E. (2017). Twelve tips for applying the science of learning to health professions education. *Medical Teacher, 39*, 26–31.

Gordon, C. J., & Hughes, V. K. (2013). Creating relevance and credibility: New approaches for bioscience education in pre-registration nursing curriculum. *International Journal of Innovation in Science and Mathematics Education, 21*(2), 53–65.

Green, R. A., Farchione, D., Hughes, D. L., & Chan, S.-P. (2014). Participation in asynchronous online discussion forums does improve student learning of gross anatomy. *Anatomical Sciences Education, 7*, 71–76.

Haagsman, M. E., Scager, K., Boonstra, J., & Koster, M. C. (2020). Pop-up questions within educational videos: Effects on students' learning. *Journal of Science Education and Technology, 29*, 713–724.

Hamilton, R. J. (1997). Effects of three types of elaboration on learning concepts from text. *Contemporary Educational Psychology, 22*(3), 299–318.

Handelsman, M. M., Briggs, W. L., Sullivan, N., & Towler, A. (2005). A measure of college student course engagement. *The Journal of Educational Research, 98*, 184–193.

Hanus, M. D., & Fox, J. (2015). Assessing the effects of gamification in the classroom: A longitudinal study on intrinsic motivation, social comparison, satisfaction, effort, and academic performance. *Computers & Education, 80*, 152–161.

Harlen, W., Crick, R. D., Broadfoot, P., Daugherty, R., Gardner, J., James, M., & Stobart, G. (2002). A systematic review of the impact of summative assessment and tests on students' motivation for learning. (EPPI-Centre Review, version 1.1*). In: *Research evidence in education library*. Issue 1. EPPI-Centre, Social Science Research Unit, Institute of Education.

Hattie, J., & Timperley, H. (2007). The power of feedback. *Review of Educational Research, 77*(1), 81–112.

Hausmann, L. R., Ye, F., Schofield, J. W., & Woods, R. L. (2009). Sense of belonging and persistence in White and African American first-year students. *Research in Higher Education, 50*(7), 649–669.

Heddy, B. C., & Sinatra, G. M. (2013). Transforming misconceptions: Using transformative experience to promote positive affect and conceptual change in students learning about biological evolution. *Science Education, 97*(5), 723–744.

Heist, P. (Ed.). (1968). *The creative college student: An unmet challenge*. Jossey-Bass.

Henderson, M., Phillips, M., Ryan, T., Boud, D., Dawson, P., Molloy, E., & Mahoney, P. (2019). Conditions that enable effective feedback. *Higher Education Research & Development, 38*(7), 1401–1416.

Herrmann, L., Beitz-Radzio, C., Bernigau, D., Birk, S., Ehlers, J. P., Pfeiffer-Morhenn, B., Preusche, I., Tipold, A., & Schaper, E. (2020). Status quo of progress testing in veterinary

medical education and lessons learned. *Frontiers in Veterinary Science, 7*, article 559. https://doi.org/10.3389/fvets.2020.00559
Hilbert, T. S., & Renkl, A. (2008). Concept mapping as a follow-up strategy to learning from texts: What characterizes good and poor mappers? *Instructional Science, 36*(1), 53–73.
Hilgard, E. R. (1969). The psychological heuristics of learning. *Proceedings of the National Academy of Sciences of the United States of America, 63*(3), 580–587.
Hirshman, E. (2001). Elaboration in memory. In N. J. Smelser & P. B. Baltes (Eds.), *International encyclopedia of the social & behavioral sciences* (pp. 4369–4374). Pergamon.
Hodges, L. C. (2020). Student engagement in active learning classes. In J. J. Mintzes & E. Walter (Eds.), *Active learning in college science* (pp. 27–41). Springer.
Hogan, M. P. (1985). Writing as punishment. *The English Journal, 74*(5), 40–42.
Horton, A. (2005). The academic achievement gap between blacks and whites: The latest version of blaming the victim? *Journal of Human Behavior in the Social Environment, 10*(2), 57–70.
Hurtado, S., & Carter, D. F. (1997). Effects of college transition and perceptions of the campus racial climate on Latino college students' sense of belonging. *Sociology of Education, 70*(4), 324–345.
Hurtado, S., & Ruiz Alvarado, A. (2015). *Discrimination and bias, underrepresentation, and sense of belonging on campus*. Higher Education Research Institute.
Idsardi, R. (2020). Evidence-based practices for the active learning classroom. In J. J. Mintzes & E. Walter (Eds.), *Active learning in college science* (pp. 13–25). Springer.
IES WWC. (2021). *What works clearinghouse*. Department of Education. https://ies.ed.gov/ncee/wwc/
Ishitani, T. T. (2006). Studying attrition and degree completion behavior among first-generation college students in the United States. *The Journal of Higher Education, 77*(5), 861–885.
Jeno, L. M., Danielsen, A. G., & Raaheim, A. (2018). A prospective investigation of students' academic achievement and dropout in higher education: A Self-Determination Theory approach. *Educational Psychology, 38*(9), 1163–1184.
Jham, B. C., Duraes, G. V., Strassler, H. E., & Sensi, L. G. (2008). Joining the podcast revolution. *Journal of Dental Education, 72*(3), 278–281.
Jovanović, J., Mirriahi, N., Gašević, D., Dawson, S., & Pardo, A. (2019). Predictive power of regularity of pre-class activities in a flipped classroom. *Computers & Education, 134*, 156–168.
Jovanović, J., Saqr, M., Joksimović, S., & Gašević, D. (2021). Students matter the most in learning analytics: The effects of internal and instructional conditions in predicting academic success. *Computers & Education*. https://doi.org/10.1016/j.compedu.2021.104251
Juvonen, J., Espinoza, G., & Knifsend, C. (2012). The role of peer relationships in student academic and extracurricular engagement. In S. L. Christenson, A. L. Reschly, & C. Wylie (Eds.), *Handbook of research on student engagement* (pp. 387–401). Springer.
Kahn, W. A. (1990). Psychological conditions of personal engagement and disengagement at work. *Academy of Management Journal, 33*(4), 692–724.
Kahu, E. R. (2013). Framing student engagement in higher education. *Studies in Higher Education, 38*(5), 758–773.
Karpicke, J. D., & Roediger, H. L., 3rd. (2007). Expanding retrieval practice promotes short-term retention, but equally spaced retrieval enhances long-term retention. *Journal of Experimental Psychology Learning Memory Cognition, 33*(4), 704–719. https://doi.org/10.1037/0278-7393.33.4.704
Kauffman, C. A., Derazin, M., Asmar, A., & Kibble, J. D. (2018). Relationship between classroom attendance and examination performance in a second-year medical pathophysiology class. *Advances in Physiology Education, 42*(4), 593–598.
Kelley, P., & Whatson, T. (2013). Making long-term memories in minutes: A spaced learning pattern from memory research in education. *Frontiers in Human Neuroscience, 7*, article 589.
Kerr, C. (2001). *The uses of the university* (5th ed.). Harvard University Press.

Kift, S., Nelson, K., & Clarke, J. (2010). Transition pedagogy: A third generation approach to FYE-A case study of policy and practice for the higher education sector. *International Journal of the First Year in higher Education, 1*(1), 1–20.

Kim, T. W., & Werbach, K. (2016). More than just a game: Ethical issues in gamification. *Ethics and Information Technology, 18*(2), 157–173.

Klein, R., Kelly, K., Sinnayah, P., & Winchester, M. (2019). The VU way: The Block Model intensive learning at University favours low achieving students. *International Journal of Innovation in Science and Mathematics Education, 27*(9), 47–59.

Kluger, A. N., & DeNisi, A. (1996). The effects of feedback interventions on performance: A historical review, a meta-analysis, and a preliminary feedback intervention theory. *Psychological Bulletin, 119*(2), 254–284.

Komarraju, M., Karau, S. J., & Schmeck, R. R. (2009). Role of the Big Five personality traits in predicting college students' academic motivation and achievement. *Learning and Individual Differences, 19*(1), 47–52.

Krieger, P. A. (2017). *A visual analogy guide to human anatomy & physiology*. Morton.

Kuh, G. D. (1993). In their own words: What students learn outside the classroom. *American Educational Research Journal, 30*(2), 277–304.

Kuh, G. D. (2001). The national survey of student engagement: Conceptual framework and overview of psychometric properties. Indiana University Center for Postsecondary Research and Planning. https://scholarworks.iu.edu/dspace/bitstream/handle/2022/24268/The%20National%20Survey%20of%20Student%20Engagement-%20Conceptual%20framework%20and%20overview%20of%20psychometric%20properties.pdf?sequence=1

Kuh, G. D. (2009a). The national survey of student engagement: Conceptual and empirical foundations. *New Directions for Institutional Research, 141*, 5–20.

Kuh, G. D. (2009b). What student affairs professionals need to know about student engagement. *Journal of College Student Development, 50*, 683–706.

Kuh, G. D., Cruce, T., Shoup, R., Kinzie, J., & Gonyea, R. M. (2008). Unmasking the effects of student engagement on college grades and persistence. *Journal of Higher Education, 79*(5), 540–563.

Kuh, G. D., Kinzie, J., Buckley, J. A., Bridges, B. K., & Hayek, J. C. (2011). *Piecing together the student success puzzle: Research, propositions, and recommendations*: ASHE higher education report 32(5). Wiley

Kuh, G. D., O'Donnell, K., & Reed, S. D. (2013). *Ensuring quality & taking high-impact practices to scale*. Association of American Colleges and Universities.

Kuh, G. D., & Umbach, P. (2004). College and character: Insights from the national survey of student engagement. *New Directions for Institutional Research, 122*, 37–54.

Kuiper, E., & de Pater-Sneep, M. (2014). Student perceptions of drill-and-practice mathematics software in primary education. *Mathematics Education Research Journal, 26*(2), 215–236.

Kurbel, S. (2014). A vector-free ECG interpretation with P, QRS & T waves as unbalanced transitions between stable configurations of the heart electric field during P-R, S-T & T-P segments. *Theoretical Biology and Medical Modelling, 11*(1), 10. https://doi.org/10.1186/1742-4682-11-10

Larsen, D. P., Butler, A. C., & Roediger, H. L., 3rd. (2008). Test-enhanced learning in medical education. *Medical Education, 42*(10), 959–966. https://doi.org/10.1111/j.1365-2923.2008.03124.x

Latessa, R., & Mouw, D. (2005). Use of an audience response system to augment interactive learning. *Family Medicine, 37*, 12–14.

Lazowski, R. A., & Hulleman, C. S. (2016). Motivation interventions in education: A meta-analytic review. *Review of Educational Research, 86*(2), 602–640.

Levine, R. E. (2008). Peer evaluation in team-based learning. Chapter 9. In L. Michaelsen, D. Parmelee, K. K. McMahon, & R. E. Levine (Eds.), *Team-based learning for health professions education: A guide to using small groups to improve learning* (pp. 103–116). Stylus.

Li, Y., & Lerner, R. M. (2011). Trajectories of school engagement during adolescence: Implications for grades, depression, delinquency, and substance use. *Developmental Psychology, 47*(1), 233–347.

Linderholm, T., Dobson, J., & Yarbrough, M. B. (2016). The benefit of self-testing and interleaving for synthesizing concepts across multiple physiology texts. *Advances in Physiology Education, 40*(3), 329–334.

Loganathan, P., Talib, C., Thoe, N., Aliyu, F., & Zawadski, R. (2019). Implementing technology infused gamification in science classroom: A systematic review and suggestions for future research. *Learning Science and Mathematics, 14*, 60–73.

Lovett, M. C. (2013). Make exams worth more than the grade: Using exam wrappers to promote metacognition. In M. Kaplan, N. Silver, D. LaVague-Manty, & D. Meizlish (Eds.), *Using reflection and metacognition to improve student learning: Across the disciplines, across the academy* (pp. 18–52). Stylus.

Majuri, J., Koivisto, J., & Hamari, J. (2018). Gamification of education and learning: A review of empirical literature. In *Proceedings of the 2nd international GamiFIN conference, GamiFIN 2018*. CEUR-WS.

Marchant, G. J. (2002). Student reading of assigned articles: Will this be on the test? *Teaching of Psychology., 29*(1), 49–51.

Mares, M. L., & Pan, Z. (2013). Effects of Sesame Street: A meta-analysis of children's learning in 15 countries. *Journal of Applied Developmental Psychology, 34*(3), 140–151.

Martin, A. J., Anderson, J., Bobis, J., Way, J., & Vellar, R. (2012). Switching on and switching off in mathematics: An ecological study of future intent and disengagement among middle school students. *Journal of Educational Psychology, 104*(1), 1–18.

Martinez, M. E., & Lipson, J. I. (1989). Assessment for learning. *Educational Leadership, 46*(7), 73–75.

Marton, F., & Säljö, R. (1976). On qualitative differences in learning: I—Outcome and process. *British Journal of Educational Psychology, 46*(1), 4–11.

Maslow, A. H. (1943). A theory of human motivation. *Psychological Review, 50*(4), 370.

Mayer, R. E., & Gallini, J. K. (1990). When is an illustration worth ten thousand words? *Journal of Educational Psychology, 82*, 715–726.

Mayer, R. E., & Moreno, R. (2003). Nine ways to reduce cognitive load in multimedia learning. *Educational Psychologist, 38*(1), 43–52.

McDaniel, M. A., & Donnelly, C. M. (1996). Learning with analogy and elaborative interrogation. *Journal of Educational Psychology, 88*(3), 508–519.

McGregor, D. (1960). *The human side of enterprise*. McGraw-Hill Higher Education.

McKeachie, W. J., Pintrich, P. R., Lin, Y. G., & Smith, D. A. (1987). *Teaching and learning in the college classroom: A review of the literature. National Center for Research to improve secondary teaching and learning*. University of Michigan Press.

Melton, A. W. (1970). The situation with respect to the spacing of repetitions and memory. *Journal of Verbal Learning and Verbal Behavior, 9*, 596–606.

Michael, J. (2006). Where's the evidence that active learning works. *Advances in Physiology Education, 30*(4), 159–167.

Michaelsen, L. K., Parmelee, D. X., McMahan, K. K., & Levine, R. E. (2008). *Team-based learning for health professions education* (1st ed.). Stylus.

Miller, R. G., Ashar, B. H., & Getz, K. J. (2003). Evaluation of an audience response system for the continuing education of health professionals. *Journal of Continuing Education in the Health Professions, 23*, 109–115.

Mintzes, J. J. (2020). From constructivism to active learning in college science. In J. J. Mintzes & E. Walter (Eds.), *Active learning in college science* (pp. 3–12). Springer.

Morisano, D., Hirsh, J. B., Peterson, J. B., Pihl, R. O., & Shore, B. M. (2010). Setting, elaborating, and reflecting on personal goals improves academic performance. *Journal of Applied Psychology, 95*(2), 255–264.

Mowrer, O. H. (1939). A stimulus-response analysis of anxiety and its role as a reinforcing agent. *Psychological Review, 46*(6), 553–565.

Munday, P. (2016). The case for using DUOLINGO as part of the language classroom experience. *RIED: Revista Iberoamericana de Educación a Distancia, 19*(1), 83–101.

Murre, J. M., & Dros, J. (2015). Replication and analysis of Ebbinghaus' forgetting curve. *PLoS One, 10*(7), e0120644.

National Research Council. (2000). *How people learn: Brain, mind, experience, and school: Expanded edition.* National Academies Press.

Newby, T. J., & Stepich, D. A. (1987). Learning abstract concepts: The use of analogies as a mediational strategy. *Journal of Instructional Development, 10*(2), 20–26. Retrieved from http://www.jstor.org/stable/30220840

Nicol, D. J., & Macfarlane-Dick, D. (2006). Formative assessment and self-regulated learning: A model and seven principles of good feedback practice. *Studies in Higher Education, 31*(2), 199–218.

Nicoll, R. A. (2017). A brief history of long-term potentiation. *Neuron, 93*, 281–290.

Norman, G. R. (1987). Current issues in the assessment of clinical competence. In I. R. Hart & H. J. Walton (Eds.), *Further developments in assessing clinical competence.* Canadian Health Publications.

Nouns, Z. M., & Georg, W. (2010). Progress testing in German speaking countries. *Medical Teacher, 32*(6), 467–470.

Ogbu, J. U. (2003). *Black American students in an affluent suburb: A study of academic disengagement.* Routledge.

Olimpo, J. T., & Esparza, D. (2020). Active learning and conceptual understanding in biology. In J. J. Mintzes & E. Walter (Eds.), *Active learning in college science* (pp. 43–57). Springer.

Olivas, M. A. (2011). If you build it, they will assess it (or, an open letter to George Kuh, with love and respect). *The Review of Higher Education, 35*(1), 1–15.

Owens, D. C., Sadler, T. D., Barlow, A. T., & Smith-Walters, C. (2020). Student motivation from and resistance to active learning rooted in essential science practices. *Research in Science Education, 50*(1), 253–277.

Page, J., Meehan-Andrews, T., Weerakkody, N., Hughes, D. L., & Rathner, J. A. (2017). Student perceptions and learning outcomes of blended learning in a massive first-year core physiology for allied health subjects. *Advances in Physiology Education, 41*, 44–55.

Paivio, A., & Csapo, K. (1973). Picture superiority in free recall: Imagery or dual coding? *Cognitive Psychology, 5*, 176–206.

Pani, J. R., Chariker, J. H., & Naaz, F. (2013). Computer-based learning: Interleaving whole and sectional representation of neuroanatomy. *Anatomical Sciences Education, 6*(1), 11–18.

Pardales, M. J., & Girod, M. (2006). Community of inquiry: Its past and present future. *Educational Philosophy and Theory, 38*(3), 299–309.

Pascarella, E. T., & Terenzini, P. T. (1980). Predicting freshman persistence and voluntary dropout decisions from a theoretical model. *The Journal of Higher Education, 51*(1), 60–75.

Paul, R., & Elder, L. (2007). Critical thinking: The art of Socratic questioning. *Journal of Developmental Education, 31*(1), 36–37.

Pechenkina, E., Laurence, D., Oates, G., Eldridge, D., & Hunter, D. (2017). Using a gamified mobile app to increase student engagement, retention and academic achievement. *International Journal of Educational Technology in Higher Education, 14*, 1–12.

Pedaste, M., Mäeots, M., Siiman, L. A., De Jong, T., Van Riesen, S. A., Kamp, E. T., Manoli, C. C., Zacharia, Z. C., & Tsourlidaki, E. (2015). Phases of inquiry-based learning: Definitions and the inquiry cycle. *Educational Research Review, 14*, 47–61.

Peters, H., Zdravkovic, M., João Costa, M., Celenza, A., Ghias, K., Klamen, D., Mossop, L., Rieder, M., Nadarajah, V. D., Wangsaturaka, D., Wohlin, M., & Weggemans, M. (2018). Twelve tips for enhancing student engagement. *Medical Teacher, 41*(6), 632–637.

Philpott, J. (2009). Revision (Chapter 5). In *Captivating your class: Effective teaching skills.* Bloomsbury.

Pieterse, V., & Thompson, L. (2010). Academic alignment to reduce the presence of 'social loafers' and 'diligent isolates' in student teams. *Teaching in Higher Education, 15*(4), 355–367.

Pike, G. R., Kuh, G. D., & Massa-McKinley, R. C. (2008). First-year students' employment, engagement, and academic achievement: Untangling the relationship between work and grades. *NASPA Journal, 45*(4), 560–582.

Prahl, K. (2017). Best practices for the think-pair-share active-learning technique. *The American Biology Teacher, 79*(1), 3–8.

Pugh, K. J., Bergstrom, C. M., Heddy, B. C., & Krob, K. E. (2017). Supporting deep engagement: The teaching for transformative experiences in science (TTES) model. *The Journal of Experimental Education, 85*(4), 629–657.

Putnam, A. L., Sungkhasettee, V. W., & Roediger, H. L. (2016). Optimizing learning in college: Tips from cognitive psychology. *Perspectives on Psychological Science, 11*, 652–660.

Rathner, J. A., Hughes, D. L., & Schuijers, J. A. (2013). Redesigning a core first year physiology subject in allied health to achieve better learning outcomes. *International Journal of Innovation in Science and Mathematics Education, 21*(2), 37–52.

Rathner, J. A., & Schier, M. A. (2020). The impact of flipped classroom andragogy on student assessment performance and perception of learning experience in two advanced physiology subjects. *Advances in Physiology Education, 44*(1), 80–92.

Reeve, J., Jang, H., Carrell, D., Jeon, S., & Barch, J. (2004). Enhancing students' engagement by increasing teachers' autonomy support. *Motivation and Emotion, 28*, 147–169. https://doi.org/10.1023/B:MOEM.0000032312.95499.6f

Reschly, A. L., & Christenson, S. L. (2012). Jingle, jangle, and conceptual haziness: Evolution and future directions of the engagement construct. In *Handbook of research on student engagement* (pp. 3–19). Springer.

Rodriguez, F., Kataoka, S., Janet Rivas, M., Kadandale, P., Nili, A., & Warschauer, M. (2021). Do spacing and self-testing predict learning outcomes? *Active Learning in Higher Education, 22*, 77–91.

Roediger, H. L., Agarwal, P. K., McDaniel, M. A., & McDermott, K. B. (2011). Test-enhanced learning in the classroom: Long-term improvements from quizzing. *Journal of Experimental Psychology: Applied, 17*(4), 382–395.

Roediger, H. L., & Butler, A. C. (2011). The critical role of retrieval practice in long-term retention. *Trends in Cognitive Sciences, 15*(1), 20–27.

Roediger, H. L., & Karpicke, J. D. (2006a). Test-enhanced learning: Taking memory tests improves long-term retention. *Psychological Science, 17*(3), 249–255. https://doi.org/10.1111/j.1467-9280.2006.01693.x

Roediger, H. L., & Karpicke, J. D. (2006b). The power of testing memory: Basic research and implications for educational practice. *Perspectives on Psychological Science, 1*(3), 181–210. https://doi.org/10.1111/j.1745-6916.2006.00012.x

Roediger, H. L., Putnam, A. L., & Smith, M. A. (2011). Chapter one—Ten benefits of testing and their applications to educational practice. In J. P. Mestre & B. H. Ross (Eds.), *Psychology of learning and motivation* (pp. 1–36). Academic Press.

Roediger, H. L., & Pyc, M. A. (2012). Inexpensive techniques to improve education: Applying cognitive psychology to enhance educational practice. *Journal of Applied Research in Memory and Cognition, 1*(4), 242–248.

Rosen, L. D., Lim, A. F., Carrier, L. M., & Cheever, N. A. (2011). An empirical examination of the educational impact of text message-induced task switching in the classroom: Educational implications and strategies to enhance learning. *Educational Psychology, 17*(2), 163–177.

Rotgans, J. I., & Schmidt, H. G. (2011). Situational interest and academic achievement in the active-learning classroom. *Learning and Instruction, 21*(1), 58–67.

Rourke, L., & Kanuka, H. (2009). Learning in communities of inquiry: A review of the literature. *International Journal of E-Learning & Distance Education/Revue internationale du e-learning et la formation à distance, 23*(1), 19–48.

Ryan, R. M., & Deci, E. L. (2000). Self-determination theory and the facilitation of intrinsic motivation, social development, and well-being. *American Psychologist, 55*, 68–78.
Ryan, R. M., & Deci, E. L. (2020). Intrinsic and extrinsic motivation from a self-determination theory perspective: Definitions, theory, practices, and future directions. *Contemporary Educational Psychology, 61*, 101860.
Ryan, R. M., & Stiller, J. (1991). The social contexts of internalization: Parent and teacher influences on autonomy, motivation, and learning. In P. R. Pintrich & M. L. Maehr (Eds.), *Advances in motivation and achievement* (Vol. 7, pp. 115–149). JAI Press.
Schmidmaier, R., Ebersbach, R., Schiller, M., Hege, I., Holzer, M., & Fischer, M. R. (2011). Using electronic flashcards to promote learning in medical students: Retesting versus restudying. *Medical Education, 45*(11), 1101–1110.
Schmidt, J. A., Rosenberg, J. M., & Beymer, P. N. (2018). A person-in-context approach to student engagement in science: Examining learning activities and choice. *Journal of Research in Science Teaching, 55*(1), 19–43.
Schofer, E., & Meyer, J. W. (2005). The worldwide expansion of higher education in the twentieth century. *American Sociological Review, 70*(6), 898–920.
Schut, S., Heeneman, S., Bierer, B., Driessen, E., van Tartwijk, J., & van Der Vleuten, C. (2020). Between trust and control: Teachers' assessment conceptualisations within programmatic assessment. *Medical Education, 54*(6), 528–537.
Schuwirth, L., van der Vleuten, C., & Durning, S. J. (2017). What programmatic assessment in medical education can learn from healthcare. *Perspectives on Medical Education, 6*(4), 211–215. https://doi.org/10.1007/s40037-017-0345-1
Schuwirth, L. W. T., & van der Vleuten, C. P. M. (2012). The use of progress testing. *Perspectives on Medical Education, 1*(1), 24–30. https://doi.org/10.1007/s40037-012-0007-2
Schwartz, D. L., & Bransford, J. D. (1998). A time for telling. *Cognition and Instruction, 16*(4), 475–5223.
Seibert Hanson, A. E., & Brown, C. M. (2020). Enhancing L2 learning through a mobile assisted spaced-repetition tool: An effective but bitter pill? *Computer Assisted Language Learning, 33*, 133–155.
Seidel, S. B., & Tanner, K. D. (2013). What if students revolt?—Considering student resistance: Origins, options, and opportunities for investigation. *CBE Life Sciences Education, 12*, 586–595.
Shekhar, P., Demonbrun, M., Borrego, M., Finelli, C., Prince, M., Henderson, C., & Waters, C. (2015). Development of an observation protocol to study undergraduate engineering student resistance to active learning. *International Journal of Engineering Education, 31*(2), 597–609.
Shernoff, D. J., & Hoogstra, L. (2001). Continuing motivation beyond the high school classroom. *New Directions for Child and Adolescent Development, 93*, 73–87.
Silverthorn, D. U. (2020). When active learning fails... and what to do about it. In J. J. Mintzes & E. Walter (Eds.), *Active learning in college science* (pp. 985–1001). Springer.
Silverthorn, D. U., Thorn, P. M., & Svinicki, M. D. (2006). It's difficult to change the way we teach: Lessons from the integrative themes in physiology curriculum module project. *Advances in Physiology Education, 30*, 204–214.
Sinatra, G. M., Heddy, B. C., & Lombardi, D. (2015). The challenges of defining and measuring student engagement in science. *Educational Psychologist, 50*(1), 1–13.
Sinatra, G. M., & Pintrich, P. R. (2002). The role of intentions in conceptual change learning. In G. M. Sinatra & P. R. Pintrich (Eds.), *Intentional conceptual change* (pp. 10–26). Routledge.
Sinclair, M. F., Christenson, S. L., Lehr, C. A., & Anderson, A. R. (2003). Facilitating student engagement: Lessons learned from Check & Connect longitudinal studies. *The California School Psychologist, 8*(1), 29–41.
Sinnayah, P., Rathner, J. A., Loton, D., Klein, R., & Hartley, P. (2019). A combination of active learning strategies improves student academic outcomes in first-year paramedic bioscience. *Advances in Physiology Education, 43*(2), 233–240. https://doi.org/10.1152/advan.00199.2018

Skinner, E., Furrer, C., Marchand, G., & Kindermann, T. (2008). Engagement and disaffection in the classroom: Part of a larger motivational dynamic? *Journal of Educational Psychology, 100*(4), 765.

Smolen, P., Zhang, Y., & Byrne, J. H. (2016). The right time to learn: Mechanisms and optimization of spaced learning. *Nature Reviews Neuroscience, 17*(2), 77–88.

Snyder, B. R. (1971). *The hidden curriculum*. MIT Press.

Söbke, H. (2018). A case study of deep gamification in higher engineering education. In *International conference on games and learning alliance* (pp. 375–386). Springer.

Soicher, R. N., & Gurung, R. A. (2017). Do exam wrappers increase metacognition and performance? A single course intervention. *Psychology Learning & Teaching, 16*(1), 64–73.

Sommer, W. G. (1990). Procrastination and cramming: How adept students ace the system. *Journal of American College Health, 39*(1), 5–10.

Stead, D. R. (2005). A review of the one-minute paper. *Active Learning in Higher Education, 6*(2), 118–131.

Stephenson, B., Craig, M., Zingaro, D., Horton, D., Heap, D., & Huynh, E. (2017). Exam wrappers: Not a silver bullet. In *Proceedings of the 2017 ACM SIGCSE technical symposium on computer science education* (pp. 573–578).

Sterling, L., Pedell, S., & Oates, G. (2020). Using motivational modelling with an app designed to increase student performance and retention. In *Early warning systems and targeted interventions for student success in online courses* (pp. 161–176). IGI Global.

Stiggins, R. J. (2005). *Student-involved assessment for learning*. Pearson Prentice Hall.

Sundrud, R. B., & Hueftle, K. (2009). Essential analogies in human anatomy & physiology. *The American Biology Teacher, 71*(9), 554–557.

Szpunar, K. K., Khan, N. Y., & Schacter, D. L. (2013). Interpolated memory tests reduce mind wandering and improve learning of online lectures. *Proceedings of the National Academy of Sciences of the United States of America, 110*(16), 6313–6317.

Tabibian, B., Upadhyay, U., De, A., Zarezade, A., Schölkopf, B., & Gomez-Rodriguez, M. (2019). Enhancing human learning via spaced repetition optimization. *Proceedings of the National Academy of Sciences of the United States of America, 116*, 3988–3993.

Talbert, R. (2017). *Flipped learning: A guide for higher education faculty*. Stylus.

Taylor, S. A., Goodwin, S. A., Melton, H., & Hunter, G. (2011). An analysis of the Student Course Engagement Questionnaire (SCEQ) in large-section marketing principles classes. *Journal on Excellence in College Teaching, 22*(1), 29–57.

Team Based Learning Collaborative. (2013). What is team based Learning: Overview. http://www.teambasedlearning.org/definition/. Accessed 28 Apr 2021.

Theobald, E. J., Hill, M. J., Tran, E., Agrawal, S., Arroyo, E. N., Behling, S., Chambwe, N., Cintrón, D. L., Cooper, J. D., Dunster, G., Grummer, J. A., Hennessey, K., Hsiao, J., Iranon, N., Jones, L., 2nd, Jordt, H., Keller, M., Lacey, M. E., Littlefield, C. E., ... Freeman, S. (2020). Active learning narrows achievement gaps for underrepresented students in undergraduate science, technology, engineering, and math. *Proceedings of the National Academy of Sciences of the United States of America, 117*(12), 6476–6483.

Thomas, L., & Jones, R. (2017). *Student engagement in the context of commuter students*. The Student Engagement Partnership & Liz Thomas Associates. https://lizthomasassociates.co.uk/projects/2018/Commuter%20student%20engagement.pdf

Tinto, V. (1975). Dropout from higher education: A theoretical synthesis of recent research. *Review of Educational Research, 45*(1), 89–125.

Tinto, V. (2003). Learning better together: The impact of learning communities on student success. *Higher Education Monograph Series, 1*(8), 1–8.

Tolman, E. C. (1938). The determiners of behavior at a choice point. *Psychological Review, 45*(1), 1.

Tomé Klock, A. C., Gasparini, I., Soares Pimenta, M., & Hamari, J. (2020). Tailored gamification: A review of literature. *International Journal of Human-Computer Studies, 144*, 102495.

Van Hoof, T. J., & Doyle, T. J. (2018). Learning science as a potential new source of understanding and improvement for continuing education and continuing professional development. *Medical Teacher, 40*(9), 880–885.

Voelkel, S. (2013). Combining the formative with the summative: The development of a two-stage online test to encourage engagement and provide personal feedback in large classes. *Research in Learning Technology, 21*(1), 1–18.

Voelkl, K. E. (2012). School identification. In S. L. Christenson, A. L. Reschly, & C. Wylie (Eds.), *Handbook of research on student engagement* (pp. 193–218). Springer.

Wahba, M. A., & Bridwell, L. G. (1976). Maslow reconsidered: A review of research on the need hierarchy theory. *Organizational Behavior and Human Performance, 15*(2), 212–240.

Waluyo, B., & Bucol, J. L. (2021). The impact of gamified vocabulary learning using quizlet on low-proficiency students. *Computer-Assisted Language Learning, 22*(1), 158–179.

Wang, M. T., & Degol, J. (2014). Staying engaged: Knowledge and research needs in student engagement. *Child Development Perspectives, 8*(3), 137–143.

Wang, M. T., & Fredricks, J. A. (2014). The reciprocal links between school engagement and youth problem behavior during adolescence. *Child Development, 85*, 722e737.

Weinstein, Y., Madan, C. R., & Sumeracki, M. A. (2018). Teaching the science of learning. *Cognitive Research: Principles and Implications, 3*, 2. https://doi.org/10.1186/s41235-017-0087-y

Wenger, E. (1999). *Communities of practice: Learning, meaning, and identity*. Cambridge University Press.

Wheeler, M. A., & Roediger, H. L. (1992). Disparate Effects of repeated testing: Reconciling Ballard's (1913) and Bartlett's (1932) results. *Psychological Science, 3*(4), 240–246. https://doi.org/10.1111/j.1467-9280.1992.tb00036.x

White, C., Bradley, E., Martindale, J., Roy, P., Patel, K., Yoon, M., & Worden, M. K. (2014). Why are medical students 'checking out' of active learning in a new curriculum? *Medical Education, 48*(3), 315–324.

Wilkinson, T. J., & Tweed, M. J. (2018). Deconstructing programmatic assessment. *Advances in Medical Education and Practice, 9*, 191–197. https://doi.org/10.2147/AMEP.S144449

Wilson, A., Howitt, S., & Higgins, D. (2016). Assessing the unassessable: Making learning visible in undergraduates' experiences of scientific research. *Assessment & Evaluation in Higher Education, 41*(6), 901–916.

Wiltbank, L. B., Williams, K. R., Marciniak, L., & Momsen, J. L. (2019). Student perceptions and use of feedback during active learning: A new model from repeated stimulated recall interviews. *Assessment & Evaluation in Higher Education, 44*(3), 431–448.

Winne, P. H. (1996). A metacognitive view of individual differences in self-regulated learning. *Learning and Individual Differences, 8*(4), 327–353.

Wissman, K. T., Rawson, K. A., & Pyc, M. A. (2012). How and when do students use flashcards? *Memory, 20*(6), 568–579.

Witchel, H., Westling, C., Tee, J., Healy, A., Needham, R., & Chockalingam, N. (2014). What does not happen: Quantifying embodied engagement using NIMI and self-adaptors. *Participations: Journal of Audience and Reception Studies, 11*(1), 304–331.

Witchel, H. J. (2013). Engagement: The inputs and the output. In *Proceedings of the 2013 inputs-outputs conference: An interdisciplinary conference on engagement in HCI and performance* (pp. 1–4).

Wittrock, M. C. (1978). The cognitive movement in instruction. *Educational Psychologist, 13*(1), 15–29.

Won, S., Wolters, C. A., & Mueller, S. A. (2018). Sense of belonging and self-regulated learning: Testing achievement goals as mediators. *The Journal of Experimental Education, 86*(3), 402–418.

Wong, M. M. (2008). Perceptions of parental involvement and autonomy support: Their relations with self-regulation, academic performance, substance use and resilience among adolescents. *North American Journal of Psychology, 10*(3), 497–518.

Wood, D. F. (2003). Problem based learning. *British Medical Journal, 326*(7384), 328–330.

Wormald, B. W., Schoeman, S., Somasunderam, A., & Penn, M. (2009). Assessment drives learning: An unavoidable truth? *Anatomical Sciences Education, 2*(5), 199–204. https://doi.org/10.1002/ase.102

Wozniak, P. (2021). SuperMemo Guru. https://supermemo.guru/wiki/SuperMemo_Guru. Accessed 30 Apr 2021.

Wyer, K. (2013). *Survey: More freshmen than ever say they go to college to get better jobs, make more money*. Higher Education Research Institute at UCLA. http://heri.ucla.edu/press-release/TheAmericanFreshman2012-PressRelease.pdf

Yeager, D. S., & Dweck, C. S. (2012). Mindsets that promote resilience: When students believe that personal characteristics can be developed. *Educational Psychologist, 47*(4), 302–314.

Zhang, Y., Haddad, E., Torres, B., & Chen, C. (2011). The reciprocal relationships among parents' expectations, adolescents' expectations, and adolescents' achievement: A two-wave longitudinal analysis of the NELS data. *Journal of Youth and Adolescence, 40*(4), 479–489.

Part II
How Educational Technologies Shape the Classroom Experience

Chapter 5
Perceptual Learning, Adaptive Learning, and Gamification: Educational Technologies for Pattern Recognition, Problem Solving, and Knowledge Retention in Medical Learning

Philip J. Kellman, Victoria Jacoby, Christine Massey, and Sally Krasne

Abstract In this chapter, we consider recent advances in the learning sciences and their potential and actual applications in medical learning. We describe emerging ideas that broaden traditional declarative and procedural emphases in learning, with a special focus on *perceptual learning*—experience-induced improvements in the pickup of information. Experience in a task or domain changes perception to be more efficient at discovering and extracting relevant information and ignoring extraneous information while also reducing effort and cognitive load. These discovery and fluency changes from perceptual learning are crucial contributors to complex task performance and advanced expertise. We also examine adaptive learning and its connections to scientific research on the testing effect and the spacing effect, as well as other examples of technology-enhanced learning (TEL), including games and gamification in learning generally and medical education specifically. New learning

Funding Announcement The authors gratefully acknowledge research support from NIH/NCI grant 5R01CA236791 to PJK, SK, and CM; NSF ECR award 1644916 to PK and CM; and support from the US Office of Naval Research Cognitive Science of Learning, US Military JPC-1/MSIS, and US Army RDECOM programs.

P. J. Kellman (✉)
Department of Psychology, University of California, Los Angeles, Los Angeles, CA, USA

Department of Surgery, UCLA David Geffen School of Medicine, University of California, Los Angeles, Los Angeles, CA, USA
e-mail: kellman@cognet.ucla.edu

V. Jacoby · C. Massey
Department of Psychology, University of California, Los Angeles, Los Angeles, CA, USA
e-mail: vjacoby@g.ucla.edu; cmassey@psych.ucla.edu

S. Krasne
Department of Physiology, UCLA David Geffen School of Medicine, University of California, Los Angeles, Los Angeles, CA, USA
e-mail: skrasne@g.ucla.edu

H. J. Witchel, M. W. Lee (eds.), *Technologies in Biomedical and Life Sciences Education*, Methods in Physiology, https://doi.org/10.1007/978-3-030-95633-2_5

technologies combining perceptual and adaptive learning make it possible to accelerate perceptual learning and rapidly advance aspects of expertise, such as pattern recognition, that have been elusive in most instructional contexts. We illustrate the efficacy of research-informed approaches to TEL in our work applying perceptual and adaptive learning modules (PALMs) in medical learning domains. We outline the scope of this research and illustrate characteristic elements and findings of PALMs using the example of electrocardiography. We also describe research that suggests that perceptual and adaptive learning in PALMs and related efforts have great potential to improve medical learning, not only for difficult perceptual classifications, but for factual learning and higher-order diagnostic skills as well.

5.1 Introduction

It is an exciting time to be involved in medical learning, either as a researcher or a practitioner seeking to improve it. Efforts in cognitive psychology and related disciplines to understand and improve learning—broadly described as the "science of learning"—have expanded dramatically in recent years. These efforts have yielded many new findings of practical importance such as spacing and interleaving in learning and the advantages of testing over studying (e.g., Carpenter et al., 2012; Pashler et al., 2007). Some of them have also begun to expand and change our very conceptions about what learning is (Kellman & Massey, 2013), a transformation that offers previously unanticipated opportunities for enhancing instruction with new methods that accelerate the development of expert pattern recognition.

In this chapter, we consider recent advances in the learning sciences and their potential and actual applications in medical learning. We describe concepts that broaden traditional declarative and procedural emphases in learning, with a special emphasis on *perceptual learning*—experience-induced improvements in the pickup (extraction and differentiation) of information (Gibson, 1969; Kellman & Garrigan, 2009). We also focus on adaptive learning and its connections to scientific research on the testing effect, the spacing effect, motivation, gaming, and mastery learning. We review efforts to apply perceptual and adaptive learning, as well as game concepts, in medical education. Following this review, we will focus in some detail on our work on adaptive perceptual learning. We use our experiments training medical students and residents to interpret ECGs to illustrate, in detail, both our approach and the types of findings we have made in a variety of areas, both published (histopathology, dermatology, echocardiography and electrocardiography) and only presented thus far at professional meetings (obstetrics and gynecology, radiology, point-of-care ultrasound, ophthalmology). We close with a discussion of issues and prospects for the further development of advanced learning technologies in medical learning.

5.2 Conceptions of Learning

When we think of what "learning" is, several prototypical examples come to mind. The storage of information in memory is one; learning a sequence of steps in a procedure is another. These declarative and procedural kinds of learning are important, but learning includes much more. One way to appreciate other crucial elements is to look backward from the long-term goals of learning. Consider, for instance, a newly minted MD's ability to interpret an abnormal ECG, detect a suspicious calcification on a mammogram, or pick up a subtle heart murmur using a stethoscope. Even the most diligent of students will nevertheless be slow and prone to making errors—in stark contrast to the seemingly effortless performance of an experienced practitioner. Although we dedicate intensive instructional effort to written materials and classroom time ostensibly to "teach" and develop competency in domains as diverse as medical diagnosis, aviation training, chemistry, and mathematics, it is accepted and obvious that when coursework and examinations end, even the student who has excelled in learning the facts and procedures is not a pilot, chemist, surgeon, or diagnostician. That functional expertise in most domains requires further extended experience, such as multi-year medical residency, speaks to other ingredients needed to fully "learn" or achieve real competence.

5.2.1 Perceptual Learning

What is missing from typical ideas of learning and most learning research? Oddly, if one looks at the different but related domain of research on expertise, one does not see this focus on declarative and procedural knowledge. Studies of expertise consistently identify as the most striking difference between experts and novices, not their storehouse of information, not their procedural knowledge, but their *information extraction and pattern recognition.* Referring to classic research on chess players by De Groot (1965), an influential review of research on learning and education (Bransford et al., 1999) put it this way:

> Experts are not simply "general problem solvers" who have learned a set of strategies that operate across all domains. The fact that experts are more likely than novices to recognize meaningful patterns of information applies in all domains, whether chess, electronics, mathematics, or classroom teaching. In DeGroot's (1965) words, a "given" problem situation is not really a given. Because of their ability to see patterns of meaningful information, experts begin problem solving at "a higher place." (DeGroot, 1965)

These changes derive from a set of learning processes called *perceptual learning*, broadly defined as experience-induced improvements in the extraction of information.

The idea that perceptual pickup of information changes with experience in a task or domain may be unfamiliar to many. That perceptual learning attunes perception to discover and extract meaningful information, often involving complex relations, may be similarly novel. Traditional views of perception as static, elementary, or providing essentially meaningless raw material unless supplemented by other cognitive processes would not suggest these ideas. However, contemporary research and theory in perception embraces a very different view of the outputs of perception. Perception is oriented toward extracting meaningful structure of objects and events; it changes dynamically to optimize task performance; and it interacts closely with other cognitive processes. (For discussion, see Kellman & Massey, 2013.)

Indeed, even the declarative and procedural aspects of learning, which might seem "post-perceptual," are deeply connected to pattern recognition skills developed through perceptual learning. Vast storehouses of declarative information, or learned procedures, are nearly useless without the learner being able to tell which facts, concepts, or procedures apply to the case or situation one is confronting. Classifying a situation, interpreting an image, and recognizing a constellation of symptoms—these are all pattern recognition problems, and the way these pattern recognition abilities develop and improve is through perceptual learning. As an example, one of the authors is a pilot who owned an aircraft with a six-cylinder piston engine. One day, as the pilot taxied in after a flight and parked at his home airport, the chief mechanic happened to be walking by. After the pilot got out of the plane, the mechanic said, "How long have you had that dead cylinder?" The pilot said "What dead cylinder?" The engine sounded normal to the pilot, but the mechanic could hear in a few seconds that one of the six cylinders had failed completely! Obviously, had the pilot noticed a significant power loss in flight and later asked the mechanic about possible causes, the troubleshooting process would have been rather greatly accelerated by the entry point provided by the mechanic's expert ear, as opposed to tearing down the whole engine trying to discover a problem. We have no doubt that medical practitioners in various specialties could recount a broad range of analogous anecdotes.

The neglect of perceptual learning in most accounts and in instructional practice relates to characteristic difficulties in teaching and learning. Perceptual learning grows gradually and often implicitly, through encountering instances and making classifications. Unlike the explicit teaching of a fact or procedure to able students, as in medical learning contexts, this form of learning is not advanced much by lectures or even demonstrations, and much of what is learned is often not verbalizable by instructors or students (e.g., Benzil, 2018). These aspects of perceptual learning account for why this kind of learning languishes under traditional approaches, and they are key to the relevance of long apprenticeship or experience to reach expert levels of pattern recognition.[1] More direct attention to perceptual learning in training has the potential to accelerate this process dramatically.

[1] It is important to note that scientific work on perceptual learning is distinctly different from sociocultural and educational theories sometimes referred to as experiential or situated learning (e.g., communities of learners (Rogoff, 1994), situated learning (Lave & Wenger, 1991), and activity theory (Engeström et al., 1999), among others). Although both perceptual learning and sociocultural experiential learning theories emphasize the importance of extended experience to promote learning, the learning processes and outcomes that they are concerned with diverge

We emphasize the pattern recognition aspects of perceptual learning because they are important and somewhat intuitive to understand. In reality, the ways that perception improves with experience are various, and together these perceptual learning effects exert even broader and more profound effects on expertise. Perceptual learning effects fall into two broad categories, which have been called *discovery* and *fluency* effects (Kellman, 2002). *Discovery* effects refer to learners coming to find and select the information that is relevant to a task—for example, the features or patterns that distinguish instances of one category from another. Learners come to extract relevant information, including new relationships or patterns to which they were originally insensitive (Gibson, 1969; Goldstone, 2000). *Fluency effects* refer to improvements in the efficiency of information extraction. Whereas the "what" (or where, etc.) in information pickup is the province of discovery, fluency involves processes that expedite the pickup of relevant information. Experience or practice leads to perceptual extraction that is faster, involves more parallel processing, picks up larger "chunks" of information, and occurs with less effort or cognitive load. This latter characteristic—greater automaticity in information extraction—allows the operator to use attentional resources for other components of complex tasks and situations. When discovery and fluency effects are considered together, it becomes clear why there is a gap between a learner's having stored relevant factual, conceptual, and procedural information and being an expert. In fact, there is more than one gap. The activation of relevant information or actions through pattern recognition is crucial in the initial grasp of a problem but also continues at each step in a procedure, as information acquired guides the choice or nuances of each new action. At the same time, new actions produce new information. This interleaving of information pickup and action has been recognized in the concept of "perception-action loops" (Gibson, 1966). The discovery aspects of perceptual learning lead to more selective pickup of relevant information and sensitivity to new relations at many points in a task. Even in tasks that may not primarily involve perception-action loops, such as medical image interpretation, growing expertise leads to better actions in terms of where to look; what to scrutinize or magnify, etc.; what to think about; and what to do next based on information acquired. In complex situations, fluency effects allow coordination of various task components, including the guiding of procedures with reduced attentional load, allowing more effective management of multiple task demands and communication, as well as better situational awareness. The apparent magic of advanced expertise in many domains (chess, music, science, medicine)

considerably. Sociocultural learning theories are primarily focused on the learning of community norms, values, and identities, as well as workplace roles and practices as they are situated in particular contexts. They emphasize interpersonal interactions and collaborative activities as the means by which learning is structured and mediated (Yardley et al., 2012). In contrast, perceptual learning involves changes in the operation of perceptual systems that lead to domain-specific selection and more rapid pickup of relevant information, extraction of relevant patterns or relations, and reductions in effort or attentional load in perception. It is likely that experiential learning and perceptual learning co-occur. Workplace or apprenticeship learning in medical education presumably provides conditions that can support both of these types of learning.

owes a great deal to the expert's specialized, advanced, and automatic information extraction, continuously interacting with other cognitive and procedural demands of complex tasks (Bryan & Harter, 1899; Kellman & Garrigan, 2009).

Broadening ordinary conceptions of learning, and specifically the understanding perceptual learning, offers clear opportunities to improve medical learning and the acceleration of expertise. This is especially true when training approaches combine perceptual learning with adaptive learning methods, to which we now turn.

5.3 Adaptive Learning

Adaptive learning may be defined broadly as the use of one or more aspects of a learner's performance to arrange learning content and events. Early work often referred to such arrangements as *adaptive instructional systems* (e.g., Atkinson, 1976), but the more or less synonymous phrase adaptive learning has become more common. There can be many forms of adaptivity, ranging from simple pre-testing to determine a learner's starting point, modification of learning content to address each learner's strengths and weaknesses, or trial-by-trial tracking of learning performance to continually decide the sequence of learning events in a learning session. While various forms of teacher-led differentiation of instruction for different learners have been in widespread use for some time (Bondie et al., 2019), possibilities and investigations of adaptive methods have blossomed with the emergence of generally available digital technologies which have the speed and capacity to arrange learning events in real time based on individual learner responses and to store large databases of learning materials.

Recently, medical education specifically has embraced emerging technology-enhanced learning or *TEL*. It has been argued that these efforts could be further enhanced by greater use of adaptive learning (Sharma et al., 2017). A number of efforts incorporating adaptive learning in medical education have been reported, spanning a range of topic areas and types of learning (Kellman & Krasne, 2018; McMahon & Drazen, 2014). One example is the NEJM Knowledge+ "Pain Management and Opioids" learning module (Hamnvik et al., 2019), which tests learners and takes both learner accuracy and rated confidence as inputs to determine gaps in knowledge and trigger presentation of learning content most needed. Hamnvik et al. (2019) cite a conference paper (Healy et al., 2019) reporting that use of the NEJM Knowledge+ review platform lowered the failure rate on the American Board of Internal Medicine (ABIM) maintenance of certification exam from 11 to 5%.

Use of the term "adaptive" or adaptive features is not a guarantee of learning effectiveness when compared to nonadaptive, business-as-usual control groups. Griff and Matter (2013) measured the effectiveness of adaptive learning in anatomy and physiology courses. Multiple schools were recruited for this study, and within each school students were assigned to either an adaptive learning or control condition. In the adaptive learning condition, students answered multiple choice, multi-select, and fill-in-the blank questions on a system called "LearnSmart" developed by

McGraw Hill Higher Education. LearnSmart uses the accuracy of the answers given in combination with each student's confidence rating to select the next questions that the student sees. Ideally, this should allow students to spend more time and receive more practice on topics that are more difficult for them. In the control condition, students were given practice questions chosen for them by their instructor from a larger question bank supplied by the class's textbook publisher. All students were given a pre-test and a post-test, both containing the same questions. The duration between the pre- and post-test varied. The study was conducted across six institutions with the pre-test occurring at the beginning of the quarter or semester, and the post-test was taken at the end of that quarter/semester.

Ultimately, the results showed that despite positive student feedback about the LearnSmart program, in most schools there was no significant difference between the adaptive and control groups in student performance gains from the pre-test to post-test. The result is consistent with a number of studies in the emerging judgments of learning (JOL) field showing that learner judgments of confidence are influenced by numerous factors that may not be predictive of actual learning or future performance (Kornell & Hausmann, 2017).

More generally, as with TEL, adding in adaptive or technological elements does not guarantee success. Spector (2014) emphasizes that there is often an eagerness to adopt the technology without thoroughly assessing its capabilities or without serious consideration for how it is integrated into a classroom. Similar comments apply to the use of games in learning. (See below.) Another issue is that many efforts in adaptive learning have been evaluated through user satisfaction measures rather than through objective measures of learning and retention.

5.3.1 Learning Principles and Adaptive Learning

We think that adaptive learning efforts are most likely to succeed when they are based on learning principles that have been clearly tested and supported in research using objective measures of learning. In this section, we briefly describe two such principles, both as examples of research-based concepts that may be valuable and as ideas that have been implemented in some successful adaptive learning approaches. These are the *testing effect* and the *spacing effect.*

5.3.1.1 The Testing Effect

Testing is usually done to assess learning. Learners typically prepare for testing by studying. It is mildly paradoxical, then, that testing events themselves have been shown to improve learning more so than studying. This idea, which has a long history (e.g., Abbott, 1909; Tulving, 1967), has come to be known as the *testing effect* (Bjork, 1975; Carrier & Pashler, 1992; Roediger & Karpicke, 2006).

The testing effect has been demonstrated in both laboratory and applied contexts (Roediger & Butler, 2011). In one example, Larsen et al. (2013) measured the testing effect in a real-world medical learning setting using residents attending an instructional conference. All residents participated in an hour-long teaching session on two topics: *status epilepticus* and *myasthenia gravis*. Following the session, half of the residents were given a test on *status epilepticus* and a review sheet to study for *myasthenia gravis*; the other half were given a test on *myasthenia gravis* and a review sheet to study for *status epilepticus*. Both the tests and review sheets covered the same information, and residents were able to check their answers following the test. This process was repeated approximately 2 weeks later, and at 6 months, all residents returned to take a final test on both topics. The results showed that despite being given the same information and amount of time devoted to each topic, residents who engaged with the material through tests significantly outperformed those who studied with the review sheet for each topic.

Why testing events are superior to studying events is not entirely settled. One likely factor is that testing events activate more attention and effort. A "retrieval effort hypothesis" suggests that more difficult (but successful) retrievals of information provide greater gains in long-term retention (Pyc & Rawson, 2009). Studying does not exercise retrieval processes in the way that testing does. Testing events with feedback may also help learners to recognize deficiencies in their current understanding of material (Roediger & Butler, 2011), whereas restudying may give illusions of familiarity that lead learners to overestimate what they know about the topic (Bjork & Bjork, 2011). Other contributors to testing effects may be greater elaboration and connections of knowledge that occur during retrieval and the idea that testing practice resembles actual test situations more so than passive studying.

Adaptive learning and the testing effect go together naturally. Adaptivity always depends in some way on assessment of a learner's status or needs. Although some adaptive efforts may use initial testing to choose a level or starting point for learning, testing and adaptive learning can be more richly combined when most of the learning itself consists of ongoing testing events (e.g., Mettler et al., 2016). With the capacities of modern computer systems, such arrangements can acquire rich and up-to-date estimates of learner performance and use the data to guide learning events in productive ways. Prominent among these is to guide spacing in learning.

5.3.2 The Spacing Effect

Considerable research in the learning sciences converges in showing that spacing the repetitions of an item in learning, rather than grouping them in close succession, enhances long-term retention. This benefit generalizes across numerous tasks, materials, and time scales and has emerged as one of the most robust effects in learning and memory research (Bjork & Bjork, 2011; Karpicke & Roediger, 2007). Although most spacing research has focused on declarative (factual) learning materials, as in foreign language vocabulary, paired associates, or learning facts from text, spacing research has been known for some time to have benefits in motor learning (Lee & Genovese, 1988), although this may vary by task (Wiseheart et al., 2017). More recently, spacing has been shown to have powerful effects in perceptual learning (Mettler & Kellman, 2014). Thus, spacing considerations are likely to be important across different types of learning, and they offer fertile opportunities for enhancing learning using instructional technology.

Almost certainly, the explanation for spacing effects involves multiple factors (Glenberg, 1979; Mettler et al., 2016). A fundamental one involves the well-studied differences between massed and distributed practice. Revisiting a learning item several times in succession offers little or no benefit for long-term learning. The problem is that in such massed practice, relevant information remains in working memory, providing no practice in retrieval from more long-term memory (Bjork & Bjork, 2011). An analogous problem occurs in more complex tasks. In perceptual-motor tasks, for example, learners retain components of performance and need not organize the full processing and response selection as would be required with more widely spaced challenges (Schmidt & Bjork, 1992). In perceptual learning, or the aspects of complex tasks that involve perceptual learning, with massing, the learner may remain attuned from one trial to the next to particular information useful for classification.

Requiring that each learning event "stands on its own," free of transient carryover from recent events, is not the only driver of spacing benefits. Spacing of items usually coincides with interleaving, such that between repetitions, not only does time elapse but other learning items occur. Interleaving exerts benefits by connecting retrieval to changing contexts and by facilitating discrimination between varying items that occur nearby in time (Kang & Pashler, 2012; Kornell & Bjork, 2008).

A considerable amount of research has attempted to establish principles for predetermined spacing schedules in learning. Some evidence suggests that expanding recurrence intervals outperform fixed ones (e.g., Landauer & Bjork, 1978). Other work suggests that equal interval practice may actually be superior to expanding practice intervals when measured after a delay (Karpicke & Roediger, 2007). On the other hand, Storm et al. (2010) reported better memory from an expanding schedule, but only when intervening items during spacing were highly related to spaced items, suggesting that expanding intervals work best for difficult material.

Mettler et al. (2016) argued that the question of the superiority of fixed vs. expanding spacing has no ultimate answer. There are many factors known to relate to optimal spacing. Most, if not all, exert their effects on an internal variable of learning strength. In general, the optimal time to practice an item is when retrieval is difficult but can still succeed. This "retrieval effort hypothesis" (Pyc & Rawson, 2009) fits a great deal of research (Bjork & Bjork, 1992; Johnston & Uhl, 1976; Pyc & Rawson, 2009; Thios & D'Agostino, 1976). Retrieval effort decreases as learning strength increases. The problem in schemes that preset spacing intervals is that fluctuations of learning strength in a learning session arise from numerous and subtle influences that cannot be captured in an a priori model. Certainly, learning strength tends to decrease over time without practice, and practice events (especially testing events) tend to increase it; but in any set of learning items, difficulty will vary by item, by learner, and by interactions of the two. Many events that occur during the course of a learning session, such as attentional fluctuations, momentary confusions between items, or resolution of confusions by patterns of interleaving and feedback, may affect a given learner's learning strength for a given item.

5.3.3 *Adaptive Learning and the ARTS System*

These issues can be addressed by adaptive methods that assess learning strength in an ongoing manner and adjust spacing intervals accordingly. Adaptive learning systems have generally relied on accuracy data alone to assess progress, but accuracy information alone is too crude to serve as the basis of optimal spacing. Several years ago, we proposed that response times, along with accuracy, can better assess underlying learning strength and serve as the basis of spacing. We developed the ARTS (Adaptive Response-Time-based Scheduling) system to do this (Kellman, 2006; Kellman & Krasne, 2018; Mettler et al., 2011, 2016). Actually ARTS adds response times to adaptive systems in two different ways: One is spacing, but the other innovation is using response time in combination with accuracy to set mastery criteria, an especially useful arrangement for many learning domains where both accuracy and fluency are goals of learning.

The ARTS system has proven to be highly effective in various domains, especially in STEM learning and medical learning, and has been shown to outperform a classic adaptive learning system (Mettler et al., 2011). These effects have been evident in both factual and perceptual learning domains, suggesting that a *successful effort hypothesis* may hold across various types of learning, generalizing the retrieval effort hypothesis to apply to types of learning that do not depend primarily on retrieval of memory items. Below we consider especially the application of ARTS to perceptual learning in perceptual-adaptive learning modules applied to a variety of medical learning domains.

With regard to efforts to optimize spacing, something more than evidence of effectiveness is required. Mettler et al. (2016) performed experiments whose results confirm the idea that optimal spacing intervals must be adaptive and based on the individual learner. They first showed the superiority of adaptive learning to a nonadaptive control condition. They then compared adaptive learning conditions to two "yoked" control conditions. In a *yoked-random* condition, a new group of participants received the same spacing intervals that had been generated by adaptive participants, but randomized across the learning items. If the benefit of adaptive learning derives, not from individually adaptive intervals per se, but from the overall distribution of spacing intervals that occur under adaptive conditions, then this group would have been expected to perform equally well as the adaptive condition. In a *yoked-item* condition, spacing schedules that had been generated by adaptive participants remained attached to the same learning items. If the benefit of adaptive conditions derives from some items consistently having greater difficulty than others, then this group would have been expected to match the performance of adaptive participants.

The adaptive learning condition outperformed both yoked control conditions. There was a trend for the yoked-item condition to be better than the yoked-random condition, suggesting that item difficulty does have an influence and that such consistent effects of difficult and easy items are appropriately picked up by ARTS and used efficaciously in spacing. Ultimately, however, adaptive intervals generated by each learner's own performance give the best learning results. These results are consistent with the idea that optimal spacing must be tied to individual learner's learning strength for learning items, and they also confirm the importance of the underlying construct of learning strength as the driver of optimal spacing. Predetermined (nonadaptive) spacing arrangements cannot achieve the same results.

5.4 Games and Gamification in Medical Learning

Earlier we noted the role of digital technology as empowering the development and implementation of adaptive learning methods. Similarly, the widespread availability of computers has opened up new opportunities for educational games and gamification. *Gamification* refers to the application of game elements to non-game contexts (Deterding et al. 2011; Hamari et al., 2014; Rutledge et al., 2018).

The motivations for making learning games, or making learning activities more like games, are several. In domains where many students struggle, such as mathematics and other STEM subjects, it has been hoped that games or gamification might increase student attention, engagement, and motivation. Instructors note with envy the degree of immersion and engagement that is evident when students play video games. If there were a way to engender such levels of attachment and commitment in academic subjects, it might boost learning outcomes. If the effect of gamification were simply to increase the time learners spend on learning tasks, it might plausibly advance learning.

An interesting preliminary consideration is that it is hard to define precisely what is meant by a game, or even game-like elements. Strikingly, the philosopher Wittgenstein used "game" as a prime example in making the case that human concepts do not have precise definitions or meanings (Wittgenstein, 1953). That said, some game-like features that have been considered to be potentially useful in learning include competition, speeded responses, feedback, markers of progress, ascent of levels with increasing success, and well-defined rewards. Obviously, some of these elements have been used in learning settings apart from explicit efforts to make game-like activities.

The idea that games might help in serious learning contexts has had remarkable momentum, especially when contrasted with the evidence base. The widespread enthusiasm for the use of learning games or gamification has been primarily driven by intuition (and, perhaps, advertising). Actual research on use of games has indicated few clearly positive results when assessed by objective measures of learning. For example, a broad review by Young et al. (2012) found no compelling evidence for learning benefits of games in STEM domains but possibly a benefit for a small number of games in social science domains. A more recent meta-analysis (Sailer & Homner, 2020) found modest positive evidence for gamification effects on learning but reported inconsistencies in which elements lead to successful outcomes.

In medical learning applications, gamification has similarly led to mixed results and conflicting views. One concern is that with many potential game elements to choose from, an overall decision to "gamify" is less meaningful than specification of particular features. Research on these specific factors seems more likely to be meaningful and transferable than the overall idea of using a game.

One element that has been investigated in medical learning settings is competition. In a study conducted by Nevin et al. (2014), medical students were given the opportunity to compete in an optional, web-based medical knowledge competition. In this competition, participants answered multiple choice questions pertaining to internal medicine. Participation was voluntary; participants had the ability to earn badges, and there was a dynamically changing leaderboard. When participants were questioned about their experience in subsequent focus groups, the students reported that the leaderboard was their most important motivator (Nevin et al., 2014). Like many studies of games in learning, the focus in data analyses was on acceptance (and in this case, attrition), and there were no measures of objective learning or comparisons that could be used to argue that games enhanced learning beyond business as usual approaches.

A study by Van Nuland et al. (2015) also tested the role of organized competition. Volunteers from an undergraduate functional anatomy course engaged in an online tournament either early or later in the term. In the tournament, participants competed by answering a series of multiple choice or matching questions on the topic of anatomy. Following the competition, students who participated in the activity reported high levels of motivation. The group that competed early in the term, but not those competing later in the term, showed higher course grades than other students in the class who did not participate. Although this study is laudable in using objective measures to compare the gaming groups to non-gaming learners, as in the study by Nevin et al., it is hard to interpret the generality of results given the self-selection of participants. Competition in particular may attract volunteers with greater confidence in their learning of particular subject matter, or gaming may actually energize learning in a subset of individuals (16.7% of the class volunteered in this study). Engagement in the competition also led to more time on task for participants. Interestingly, the authors characterized the intervention as game-like due to competition but attributed the learning gains to the testing effect. Competition and testing can be related, especially in that competition would seem necessarily to involve testing, but they are conceptually separable factors, and testing can be, and often is, implemented without competition.

Perhaps because it has been difficult to show clear learning gains from games, motivational measures are more often reported (Hamari et al., 2014). Here too, the particular elements responsible for group differences may be hard to pin down. Also, enhanced motivation and acceptance presumably have value in learning contexts due to their potential for increasing learning. Studies that relate motivational measures to learning outcomes are likely to be most valuable. Even apart from objective learning measures, it also seems wise to include objective data on motivation such as increased time on task or some other objective measure of participation or persistence, as has been done in some studies, rather than rely exclusively on self-report measures.

Recognition of levels of progress is another potentially valuable element often used in games or gamification. In the same study mentioned above by Nevin et al. (2014), they reported that students who earned badges throughout the competition went on to answer more questions in the activity than those who did not earn badges. That indications of progress might be helpful in increasing motivation and learning is consistent with many different psychological and educational ideas, including reinforcement theory, mastery learning, and goal-directed behavior.

As has been often recognized with the use of standardized tests in schools, it may be important for instructional designers to seek to avoid having the game, rather than the learning content, become the goal. Ng and Bereiter (1991) distinguished three kinds of goals in self-directed learning: task-completion goals, instructional goals, and personal knowledge-building goals. In the learning of basic computer programming, they found that knowledge-building goals, relating to personal interests in learning, were the least frequently expressed, but that students who did express them showed superior learning outcomes in several respects.

Recognition of levels of progress as a learning element readily connects to all three of these goal types. Students' motivation to satisfy the instructional goals (i.e., perform well in the game or on a test) or the task-completion goals, however, may actually compete with knowledge-building (Ng & Bereiter, 1991). We have seen this effect in sequences of trials in adaptive learning where the task and instructional goals of getting to the next trial or finishing a module may cause a learner to speed through and miss valuable trial feedback. A separate issue is that progress indicators may backfire if a learning task is too hard, as participation for long periods without obvious progress undercuts motivation. On balance, however, levels of progress indications seem to have strong potential to maintain motivation. It is worth careful effort, however, to align task and instructional goals with the deeper learning objectives, as game elements, such as speeded response and competition to finish quickly, can act at cross purposes to learning goals.

In STEM and social science subjects, games for primary and secondary school students have often been built around fictional stories, where progress toward the treasure or defeating the space aliens occurs through answering questions or other challenges involving learning items. These approaches have not often been used in medical learning, where the commitment and motivation of learners tends to be higher and more consistent.

5.4.1 Gaming Prospects and Pitfalls

Although more successful versions of learning games could be developed, there are sound scientific reasons that suggest serious limitations on how much games might ultimately enhance meaningful learning. Put another way, these are concerns that identify issues that would have to be addressed or circumvented to derive greater learning benefits from games.

One well-known concern has been called the *bells and whistles* problem (e.g., Moreno & Mayer, 2000). Since the emergence of enhanced graphics, sound, and animation capabilities in ordinary computers, there has been widespread enthusiasm for adding multimedia features, story lines, scenery, characters, sounds, animations, etc. as features in learning software and games in particular. These additions often fail to produce learning benefits and may instead lead to poorer performance. A deep reason is that such additions almost always add extraneous cognitive load (Chandler & Sweller, 1991), which can impair learning of relevant material. Extraneous cognitive load refers to information or inputs that attract processing resources but are not inherently part of the learning content. This kind of load contrasts with intrinsic and/or germane load, which relate to the learning content and associated problem solving efforts (Van Merriënboer & Sweller, 2009); these are relevant to the targeted learning content, and coping with increased load of these types may often be an important factor in learning, as in the "desirable difficulties" framework (Bjork & Bjork, 2006, 2011).

A related problem is what we call the *spoonful of sugar* problem. Embedding mathematics or science learning in a game has often been done on the assumption that the game will be more fun than the actual content of these disciplines. Thus, the engaging storyline, action, or adventure are the "sugar," whereas the mathematics or science can be slipped into the mix as the medicine, and all will be well. The problem with this reasoning is that students tend to learn the story or game but not so much the math or science (Young et al., 2012). An increase in time and attention will not benefit learning goals if it is primarily dedicated to game elements that are not intrinsically related to the intended academic content. This is not an incidental finding; it is exactly what would be expected on the grounds of limited capacity in human information processing and the extraneous load presented by a story or game. Yet another difficulty is that in any learning domain, some of the material is more amenable to gamification than other parts. Thus, much research on games and learning has essentially "cherry-picked" parts of learning that seem suitable for a game; as a result, there are numerous games in which students advance by correctly doing arithmetic but few that teach quantum mechanics. It remains unclear how general a strategy gamification might be.

Both the bells and whistles problem and the spoonful of sugar problems are related to what has been called the problem or paradox of *seductive details*. Seductive details refer to interesting pieces of information, stories, or images that are only tangentially related to the topic being learned, but are not oriented toward the specific learning objectives (Harp & Mayer, 1998). Such details are added most often in an attempt to make the topic more interesting and relatable for students. However, inclusion of such irrelevant information has been found to actually disrupt learning (Rey, 2012). Moreno and Mayer (2000) discuss seductive details in relation to what they call the *coherency principle*—that multimedia presentations should use minimal extraneous words or images.

Finally, efforts to improve motivation and engagement using games seem to have made little contact with an extensive literature in animal and human behavior about explicit rewards in learning (e.g., Hidi, 2015; Schwartz, 1982). The explicit rewards in a game may lead to short-term engagement but may actually undermine subsequent interest, participation, and learning in that domain when no explicit rewards are offered. We know of no studies that have actually examined long-term effects of game-based learning in this regard.

Medical learning may be in something of a privileged position regarding some of these factors. Medical learners are usually self-selected, highly intelligent, and strongly motivated. Couching learning in a story about fighting off space aliens is usually unnecessary. And competition may be less off-putting than it would be to many students in middle school mathematics. Thus, having a game-like flow, levels of progress, and even competition may have useful applications. These factors may increase motivation, attention, or time on task, and, as mentioned, competition almost always incorporates benefits of the testing effect. Not much research has attempted to disentangle separable effects of these potentially relevant factors.

On balance, continued efforts may lead to development of more valuable gamification efforts. Crucial in such developments will be use of objective measures in learning, as well as avoidance of clear pitfalls of extraneous cognitive load. Most of all, efforts to identify and test rigorously specific game-like elements will have the clearest implications for generalizable efforts.

5.5 Perceptual and Adaptive Learning in Medical Domains

In this section, we consider applications of perceptual and adaptive learning in medical domains. The work in this section focuses on our program of research, which has combined perceptual learning and adaptive learning in efforts to accelerate expertise in a variety of medical learning domains. For brevity, we give some of the general aims and characteristics of PALMs, and we elaborate these in describing PALMs in electrocardiography.

This combination has wide application, as much of medical diagnosis is based upon perceiving and interpreting patterns, be they the details of the squiggles in a 12-lead electrocardiogram (ECG), the sounds transmitted via stethoscope of a beating heart, or specific features in patients' descriptions of why they are consulting a physician. Abilities to pick up the relevant features embedded in each of these, and other, clinical recordings and presentations are not readily mastered but require years of experience, extending from medical school through residency and even throughout the early years of clinical practice. The fact that many patients seek out clinicians who have "been around awhile" rather than newly minted physicians is commonplace, but revealing. Recently trained physicians may have up-to-date factual knowledge and training in the latest techniques, but hard-won expertise in pattern recognition is found in more experienced practitioners.

Although most applications have involved classification of complex clinical displays and presentations, we discuss at the end applications of the core concepts to two other areas in medical learning: factual and conceptual learning that may be greatly improved by adaptive learning methods such as ARTS and higher-order pattern recognition in diagnosis integrating across multiple sources of information. Some work indicates that adaptive and perceptual learning methods also have great potential to enhance these aspects of learning.

5.5.1 General Aims of Perceptual-Adaptive Learning Modules (PALMs)

Why are pattern recognition and diagnostic skills at various levels of medical practice so elusive? Why do they take so long to develop? These questions actually have answers. In fact, there are two related answers. First, as we have already mentioned, the relevant mechanisms involved in these attainments are not much engaged by typical declarative and procedural instructional methods. Much of the relevant learning cannot be well verbalized, and seeing a couple of examples in a textbook or lecture does little to advance perceptual learning.

The second, related, answer is that until recently there has not been much in the way of instructional methods that effectively target pattern recognition. The relevant perceptual learning of classifications, distinguishing features, and the seeing of complex relationships develops gradually and implicitly through encounters with many examples, efforts to classify them (in other words, having a task), and receiving feedback (Kellman & Garrigan, 2009). The instances in any learning domain also need to cover the space in important ways. For a set of diagnostic categories in a domain, for example, training must include ample instances that allow learning mechanisms to sort diagnostic properties that determine membership in one category vs. another, and there must be ample exemplars that show irrelevant variation that spans both categories. Incidental or spurious properties are the bane of effective perceptual category learning. Internship, residency, and actual practice do much to advance the relevant skills, but even there, some phenomena appear rarely, some may be incidentally correlated with other characteristics, and the overall flow of encounters across a learning domain is haphazard.

PALMs were developed to address these problems, based on an emerging understanding of mechanisms of perceptual learning, its role in expertise, and the conditions needed to optimize it. The goals include accelerating the growth of expertise and making learning more comprehensive. The adaptive features of PALMs leverage a number of now well-established learning principles, such as testing and spacing, and use objective data both to guide learning and to certify mastery. The available evidence (see below) suggests that the combined approach can accelerate expertise and produce more comprehensive learning in a variety of domains.

Across various domains, PALMs use the same core features. Trials (learning events) in PALMs consist of classification episodes in which one or more displays are presented and learners must make a response, such as classifying a display, comparing displays, or distinguishing some feature or property. Accuracy and response times are used for spacing, to determine progress and eventually mastery, based on both accuracy and fluency. Feedback occurs at the trial level, after blocks of trials, and in terms of levels of progress indicated periodically. Usually, mastery criteria are implemented, such that learning continues until all learning categories have been mastered. Progress tracking spans sessions and allows learns to return and continue learning where they left off. In research investigations of PALMs with real medical learners (medical students, residents, or practitioners), we usually employ designs in which assessments are administered at pre-test and post-test and in a delayed post-test, administered weeks, or sometimes months, later. Active control groups are sometimes used, but comparisons to non-users (business-as-usual control) and comparisons of PALM users to more advanced practitioners (e.g., comparing 3rd-year medical students to 2nd-year residents) have often been a more practical methodology than subjecting a separate group of learners to some other intervention.

We have created and investigated PALMs in the areas of anatomy, cardiology, dermatology, histopathology, obstetrics and gynecology, ophthalmology, radiology, and ultrasound. Details for many of these interventions have been published or are under review (Krasne et al., 2013, 2020; Rimoin et al., 2015; Romito et al., 2016; Slaught et al., 2021).

5.5.2 *Electrocardiography PALMs*

Learning to discriminate pathologies underlying the patterns produced in 12-lead electrocardiograph recordings used in the PALMs relies very heavily on perceptual learning, since there are large variations among individual patients for both normal patterns and in features associated with a given pathology, and relevant features are embedded in varying degrees of "noise." Thus, beyond knowing the names of the categories and the general features associated with each, mastery primarily involves more selective extraction of subtle features and detection of important relations, as well as quicker and more effortless performance. Some of the relevant complexity is illustrated by the electrocardiogram (ECG) traces in Fig. 5.1.

Here is a bit of background for understanding how to view a 12-lead ECG, such as those shown in Fig. 5.1. Each of the 12 panels display 2.5 s traces of electrical signals arising from different combinations of electrodes placed on the body in a specified pattern. The vertical axis of the traces displays the electrical signal in millivolts, and the horizontal axis displays time in seconds. Three electrical signal traces are displayed at a time for 2.5 s before switching to another set of three traces. The labels on the traces are standard, referring to the particular combination of electrodes from which the recording is derived.

ECGs from two different normal patients are shown in Fig. 5.1a, b, whereas Fig. 5.1c, d displays two patients with an acutely occurring type of heart attack

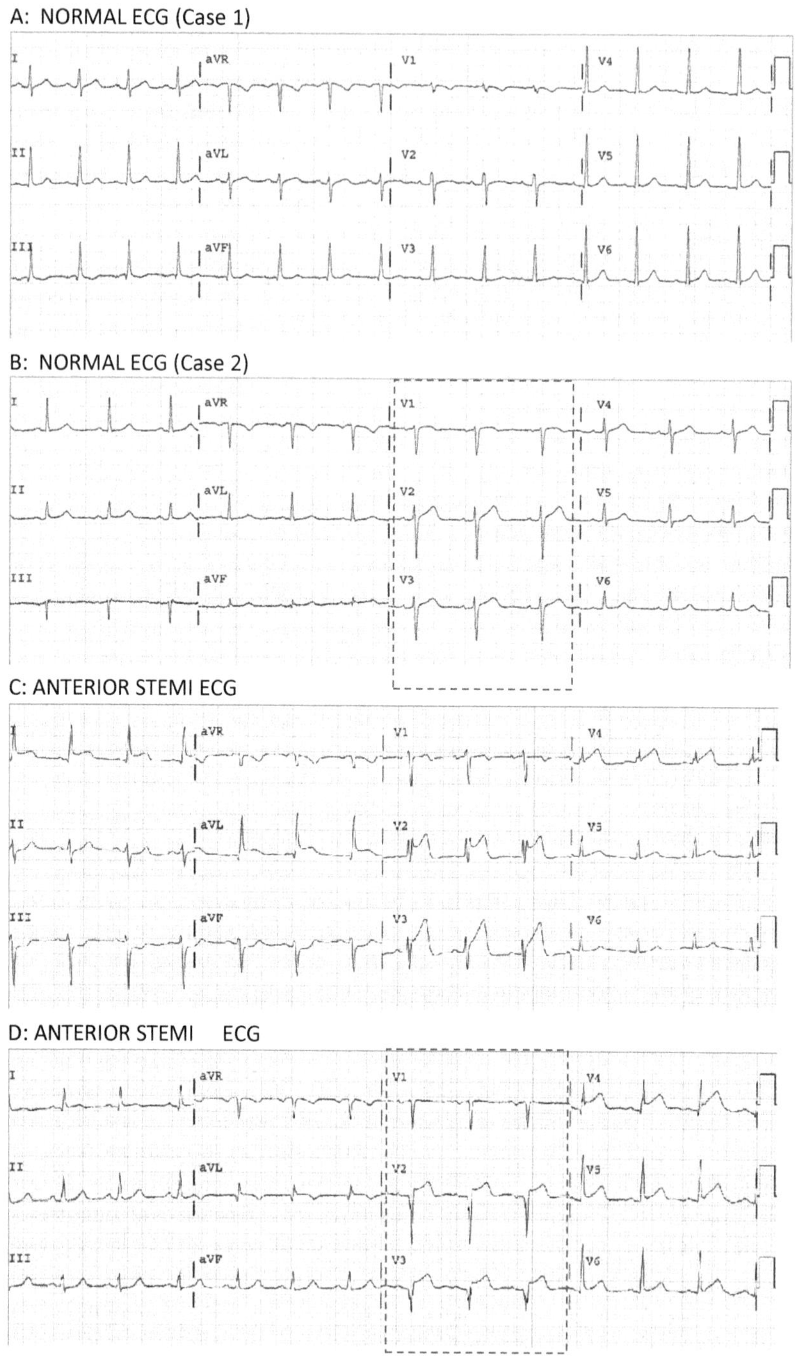

Fig. 5.1 Examples of 12-lead echocardiograms (ECGs), as used in the ECG Morphology PALM. (**a**) and (**b**) show two different normal cases. (**c**) and (**d**) show two different patients with anterior STEMI. Note the complexity of the information displayed and the variation for cases within each

arising in the anterior portion of the left ventricle, referred to as an anterior ST-elevated myocardial infarction (STEMI).

These records illustrate two of the problems inherent in classifying ECGs as well as in other medical classification tasks: Within any given diagnostic category, there is a large range of variation (as can be seen in Fig. 5.1a vs. b and Fig. 5.1c vs. d); and between diagnostic categories, there is a large overlap in appearance (Fig. 5.1b vs. d). Of particular relevance in distinguishing the type of heart attack shown are the traces within the dashed rectangles superimposed on each recording. Although these traces appear quite similar, a key difference in determining that Fig. 5.1d, but not b, represents an acutely occurring heart attack is that the upswing following the large dip in traces V2 and V3 of d overshoots each trace's baseline by at least two of the small horizontal voltage lines prior to bending, whereas those in the equivalent traces of Fig. 5.1b begin bending when they reach the baseline. Diagnoses of a large number of cardiac disorders are made based on small differences such as these or even less obvious ones, occurring in one or more traces, and expertise in picking out such differences requires many years of experience.

Because interpreting ECGs is difficult and critical to making rapid and accurate diagnoses of cardiac problems, we developed PALMs to improve the early mastery of this clinical tool. Most extensively, we studied, over a broad range of educational levels and retention times, outcomes associated with an *ECG Morphology PALM* that focuses on the shapes of the electrical signals, rather than their rhythmicity. We use this specific PALM to illustrate some features of medically related PALMs and their performance. Further detail and all of the data can be found in Krasne et al. (2020).

In our experiments on the ECG Morphology PALM, we evaluated its effectiveness by comparing performance on pre-test assessment prior to PALM training and on a post-test following PALM training. The ECG exemplars used in assessments had not been seen previously and were different for pre- and post-tests. Effects of the PALM intervention on classifying 12-lead ECGs into 15 different diagnostic categories were examined for third- and fourth-year medical students (MS3 and MS4, respectively) and emergency medicine residents in their first, second, and third years (EMR1, EMR2, and EMR3). Individual progress through the PALM was tracked for each individual in terms of accuracy and response times for instances of each diagnostic category. Mastery criteria including accuracy, speed, and success across spaced delays were applied to each diagnostic category, such that learning in the PALM by a given learner continued until mastery was attained for each of the learning categories. The durations of participation required to reach mastery varied by educational level, ranging from 88 $\pm$ 32 min for MS3 students to 46 $\pm$ 24 min for EMR3 residents.

Figure 5.2 shows results of the ECG Morphology PALM for a range of participants from MS3s to EMR3s based on pre-test and post-test performance on an

Fig. 5.1 (continued) category. Becoming able to accurately choose a diagnostic category from these displays requires perceptual learning to select relevant information, detect subtle features, and extract important relations, as well as ignore irrelevant variations (see text)

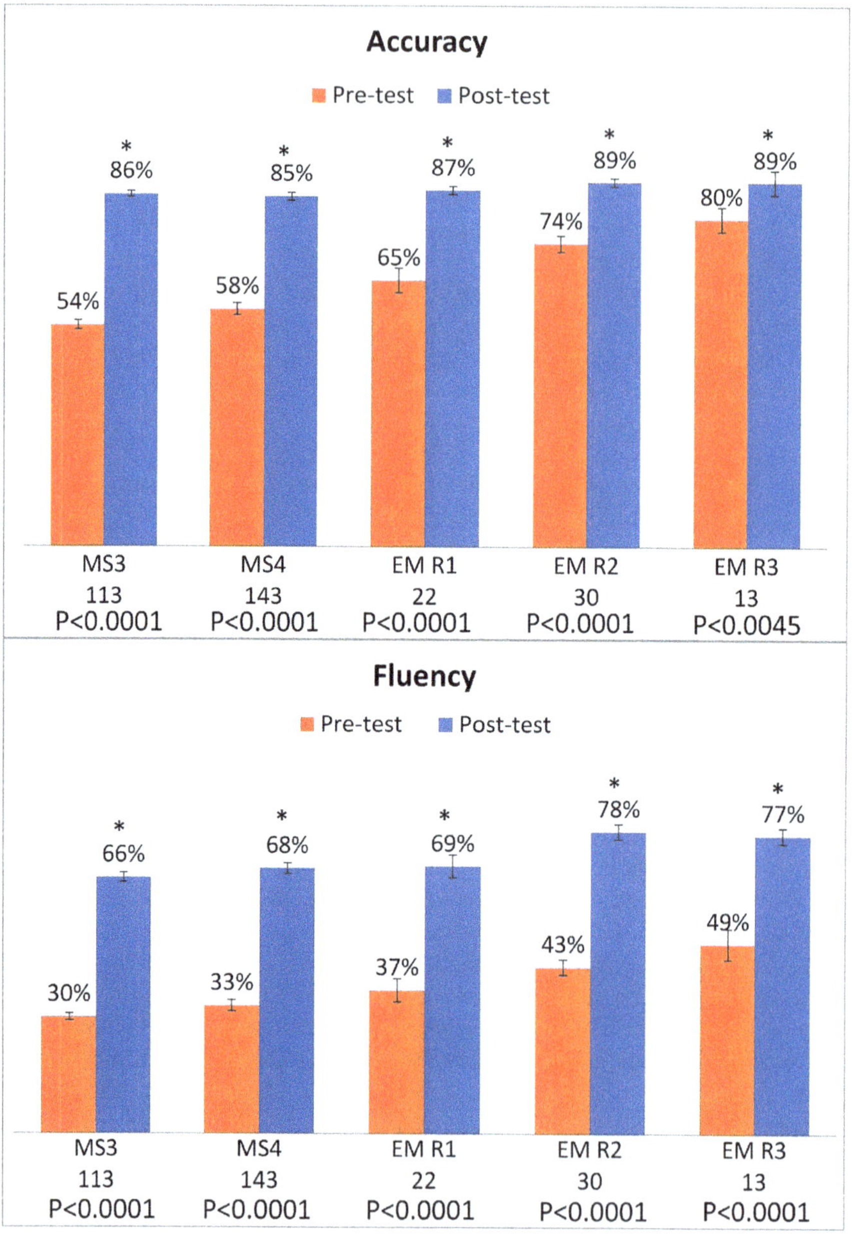

Fig. 5.2 Effect of the ECG Morphology PALM on ECG interpretation skill. *Upper panel*: Percent accurate classification shown as a function of educational level for ECG interpretation before (orange bars or gray in grayscale version) and after (blue bars or black in grayscale version) completing the ECG Morphology PALM for the first time. MS3 and MS4 were medical students in years 3 and 4; EM R1, EM R2, and EM R3 were emergency medicine residents in years 1, 2, and 3. *Lower panel*: Fluency, defined as accurate responding on assessment ECGs with a response time ≤15 s is shown as a function of educational level. For both panels, numbers of participants are

assessment of ECG interpretation. We used two primary measures: *accuracy*, defined as proportion of correct responses made within the 30 s allotted time period for each trial, and *fluency*, which measures the fraction of responses that were both accurate and completed within 15 s. By examining responses made within a shorter target time, the fluency measure aims to capture changes in perceptual extraction of information and progress toward automaticity rather than simply the ability to consciously analyze the ECG.

Initial (pre-test) performance reflected educational level. Performance following the PALM training greatly improved and was much more equalized across the different training levels. Besides the robust improvement for participants at all stages of medical training, the comparisons across educational levels provide evidence that PALM training leads to acceleration of proficiency in these skills relative to conventional training. Note that 3rd- and 4th-year medical students at the end of PALM use achieved levels of performance that exceeded pre-test performance of emergency medicine residents in years 1, 2, and 3. Interpreting ECGs is an important part of emergency medicine training, and residency provides considerable exposure to ECGs and opportunities to develop these skills. The monotonic increase in pre-test scores for EM1 through EM3 suggests that proficiency does increase across years of residency. By comparison, however, only 1–2 h of the ECG Morphology PALM produced much larger gains, such that PALM-trained medical students not only exceeded residents' pre-test accuracy in classifying ECGs, but did not differ much from residents' post-test accuracy scores.

In the fluency data, some of these patterns were even more pronounced. There, 3rd- and 4th-year medical students at post-test substantially outperformed residents in any year who had not undergone PALM training. It is also clear, however, that residents' fluency benefitted greatly from the PALM, with EM2s and EM3s going from 43 and 49% to 78 and 77%, respectively. As with the accuracy data, there was a clear increase in pre-test fluency across years of residency, suggesting that conventional training does improve these skills, but these gains were somewhat subtle (37–49% from EM1 to EM3) compared to the gains from the 1–2 h PALM (which led to gains in fluency of 32, 35, and 28 percentage points for the EM1, EM2, and EM3 groups, respectively). The gains in both fluency and accuracy suggest that this adaptive perceptual learning intervention effectively addressed both discovery and fluency aspects of perceptual learning (Kellman, 2002). The especially large gains in fluency suggest that PALM training may specifically impact aspects of learning that are relatively poorly addressed through conventional instruction or apprenticeship (although it is also possible that fluency gains appear to exceed accuracy gains due to a ceiling effect in the accuracy data).

Fig. 5.2 (continued) shown below the training level on the abscissa. For all groups and measurements, the post-test performance was statistically better than the pre-test performance with $P < 0.0001$ for all pre-/post-test comparisons except for R3 Accuracy, for which $P = 0.0045$

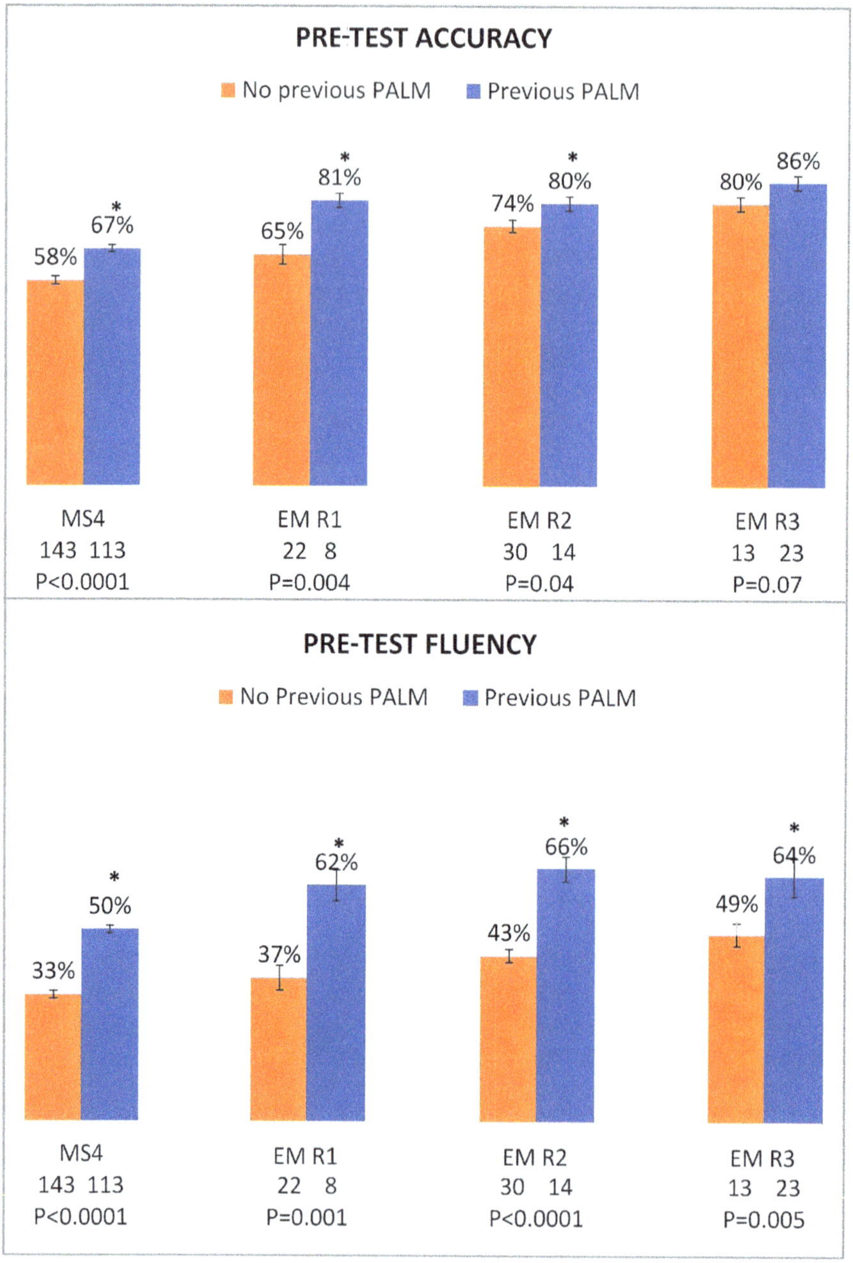

Fig. 5.3 Delayed test effects of ECG Morphology PALM training on ECG assessment performance. Left-hand bars in each pair (orange in color version of the figure; gray in grayscale version) represent trainees at several educational levels who had no previous ECG PALM experience. Right-hand bars in each pair (blue in color version of the figure; black in grayscale version) represent current performance on the same assessment for trainees at the same educational level who had completed the ECG PALM approximately 1 year previously. Numbers of participants are shown

We were able to assess some of the long-term effects of completion of the ECG Morphology PALM by testing a year later participants of equivalent educational level who had or had not previously used the PALM. These data can be seen in Fig. 5.3. (The data are labeled as "pre-test" scores because these learners went on to complete the PALM afterward.) At each level, participants who had completed the 1–2 h PALM a year or so earlier outperformed peers who had not done so. With the exception of the accuracy performance of R3 residents (for which $P = 0.07$), both accuracy and fluency scores for trainees who had completed the PALM approximately 1 year earlier were statistically significantly higher than those for trainees at the same educational level but with no previous ECG PALM training. Effect sizes were large for all fluency comparisons and were medium to large for accuracy comparisons. Another comparison that can be made from the data of Fig. 5.3 is the relative effectiveness of the PALM training versus that of the clinical experience of the emergency medicine residents during the previous year. The accuracy performance of residents completing ECG PALM training 1 year previously is comparable to or greater than that of residents 1 year more advanced who had spent the previous year in the clinic but without PALM training (compare right hand [black or orange] bars for MS4, EMR1, and EMR2 with left hand [gray or blue] bars for EMR1, EMR2, and EMR3, respectively). In comparing fluency performance, the effectiveness of the PALM is even greater than that of the additional year of clinical experience of residents in the next highest year of training. These results indicate substantial lasting gains from PALM training. We know of no other learning interventions that have been shown to produce meaningful and enduring learning gains a year later from a 1–2 h activity.

5.5.3 PALMs in Other Medical Learning Domains

The learning principles and algorithms in PALMs have broad applicability, and they have proven to be useful in many domains, including mathematics and STEM learning (Kellman et al., 2010; Mettler, Massey, et al., 2020; Unuma et al., 2016), chemistry (Mettler, El-Ashmawy, et al., 2020), music (Bufford et al., 2016) and aviation training (Kellman & Kaiser, 1994). As mentioned earlier, we have developed and tested PALMs to improve the learning of difficult perceptual classifications in a variety of other medical learning domains, including histopathology (Krasne et al., 2013), dermatology (Rimoin et al., 2015; Slaught et al., 2021), and trans-

Fig. 5.3 (continued) below the educational level on the abscissa along with P values (two-sided) for assessment performance differences between those with and those without the ECG PALM experience the previous year. For accuracy, effect sizes were $d = 0.7$, 1.7, 0.5, and 0.7, and for fluency, they were $d = 1.3$, 1.9, 1.8, and 0.8 for MS4, R1, R2, and R3 trainees, respectively

esophageal echocardiography (Romito et al., 2016). Uniformly, we have found that PALM training appears to produce large, significant, durable improvements in performance for a variety of clinical tests and presentations. PALMs or elements of PALMS are increasingly being applied by others interested in improving medical learning (e.g., Guerlain et al., 2004; Sha et al., 2020; Perham et al., 2015), have been evaluated in a systematic review (Guegan et al., 2020), and have been the subject of favorable commentaries in medical journals (e.g., Evered, 2018; Waite et al., 2020; Weller, 2016).

In addition, surveys that we implemented early on in conjunction with our first medical PALM applications, those for training basic dermatology features and histopathology processes, demonstrated strong support for PALM implementation in the curriculum, with MS1 students averaging 4.5/5 on a Likert scale evaluating whether there should be comparable modules for other areas of the curriculum and MS2 students averaging 4.6/5 for the same issue (Krasne et al., 2013; Rimoin et al., 2015).

Finally, the gamification features discussed earlier in this chapter were clearly recognized and influenced the positive response to the PALM modules, with students spontaneously including characterizations such as the following:

> . . . really nice...and made it seem like a game, which I enjoyed and wish we had more of.
> . . . lots of examples, game format made it fun
> The derm and histo 'games' were extremely helpful . . .
> . . . it was fun and stress-free, more like a game as opposed to a test.
> The game-type set up made the exercise fun
> —repetition—use of images—"game" format—diagnosis—interactive-ness of the program

5.5.3.1 Factual Learning, Simulation, and Higher-Order Pattern Recognition in PALMs

Although we have emphasized applications to perceptual classifications, PALMs also have considerable potential to enhance other kinds of learning. The adaptive learning algorithms in PALMs are well-suited to learning of factual material, and evidence suggests that the ARTS system in particular offers advantages over other spacing schemes (Mettler et al., 2016). Many aspects of medical learning involve substantial learning of facts, concepts, and procedures, all of which can be enhanced by the testing, spacing, and mastery components of ARTS. Yet another difficult but common learning challenge involves integration of various signs, symptoms, patient characteristics, and test results in medical diagnosis. This can be considered a higher-order pattern recognition problem, which, although it may involve perceptual components, is not primarily perceptual. The formal similarity to perceptual learning, in terms of discovering distinguishing features and important relations, and acquiring fluency, however, has suggested that a PALM approach may be capable of accelerating learning in diagnostic domains.

Some work has tested this hypothesis. Kellman and Mettler (2011) studied PALMs in a US Army project aimed at teaching information and diagnostic skills at different levels. The focus was on chief complaints that might present among military personnel and their families at military installations. Adaptive factual learning modules (FLMs) were developed to teach the basics of specific medical conditions that might lead to a particular chief complaint (e.g., meningitis, systematic viral infection, subarachnoid hemorrhage, migraine, or tension headache for the complaint of headache). Integrated factual learning modules (IFLMs) mixed learning across medical conditions with the aim of sharpening long-term retention and understanding of distinctions. Finally, cognitive task modules (CTMs) presented novel cases that trained both the procedural aspects of treatment and investigatory steps related to different specific conditions as well as differential diagnosis based on sets of characteristics, signs, symptoms, and test results. Modules at all levels used the ARTS system for adaptive learning and mastery. Undergraduate participants with health-related career interests served as participants. Results showed substantial advantages of adaptive learning modules over conventional study materials, and they also showed that the sequence from FLMs to CTMs added considerably to mastery and later diagnostic accuracy. PALMs have also shown great utility in anatomy learning, which is a combination of factual and perceptual learning. Suites of anatomy learning focused on specific body areas or for different purposes (e.g., surgery vs. physical therapy training) have been developed by Insight Learning Technology, Inc. and Primal Pictures, Inc., a leading distributor of high-quality anatomy graphics online.

In another project (Benharash et al., 2015), PALMs were applied to simulation in trauma care. The goal of the project was to develop an integrated simulation and training platform based on perceptual and adaptive learning principles, focused on training combat medics in the factual and conceptual underpinnings, diagnosis, and emergency treatment in hemorrhage control. The trauma care simulator displayed vital signs of a patient, presented other information, and could present queries or cases. Training interleaved several levels of interactive learning. Recognition trials trained learners to fluently recognize important indications, such as a vital sign out of normal limits. Some examples included blood pressure, respiration rate, and cardiac arrhythmias. Diagnostic trials involved higher-order pattern recognition from combined simulator and case information inputs. Examples included hypovolemic shock, neurogenic shock, tension pneumothorax, cardiac tamponade, and traumatic brain injury. Treatment trials involved administration of medications, electric shock, airway interventions, etc. and assessment of their consequences in the simulator. Some examples included pressure or tourniquet for hemorrhage, shock for ventricular tachyarrhythmias, morphine administration, and resolution of airway issues through intubation, O_2 administration, or cricoidotomy. Finally, sets of factual learning points were included to teach the MARCH algorithm, which gives an order of priority for initial assessment and treatment for trauma victims (M, massive bleeding; A, airway; R, respirations; C, circulation; and H, head).

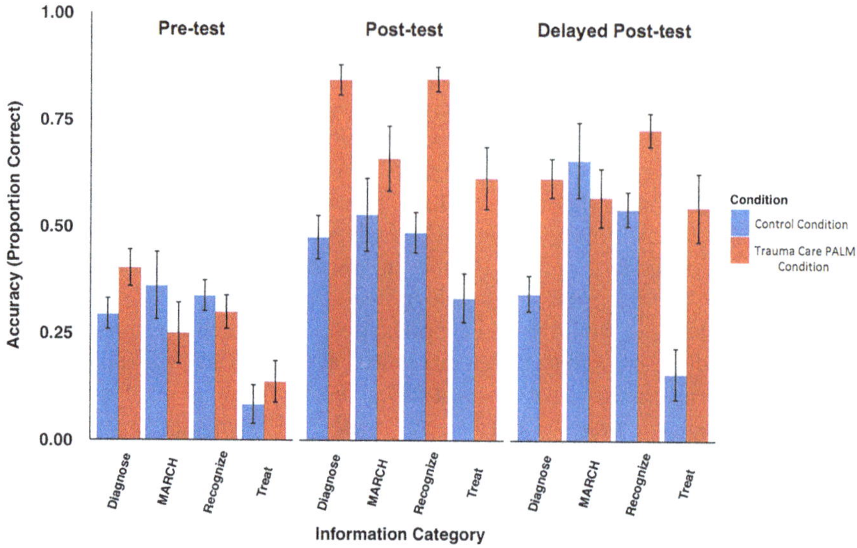

Fig. 5.4 Assessment data by Trial Type and Assessment Phase for trauma care training. Blue bars (black in grayscale version) in each pair show control group data. Red bars (gray in grayscale version) show data from the Trauma Care Simulation PALM. Pre-test, post-test, and delayed post-test data are shown for the four types of learning trials: diagnostic trials, MARCH algorithm trials, recognition trials, and treatment trials. Error bars show ± standard error of the mean

One group learned entirely in the Trauma Care Simulator PALM, and a control group learned from actual study materials used in basic military medical training. The learning content was the same in both groups.

Assessments were given half within the simulator environment and half as paper and pencil tests. The Trauma Care Simulation PALM produced substantially better accuracy in performance on both types of tests, both in immediate post-tests and in a delayed post-test 1 week later. Figure 5.4 shows the patterns of data for the simulation-based assessments. The four learning areas (diagnosis, MARCH algorithm, recognition, and treatment) are shown from left to right for the pre-test, post-test, and delayed post-test, given 1 week later. There are conspicuous advantages for the PALM group in diagnostic accuracy, recognition, and selection of treatment actions. These are evident at immediate post-test and also after a delay. For the MARCH algorithm, the group trained using simulation and adaptive learning showed a trend toward better learning at immediate post-test, especially given an advantage for the control group at pre-test, but the difference in learning gains was not different between groups at delayed post-test. The treatment category appeared to be the hardest to learn, in terms of immediate post-test scores, but what was learned declined very little between immediate and delayed post-tests for the simulation group, especially in comparison with the control group. This might indicate greater incorporation of pattern recognition components when the learning occurred in the PALM format. Oddly, the PALM group outperformed the control group on MARCH algorithm content, which might be considered the most purely declarative and procedural learning content in the study, when tested in paper and pencil format (as well as in all of the other learning categories).

These results suggest potential for much broader application of perceptual and adaptive learning technologies in both factual learning in medicine and in more complex settings, including simulation, integration of information in diagnosis, and the application and assessment of treatments.

5.6 Conclusion

Instruction and learning in medical domains can be improved by application of findings from the science of learning and by advances in technology. For new approaches or enhancements to be effective requires fruitful combinations of these. We have considered a number of current and emerging ideas with already realized or potential value in medical education. Scientific research and technological implementations involving perceptual learning add value both by broadening our conceptions of what learning includes and how it can be systematically advanced. Advances in digital technology and principles of learning make possible adaptive methods that can personalize learning and make it more efficient, as well as track progress and ensure more comprehensive mastery. Efforts in these areas, basic and applied, are also expanding ideas of mastery to include fluency, automaticity, and transfer to new situations. Adaptive learning technologies have powerful applications when combined with perceptual learning, as we have illustrated, and they also have compelling potential to improve factual and procedural learning. Games and gamification may be best considered in terms of useful elements; there is no magic for learning that accrues just because something is a game. Specific components of gaming may have value in learning, especially those that have been or could be subjected to experimental work using objective measures of learning that can test their effectiveness and reveal best practices. Promising elements include progress tracking, feedback, and potential effects on motivation, whereas challenges include avoidance of distracting seductive details, extraneous cognitive load, and fostering intrinsic motivation that might persist outside of the gamified learning environment. The continuing emergence of findings from scientific research on learning and new technologies in which they can be instantiated promises a bright future for evidence-based advances in instruction, but these will also require continuing, committed, and collaborative efforts from domain experts and learning scientists to devise, test, and implement new methods that truly add value.

References

Abbott, E. (1909). On the analysis of the factors of recall in the learning process. *Psychological Monographs: General and Applied, 11*(1), 159–177.

Atkinson, R. C. (1976). Adaptive instructional systems: Some attempts to optimize the learning process. In D. Klahr (Ed.), *Cognition and instruction*. Wiley.

Benharash, P., Older, R., Mettler, E., Genovese, B., Frank, P., Dutson, E., & Kellman, P. (2015). Perceptual and adaptive learning and simulation in training combat medics in trauma care. *UCLA human perception laboratory technical report to ONR cognitive science of learning program review*, Sept., 2016.

Benzil, D. L. (2018). See one, do one, teach one. *AANS Neurosurgeon, 27*(3).

Bjork, E. L., & Bjork, R. A. (2011). Making things hard on yourself, but in a good way: Creating desirable difficulties to enhance learning. *Psychology and the Real World: Essays Illustrating Fundamental Contributions to Society, 2*, 59–68.

Bjork, R. A. (1975). Retrieval as a memory modifier. In R. Solso (Ed.), *Information processing and cognition: The Loyola Symposium* (pp. 123–144). Lawrence Erlbaum.

Bjork, R. A., & Bjork, E. L. (1992). A new theory of disuse and an old theory of stimulus fluctuation. *From Learning Processes to Cognitive Processes: Essays in Honor of William K. Estes, 2*, 35–67.

Bjork, R. A., & Bjork, E. L. (2006). Optimizing treatment and instruction: Implications of a new theory of disuse. In L.-G. Nilsson & N. Ohta (Eds.), *Memory and society: Psychological perspectives* (pp. 109–133). Psychology Press.

Bondie, R. S., Dahnke, C., & Zusho, A. (2019). How does changing "one-size-fits-all" to differentiated instruction affect teaching? *Review of Research in Education, 43*, 336–362. https://doi.org/10.3102/0091732X18821130

Bransford, J. D., Brown, A. L., & Cocking, R. R. (1999). *How people learn: Brain, mind, experience, and school.* National Academies Press.

Bryan, W. L., & Harter, N. (1899). Studies on the telegraphic language: The acquisition of a hierarchy of habits. *Psychological Review, 6*, 345–375.

Bufford, C. A., Thai, K. P., Ho, J., Xiong, C., Hines, C. A., & Kellman, P. J. (2016). Perceptual learning of abstract musical patterns: Recognizing composer style. In T. Zanto, T. Fujioka, P. Janata, J. Johnson, J. Berger, J. Slater, & C. Chafe (Eds.), *Proceedings of the 14th annual international conference for music perception and cognition* (pp. 8–12).

Carpenter, S. K., Cepeda, N. J., Rohrer, D., Kang, S. H., & Pashler, H. (2012). Using spacing to enhance diverse forms of learning: Review of recent research and implications for instruction. *Educational Psychology Review, 24*(3), 369–378.

Carrier, M., & Pashler, H. (1992). The influence of retrieval on retention. *Memory & Cognition, 20*(6), 632–642. https://doi.org/10.3758/bf03202713. PMID 1435266. S2CID 15893469.

Chandler, P., & Sweller, J. (1991). Cognitive load theory and the format of instruction. *Cognition and Instruction, 8*(4), 293–332.

De Groot, A. D. (1965). *Thought and choice in chess*. Mouton.

Deterding, S., Dixon, D., Khaled, R., & Nacke, L. (2011). From game design elements to gamefulness: Defining "gamification." In *Proceedings of the 15th international academic MindTrek conference: Envisioning future media environments* (pp. 9–15).

Engeström, Y., Miettinen, R., & Punamäki, R. (Eds.). (1999). *Perspectives on activity theory*. Cambridge University Press.

Evered, A. (2018). Perceptual and adaptive learning modules and their potential to transform cytology training programmes. *Cytopathology, 29*(4), 371–374.

Gibson, E. J. (1969). *Principles of perceptual learning and development*. Appleton-Century-Crofts.

Gibson, J. J. (1966). *The senses considered as perceptual systems*. Houghton-Mifflin.

Glenberg, A. M. (1979). Component-levels theory of the effects of spacing of repetitions on recall and recognition. *Memory & Cognition, 7*(2), 95–112.

Goldstone, R. L. (2000). Unitization during category learning. *Journal of Experimental Psychology: Human Perception and Performance, 26*, 86–112.

Griff, E. R., & Matter, S. F. (2013). Evaluation of an adaptive online learning system. *British Journal of Educational Technology, 44*(1), 170–176.

Guegan, S., Steichen, O., & Soriace, A. (2020). Literature review of perceptual learning modules in medical education: What can we conclude regarding dermatology? *Annales de Dermatologie et de Vénéréologie*. Available online 9 July 2020. https://doi.org/10.1016/j.annder.2020.01.023

Guerlain, S., Green, K. B., LaFollette, M., Mersch, T. C., Mitchell, B. A., Poole, G. R., Calland, J. F., Lv J.,, & Chekan, E. G. (2004). Improving surgical pattern recognition through repetitive viewing of video clips. *IEEE Transactions on Systems, Man, and Cybernetics-Part A: Systems and Humans, 34*(6), 699–707.

Hamari, J., Koivisto, J., & Sarsa, H. (2014). Does gamification work?—A literature review of empirical studies on gamification. In *2014 47th Hawaii international conference on system sciences* (pp. 3025–3034). IEEE.

Hamnvik, O. R., Alford, D. P., Ryan, C. T., Hardesty, I. T., & Drazen, J. M. (2019). NEJM knowledge+ pain management and opioids—A new adaptive learning module. *New England Journal of Medicine, 380*, 1576–1577.

Harp, S. F., & Mayer, R. E. (1998). How seductive details do their damage: A theory of cognitive interest in science learning. *Journal of Educational Psychology, 90*(3), 414.

Healy, M. G., Phitayakorn, R., Axelsson, C. G., O'Rourke, M., Duhigg, L., & Gray, B. A. (2019, February 28). *Novel adaptive e-learning product helping general internists pass their ABIM IM-MOC exam*. Presented at the Society for Academic Continuing Medical Education annual meeting, Charleston, SC.

Hidi, S. (2015). Revisiting the role of rewards in motivation and learning: Implications of neuroscientific research. *Educational Psychology Review, 28*(1), 61–93. https://doi.org/10.1007/s10648-015-9307-5

Johnston, W. A., & Uhl, C. N. (1976). The contributions of encoding effort and variability to the spacing effect on free recall. *Journal of Experimental Psychology: Learning, Memory, and Cognition, 2*, 153–160.

Kang, S. H. K., & Pashler, H. (2012). Learning painting styles: Spacing is advantageous when it promotes discriminative contrast. *Applied Cognitive Psychology, 26*(1), 97–103.

Karpicke, J. D., & Roediger, H. L., III. (2007). Expanding retrieval practice promotes short-term retention, but equally spaced retrieval enhances long-term retention. *Journal of Experimental Psychology: Learning, Memory, and Cognition, 33*, 704–719.

Kellman, P. J. (2002). Perceptual learning. In H. Pashler & R. Gallistel (Eds.), *Stevens' handbook of experimental psychology: Learning, motivation, and emotion*. Wiley.

Kellman, P. J. (2006). *System and method for adaptive learning*. Inventor: P. Kellman (US PTO #7052277, awarded 4/2006).

Kellman, P. J., & Garrigan, P. (2009). Perceptual learning and human expertise. *Physics of Life Reviews, 6*, 53–84.

Kellman, P. J., & Kaiser, M. K. (1994). Perceptual learning modules in flight training. *Proceedings of the Human Factors and Ergonomics Society Annual Meeting, 38*, 1183–1187.

Kellman, P. J., & Krasne, S. (2018). Accelerating expertise: Perceptual and adaptive learning technology in medical learning. *Medical Teacher, 40*(8), 797–802.

Kellman, P. J., & Massey, C. M. (2013). Perceptual learning, cognition, and expertise. In *Psychology of learning and motivation* (Vol. 58, pp. 117–165). Academic Press.

Kellman, P. J., Massey, C. M., & Son, J. (2010). Perceptual learning modules in mathematics: Enhancing students' pattern recognition, structure extraction, and fluency. *Topics in Cognitive Science (Special Issue on Perceptual Learning), 2*(2), 285–305.

Kellman, P. J., & Mettler, E. M. (2011). *ACAT technology for optimal sick call performance*. UCLA Human Perception Laboratory Technical Report, US Army RDECOM Contract No: W91CRB-09-C-0093.

Kornell, N., & Bjork, R. A. (2008). Learning concepts and categories: Is spacing the "enemy of induction"? *Psychological Science, 19*(6), 585–592.

Kornell, N., & Hausmann, H. (2017). Performance bias: Why judgments of learning are not affected by learning. *Memory and Cognition, 45*, 1270–1280.

Krasne, S., Hillman, J. D., Kellman, P. J., & Drake, T. A. (2013). Applying perceptual and adaptive learning techniques for teaching introductory histopathology. *Journal of Pathology Informatics, 4*, 34. https://doi.org/10.4103/2153-3539.123991

Krasne, S., Stevens, C. D., Kellman, P. J., & Niemann, J. T. (2020). Mastering electrocardiogram interpretation skills through a perceptual and adaptive learning module. *Academic Emergency Medicine Education and Training, 5*(2), *e10454*. https://doi.org/10.1002/aet2.10454. eCollection 2021 Apr. PMID: 33796803 1–11.

Landauer, T. K., & Bjork, R. A. (1978). Optimum rehearsal patterns and name learning. In M. M. Gruneberg et al. (Eds.), *Practical aspects of memory* (pp. 625–632). Academic Press.

Larsen, D. P., Butler, A. C., & Roediger, H. L., III. (2013). Comparative effects of test-enhanced learning and self-explanation on long-term retention. *Medical Education, 47*(7), 674–682.

Lave, J., & Wenger, E. (1991). *Situated learning: Legitimate peripheral participation*. Cambridge University Press.

Lee, T. D., & Genovese, E. D. (1988). Distribution of practice in motor skill acquisition: Learning and performance effects reconsidered. *Research Quarterly for Exercise and Sport, 59*, 277–287.

McMahon, G. T., & Drazen, J. M. (2014). Introducing NEJM Knowledge+ and its adaptive personalized learning. *New England Journal of Medicine, 370*(17), 1648–1649.

Mettler, E., El-Ashmawy, A. K., Massey, C. M., & Kellman, P. J. (2020). Adaptive vs. fixed spacing of learning items: Evidence from studies of learning and transfer in chemistry education. In *Proceedings of the 42nd annual conference of the Cognitive Science Society* (pp. 1598–1604). Cognitive Science Society.

Mettler, E., & Kellman, P. J. (2014). Adaptive response-time-based category sequencing in perceptual learning. *Vision Research, 99*, 111–123.

Mettler, E., Massey, C., Burke, T., & Kellman, P. J. (2020). Comparing adaptive and random spacing schedules during learning to mastery criteria. In *Proceedings of the 42nd annual conference of the Cognitive Science Society* (pp. 773–779). Cognitive Science Society.

Mettler, E., Massey, C., & Kellman, P. J. (2011). Improving adaptive learning technology through the use of response times. In L. Carlson, C. Holscher, & T. Shipley (Eds.), *Proceedings of the 33rd annual conference of the Cognitive Science Society* (pp. 2532–2537). Cognitive Science Society.

Mettler, E., Massey, C. M., & Kellman, P. J. (2016). A comparison of adaptive and fixed schedules of practice. *Journal of Experimental Psychology: General, 145*(7), 897–917.

Moreno, R., & Mayer, R. (2000). A coherence effect in multimedia learning: The case for minimizing irrelevant sounds in the design of multimedia instructional messages. *Journal of Educational Psychology, 92*(1), 117–125.

Nevin, C. R., Westfall, A. O., Rodriguez, J. M., Dempsey, D. M., Cherrington, A., Roy, B., Patel, M., & Willig, J. H. (2014). Gamification as a tool for enhancing graduate medical education. *Postgraduate Medical Journal, 90*(1070), 685–693.

Ng, E., & Bereiter, C. (1991). Three levels of goal orientation in learning. *Journal of the Learning Sciences, 1*(3,4), 243–271.

Pashler, H., Bain, P. M., Bottge, B. A., Graesser, A., Koedinger, K., McDaniel, M., & Metcalfe, J. (2007). Organizing instruction and study to improve student learning. IES practice guide. *NCER 2007–2004*. National Center for Education Research.

Perham, N., Evered, A., Walker, D., & Watt, A. (2015). Visual perceptual learning in cytopathology. In T. Heinen (Ed.), *Advances in visual perception* (pp. 203–232). Nova Science.

Pyc, M. A., & Rawson, K. A. (2009). Testing the retrieval effort hypothesis: Does greater difficulty correctly recalling information lead to higher levels of memory? *Journal of Memory and Language, 60*, 437–447.

Rey, G. D. (2012). A review of research and a meta-analysis of the seductive detail effect. *Educational Research Review, 7*(3), 216–237.

Rimoin, L., Altieri, L., Craft, N., Krasne, S., & Kellman, P. J. (2015). Training pattern recognition of skin lesion morphology, configuration, and distribution. *Journal of the American Academy of Dermatology, 72*(3), 489–495.

Roediger, H. L., III, & Butler, A. C. (2011). The critical role of retrieval practice in long-term retention. *Trends in Cognitive Sciences, 15(1)*, 20–27.

Roediger, H. L., III, & Karpicke, J. D. (2006). The power of testing memory: Basic research and implications for educational practice. *Perspectives on Psychological Science, 1*, 181–210.

Rogoff, B. (1994). Developing understanding of the idea of communities of learners. *Mind, Culture and Activity, 1*(4), 209–229. https://doi.org/10.1080/10749039409524673

Romito, B. T., Krasne, S., Kellman, P. J., & Dhillon, A. (2016). The impact of a perceptual and adaptive learning module on transoesophageal echocardiography interpretation by anaesthesiology residents. *British Journal of Anaesthesia, 117*(4), 477–481.

Rutledge, C., Walsh, C. M., Swinger, N., Auerbach, M., Castro, D., Dewan, M., Khattab, M., Rake, A., Harwayne-Gidansky, I., Raymond, T., Maa, T., & Chang, T. (2018). Gamification in action: Theoretical and practical considerations for medical educators. *Academic Medicine, 93*(7), 1014–1020.

Sailer, M., & Homner, L. (2020). The gamification of learning: A meta-analysis. *Educational Psychology Review, 32(1)*, 77–112. https://doi.org/10.1007/s10648-019-09498-w

Schmidt, R. A., & Bjork, R. A. (1992). New conceptualizations of practice: Common principles in three paradigms suggest new concepts for training. *Perspectives on Psychological Science, 13*(2), 146–148.

Schwartz, B. (1982). Reinforcement-induced behavioral stereotypy: How not to teach people to discover rules. *Journal of Experimental Psychology: General, 111*(1), 23–59.

Sha, L. Z., Toh, Y. N., Remington, R. W., & Jiang, Y. (2020). Perceptual learning in the identification of lung cancer in chest radiographs. *Cognitive Research: Principles and Implications, 5*(4).

Sharma, N., Doherty, I., & Dong, C. (2017). Adaptive learning in medical education: The final piece of technology enhanced learning? *The Ulster Medical Journal, 86*(3), 198.

Slaught, C., Madu, P., Chang, A., Williams, V., Kebaetse, M., Nkomazana, O., Molefe-Baikai, J., Bekele, N., Omech, B., Kellman, P., Krasne, S., & Kovarik, C. (2021). Novel education modules addressing the underrepresentation of skin of color in dermatology training. *Journal of Cutaneous Medicine & Surgery*. Advanced online publication. https://doi.org/10.1177/12034754211035093

Spector, J. M. (2014). Conceptualizing the emerging field of smart learning environments. *Smart Learning Environments, 1*(1), 1–10.

Storm, B. C., Bjork, R. A., & Storm, J. C. (2010). Optimizing retrieval as a learning event: When and why expanding retrieval practice enhances long-term retention. *Memory & Cognition, 38*, 244–253.

Thios, S. J., & D'Agostino, P. R. (1976). Effects of repetition as a function of study-phase retrieval. *Journal of Verbal Learning & Verbal Behavior, 15*(5), 529–536.

Tulving, E. (1967). The effects of presentation and recall of material in free-recall learning. *Journal of Verbal Learning and Verbal Behavior, 6*, 175–184.

Unuma, H., Hasegawa, H., & Kellman, P. J. (2016). Perceptual learning facilitates precise mental representations of fractions. *The Journal of Kawamura Gakuen Women's University, 27*(1), 35–49.

Van Merriënboer, J. J. G., & Sweller, J. (2009). Cognitive load theory in health professional education: Design principles and strategies. *Medical Education, 44*(85-93), 16.

Van Nuland, S. E., Roach, V. A., Wilson, T. D., & Belliveau, D. J. (2015). Head to head: The role of academic competition in undergraduate anatomical education. *Anatomical Sciences Education, 8*(5), 404–412.

Waite, S., Farooq, Z., Grigorian, A., Sistrom, C., Kolla, S., Mancuso, A., Martinez-Conde, S., Alexander, R. G., Kantor, A., & Macknik, S. L. (2020). A review of perceptual expertise in radiology: How it develops, how we can test it, and why humans still matter in the era of artificial intelligence. *Academic Radiology, 27*(1), 26–38.

Weller, J. M. (2016). New approaches to the development of expertise in anaesthesia. *British Journal of Anaesthesia, 117*(5), 545–546. https://doi.org/10.1093/bja/aew329

Wiseheart, M., D'Souza, A. A., & Chae, J. (2017). Lack of spacing effects during piano learning. *PLoS One, 12*(8), e0182986.

Wittgenstein, L. (1953). *Philosophical investigations*. Blackwell.

Yardley, S., Teunissen, P. W., & Dornan, T. (2012). Experiential learning: Transforming theory into practice. *Medical Teacher, 34*(2), 161–164. https://doi.org/10.3109/0142159X.2012.643264

Young, M. F., Slota, S. T., Cutter, A., Jalette, G. C., Mullin, G., Lai, B., Simeoni, Z., Tran, M., & Yukhymenko, M. (2012). Our princess is in another castle: A review of trends in serious gaming for education. *Review of Educational Research, 82*(1), 61–89. https://doi.org/10.3102/0034654312436980

Chapter 6
The Flipped Classroom: A Guide to Making Evidence-Based Decisions About Implementation

Michael W. Lee and Andrew C. Butler

Abstract Flipped classroom is a loosely defined pedagogical method that is gaining popularity in health and life sciences education. The shift of lecture-based material to outside the classroom frees up time in class for active learning that facilitates the construction of knowledge and development of understanding. However, there is a great degree of heterogeneity in how flipped classrooms are implemented in terms of the format and structure of pre-class activities and in-class activities. Pre-class knowledge is often acquired by students via media such as videos and knowledge application in-class is structured around active learning activities. Here we examine and synthesize the literature in an effort to distill out best practices for implementing the flipped classroom model that will enable teachers to make decisions about how to design the most effective experience for their particular students and classroom context. Covered are the merits of technology in flipped learning, the role of student preparation, and how to best leverage assessment. Also explored is the impact that flipped classrooms have on student engagement, motivation, self-efficacy, and critical thinking. The conclusion offers a model that describes and connects some of the key features that teachers and students need to consider to gain the most from this teaching modality.

6.1 Introduction

The concept of the flipped classroom has grown in popularity over the past 20 years from an obscure model introduced in conference papers to a dominant pedagogical approach that transcends academic disciplines and levels of education. Interest in the

M. W. Lee (✉)
Department of Medical Education, Geisel School of Medicine, Dartmouth College, Hanover, NH, USA
e-mail: Michael.W.Lee@dartmouth.edu

A. C. Butler
Department of Education and Department of Psychological & Brain Sciences, Washington University in St. Louis, St. Louis, MO, USA

H. J. Witchel, M. W. Lee (eds.), *Technologies in Biomedical and Life Sciences Education*, Methods in Physiology, https://doi.org/10.1007/978-3-030-95633-2_6

flipped classroom model of teaching can be linked to the growing recognition among faculty that "transmissionist" teaching methods are much less effective than teaching methods that promote active learning and resulting paradigm shift from teacher-centered to student-centered learning (Ambrose et al., 2010; Bonwell & Eison, 1991; Braxton et al., 2000; Freeman et al., 2014; National Academies of Sciences, Engineering, and Medicine, 2018; National Research Council, 2000). The watershed moment for this progressive constructivist teaching method might be the publication of *Flip Your Classroom: Reach Every Student in Every Class Every Day* by Aaron Sams and Jonathan Bergmann (Sams & Bergmann, 2012; Talbert, 2017), which helped to introduce it to broader audience. Regardless of the specific tipping point, the term "flipping the classroom" is now firmly established in the educational lexicon.

Despite the popularity of the flipped classroom model, there is still much debate about its definition. Numerous attempts have been made to identify the critical features that define the flipped classroom. For example, Talbert (2017) defined flipped learning as "a pedagogical approach in which first contact with new concepts moves from the group learning space to the individual learning space in the form of structured activity, and the resulting group space is transformed into a dynamic, interactive learning environment where the educator guides students as they apply concepts and engage creatively in the subject matter" (p. 20) (Talbert, 2017). By contrast, DeLozier and Rhodes (2017) described flipped learning as being "characterized by course structure: instructional content (i.e., prerecorded class lectures) is assigned as homework before coming to class and in-class time is then spent working on problems, advancing concepts, and engaging in collaborative learning" (p. 141). One reason that the flipped classroom model remains somewhat amorphous is that it has been implemented in so many different ways (Talbert, 2017). Indeed, even the term "flipped classroom" is not consistently used by researchers and teachers, who have alternated used other terms such as "inverted classroom" and "reverse instruction" to refer to the same basic model.

Resolving these definitional issues is beyond the scope of this chapter, so we have opted to use the conceptualization of the flipped classroom model put forth by Abeysekera and Dawson (2015). They identify three key aspects of the model: (1) most information transmission occurs outside the classroom, (2) time in the classroom is devoted to active and social learning, and (3) assessments are completed before and after class. Although some researchers have suggested that the use of technology to support activities outside of class is critical to constraining the definition (Bernard, 2015; Bishop & Verleger, 2013), we purposefully do not take a firm stance in this particular debate. The argument for needing to constrain the definition is strong, but it is unclear that the inclusion of technology use in the definition is the best way to accomplish this goal. Our rationale is that there is much agreement about the three key aspects of the model laid out above (Abeysekera & Dawson, 2015) and technology is not necessary to accomplish them. That said, as we discuss below, technology is quite useful in accomplishing them and might thus be better thought of as a moderator variable that enhances the effectiveness of the model.

The purpose of this chapter is to provide a resource for teachers who want to implement the flipped classroom model to guide their decision-making. In the following sections, we first delve into the theoretical and empirical rationale for flipping the classroom by describing how the science of learning supports the model. Next, we evaluate the evidence that exists to support the effectiveness of flipping the classroom when compared with the traditional lecture model while also exploring the generalizability of the findings across variables such as student characteristics and educational outcomes. After establishing the efficacy of the flipped classroom model, we go through the major decisions that teachers must make when implementing it and discuss the benefits and costs of different options. Finally, we grapple with some challenges and additional considerations in flipping the classroom before offering a conclusion and some future directions.

6.2 The Theoretical and Empirical Rationale for the Flipped Classroom

In this section, we review theory and empirical findings from the science of learning that support the potential efficacy of the flipped classroom. At the core of the model is the radical (at least by the standards of academia) shift from a teacher-centered, transmissionist approach to a student-centered, constructivist approach. Taking a student-centered, constructivist approach leads to numerous important changes to the way students experience a course. Faculty must redesign their course, which often results in better defined learning objectives (Wiggins & McTighe, 2005), the implementation of structure to scaffold student learning (Reiser, 2004), the use of dynamic activities that promote deeper learning (Chi, 2009), and increased opportunities for formative and summative assessment (Black & Wiliam, 1998; Sams & Bergmann, 2012; Talbert, 2017). As a result of this redesign, students are more likely to keep up with material outside of class (Mawhinney et al., 1971) and to engage in active learning during class time. Active learning is a process by which learners construct knowledge rather than passively receiving it, thereby facilitating self-assessment of understanding and a sense of agency (National Research Council, 2000). A wealth of evidence supports the conclusion that increasing the amount of active learning in a course improves student learning and academic performance (Freeman et al., 2014; Richardson et al., 2012). By contrast, much of what is transmitted to students during a traditional lecture is forgotten relatively quickly (Custers, 2010; Ling et al., 2008). Beyond the general shift to active learning, the flipped classroom model would be predicted to improve student learning in several ways.

First, the flipped classroom provides students with greater control over the initial process of knowledge acquisition by shifting to self-paced engagement with instructional material outside of class. Providing students control of the pace of flipped classroom pre-work (controlling video or reading speed, rewinding/fast forwarding, etc.) reduces extraneous cognitive load relative to listening to a lecture (Clark et al., 2006). Cognitive load relates to the amount of information a learner is working with at any given time. Given the limited capacity of working memory, too much cognitive load can hinder the learner's ability to process the information with negative effects for understanding and retention. Indeed, building in additional structure with the pre-work such as quiz questions or reflection likely prompts further reductions in cognitive load (Haak et al., 2011; Wilton et al., 2019). Finally, the greater control given to students allows them to choose how much pre-work they cover in line with their prior knowledge.

Second, the in-class learning activities enable students to work with the new knowledge that they have acquired from the pre-work and develop a greater understanding. Chi and Wylie (2014) provide a useful framework for conceptualizing the types of cognitive processing induced by different learning activities and how they relate to learning outcomes. The ICAP framework distinguishes among four different types of processing: *passive* processing refers to receiving information by paying attention (e.g., listening to a lecture); *active* processing involves manipulating information without adding new information (e.g., taking verbatim notes); *constructive* processing requires students to integrate new information with existing knowledge to construct deeper understanding (e.g., generating an explanation); and interactive processing refers to engaging with information through collaboration with another student. Research suggests that as learners move from passive to active to constructive to interactive, the quantity, quality, and durability of their learning will improve (Chi, 2009; Chi & Wylie, 2014). Mapping this framework on to the shift from traditional lecture to the flipped classroom, learners are moving from engaging in passive and active processing to constructive and interactive, which should yield substantial benefits. More generally, there is much evidence to support the conclusion that once students have acquired some initial knowledge, it is better for them to engage in activities that require them to retrieve and work with that knowledge (Dunlosky et al., 2013; Larsen et al., 2013; Roediger & Butler, 2011).

Third, when students engage in learning activities inside and outside the classroom, it creates the opportunity for copious amounts of feedback. That is, even though these activities facilitate learning, they also serve a formative (and potentially summative) assessment purpose (Black & Wiliam, 1998). Feedback is critical to student learning because it enables the self-monitoring of current understanding and provides actionable information to close any existing gaps (Bangert-Drowns et al., 1991; Hattie & Timperley, 2007). Even more important than the feedback provided to the student is the feedback provided to the teacher (Hattie, 2009). In contrast with the traditional lecture model, which offers minimal opportunities for teachers to check student understanding, the flipped classroom model provides teachers with an accurate sense of how learning is progressing. This formative feedback to teachers is critical to enabling them to modify or differentiate instruction, address misunderstandings, and provide appropriate challenge for learners. For example, such feedback has been used to identify pre-work misconceptions which can guide in-class discussion and feedback sessions in tandem with application exercises (Steinel et al., 2019). To the extent that students are learning in groups within the flipped classroom, there is also opportunity for feedback from student-to-student, which is a valuable aspect of collaborative learning (Nokes-Malach et al., 2015).

Fourth and finally, the student-centered nature of the flipped classroom model supports student motivation, persistence, and engagement. The shift to active, collaborative learning helps to fulfill the fundamental student needs of autonomy, competence, and relatedness as identified within self-determination theory (Ryan & Deci, 2000). The challenge of active learning can encourage students to adopt productive beliefs about the benefits of effort and strategic practice (i.e., a growth mindset) (Yeager & Dweck, 2012), which has been shown to produce greater resilience and higher achievement (Blackwell et al., 2007). To the extent that the teacher communicates the purpose of flipping the classroom and design of the course, students may also find greater value in what they are learning and the activities in which they are engaging, which will increase motivation (Cohen & Sherman, 2014; Harackiewicz & Hulleman, 2010). If the in-class group work and broader culture of the course is designed to create a cooperative learning environment rather than a competitive one, then students will be more likely adopt mastery learning goals (i.e., rather than performance goals) and engage in deep learning (Ames, 1992; Johnson et al., 1981). In terms of extrinsic motivation, the flipped classroom also provides greater opportunities for both reward and fulfillment of obligations to the course, instructor, and peers through successful completion of their work (Abeysekera & Dawson, 2015). The size and environment in the classroom may also play a role in increasing extrinsic motivation by enhancing peer-peer contact and relatedness (Abeysekera & Dawson, 2015; Silverthorn et al., 2020).

6.3 Evidence Supporting the Effectiveness of the Flipped Classroom

Numerous studies have been published that compare the overall effect of flipped classroom teaching to traditional teaching. The literature has grown so rapidly that it has spawned a boom in secondary research on the topic with at least eight literature reviews (Betihavas et al., 2016; Chen et al., 2017; DeLozier & Rhodes, 2017; Lundin et al., 2018; O'Flaherty & Phillips, 2015; Tang et al., 2018; Velegol et al., 2015; Ward et al., 2018), 19 meta-analyses (Algarni, 2018; Chen et al., 2018; Cheng et al., 2019; Gillette et al., 2018; Hew & Lo, 2018; Hu et al., 2018; Karagöl & Esen, 2018; Låg & Sæle, 2019; Liu et al., 2018; Lo et al., 2017; Lo & Hew, 2019; Shi et al., 2019; Tan et al., 2017; van Alten et al., 2019; Xu et al., 2019; Zhang, 2018; Zhu et al., 2019), and a recent review of the methodology of these meta-analyses (Hew et al., 2021). Three general conclusions can be drawn from these various attempts to synthesize the literature. First, almost all studies show a benefit of flipping the classroom, and it is relatively rare to find a negative effect in a meta-analysis (Cheng et al., 2019). Second, the meta-analyses that test for the presence of publication bias find no bias meaning that there is unlikely to be a hidden "file drawer"

Table 6.1 Comparison of effect sizes from recent representative flipped classroom meta-analyses. A summary of the publication year, overall effect sizes, total number of studies meeting inclusion criteria, number of life and health science studies meeting inclusion criteria, and significant/notable moderators reported in the different meta-analyses discussed in the text

	Hew et al	Cheng et al	Strelan et al	Chen et al	van Alten et al
Year:	2018	2019	2020	2018	2019
Effect Size:	0.33 (SMD)	0.193 (g)	0.5 (g)	0.47 (SMD)	0.36 (g)
Total studies (n):	28	55	198	46	114
Life and health science studies (n):	26	17	53	32	51
Significant and notable moderators:	• disciplines • availability of pre-class assessments, exercises, pre-class readings, and notes • quiz at the start of the in-class session	• disciplines • duration of flipped classroom use	• education level • type of design (full vs partial flip) • types of pre-class and in-class activities • type and timing of assessment	• type of assessment	• addition of quizzes to flipped classroom • incorporation of micro-lectures into in-class sessions

literature of unpublished negative effects. Third, although there are a few moderators that influence the magnitude of the positive effect of flipping the classroom (for a summary, see Table 2 in Hew et al. (2021)), most of the moderators analyzed do produce a significant difference in effect size, and none produce negative effect sizes. Taken as a whole, it suggests that as long as some minimum criteria are met, almost all variations of the flipped classroom model are effective. In addition, it suggests that there is limited utility in continuing to conduct studies comparing the flipped classroom model to a lecture-only control condition because it is so easy to outperform. Indeed, recent reviews and meta-analyses have concluded with calls to move beyond this basic comparison to investigate the various components of the flipped classroom model and how they contribute its effectiveness (Hew et al., 2021).

To provide the reader with a better sense of the literature supporting the effectiveness of the flipped classroom approach, we briefly summarize five recent representative meta-analyses that include a sizeable number of studies from the health and life sciences (see Table 6.1) (Chen et al., 2018; Cheng et al., 2019; Gillette et al., 2018; Hew & Lo, 2018; Strelan et al., 2020; van Alten et al., 2019). Hew and Lo (2018) performed a meta-analysis comparing student assessment outcomes between flipped learning (using pre-recorded lectures as pre-class work) to traditional learning (defined by mandatory class attendance where teachers use varying pedagogical approaches, followed by students completing homework at home) and reported a standard mean difference (SMD) of 0.33 (95% CI 0.21–0.46, $p < 0.001$) for 28 health profession-focused studies (i.e., medical students, residents, doctors, nurses, and students in dental, pharmacy, and environmental and occupational health disciplines) (Table 6.1). This SMD, according to Cohen, is between a small and medium effect size (Cohen, 1988). In addition to the small number of studies included in the analysis, Hew and Lo (2018) also report a very high I^2 (75.6%) indicating significant heterogeneity among the studies. As discussed later, they identified several potential moderators such as student and instructor equivalence, research study designs, disciplines, availability of pre-class assessments/exercises, availability of pre-class readings/notes, and availability of a quiz at the start of the in-class session.

Chen et al. (2018) also published a meta-analysis that focused more on health science disciplines and included 46 studies overall with 32 studies from health science disciplines and 14 studies from non-health science disciplines. They looked across health disciplines including medical, pharmacy, and nursing comparing outcomes for course grades, examination scores, and objective structured clinical examinations (OSCEs) between flipped classroom-based sessions and lecture-based sessions with an overall effect size SMD = 0.476 US studies (n = 34) and SMD = 0.435 for non-US studies (n = 7) (Table 6.1). For each of these outcomes, they calculated pooled effect sizes: examination scores (SMD = 0.468, n = 41), pre- and post-test changes (SMD = 0.454, n = 7), course grades (SMD = 0.354, n = 9), and OSCEs (SMD = 3.116, n = 2). Looking specifically at health science studies (n = 32), they found slightly higher effect sizes for most outcomes: examination scores (SMD = 0.451, n = 28), pre- and post-test changes (SMD = 0.603, n = 5), course grades (SMD = 0.440, n = 5), and OSCEs (SMD = 3.116, n = 2). Parsing the studies further, they looked at specific professional programs within health science studies. These included medical education, pharmacy education, nursing education, and others. Of these, medical education studies were the most well represented (n = 14) and yielded the highest effect sizes for examination scores (SMD = 0.53, n = 14), pre/post-test changes (SMD = 0.70, n = 4), and OSCEs (SMD = 5.86, n = 1). Pharmacy had an effect size of 0.53 for both examination scores (n = 7) and course grades (n = 3), whereas nursing had an effect size of 0.20 for examination scores (n = 2), 0.05 for course grades (n = 1), and 0.41 for OSCEs (n = 1). Similar to the meta-analysis by Hew and Lo (2018), the test for heterogeneity (I^2) was high for most outcomes measured in this meta-analysis, regardless of discipline with the exception of nursing, which was slightly lower than medicine, pharmacy, and other health sciences (Chen et al., 2018).

A more recent meta-analysis by Cheng et al. (2019) included 55 studies with a total of n = 7912 students comparing flipped classroom learning (n = 3727) with traditional classroom learning (n = 4185). A variety of disciplines were represented in these studies. Of the total number of students represented here, n = 2358 students were from studies in health or science disciplines. When examining cognitive outcomes including exam scores, the authors found a low, near trivial, overall effect size of g = 0.193 for flipped classroom learning effects on student learning compared to traditional classroom learning. The authors reported a moderate level of heterogeneity (I^2) of 59.88% suggesting that some moderators may be influencing variance among the sample studies rather than sampling error. Moderators examined included level of education, discipline, publication type, and study duration. Interestingly, Cheng and colleagues found that studies using flipped classroom for less than a semester had a higher effect size ($g = 0.348$) compared to studies that used flipped classroom for longer than a semester ($g = 0.153$), although the results were not statistically significant. Despite the lack of statistical significance, this observation warrants consideration. Does the impact of flipped classroom learning on cognitive learning outcomes fluctuate with prolonged use? It may be possible that students become fatigued with the flipped learning format over time or with prolonged use and, in turn, alter their participation in certain components of the flipped approach (i.e., stop doing pre-work or don't come to class for application exercises if attendance is not mandatory).

A meta-analysis by van Alten et al. (2019) compared effect sizes for assessed learning outcomes (derived from assessments of students) and perceived learning outcomes (derived from student perceptions of learning gains). Here they analyzed 114 studies including 51 studies from natural sciences, 9 studies from humanities, 15 studies from social sciences, and 39 studies from formal sciences (van Alten et al., 2019). They reported a significant average effect size for assessed learning outcomes ($g = 0.36, p < 0.001$) compared to a non-significant average effect size for perceived learning outcomes ($g = 0.36, p = 0.13$) (van Alten et al., 2019). They attribute this finding to differences in the ability to detect the smallest important effects within the sets of studies comprising each of these groups (van Alten et al., 2019).

Finally, Strelan et al. (2020) published a meta-analysis investigating 198 studies from a range of disciplines (including medical and health sciences) with a total of 33,678 students yielding an overall effect size of $g = 0.5$ compared to traditional classroom learning. They report a larger effect size for courses where part of the course/subject was flipped ($g = 0.77$) compared to the entire course/subject ($g = 0.42$). This provides some support for the role of flipped classroom teaching as a component of a blended learning approach in a course that incorporates other pedagogical techniques (Cheng et al., 2019; DeLozier & Rhodes, 2017; Flynn, 2015), although this contrasts with the findings of others who compared partially flipped course sections (part of the class time was devoted to active learning exercises) to fully flipped course sections (where the entire class time was devoted to active learning exercises) and found no statistically difference in the performance of the two sections (Adams et al., 2016; Jensen et al., 2015). The underlying reason for the differences between these studies remains unclear. Nevertheless, incorporating an active component into a course at some threshold leads to greater gains for the student (Adams et al., 2016; Jensen et al., 2015). Strelan et al. (2020) also address a key question we will explore and that is: Are all of the components of flipped classroom necessary to derive benefit from this pedagogical method? They report that the difference in effect size between classically flipped (incorporating pre-class preparation of any type, some form of pre-class engagement measure, and student activities in class with or without in-class lectures) and minimally flipped (*does not* incorporate all three *classic* elements: pre-class preparation, pre-class engagement measure, and in-class activities) is marginal with a slight advantage for the classically flipped model ($g = 0.52$ and $g = 0.45$, respectively).

Thus, a driving force of the flipped classroom approach (both inside and outside the classroom) is active learning (Flynn, 2015; Jensen et al., 2015). Active learning has been shown to yield superior assessment outcomes and knowledge retention compared to passive lecturing (Freeman et al., 2014). Although not all studies agree with the premise that active learning enhances student learning, active learning may be especially useful in helping to resolve misconceptions, likely through feedback from peers and instructors (Andrews et al., 2011). Both student interest in the subject ($p = 0.004$) and perception of course difficulty ($p = 0.021$) are also positively associated with performance in an active classroom (Andrews et al., 2011). This further bolsters the argument that increased structure and rich pre-class and in-class exercises can enhance student learning compared to traditional instruction. It also supports the notion that curiosity and struggle play important parts in motivation and performance which could enhance engagement.

6.4 Evidence-Based Guidance for Implementing Flipped Classroom Teaching

In this section, we will examine studies that assess the principal components of flipped classroom teaching that mediate learning gains inside and outside the classroom. There are a number of critical questions that need to be considered as we review the published evidence supporting the flipped classroom approach:

- Does the type of pre-class material (i.e., reading, videos, audio, computer simulations, etc.) matter for student outcomes or performance?
- Does the type of in-class application exercise matter?
- If students do not complete the pre-work (i.e., read or watch the videos), does this impact their performance in the in-class application exercise?
- Do certain types of in-class activity optimally synergize with certain types of pre-class materials?
- Does attendance at in-class application sessions impact students' outcomes with flipped classroom learning?
- Do certain types of students stand to benefit more from flipped classroom learning than others?

These and other questions will be considered in the following sections with the goal of understanding how to optimally develop and run a flipped classroom to maximize student success.

6.4.1 Outside the Classroom

As discussed in Sect. 6.2, one of the advantages of flipped classroom learning is the control students have over their initial learning outside the classroom in terms of both the amount of material they cover and their pace of coverage. Here we consider the format and types of materials a student can use as part of their preparation and their relative effectiveness on performance.

6.4.1.1 Technology

Flipped classroom learning is classically associated with the use of technology such as short, recorded videos for pre-class learning rather than other types of activities that do not depend upon technology, such as reading. Indeed, many definitions of flipped classroom do not include reading as a form of pre-class work. It has been suggested that pre-reading is inferior to videos because a book does not contain instructor explanations or elaboration whereas videos do, but the data for such an argument is sparse (Hew & Lo, 2018). In the meta-analysis by Hew and Lo (2018), they compared studies which used pre-class readings or notes (SMD = 0.30) to studies which did not provide access to pre-reading or notes (SMD = 0.35). They state that, due to a lack of heterogeneity among these subgroups, there were no performance gains by providing notes or pre-readings. Likewise, Hew and Lo state that including an online assessment or exercise *prior* to attending face-to-face sessions did not appear to moderate performance gains, again due to a lack of heterogeneity among the subgroups (Hew & Lo, 2018). This suggests that there may not be any added value in providing additional material along with the pre-recorded videos. What about if alternate pre-work is used *in lieu* of pre-recorded videos?

An early study which pre-dated the widespread use of the term flipped classroom addressed this very question. Moravec et al. (2010) measured the impact of migrating a portion of lecture material in an introductory biology class to either narrated PowerPoints or worksheets delivered online and compared their effectiveness with a traditional, unmodified form of the course. They dubbed this process learn before lecture (LBL) (Moravec et al., 2010). By moving some of the former lecture content online they were able to increase active learning time during in-class sessions (Moravec et al., 2010). An important distinction here is that both types of pre-work were delivered digitally, and the worksheets, which perhaps come the closest to "pre-reading," were active in the sense that students must complete them rather than simply read them. They found that this subtle shift of material led to an increase in student performance on LBL-associated exam questions ($p < 0.001$) with an average increase of 21% for these questions compared to prior non-LBL sections of the course (Moravec et al., 2010). Importantly, they found no statistically significant performance differences between the video-based pre-class material and worksheets (Moravec et al., 2010).

What do students actually use if given the choice, reading, or technology (i.e., video)? Bassett et al. (2020) compared student use and preference of reading and/or video viewing on performance in a partially flipped course. Here they assigned 10–20 pages of pre-reading in a textbook for the traditional lecture sessions, held 2 days per week, and online videos between 8 and 13 min in length for the flipped session, occurring 1 day/week (Bassett et al., 2020). The difference with this study is that pre-class material was accompanied by low-stakes assessments: in-class clicker questions for the pre-reading and online quizzes for the videos (Bassett et al., 2020). Completion of the online quizzes was required and comprised 10% of the final grade, whereas the in-class clicker questions were available to students who attended class and comprised 5% of the final course grade (Bassett et al., 2020). Based on their analysis, both reading and video viewing had positive effects on exam grades (Bassett et al., 2020). Interestingly, most students did a combination of both reading and video viewing. This student behavior likely arises from the fact that both pre-class material formats were linked to assessments that comprised a percentage of the final grade. Students may have also viewed having two types of preparatory materials and assessments as just that: additional opportunities for preparation and practice. However, students did report both a stronger preference for videos, and they also exhibited greater compliance with video viewing compared with reading (Bassett et al., 2020).

When taken together, these results suggest that it is neither the format of the pre-work nor the use of technology per se that has a substantial impact on student learning. Rather student engagement and compliance with the pre-work may be more important (DeLozier & Rhodes, 2017; Stephenson et al., 2008; Zhang et al., 2006). In addition, pairing low-stakes assessment with pre-work regardless of the format (i.e., technology or not) may be more important than the type of pre-work itself (see Sect. 6.4.3 below) because assessment can actually increase accountability and motivate students to engage, prepare, and/or attend the in-class active learning sessions and thus have a significant impact on assessment outcomes.

6.4.1.2 Activities

Compared to traditional passive lecturing, teaching and learning in a flipped classroom model require more deliberate, engaged, and persistent preparation. While students initially appear to put more time into preparing for flipped classroom sessions than lecture-based sessions, their participation wanes over time (Bouwmeester et al., 2019). Thus, constructing pre-class activities that support engagement or incentivize preparation is key. For example, Strelan et al. (2020) found that pre-class work that engages students in discussion has a higher effect size ($g = 0.78$) compared to pre-class work where students simply reflect on ($g = 0.35$) and that scoring the pre-class work also has a measurable effect ($g = 0.53$). However, other factors come into play that determine a student's engagement and level of participation beyond the type of activity. The *amount of pre-class preparation* plays an important role in moderating assessment outcomes.

Gross et al. (2015) assessed the impact of pre-class content engagement on final exam performance in a flipped version of an undergraduate physical chemistry course compared to a traditional version of this course. Here pre-class activities in both the traditional and flipped versions of the course consisted of online homework that emphasized problem solving together with an online interactive book replete with background readings, interactive illustrations, and example problem sets more complex than the homework (Gross et al., 2015). In addition to this pre-work, the flipped class also had online supplemental lectures (e.g., videos). The in-class sessions of the flipped classroom section were made active using clicker questions, peer learning, and team-based learning (TBL) exercises focused on difficult problems, together with instructor feedback. The written exams used in each of the courses (traditional and flipped) were "identical in structure and very similar in content." Students in the flipped classroom sections ($n = 244$) engaged with pre-work more readily than students in a traditionally taught version of the course ($n = 245$). Similar to the report by Jensen et al. (2015), Gross et al. (2015) also observed that students in the flipped classroom section engaged with their homework in a more timely manner and were more accurate in their responses compared to the students in the traditional course. Significantly, the number of pre-class video accesses was positively correlated with both exam average for the course ($r(235) = 0.22$, $p < 0.001$) and with GPA ($r(235) = 0.24$, $p < 0.001$) (Gross et al., 2015). This finding suggests what one would expect and that is students with higher GPAs would watch more of the pre-class videos and score better on the exam. However, the results are a bit more nuanced because it depends on the timing of the video viewing relative to the exam. It turns out that the students with the lowest GPA tended to access and view pre-work videos more, but in close temporal proximity to the exam, whereas students with higher GPAs accessed and viewed the videos slightly less, but in a more even and consistent fashion across time (Gross et al., 2015). This result may be explained by the observation that higher performing students have higher degrees of self-efficacy and exhibit better resource management and utilization (Kauffman et al., 2018; Stegers-Jager et al., 2012). This relationship may extend into outcomes as well. Bassett et al. (2020) reveal, via a multiple linear regression model, a large positive effect of GPA on exam scores and that GPA plays a role in prediction of student performance in flipped classroom learning (Bassett et al., 2020). Thus, in the face of these results, it stands to reason that students who are more *strategically* engage with pre-work come to class more prepared and derive more benefit from in-class active learning activities. However, as we will discuss later on in Sect. 6.4.3, engagement with pre-work, self-regulated learning skills, and attendance interact in a complicated admixture that may not manifest on all assessment item or question types (White et al., 2017).

In conclusion, these findings may provide some small comfort to instructors who are concerned about choosing the "right type" of pre-work to use or whether students who deviate from assigned pre-work will not do as well. It seems participating and engaging with pre-class materials or activities will lead to improved performance so long as there is an explicit connection with in-class activities that justify a student's time investment with this material (Fig. 6.1).

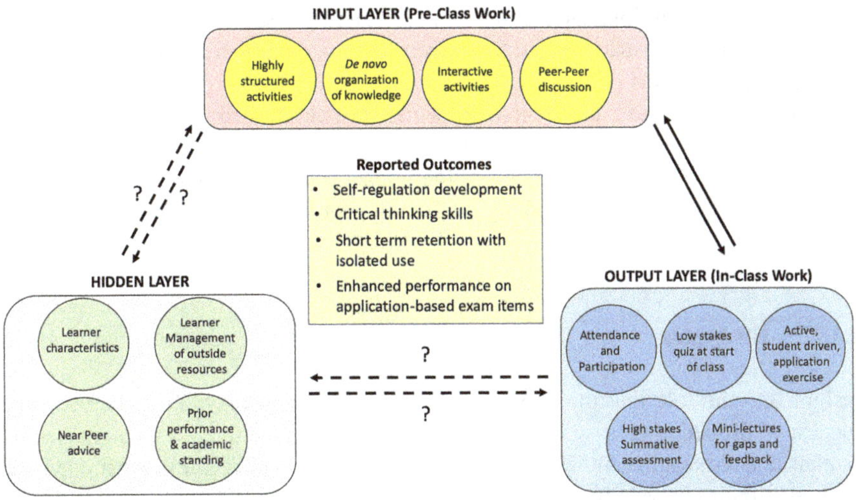

Fig. 6.1 Proposed model of factors that mediate optimal delivery of flipped classroom sessions. Documented and inferred relationships between the different elements of flipped learning and "extra-instructional" factors (hidden layer). This model is an adaptation of a neural network model that synthesizes what has been reported for flipped classroom as well as studies focused on flipped elements such as self-directed learning (Murad & Varkey, 2008), motivation, and resource management (Stegers-Jager et al., 2012). Unlike what is found with a traditional neural network diagram, here there is a reciprocal relationship between hidden layer elements and the input and output layers such that hidden layer can both shape, and be shaped by, the input and output layers. In this model, the hidden layer represents "extra-instructional" attributes, behaviors, and characteristics of students that lie outside of the direct view of the instructor designing and delivering the flipped classroom sessions, which warrant further investigation (see text for details on individual elements systematized within each layer)

6.4.2 *Inside the Classroom*

In contrast to what happens outside of class in a flipped classroom format, what happens inside the classroom is, by definition, active. Students work to apply knowledge they acquired from the pre-work in order to strengthen their understanding and information retention.

6.4.2.1 Activities

The active learning exercises used in in-class flipped classroom sessions may take any number of forms and may be accompanied by mini lectures and assessments (DeLozier & Rhodes, 2017). Frequently employed active learning strategies span a range of interactivity and include clicker questions using an audience response system, open-ended questions, think-pair-share activities, case-based learning, team-based learning, games, simulations, and others (DeLozier & Rhodes, 2017). However, as any practitioner of active learning would quickly admit, building and delivering active sessions take significant effort and preparation on the part of both the student and the instructor, and they cover less material than a typical lecture. For this reason, the effects of varying the amount of active learning, the types of active learning exercises, and the degree to which pre-class work and in-class active learning sessions are linked need to be carefully considered and understood. We discuss these considerations as well as the impact of flipped classroom learning on the development of critical thinking skills, self-regulated learning, and social/non-cognitive behaviors.

6.4.2.2 Structure

Insight into the role that the structure of in-class learning plays in moderating the success of flipped in-class sessions comes in part from a meta-analysis published by Strelan and colleagues in 2020 (Strelan et al., 2020). Here they partitioned the studies in their analysis into four groups: standard didactic (lecture only), predominately teacher guided, predominately student centered, or not reported. The predominately teacher-guided group may involve some discussion or question and answer with or without lectures, but it does not have student-led in-class activities. On the other hand, the predominately student-centered group includes studies where students drive or generate content and engage in problem solving exercises, which may or may not operate in tandem with teacher discussions, question and answer sessions, or mini lectures. Analysis of the moderating effects of in-class activities for flipped classroom studies compared to traditional classroom teaching revealed that predominately student-centered flipped classroom sessions had a larger effect size ($g = 0.52$, $p = 0.00$, $n = 184$) than predominately teacher centered ($g = 0.06$, $p = 0.33$, $n = 4$) or studies where the in-class activity was not reported ($g = 0.32$, $p = 0.16$, $n = 9$). They do note, however, that the presence of a didactic lecture in the in-class sessions had a significant effect based on a small number of studies ($g = 0.33$).

Jensen et al. (2015) looked more closely at the role that both student-centered active learning and instructors play in flipped learning. Here they compared shifting the role of the instructor from facilitating first contact with material (active content attainment) that was followed by student-led concept application after class (non-flipped condition) against a condition where instructors facilitated in-class concept application where students did pre-work before class (flipped condition). The role of the instructor in both conditions was relegated to preparing materials and facilitating discussion, whereas the role of the student was to engage in group work in-class. They observed no differences in outcome measures between the two conditions. The only difference they reported between the two conditions is that students in the flipped classroom had more accurate responses to their homework questions than the traditional condition.

In a more recent study by Bouwmeester et al. (2019), they determined that when they flipped a hematology and oncology course at the Utrecht University Medical School, instructors and students interacted differently with the altered course structure. In the flipped classroom sessions, instructors tended to ask more questions of students, provide more answers, and explain more in-depth information compared to traditional class sessions (Bouwmeester et al., 2019). This led to more participation in these sessions. However, like Jensen et al. (2015), Bouwmeester et al. (2019) found that performances on both knowledge-based questions immediately after the course ($M = 48.8$ (SD $= 13.3$)) and also 10 months later ($M = 44.1$ (SD $= 11.4$)) were not significantly different ($t(36) = 1.23, p = 0.23$) between the traditional and flipped classes (Bouwmeester et al., 2019). Nor were application-based questions immediately after the course ($M = 40.9$ (SD $= 6.0$)) or 10 months later ($M = 41.3$ (SD $= 6.9$)) significantly different ($t(41) = -0.18, p = 0.86$) between the courses (Bouwmeester et al., 2019). Thus, at a minimum, incorporating instructor input and student-directed group activity in class is not harming students or diminishing performance. Similar to Bouwmeester et al. (2019), Jensen et al. (2015) observed no differences in scientific thinking skills or reasoning between the flipped and non-flipped classes as measured by Lawson's Classroom Test of Scientific Reasoning (LCTSR). The authors also found that students performed equally well on both low-level and high-level items between the two conditions. Thus, if the temporal order instructor and student participation in active learning do not matter for overall performance or attainment of critical thinking skills, perhaps the dosage of active learning in the flipped sessions impacts critical thinking.

In contrast to what was shown by Jensen et al. (2015), Styers et al. (2018) reported that the dosage of active learning in flipped session may impact critical thinking skills performance. Styers et al. (2018) studied three undergraduate level science courses (a molecular biology course, $n = 52$; an evolutionary ecology course, $n = 27$; and a biochemistry course, $n = 21$) that were partially flipped with selective incorporation of active learning sessions together with traditional lecture. The percentage of class time dedicated to active learning varied by course: ~20% in the molecular biology course, ~75% in the evolutionary ecology course, and ~90% in the biochemistry course (Styers et al., 2018). In each of these courses, the active learning sessions were paired with in-class quizzes where ~67% of the items evaluated higher-order cognitive skills and covered the pre-work videos and computer simulations (which as we have seen from the meta-analysis by Hew and Lo (2018) leads to improved performance) (Styers et al., 2018). The in-class active learning component of the flipped-sessions included multiple pedagogical methods such as clicker questions, think-pair-share, group problem solving, and hands-on model building (Styers et al., 2018). Critical thinking was measured using a validated instrument known as the critical thinking assessment (CAT) test in a pre- and post-fashion at the beginning of the course. The CAT contains questions that link to four skill domains, evaluate and interpret information (E), problem solving (P), creative thinking (C), and effective communication (Co). Results of the CAT showed a significant increase in overall CAT scores from pre- to post-administration for the two courses with the highest amount of flipped classroom sessions, the evolutionary ecology course with ~75% flipped sessions and the biochemistry course with ~95% flipped sessions ($\delta = 3.56$, effect size $= 0.52$ and $\delta = 2.74$, effect size $= 0.46$, respectively). Statistically significant gains for both of these courses were clustered around pre- to post-improvements in E and P, although several questions related to C and Co were also significantly improved.

6.4.2.3 Group Size and Attendance

Other factors that may mediate the success of flipped classroom sessions are group size and attendance. Strelan et al. (2020) found a difference in effect sizes between in-class activities that were done as small groups ($g = 0.46$) or work individually ($g = 0.59$). In general, when specific group number data was available, there appeared to be a steadily decreasing effect size as group numbers increased. Although this overall result appears to support the notion that working individually might be marginally better, given the other benefits of group work, including deeper engagement with material as a result of peer-learning, there is a role for having both a group and individual component with the in-class sessions (Steinel et al., 2019). This conclusion suggests that, preliminarily, the way the in-class active learning component is structured, the size of the group, and whether or not students are afforded the opportunity to work individually may shape learning outcomes.

What if students do not attend the in-class sessions? Clearly group work matters and a lack of attendance can hinder group formation and reduce the richness of group composition. As a result, it stands to reason that students need to attend flipped classroom sessions in order to derive any benefit from the active learning and peer learning. Yet the role of attendance in influencing performance in in-class sessions is a contentious area of health science education research rife with conflicting data and results (see Chaps. 11 and 12 in this volume) (Doggrell, 2020; Kauffman et al., 2018). As a result several questions arise here: (1) Does flipped teaching promote increased attendance and in-class engagement and does it matter if attendance is voluntary or mandatory? and (2) One would expect that attendance would have a profound effect on the success of flipped learning because the active learning portion of the flipped classroom appears to have the greatest impact (Jensen et al., 2015). Indeed, some have found a positive association between voluntary attendance at flipped classroom sessions and performance on both summative assessments and high-stakes licensing exams (Laird-Fick et al., 2018), whereas others have not found a correlation between voluntary lecture attendance and performance (Doggrell, 2020). Of course, it may vary by course and the types of students.

Likewise, Kauffman et al. (2018) argue that voluntary in-class attendance does not correlate with performance on summative assessments (Chap. 11 in this volume). Furthermore, they found that in their cohort of medical students who participated in their study ($n = 78$), those who attended "greater than zero classes" ($n = 52$) were the historically lower academically performing students who had lower self-efficacy and were less self-confident (as assessed by results from the MSLQ). On the other hand, the students who did not attend class ($n = 26$) had statistically higher MCAT scores and stronger academic performance prior to coming to class, compared to the students who attended, and high self-efficacy based on MSLQ results. In addition, they report that students who attended class had a greater orientation toward peer learning, help seeking, and organization but had a low capacity for effort regulation. This suggests that students who are more motivated and can self-regulate their own learning and study habits are utilizing learning resources in a self-determined way outside of class, compared to students who lack such self-regulation skills. It may be then that attendance at flipped classroom sessions is in fact more beneficial to the student with low self-efficacy and low resource management (Day, 2018). This aligns with the observations of Lai and Hwang (2016) discussed above. When taken together, it would seem that we want students to become self-regulated learners but not so much they stop attending active learning sessions, the flipped classroom, or the formal curriculum for that matter. Indeed, by this logic, if attendance and engagement in flipped classroom sessions help students develop self-regulated learning behaviors, then it could be predicted that enhanced engagement of poor performing students with low self-efficacy in flipped sessions could convert them to non-attending self-regulated learners! While this hypothesis needs further validation, it begs the question: should attendance at in-class flipped classroom application sessions be required and what effect does this have on student performance and preparation? (For more discussion, see Chaps. 11 and 12 in this volume.)

6.4.3 Assessment

In this section, we focus on *how* assessment can best be used to *improve* performance and engagement with the flipped classroom model, as opposed to simply measuring performance. More specifically, we will look at studies addressing which types of assessments (i.e., formative and summative) and assessment items (i.e., multiple choice, essay, short answer, etc.) facilitate optimal learning in a flipped classroom, the role of feedback, and the optimal timing for assessment delivery during a flipped learning cycle.

6.4.3.1 Pre-class and Post-class Assessments

The timing and types of assessments used to support flipped classroom learning vary widely by study. Some studies administer quizzes prior to class, others in class, some at the end of the week, or not at all. What is the best practice for using assessments to measure learning in a flipped classroom setting? Can formative assessment be used to drive compliance with pre-class learning? What is the role for feedback in formative assessment for flipped learning? Assessing pre-class learning appears to reinforce and support the role of pre-class activities in the overall success of the flipped classroom approach and improves performance (van Alten et al., 2019). However, the timing of the quiz administration makes a difference. Hew and Lo (2018) reported an effect size of $g = 0.56$ for studies where pre-class material was assessed with a quiz at the start of class, compared to $g = 0.26$ where there was no quiz. Thus, it appears that delivering a quiz in-class, before the in-class session, has a greater impact than delivering it outside of class, insofar as the class attendance is mandatory (Hew & Lo, 2018). These authors note that the explanation for this finding may relate to the ability of students and instructors to identify and discuss unclear concepts or misconceptions from the pre-work prior to the start of the in-class activity (Hew & Lo, 2018). In other words, it affords instructors the opportunity to provide feedback to students prior to the application phase. It also holds students accountable for completion of the pre-class material. Even scoring pre-class materials increases student performance ($g = 0.53$) (Strelan et al., 2020). In many ways then, this closely mirrors that approach taken in team-based learning (TBL) which has an individual readiness assurance test (IRAT) that tests pre-work and an application exercise (Michaelsen et al., 2002).

Timing also plays a role with summative exams as well. The impact of flipped learning on performance is strongest when students are immediately assessed right after the session ($g = 0.77$), whereas the learning gains appear to fade when assessed later, on a summative final exam ($g = 0.28$) (Strelan et al., 2020). Clearly, more study is needed that looks at the durable long-term retention of knowledge and reasoning skills obtained from flipped classroom sessions.

6.4.3.2 Types and Cognitive Level of Assessment Items

In the aforementioned meta-analysis by Chen et al. (2018), they note that the types of assessment questions used appear to impact outcomes for flipped classroom learning with essay questions exhibiting a larger pooled effect size (SMD = 1.186, $n = 3$ studies) compared to the more frequently employed MCQ style items (SMD = 0.394, $n = 20$ studies), or a combination of item types (SMD = 0.378, $n = 5$), and unknown (SMD = 0.451, $n = 13$). However, heterogeneity (I^2) was very high (ranging from 51.6 to 93.9 depending on the format), so further study is needed to determine if a particular type of exam item format influences outcomes of flipped learning. Though given what was described in the study by White et al. (2017), certain item types may be more reflective of true student gains from participating in flipped classroom sessions than other item types. White et al. (2017) suggest that the combination of pre-class preparation, attendance, and learner characteristics may impact performance with particular types of assessment items. Specifically, they saw a stronger correlation between attendance and performance on novel scenario-based summative exam MCQ items (pooled data for two sections of students: $r = 0.44, p = 0.00$) as compared to simple knowledge-based exam MCQ items (pooled data for two sections of students: $r = 0.19, p = 0.02$). When taken together with preparation data, they observed that students who both *prepared* and *attended* performed significantly better on novel scenario-based exam items than those who did one or the other (59% of the total pooled cohort of students). Predictably, students who neither prepared nor attended performed significantly worse on knowledge-based questions, and, although not statistically significant, they tended to perform worse on novel scenario-based questions.

In addition to the item format, the cognitive level of assessment items may impact both learning *and* the ability to detect learning from flipped classroom teaching. In contrast to other studies, Barral et al. (2018) reported that students in an accelerated 4-week-long course using flipped learning ($n = 88$) actually performed better on lower-level Bloom's taxonomy mapped multiple-choice questions compared to students in a traditionally taught condition ($n = 54$). The two groups of students (flipped and traditional) were equivalent in terms of gender, sex, ethnicity, age, and GPA (although the latter variable is not shown in the table). For both flipped and traditional groups, two quizzes were administered, using the same questions for both, to gage a change in performance over time. They found no significant performance differences between the two classes for high-level Bloom's mapped questions.

These results raise some important questions that need to be addressed in future research around the type of learning purportedly driven by flipped classroom teaching (application), the type of assessment (formative versus summative), the type of item/question (i.e., MCQ, essay, short answer, etc.), and the cognitive level of the assessment items/questions. Indeed, when considering that flipped classroom is ultimately focused on knowledge application, it is surprising that most examples of flipped classroom teaching discussed thus far use MCQ assessments. In future studies more focus should be devoted to comparing different flipped classroom conditions with distinct assessment formats. This way, rather than comparing flipped classroom to traditional teaching, a true conclusion can be made on the role assessment types play in moderating learning.

In conclusion, achieving optimal impact from in-class learning in a flipped classroom may depend, in part, on the dosage of active learning, attendance, engagement, and types of assessment items, more so than the types or order of activities (Fig. 6.1).

6.5 The Key Ingredients for a Successful Flipped Classroom: The State of the Instructors and Students

Up to this point, we have focused primarily on the process and components of flipped classroom teaching. Clearly student preparation is important, regardless of the method by which they prepare. Having an active element in a flipped classroom is also vital. However, as with any active learning-based pedagogy, arguably the most important part is the knowledge and commitment of the instructor in delivering the activity and the receptiveness of the student to "buy in," participate, and engage with the process.

6.5.1 Instructors

In order for an instructor to successfully deliver a flipped classroom session, they need to be committed to taking risks but also be confident and knowledgeable about the topic area and the active learning process (Wang, 2017). Simply throwing material online for students to consume and then having them discuss a case in class is not enough. The pre-class material should be carefully devised and prepared so that it is both achievable prior to class (both in terms of volume and complexity) and that it nests with the active application session the following class period (Silverthorn et al., 2020). The in-class active application session, ideally, will be a challenge that goes beyond simple regurgitation of memorized pre-work and instead has them use the pre-class information to construct something or solve a problem. This way at the end of the suite of pre- and in-class activities, the students remember information in the context of its use as opposed to remembering it because we told them to do it.

This requires that instructors overcome reluctances to move past teaching approaches and embrace technology or innovative techniques if warranted (Wang, 2017). Many instructors are interested in active learning but have little real experience with it, and this can interfere with the delivery and execution of a successful active session (Andrews et al., 2011; Silverthorn, 2020). Active learning activities require time: time to develop materials, time to create appropriate activities, and time to deliver in class (Silverthorn et al., 2020). Resources and support, both materially and administratively, are potential barriers as well. Finally, learners resist when challenged so having the resiliency and administrative/institutional support to weather this resistance is key. Using active learning can lead to negative course and instructor evaluations, particularly if other instructors or courses are using traditional lecturing (Silverthorn, 2020).

Constructing pre- and in-class materials with learning theory in mind or using a framework can help ensure standardization of the flipped sessions (Quesnelle et al., 2021a). Working in collaboration with colleagues (within and between institutions) can also be very helpful for creating active learning activities like flipped classroom (Quesnelle et al., 2021b).

6.5.2 Students

Students often resist active learning as a consequence of the greater work load and self-directed learning that it demands (Silverthorn, 2020). Students need to have clear expectations of the work and effort that active learning entails. They also need to understand the longer-term benefits it affords them. What characteristics, training, or experience do students need to succeed in active learning and flipped classroom sessions? Abeysekera and Dawson (2015) astutely note that the success of a flipped classroom session is integrally dependent on student motivation because it is necessary to engage with, and complete, the pre-class work as well as participate in the in-class application exercise. They adopt Baron and Corbin's definition of engagement: "the engaged student is the student who has a positive, fulfilling, and work-related state of mind that is characterized by vigor, dedication and absorption and who views him or herself as belonging to, and an active participant in, his or her learning communities" (Baron & Corbin, 2012). However, many students may need help learning *how* to study and engage with active learning approaches. More specifically, certain "learner characteristics" appear to be more important for success in active learning environments.

6.5.2.1 Student Participation and Self-Regulated Learning Skills

Insight into how learner characteristics and participation can impact performance in a flipped classroom setting, and vice versa, can be derived from a study by Stegers-Jager et al. (2012) who administered the Motivated Strategies for Learning Questionnaire (MSLQ) to two 1st year classes of medical students ($n = 817$) at the Erasmus MC Medical School in Rotterdam in the Netherlands. Using subscales of the MSLQ linked to motivation (internal goal orientation, task value, and academic self-efficacy), deep learning (cognitive and metacognitive strategies), and elaboration (organization and metacognitive self-regulation), the authors sought to link learner characteristic with successful performance on summative assessments and course grades. Correlational analysis of self-reported data from the MSLQ with performance data revealed that both student's participation ($\beta = 0.77$) and their self-efficacy ($\beta = 0.19$) were directly and positively associated with performance. They define participation to include lecture attendance, skills training attendance, guided individual study, and homework completion. In addition, deep learning strategies, such as elaboration or organization, were directly and negatively associated with performance ($\beta = -0.31$) but appear to have a positive association with resource management (i.e., time management and effort regulation) ($\beta = 0.77$) which, in turn, is positively associated with participation ($\beta = 0.79$). Thus, deep learning strategies on their own do not improve performance, only when combined with good resource management and participation. The authors posit that good resource management strategies may drive participation, highlighting an area for intervention with highly structured exercises that stimulate curiosity.

To assess the durability of learner characteristics, such as motivation and self-efficacy, Bouwmeester et al. (2019) measured learner characteristics at the end of a flipped and non-flipped course and then again 10 months after the end of these courses. Changes in students' self-reported motivation and self-efficacy (i.e. autonomous and controlled reasons for learning, competence) were examined by a previously reported self-regulated learning questionnaire (Black & Deci, 2000). Autonomous and controlled reasons for learning were found to be equivalent for both the traditional and flipped course ($t(41) = -0.28$, $p = 0.78$ and $t(37) = -1.66$, $p = 0.11$, respectively) (Bouwmeester et al., 2019). Self-reported self-efficacy (perceived competence) was found to be higher in the flipped course ($M = 64.9\%$

(SD = 10.0)) after the exam compared to the traditional course (M = 58.9% (SD = 10.4)) ($t(41) = -2.09$, $p = 0.04$, $\eta 2 = 0.580$), but then declined over the 10 months until a retention test was administered at which point it was equivalent to that reported for the traditional course (traditional, $M = 37.3$ (SD = 13.5); flipped, $M = 34.2$ (SD = 14.3), $t(43) = 0.76$, $p = 0.45$). The authors note that students may have felt less competent in reporting self-efficacy after a 10-month break from the initial exposure to the information and that perception of self-efficacy is task-related and declines with time. Thus, they conclude that flipped classroom may improve self-efficacy, which plays a part in self-regulated learning, but these effects are not durable. This conclusion is also supported by van Vliet et al. (2015) who examined learner characteristics in two courses taught in tandem, a traditionally taught (lecture based) course followed by a partially flipped (a mixture of lectures and flipped classroom teaching) using the MSLQ survey in a pre- and post-test fashion. Despite being similar at the outset, meta-cognitive and collaborative learning characteristics increased during the partially flipped course, whereas self-efficacy increased in both courses (van Vliet et al., 2015). This suggests that, although the content of the courses was independent of each other, the students' experience in managing the information load on their own in the traditional course with less structure led them to develop more self-efficacy. Like Bouwmeester et al. (2019), the durability of the changes observed on the MSLQ appear to dissipate over time after the student's exposure to flipped learning (van Vliet et al., 2015).

Yet another possibility is that simply participating in flipped classroom sessions alone may not be enough to build durable SRL skills in students. Rather than comparing flipped learning to traditional teaching, Lai and Hwang (2016) compared a flipped classroom group to a flipped classroom condition where SRL was intentionally incorporated and measured. In the SRL flipped class, the authors had the students set goals *before* accessing the pre-work and perform self-evaluation and self-reflection *after* the in-class application session. The SRL flipped class not only scored better on performance metrics but also showed more self-efficacy. Despite improving self-efficacy across the SRL flipped cohort, students with high SRL derived more benefit from the SRL flipped classroom approach and performed better on assessments, compared to students with low SRL skills. This difference was not observed in the regular flipped classroom condition suggesting that incorporating an SRL component in flipped classroom teaching strengthens all student learning and performance but to a degree dependent on baseline SRL skills at the outset. The caveats to this study are twofold: the study population was elementary school mathematics students, and the intervention was over 4 weeks. Open questions remain as to (1) how the phenomena observed in this study may apply to older or more experienced learners and (2) whether the changes observed are durable. Indeed, others have shown that higher levels of self-efficacy do not correlate with performance in flipped classroom learning in a health profession context (White et al., 2017).

Implementing efforts to increase student's self-efficacy may also improve participation (Stegers-Jager et al., 2012). Deliberate and targeted feedback and elaboration on a student's learning have the potential to improve self-efficacy, performance, and long-term retention of information (Schell & Butler, 2018; Stegers-Jager et al., 2012). A number of other studies have also shown a positive role for feedback in modulating learner characteristics (Deslauriers et al., 2011; Persky & McLaughlin, 2017; Ryan & Deci, 2000).

6.5.2.2 Student Perceptions

Lastly, how students perceive flipped classroom teaching and the implications this has on the success of this teaching modality bear consideration. Overall, students appear to prefer flipped classroom learning over traditional classroom teaching. Hew and Lo (2018) aggregated data from 5 studies including a total of 320 students and found that on average 71% of the respondents preferred flipped classroom to traditional classroom teaching. The study by van Vliet et al. (2015) also reported a marked divergence in student favorability of flipped classroom learning (8.0 ± 0.7 out of 10) compared to traditional lecture-based learning (6.8 ± 0.5 out of 10) (van Vliet et al., 2015). On the other hand, Jensen et al. (2015) looked more specifically at student perceptions and found that students in the flipped classroom condition had a more negative perception of the helpfulness of technology-facilitated activities compared to the non-flipped version. However, they observed students felt that the organization of the flipped course was better and each activity had a clear purpose compared to a non-flipped course. Nevertheless, there was no difference in the overall evaluation of the traditional or flipped versions of the course.

Other studies have also found mixed student perceptions of flipped classroom-associated technology including the value of the pre-class videos (Bassett et al., 2020). Here the students surveyed appeared to like the control they had over videos in terms of pace regulation, and they found the in-class application component of the flipped sessions useful; however students' perceptions of how they studied and how prepared for the exam did not have a statistically significant impact on their actual performance (Bassett et al., 2020). Likewise, based on self-reported survey results on students' perceptions of their learning, Styers et al. (2018) discovered that many students in their study overestimated their own learning gains in critical thinking.

This latter observation fits into a recurring trend seen in many of these studies and that is the disconnect between student perceptions and their actual performance. This may be related to the observations of Dunning and Kruger on incomplete understanding and misperceptions of mastery, the Dunning-Kruger effect (Kruger & Dunning, 1999). Ignorance, such as seen with a novice, often makes it difficult to recognize one's own incompetence and can lead to exaggerated self-confidence about knowledge and understanding (Kruger & Dunning, 1999). This is supported, in part, in the meta-analysis by van Alten et al. (2019) where they observed a non-significant effect size for perceived learning outcomes together with a trivial, non-significant effect size of $g = 0.05$ for student satisfaction. This led to the conclusion that while students may overestimate their own performance, they are at least as satisfied with flipped classroom learning as compared to traditional lecture-based teaching (van Alten et al., 2019). It is possible that the heavy independent learning element of flipped learning may create an environment that is ripe for developing incomplete assumptions, fostering premature closure on knowledge, and erroneous or faulty understanding of concepts or processes built on hastily researched information from non-assigned/non-recommended resources (i.e., relying on simple Google searches). This is an area of flipped classroom teaching that should be a priority for research with the aim of developing interventions to close the gap between student perceptions of success and reality.

6.6 Conclusion and Future Directions

What has emerged from empirical studies and meta-analyses discussed in this chapter is that for flipped classroom teaching to be successful, there needs to be a compatible pairing of learner characteristics with thoughtfully structured, pre-class activities and in-class active learning exercises. Assimilating all of the lessons learned from the data discussed thus far, we have created a model articulating optimal components for the successful use of flipped classroom teaching together with their complicated and reciprocal relationships (Fig. 6.1).

6.6.1 Model of Factors Influencing Flipped Classroom Success

6.6.1.1 The Input Layer

As discussed, highly structured pre-class activities can improve engagement, potentially increase performance, may reduce extraneous cognitive load, promote development of learner characteristics (e.g., self-efficacy), and facilitate development of self-regulated learning skills. De novo organization of knowledge describes activities that require students to organize knowledge they are obtaining from resources associated with the pre-class work in a way that drives understanding of interrelationships among knowledge (i.e., concept maps). These types of activities appear to aid in retention of knowledge and enhanced performance on application-based exam items. Interactive activities could describe pre-work that requires the students to do something rather than simply watch or read in the absence of accompanying questions (which would serve to elevate these passive activities toward interactive): interactive videos, computer simulations, serious games, and similar types of tools. These types of activities will increase engagement, improve completion of pre-work, and strengthen retention of knowledge for the in-class sessions. Finally, intentionally incorporating *peer-peer discussion* into pre-class activities has been shown by several studies noted above to improve performance and may improve participation.

6.6.1.2 The Output Layer

There are myriad reasons why a student would choose not to attend an in-class session. Failure to complete the pre-work may be a factor in guiding a student's decision to attend a flipped classroom session. Indeed, the increased pre-class workload and in-class active learning (which can expose student gaps in knowledge) may serve as barriers for student attendance and engagement. Including a quiz at the beginning of the in-class session and making attendance mandatory may both lead to improved performance and compliance, likewise developing activities and messaging that demonstrate the value of using flipped classroom teaching to students. Considering that the in-class active learning component is the major driver of flipped learning, attendance and participation are a critical area for intervention and study. Attendance and participation are also critical considering that the optimal in-class active learning session should be student driven, paired with small mini lectures to provide feedback.

6.6.1.3 The Hidden Layer

Moreover, there are a number of factors that are not routinely considered or incorporated into the design of flipped classroom sessions because they are not readily available or, in some cases, even known to the instructor. These factors make up the hidden layer in our flipped classroom model patterned after a neural network model (Fig. 6.1). They include learner characteristics (encompassing traits such as self-efficacy, self-regulated learning skills, others). Prior performance, GPA, and academic standing are also significant factors that appear to predict student performance in a flipped classroom and have pervasive connections among the different elements in our model. However, given that many health science programs are going to pass/fail systems, the role of GPA and prior performance needs to be more thoroughly evaluated in these contexts. Near peer advice is the insight provided to learners from other learners who have already taken the course. This represents a potent and significant factor that influences student's behaviors and study habits. Lastly, the specific contribution that pre-class peer discussion plays in the overall performance of students remains opaque to the instructor unless a system is used to monitor it (e.g., Perusall®) (Strelan et al., 2020; White et al., 2017). It may serve to resolve uncertainties or misconceptions, such as would be encountered during a lecture with a content expert. Indeed, asking questions before class or in-class during a lecture is a real-time way to correct mental models as the learning is happening, whereas watching a pre-recorded lecture or doing an activity the night beforehand may mean that questions need to be saved or possibly go unanswered (or even forgotten) until the following day. In theory, this latter scenario provides valuable training in development of self-regulated learning skills and task management. It may also mean that an important piece of the learning process is lost between the outside and inside of the flipped classroom without peer-to-peer discussion.

6.6.2 Conclusions and Future Directions

In a progressively developing medical or health science curriculum, where knowledge compounds and builds on itself over time, it stands to reason that flipped classroom may benefit students earlier on in their learning while they are still developing their learning habits and study approaches and have less robust mental models of key concepts, whereas later in a curriculum, in-class application sessions might be less impactful once students have a base of knowledge and know how to apply it, particularly in the face of the copious amounts of formative assessment available both in house (faculty derived) and online (third party resources). This shift could explain the result noted earlier in the meta-analysis by Cheng et al. (2019) that showed a higher effect size for flipped classroom teaching on performance based on the duration of use.

Nevertheless, the trick seems to be convincing students that they should embrace and engage with exercises or teaching/learning approaches that intentionally cause them to struggle. Finding ways to harness innate curiosity is one of the most obvious ways to accomplish this. Coupling curiosity by focusing on problems or application exercises that truly use the details we expect students to learn from the pre-work may help drive compliance with pre-work and change study behaviors. Thus, future studies should consider addressing the proper temporal use of flipped learning by taking into consideration learner characteristics, experiences, and interests.

Finally, taking time to clearly explain and articulate the process of flipped classroom learning and the benefits it has for students at the outset of the course, with frequent reminders, may increase student buy-in, engagement, and perception although this needs to be studied. Having faculty members who are knowledgeable and versed in the literature and practices of active learning and flipped classroom teaching may actually make an impact in student's success with this pedagogy. This is hinted at in the work by Andrews et al. who reported that active learning did not lead to gains in student learning but that instructors who understand the process can improve its success (Andrews et al., 2011). This too is an aspect of flipped classroom teaching that needs to be rigorously investigated.

References

Abeysekera, L., & Dawson, P. (2015). Motivation and cognitive load in the flipped classroom: Definition, rationale and a call for research. *Higher Education Research and Development, 34*(1), 1–14. https://doi.org/10.1080/07294360.2014.934336

Adams, A. E., Garcia, J., & Traustadottir, T. (2016). A quasi experiment to determine the effectiveness of a "partially flipped" versus "fully flipped" undergraduate class in genetics and evolution. *CBE—Life Sciences Education, 15*(2). https://doi.org/10.1187/cbe.15-07-0157

Algarni, B. (2018, July). A meta-analysis on the effectiveness of flipped classroom in mathematics education. In EDULEARN18 proceedings (pp. 7970–7976). https://doi.org/10.21125/edulearn.2018.1852

Ambrose, S. A., Bridges, M. W., DiPietro, M., Lovett, M. C., & Norman, M. K. (2010). *How learning works*. Wiley.

Ames, C. (1992). Achievement goals and the classroom motivational climate. In *Student perceptions in the classroom* (pp. 327–348). Lawrence Erlbaum.

Andrews, T. M., Leonard, M. J., Colgrove, C. A., & Kalinowski, S. T. (2011). Active learning not associated with student learning in a random sample of college biology courses. *CBE—Life Sciences Education, 10*(4), 394–405. https://doi.org/10.1187/cbe.11-07-0061

Bangert-Drowns, R. L., Kulik, C. C., Kulik, J. A., & Morgan, M. (1991). The instructional effect of feedback in test-like events. *Review of Educational Research, 61*, 213–238. https://doi.org/10.3102/00346543061002213

Baron, P., & Corbin, L. (2012). Student engagement: Rhetoric and reality. *Higher Education Research and Development, 31*(6), 759–772. https://doi.org/10.1080/07294360.2012.655711

Barral, A. M., Ardi-Pastores, V. C., & Simmons, R. E. (2018). Student learning in an accelerated introductory biology course is significantly enhanced by a flipped-learning environment. *CBE—Life Sciences Education, 17*(3), ar38. https://doi.org/10.1187/cbe.17-07-0129

Bassett, K., Olbricht, G. R., & Shannon, K. B. (2020). Student preclass preparation by both reading the textbook and watching videos online improves exam performance in a partially flipped course. *CBE—Life Sciences Education, 19*(3), ar32. https://doi.org/10.1187/cbe.19-05-0094

Bernard, J. S. (2015). The flipped classroom: Fertile ground for nursing education research. *International Journal of Nursing Education Scholarship, 12*. https://doi.org/10.1515/ijnes-2015-0005

Betihavas, V., Bridgman, H., Kornhaber, R., & Cross, M. (2016). The evidence for 'flipping out': A systematic review of the flipped classroom in nursing education. *Nurse Education Today, 38*, 15–21. https://doi.org/10.1016/j.nedt.2015.12.010

Bishop, J., & Verleger, M. A. (2013, June). *The flipped classroom: A survey of the research*. Paper presented at 2013 ASEE annual conference & exposition, Atlanta, Georgia. https://doi.org/10.18260/1-2—22585

Black, A. E., & Deci, E. L. (2000). The effects of instructors' autonomy support and students' autonomous motivation on learning organic chemistry: A self-determination theory perspective. *Science Education, 84*(6), 740–756. https://doi.org/10.1002/1098-237X(200011)84:6<740::AID-SCE4>3.0.CO;2-3

Black, P., & Wiliam, D. (1998). Assessment and classroom learning. *Assessment in Education: Principles, Policy & Practice, 5*(1), 7–74. https://doi.org/10.1080/0969595980050102

Blackwell, L. S., Trzesniewski, K. H., & Dweck, C. S. (2007). Implicit theories of intelligence predict achievement across an adolescent transition: A longitudinal study and an intervention. *Child Development, 78*(1), 246–263. https://doi.org/10.1111/j.1467-8624.2007.00995.x

Bonwell, C. C., & Eison, J. A. (1991). *Active learning: Creating excitement in the classroom. ASHE-ERIC higher education report*. School of Education and Human Development, George Washington University.

Bouwmeester, R. A. M., de Kleijn, R. A. M., van den Berg, I. E. T., ten Cate, O. T. J., van Rijen, H. V. M., & Westerveld, H. E. (2019). Flipping the medical classroom: Effect on workload, interactivity, motivation and retention of knowledge. *Computers & Education, 139*, 118–128. https://doi.org/10.1016/j.compedu.2019.05.002

Braxton, J., Milem, J., & Sullivan, A. (2000). The influence of active learning on the college student departure process: Toward a revision of Tinto's theory. *The Journal of Higher Education, 71*, 569–590. https://doi.org/10.1080/00221546.2000.11778853

Chen, F., Lui, A. M., & Martinelli, S. M. (2017). A systematic review of the effectiveness of flipped classrooms in medical education. *Medical Education, 51*(6), 585–597. https://doi.org/10.1111/medu.13272

Chen, K. S., Monrouxe, L., Lu, Y. H., Jenq, C. C., Chang, Y. J., Chang, Y. C., & Chai, P. Y. (2018). Academic outcomes of flipped classroom learning: A meta-analysis. *Medical Education*. https://doi.org/10.1111/medu.13616

Cheng, L., Ritzhaupt, A. D., & Antonenko, P. (2019). Effects of the flipped classroom instructional strategy on students' learning outcomes: A meta-analysis. *Educational Technology Research and Development, 67*(4), 793–824. https://doi.org/10.1007/s11423-018-9633-7

Chi, M. T. H. (2009). Active-constructive-interactive: A conceptual framework for differentiating learning activities. *Topics in Cognitive Science, 1*, 73–105.

Chi, M. T. H., & Wylie, R. (2014). The ICAP framework: Linking cognitive engagement to active learning outcomes. *Educational Psychologist, 49*(4), 219–243. https://doi.org/10.1080/00461520.2014.965823

Clark, R., Nguyen, F., Sweller, J., & Baddeley, M. (2006). Efficiency in learning: Evidence-based guidelines to manage cognitive load. *Performance Improvement, 45*. https://doi.org/10.1002/pfi.4930450920

Cohen, G. L., & Sherman, D. K. (2014). The psychology of change: Self-affirmation and social psychological intervention. *Annual Review of Psychology, 65*(1), 333–371. https://doi.org/10.1146/annurev-psych-010213-115137

Cohen, J. (1988). *Statistical power analysis for the behavioral sciences* (2nd ed.). Lawrence Erlbaum.

Custers, E. (2010). Long-term retention of basic science knowledge: A review study. *Advances in Health Sciences Education: Theory and Practice, 15*(1), 109–128. https://doi.org/10.1007/s10459-008-9101-y

Day, L. J. (2018). A gross anatomy flipped classroom effects performance, retention, and higher-level thinking in lower performing students. *Anatomical Sciences Education, 11*(6), 565–574. https://doi.org/10.1002/ase.1772

DeLozier, S. J., & Rhodes, M. G. (2017). Flipped classrooms: A review of key ideas and recommendations for practice. *Educational Psychology Review, 29*(1), 141–151. https://doi.org/10.1007/s10648-015-9356-9

Deslauriers, L., Schelew, E., & Wieman, C. (2011). Improved learning in a large-enrollment physics class. *Science, 332*(6031), 862–864. https://doi.org/10.1126/science.1201783

Doggrell, S. A. (2020). No apparent association between lecture attendance or accessing lecture recordings and academic outcomes in a medical laboratory science course. *BMC Medical Education, 20*(1), 207. https://doi.org/10.1186/s12909-020-02066-9

Dunlosky, J., Rawson, K. A., Marsh, E. J., Nathan, M. J., & Willingham, D. T. (2013). Improving students' learning with effective learning techniques: Promising directions from cognitive and educational psychology. *Psychological Science in the Public Interest, 14*(1), 4–58. https://doi.org/10.1177/1529100612453266

Flynn, A. B. (2015). Structure and evaluation of flipped chemistry courses: Organic & spectroscopy, large and small, first to third year, English and French. *Chemistry Education Research and Practice, 16*(2), 198–211. https://doi.org/10.1039/C4RP00224E

Freeman, S., Eddy, S. L., McDonough, M., Smith, M. K., Okoroafor, N., Jordt, H., & Wenderoth, M. P. (2014). Active learning increases student performance in science, engineering, and mathematics. *Proceedings of the National Academy of Sciences of the United States of America, 111*(23), 8410. https://doi.org/10.1073/pnas.1319030111

Gillette, C., Rudolph, M., Kimble, C., Rockich-Winston, N., Smith, L., & Broedel-Zaugg, K. (2018). A meta-analysis of outcomes comparing flipped classroom and lecture. *American Journal of Pharmaceutical Education, 82*(5), 6898–6898. https://doi.org/10.5688/ajpe6898

Gross, D., Pietri, E. S., Anderson, G., Moyano-Camihort, K., & Graham, M. J. (2015). Increased preclass preparation underlies student outcome improvement in the flipped classroom. *CBE—Life Sciences Education, 14*(4), ar36. https://doi.org/10.1187/cbe.15-02-0040

Haak, D. C., HilleRisLambers, J., Pitre, E., & Freeman, S. (2011). Increased structure and active learning reduce the achievement gap in introductory biology. *Science, 332*(6034), 1213–1216. https://doi.org/10.1126/science.1204820

Harackiewicz, J. M., & Hulleman, C. S. (2010). The importance of interest: The role of achievement goals and task values in promoting the development of interest. *Social and Personality Psychology Compass, 4*(1), 42–52. https://doi.org/10.1111/j.1751-9004.2009.00207.x

Hattie, J. (2009). *Visible learning: A synthesis of over 800 meta-analyses relating to achievement.* Routledge.

Hattie, J., & Timperley, H. (2007). The power of feedback. *Review of Educational Research, 77*, 81–112.

Hew, K. F., Bai, S., Dawson, P., & Lo, C. K. (2021). Meta-analyses of flipped classroom studies: A review of methodology. *Educational Research Review, 33*, 100393. https://doi.org/10.1016/j.edurev.2021.100393

Hew, K. F., & Lo, C. K. (2018). Flipped classroom improves student learning in health professions education: A meta-analysis. *BMC Medical Education, 18*(1), 38. https://doi.org/10.1186/s12909-018-1144-z

Hu, R., Gao, H., Ye, Y., Ni, Z., Jiang, N., & Jiang, X. (2018). Effectiveness of flipped classrooms in Chinese baccalaureate nursing education: A meta-analysis of randomized controlled trials. *International Journal of Nursing Studies, 79*, 94–103. https://doi.org/10.1016/j.ijnurstu.2017.11.012

Jensen, J. L., Kummer, T. A., & Godoy, P. D. d. M. (2015). Improvements from a flipped classroom may simply be the fruits of active learning. *CBE—Life Sciences Education, 14*(1), ar5. https://doi.org/10.1187/cbe.14-08-0129

Johnson, D., Murayama, G., Johnson, R., Nelson, D., & Skon, L. (1981). Effects of cooperative, competitive, and individualistic goal structures on achievement: A meta-analysis. *Psychological Bulletin, 89*, 47–62. https://doi.org/10.1037/0033-2909.89.1.47

Karagöl, I., & Esen, E. (2018). The effect of flipped learning approach on academic achievement: A meta-analysis study. *Hacettepe Universitesi Egitim Fakultesi Dergisi-hacettepe University Journal of Education, 34*, 708–727.

Kauffman, C. A., Derazin, M., Asmar, A., & Kibble, J. D. (2018). Relationship between classroom attendance and examination performance in a second-year medical pathophysiology class. *Advances in Physiology Education, 42*(4), 593–598. https://doi.org/10.1152/advan.00123.2018

Kruger, J., & Dunning, D. (1999). Unskilled and unaware of it: How difficulties in recognizing one's own incompetence lead to inflated self-assessments. *Journal of Personality and Social Psychology, 77*(6), 1121–1134. https://doi.org/10.1037/0022-3514.77.6.1121

Låg, T., & Sæle, R. (2019). Does the flipped classroom improve student learning and satisfaction? A systematic review and meta-analysis. *AERA Open, 5*, 233285841987048. https://doi.org/10.1177/2332858419870489

Lai, C.-L., & Hwang, G.-J. (2016). A self-regulated flipped classroom approach to improving students' learning performance in a mathematics course. *Computers & Education, 100*, 126–140. https://doi.org/10.1016/j.compedu.2016.05.006

Laird-Fick, H. S., Solomon, D. J., Parker, C. J., & Wang, L. (2018). Attendance, engagement and performance in a medical school curriculum: Early findings from competency-based progress testing in a new medical school curriculum. *PeerJ, 6*, e5283. https://doi.org/10.7717/peerj.5283

Larsen, D. P., Butler, A. C., & Roediger, H. L., 3rd. (2013). Comparative effects of test-enhanced learning and self-explanation on long-term retention. *Medical Education, 47*(7), 674–682. https://doi.org/10.1111/medu.12141

Ling, Y., Swanson, D. B., Holtzman, K., & Bucak, S. D. (2008). Retention of basic science information by senior medical students. *Academic Medicine, 83*(10 Suppl), S82–S85. https://doi.org/10.1097/ACM.0b013e318183e2fc

Liu, Y.-Q., Li, Y.-F., Lei, M.-J., Liu, P.-X., Theobald, J., Meng, L.-N., Liu, T.-T., Zhang, C.-M., & Jin, C.-D. (2018). Effectiveness of the flipped classroom on the development of self-directed learning in nursing education: A meta-analysis. *Frontiers of Nursing, 5*, 317–329. https://doi.org/10.1515/fon-2018-0032

Lo, C. K., & Hew, K. F. (2019). The impact of flipped classrooms on student achievement in engineering education: A meta-analysis of 10 years of research. *Journal of Engineering Education, 108*(4), 523–546. https://doi.org/10.1002/jee.20293

Lo, C. K., Hew, K. F., & Chen, G. (2017). Toward a set of design principles for mathematics flipped classrooms: A synthesis of research in mathematics education. *Educational Research Review, 22*, 50–73. https://doi.org/10.1016/j.edurev.2017.08.002

Lundin, M., Bergviken Rensfeldt, A., Hillman, T., Lantz-Andersson, A., & Peterson, L. (2018). Higher education dominance and siloed knowledge: A systematic review of flipped classroom research. *International Journal of Educational Technology in Higher Education, 15*(1), 20. https://doi.org/10.1186/s41239-018-0101-6

Mawhinney, V. T., Bostow, D. E., Laws, D. R., Blumenfeld, G. J., & Hopkins, B. L. (1971). A comparison of students studying-behavior produced by daily, weekly, and three-week testing schedules. *Journal of Applied Behavior Analysis, 4*(4), 257–264. https://doi.org/10.1901/jaba.1971.4-257

Michaelsen, L., Knight, A. B., & Fink, L. (2002). *Team-based learning: A transformative use of small groups in college teaching*. Stylus.

Moravec, M., Williams, A., Aguilar-Roca, N., & O'Dowd, D. K. (2010). Learn before lecture: A strategy that improves learning outcomes in a large introductory biology class. *CBE—Life Sciences Education, 9*(4), 473–481. https://doi.org/10.1187/cbe.10-04-0063

Murad, M. H., & Varkey, P. (2008). Self-directed learning in health professions education. *Annals Academy of Medicine Singapore, 37*(7), 580.

National Academies of Sciences, Engineering, & Medicine. (2018). *How people learn II: Learners, contexts, and cultures*. The National Academies Press.

National Research Council. (2000). *How people learn: Brain, mind, experience, and school: Expanded edition*. The National Academies Press.

Nokes-Malach, T. J., Richey, J. E., & Gadgil, S. (2015). When is it better to learn together? Insights from research on collaborative learning. *Educational Psychology Review, 27*(4), 645–656. https://doi.org/10.1007/s10648-015-9312-8

O'Flaherty, J., & Phillips, C. (2015). The use of flipped classrooms in higher education: A scoping review. *The Internet and Higher Education, 25*, 85–95. https://doi.org/10.1016/j.iheduc.2015.02.002

Persky, A. M., & McLaughlin, J. E. (2017). The flipped classroom—From theory to practice in health professional education. *American Journal of Pharmaceutical Education, 81*(6), 118. https://doi.org/10.5688/ajpe816118

Quesnelle, K. M., Zaveri, N. T., Schneid, S. D., Blumer, J. B., Szarek, J. L., Kruidering, M., & Lee, M. W. (2021a). Design of a foundational sciences curriculum: Applying the ICAP framework to pharmacology education in integrated medical curricula. *Pharmacology Research & Perspectives, 9*(3), e00762. https://doi.org/10.1002/prp2.762

Quesnelle, K. M., Zaveri, N. T., Schneid, S. D., Blumer, J. B., Szarek, J. L., Kruidering, M., & Lee, M. W. (2021b). The importance of collaboratively designing pharmacology education programs. *Pharmacology Research & Perspectives, 9*(3), e00773. https://doi.org/10.1002/prp2.773

Reiser, B. J. (2004). Scaffolding complex learning: The mechanisms of structuring and problematizing student work. *The Journal of the Learning Sciences, 13*(3), 273–304. https://doi.org/10.1207/s15327809jls1303_2

Richardson, M., Abraham, C., & Bond, R. (2012). Psychological correlates of university students' academic performance: A systematic review and meta-analysis. *Psychological Bulletin, 138*(2), 353–387. https://doi.org/10.1037/a0026838

Roediger, H. L., 3rd, & Butler, A. C. (2011). The critical role of retrieval practice in long-term retention. *Trends in Cognitive Sciences, 15*(1), 20–27. https://doi.org/10.1016/j.tics.2010.09.003

Ryan, R. M., & Deci, E. L. (2000). Self-determination theory and the facilitation of intrinsic motivation, social development, and well-being. *The American Psychologist, 55*(1), 68–78. https://doi.org/10.1037//0003-066x.55.1.68

Sams, A., & Bergmann, J. (2012). *Flip your classroom: Reach every student in every class every day*. International Society for Technology in Education.

Schell, J. A., & Butler, A. C. (2018). Insights from the science of learning can inform evidence-based implementation of peer instruction. *Frontiers in Education, 3*(33). https://doi.org/10.3389/feduc.2018.00033

Shi, Y., Ma, Y., MacLeod, J., & Yang, H. H. (2019). College students' cognitive learning outcomes in flipped classroom instruction: A meta-analysis of the empirical literature. *Journal of Computers in Education, 7*. https://doi.org/10.1007/s40692-019-00142-8

Silverthorn, D. U. (2020). When active learning fails... and what to do about it. In J. J. Mintzes & E. M. Walter (Eds.), *Active learning in college science: The case for evidence-based practice* (pp. 985–1001). Springer.

Silverthorn, D. U., Lee, M. W., Corliss, S. B., Nelson, E. A., & Bergemann, A. D. (2020). Words of advice: Preparing to teach. *The FEBS Journal, 287*(3), 443–451. https://doi.org/10.1111/febs.15184

Stegers-Jager, K. M., Cohen-Schotanus, J., & Themmen, A. P. (2012). Motivation, learning strategies, participation and medical school performance. *Medical Education, 46*(7), 678–688. https://doi.org/10.1111/j.1365-2923.2012.04284.x

Steinel, N., Palmer, G. C., Nowicki, E., Lee, E., Nelson, E., Whiteley, M., & Lee, M. W. (2019). Integration of microbiology, pharmacology, immunology, and infectious disease using active teaching and self-directed learning. *Medical Science Educator, 29*(1), 315–324. https://doi.org/10.1007/s40670-018-00689-8

Stephenson, J. E., Brown, C., & Griffin, D. K. (2008). Electronic delivery of lectures in the university environment: An empirical comparison of three delivery styles. *Computers & Education, 50*(3), 640–651. https://doi.org/10.1016/j.compedu.2006.08.007

Strelan, P., Osborn, A., & Palmer, E. (2020). The flipped classroom: A meta-analysis of effects on student performance across disciplines and education levels. *Educational Research Review, 30*, 100314. https://doi.org/10.1016/j.edurev.2020.100314

Styers, M. L., Van Zandt, P. A., & Hayden, K. L. (2018). Active learning in flipped life science courses promotes development of critical thinking skills. *CBE—Life Sciences Education, 17*(3), ar39. https://doi.org/10.1187/cbe.16-11-0332

Talbert, R. (2017). *Flipped learning : A guide for higher education faculty*. Stylus.

Tan, C., Yue, W.-G., & Fu, Y. (2017). Effectiveness of flipped classrooms in nursing education: Systematic review and meta-analysis. *Chinese Nursing Research, 4*(4), 192–200. https://doi.org/10.1016/j.cnre.2017.10.006

Tang, B., Coret, A., Qureshi, A., Barron, H., Ayala, A. P., & Law, M. (2018). Online lectures in undergraduate medical education: Scoping review. *JMIR Medical Education, 4*(1), e11. https://doi.org/10.2196/mededu.9091

van Alten, D. C. D., Phielix, C., Janssen, J., & Kester, L. (2019). Effects of flipping the classroom on learning outcomes and satisfaction: A meta-analysis. *Educational Research Review, 28*, 100281. https://doi.org/10.1016/j.edurev.2019.05.003

van Vliet, E. A., Winnips, J. C., & Brouwer, N. (2015). Flipped-class pedagogy enhances student metacognition and collaborative-learning strategies in higher education but effect does not persist. *CBE—Life Sciences Education, 14*(3). https://doi.org/10.1187/cbe.14-09-0141

Velegol, S., Zappe, S. E., & Mahoney, E. (2015). The evolution of a flipped classroom: Evidence-based recommendations. *Advances in Engineering Education, 4*.

Wang, T. (2017). Overcoming barriers to 'flip': Building teacher's capacity for the adoption of flipped classroom in Hong Kong secondary schools. *Research and Practice in Technology Enhanced Learning, 12*(1), 6. https://doi.org/10.1186/s41039-017-0047-7

Ward, M., Knowlton, M. C., & Laney, C. W. (2018). The flip side of traditional nursing education: A literature review. *Nurse Education in Practice, 29*, 163–171. https://doi.org/10.1016/j.nepr.2018.01.003

White, P. J., Naidu, S., Yuriev, E., Short, J. L., McLaughlin, J. E., & Larson, I. C. (2017). Student engagement with a flipped classroom teaching design affects pharmacology examination performance in a manner dependent on question type. *American Journal of Pharmaceutical Education, 81*(9), 5931. https://doi.org/10.5688/ajpe5931

Wiggins, G., & McTighe, J. (2005). *Understanding by design, expanded* (2nd ed.). Association for Supervision and Curriculum Development.

Wilton, M., Gonzalez-Nino, E., McPartlan, P., Terner, Z., Christoffersen, R. E., & Rothman, J. H. (2019). Improving academic performance, belonging, and retention through increasing structure of an introductory biology course. *CBE—Life Sciences Education, 18*(4), ar53. https://doi.org/10.1187/cbe.18-08-0155

Xu, P., Chen, Y., Nie, W., Wang, Y., Song, T., Li, H., Li, J., Yi, J., & Zhao, L. (2019). The effectiveness of a flipped classroom on the development of Chinese nursing students' skill competence: A systematic review and meta-analysis. *Nurse Education Today, 80*, 67–77. https://doi.org/10.1016/j.nedt.2019.06.005

Yeager, D., & Dweck, C. (2012). Mindsets that promote resilience: When students believe that personal characteristics can be developed. *Educational Psychologist, 47*. https://doi.org/10.1080/00461520.2012.722805

Zhang, D., Zhou, L., Briggs, R. O., & Nunamaker, J. F. (2006). Instructional video in e-learning: Assessing the impact of interactive video on learning effectiveness. *Information & Management, 43*(1), 15–27. https://doi.org/10.1016/j.im.2005.01.004

Zhang, L. (2018). English flipped classroom teaching model based on cooperative learning. *Educational Sciences: Theory & Practice, 18*(6), 3652–3661. https://doi.org/10.12738/estp.2018.6.278

Zhu, G., Thompson, C., Suarez, M., & Peng, Z. (2019, April 5–9). *A meta-analysis on the effect of flipped instruction on K-12 students' academic achievement*. Paper presentation at the American Educational Research Association annual meeting, Toronto, Canada.

Chapter 7
Supplementary Videos in the Biosciences: How Stakeholders Can Reinforce Complex Concepts for Self-Directed Learners

Richard Guy and Harry J. Witchel

Abstract Supplementary videos for learning are those that are accessed by students as an original adjunct to other learning materials. Supplementary videos are very popular, and YouTube view numbers suggest extraordinary engagement, to the point where students overvalue such videos versus textbooks. Thus, one rationale for instructors and institutions to engage with educational video media is the increased student engagement, which may result in a virtuous cycle of further student engagement. The unique audiovisual content enabled by videos may also create more memorable learning experiences or unique explanatory opportunities, particularly for 3-D or procedural/sequential information. We review the elements that enable effective video learning of the biosciences. Much research has produced contradictory results, although there is agreement that active learning during videos that ask students to stop the video to answer questions or perform other tasks results in better outcomes. The disadvantages of using videos for education are inaccurate content, profit-making search-engine curation, poor student insight, difficulty in editing/updating, and the high cost and labor of production. The proliferation of video learning materials has been contemporaneously accompanied by increasing student numbers and institutional efficiencies, lessening students' contact with instructors for adaptive learning experiences regardless. Students, currently guided by opaque search algorithms, need to become truly self-directed learners able to find and evaluate information online that is poorly curated and may be oversimplified, misleading, or false. Scientific organizations and learned societies could certify videos or producers, helping videos recoup some of the learning losses driven by the efficiencies occurring in higher education.

R. Guy
Victoria University, Melbourne, VIC, Australia

H. J. Witchel (✉)
Department of Neuroscience, Brighton and Sussex Medical School, Brighton, UK
e-mail: h.witchel@bsms.ac.uk

H. J. Witchel, M. W. Lee (eds.), *Technologies in Biomedical and Life Sciences Education*, Methods in Physiology, https://doi.org/10.1007/978-3-030-95633-2_7

7.1 Introduction: The Educational Case for Supplementary Videos

7.1.1 Remit of This Chapter

How do we reap the potential value of educational videos in the biosciences, and are there approaches to sidestep the obvious pitfalls? In this chapter, we investigate the different roles of supplementary video, effective approaches to support these roles, and the future of this learning method. We also attempt to assess the extent to which a resource of this type meets its potential (i.e., does it reliably improve learning outcomes?). We reaffirm the concerns of previous authors who have highlighted that video producers need to consider (1) how to maximize engagement with such a video, (2) how to manage the cognitive load demanded by the video, and (3) ways to promote active learning with the video (Brame, 2016). To address the pitfalls, we revisit the illusion of understanding as well as the trustworthiness problem inherent in social media (Schwartz, 2013; Thompson et al., 2020). We conclude that science educators (and their learned societies) need to organize the effective curation of (and access to) those supplementary videos that manage to balance credibility and accuracy with engagement (Celentano et al., 2020); if we do not, then this curation will be managed for us by opaque search algorithms (Gorwa et al., 2020).

Supplementary videos for learning are those that are provided to students as an original adjunct to other learning materials; they are sometimes referred to as video podcasts, and the most common varieties are short screencasts; a screencast is an instructional video made by recording what appears on your computer's screen, and it usually includes audio narration. In this chapter we do not rate individual videos, nor do we compare video producers; because the videos are too numerous, the poor overlap in their learning outcomes makes them impossible to compare. Furthermore, the methods for assessing learning from videos are not standardized. Instead, we propose approaches to address those issues and provide some suggestions for best practices; whereas the literature will eventually compare the extraordinary breadth of videos (or instructors, see Emanuel (2020)), right now this is not possible or fair. This would be a task for an international organization or a learned society, not an individual researcher.

Also, in this chapter we do not consider recordings or translations of other learning materials into video formats such as lecture capture recordings, which are covered elsewhere (see Chaps. 11 and 12 on Classroom Attendance in this volume). Although this chapter focuses exclusively on English-language videos, there are extensive educational video resources in other languages that use many of the same processes and platforms, including YouTube (Ramos-Rincón et al., 2017); ultimately, it is our hope and expectation that the educational benefits of videos in all languages would have many commonalities and that the research on educational videos from all languages would mutually inform one another.

7.1.2 Environment for the Use of Videos

Educational video has become a ubiquitous medium for the recording and playback of moving images (mainly with accompanying audio track). Currently the term relates to discrete packages of digital information that can be viewed within lectures/laboratories, accessed using institutional computers, or used online, or downloaded to a personal electronic device for viewing. Video can also be used as an adjunct to face-to-face classes in blended learning or flipped classes (see Chap. 6 on Flipped Classrooms in this volume).

The combination of ease of recording, storage, and access has initiated a modest revolution in teaching methods. We believe that this is only the beginning of a much more substantial change in teaching. To this end we investigate the different roles of supplementary video, the student perspective, and, most importantly, the impact on learning.

The first indicators of this teaching revolution came when universities started to install lecture capture recording systems (LCRs) where recordings (of whole lectures) were made available to students, often via the relevant learning management system. Initially this was done without any supporting research, but the general approach was that recorded lectures would provide a useful backup to face-to-face lectures. However, the early literature was much more focused on whether or not the learning impact of recorded lectures was better, worse, or equal to the face-to-face equivalent and also the extent to which availability of recorded lectures impacted on student attendance to face-to-face lectures (Banerjee, 2021; O'Callaghan et al. 2017). Thus, the issue of supplementation was in the background. This situation has continued to the present day, where many publications simply regard students using recorded lectures as one uniform group without separating them into (a) those that substitute and (b) those that supplement face-to-face lectures. The role of LCRs as a supplement to learning is covered in Chaps. 11 and 12 on Attendance in this volume.

Following the implementation of lecture capture, there was an increase in both the institutional generation of short videos for teaching and, with the advances in personal computing power, the generation of short teaching videos by academic staff (Kay, 2012). We also investigate the design, use, and learning outcomes of this type of learning tool.

Hand in hand with the educational use of video came the development of YouTube, where anyone could upload short videos on practically any topic. This platform also became an alternative place where videos of an academic nature could also be uploaded. We also investigate the occurrence, reliability, and learning potential of videos made available in this context (Curran et al., 2020; Jaffar, 2012).

The online environment poses a number of challenges, one of which is the (relative) absence of an instructor's guidance (Guy et al., 2014). The availability of online resources and the shift toward self-directed learning have generally been regarded as an advance relative to the more traditional didactic face-to-face lecture. However, when one compares the two situations, it becomes obvious that this new frontier may lead to two related problems; together these problems may result in students learning incorrect information. (1) Students have less personalized/adaptive teaching from instructors. Before the advent of online resources, a student might well approach an academic for assistance with a difficult concept. We are not suggesting that this no longer occurs; however, increasing numbers of student and staff responsibilities sometimes make it difficult for this type of interaction to occur. There is also evidence that most students prefer the instant option of trying to work out their questions using online resources rather than booking a time to see an academic (O'Malley et al., 2019). (2) Students increasingly need to be able to evaluate both unvetted sources of information and their own ability to use them. This preference for online answers raises questions about learning: how good is the information that the student accesses online, how good are they at finding accurate online material, and how effective is the answering of their query?

Face-to-face learning does provide the opportunity for interaction and adaptive teaching, so its replacement by online resources (including videos) is a potential problem. Efficiencies in education as a whole demand that students become self-directed learners, who are able to both find and evaluate online information. In particular, students rely upon opaque search algorithms to find both relevant and accurate information. By contrast, worldwide there are a multitude of physiology educators with a wealth of experience, which is, to some extent, tied to the individual. Experience, which translates into providing effective answers for students, may, to some extent, be lost as we progress into a digital-heavy future. In this chapter, we consider how supplementary videos may benefit students in an educational environment where students receive diminishing levels of one-to-one tuition.

7.2 Formats of Educational Videos

About 10 years ago, digital video capture of lectures (lecture capture recording) started to become popular, and it was adopted by many universities (O'Callaghan et al., 2017). The recorded lectures were made available online via the institutional intranet. At that point in time, adoption preceded any significant amount of research into whether or not this new resource actually had any learning benefits. Apart from the presumed benefit of this supplementary resource, lecture capture was also considered as a replacement for face-to-face lectures (i.e., to become a core component in its own right) (Bassili, 2008).

The production of lecture capture videos tended to be standardized and regulated within institutions (mainly due to cost limitations). However, the advent of desktop computers with video cameras and video-editing software facilitated the "Do-It-Yourself" approach and accelerated the development of video resources by individual academics. There are many different software approaches for recording, editing, and presenting videos to students (Table 7.1). Whereas many institutions will have video content creation and management solution (in which recording, basic editing, storage, and playback to students are managed from inside the platform), some

Table 7.1 Software categories used for making supplementary video experiences

Software category	Record	Edit	Audio	Listing of software on Wikipedia: https://en.wikipedia.org/wiki/ +
Screencasting	✓	✓	✓	Comparison_of_screencasting_software
Webcam recording	✓		✓	Comparison_of_webcam_software
Video editing		✓		Comparison_of_video_editing_software
Mix >2 sources	✓	[a]	✓	Software_vision_mixer
Online solutions	✓	✓	✓	
Panopto	✓	[b]	✓	
Echo 360	✓	[b]	✓	
Kaltura	✓	[b]	✓	
Vimeo Create	✓	✓	✓	

Audio refers to recording of live audio onto the experience
[a]Software vision mixers edit and transition, but livestream
[b]Basic edit features, but include many extra educational features

instructors will be attracted to the simplest approach, which is screencasting software. This allows the authors to record directly from their computer screen (e.g., PowerPoint slide show) while also overdubbing a voice "as you record," resulting in a video file (e.g., avi or mpeg) that can be played on standard media players (e.g., VLC or QuickTime) and uploaded to video platforms such as YouTube. Most screencasting software comes with basic editing features, which allows for ease of use rather than professional video editing.

Many researchers have considered at length the format of the video in terms of where or how the instructor appears (Choe et al., 2019). These tests of student preference, like most market research, are subject to fads that seem to be driven by novelty, as seen by the progression from Khan Academy to "Learning Glass" videos. We will focus on two formatting issues that are unlikely to change: video duration and the type of information being learned.

The duration of each video is a topic of some controversy. Shorter videos are more engaging and more likely to be watched to completion (Brame, 2016). By contrast, complex topics require time to explain and contextualize, with the implication that shorter videos are superficial and do little more than misleadingly reassuring students rather than communicating complex concepts (Paik & Schraw, 2013). The compromise position is to break apart long explanations into a series of shorter videos that allow the student to dip in and out or to search for the precise problem they need to address. Generally, most supplementary videos tend to be short (less than 20 min).

Thus far, a diverse range of supplementary videos has been generated (including institutionally made material) with the intention of providing an improved learning experience for students over and above their use of core course materials (both face-to-face and online) (Kay, 2012). The term instructional video is often used to cover both procedural and concept-based videos. Procedural videos are those where some form of manual task is accomplished (e.g., dissection or laboratory preparation (Croker et al., 2010)) or where a technique is demonstrated (e.g., clinical

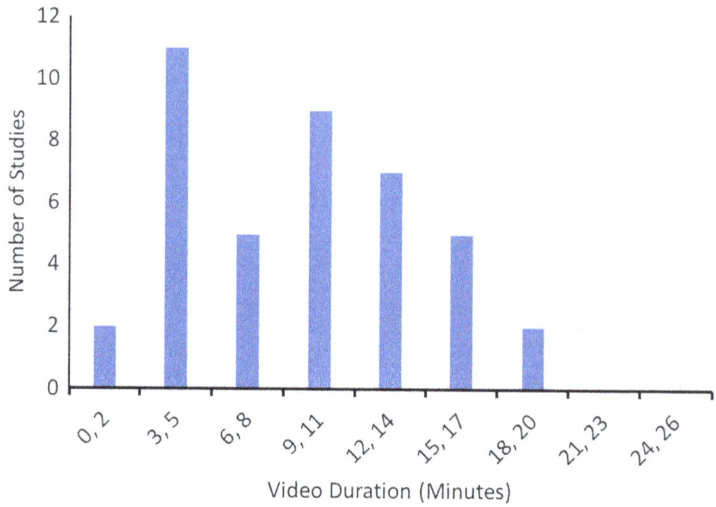

Fig. 7.1 Video durations from all the studies referred to in this chapter (where values are given). Where a study has multiple videos, a mean or median value is used. For studies that include different types of video, the data is plotted separately. Four studies with videos longer than 26 min are not shown. For further details, see Sect. 7.5.4

examination). With concept-based videos, the aim is to assist students with difficult course content. Although we have, for the sake of clarity, attempted to deal with these categories of video separately, a given video may contain any mix of approaches, depending on the desired learning outcome. In this chapter we will only focus on videos that are designed to assist student learning and will not consider other uses, e.g., providing information about course structure, assessment procedures, etc. However, we have included a number of studies that can be classified as experimental, where best practice approaches for video design are established. Although these videos are relevant to any teaching video, they do provide guidelines for the development of supplementary videos and are thus included within this chapter. The durations of all videos from all the studies referred to in this chapter (procedural, conceptual, and experimental) are shown in Fig. 7.1. Because of the well-known preference of some viewers for short videos (Brame, 2016), there are fewer studies on longer videos, despite their potential utility for complex topics.

7.3 The Rationale for Videos: Learning, Engagement, and Satisfaction

7.3.1 Instructor Viewpoint

Before proceeding with a detailed analysis of the use and design of videos, it is important to consider whether there is an advantage of using video compared to an equivalent text/static image format. Of course, text-based materials have been used for many years and are still an important part of education. However, several studies have made direct comparisons between video and text to answer the above question. In a study by Stockwell et al. (2015) (biochemistry), a problem-solving concept video was developed for use prior to in-lecture problems, and this was compared with an equivalent text version. The pre-lecture video preparation group did not have a significantly different exam performance compared to the text-only preparation group. However, the students were significantly more satisfied with the video material compared to the text material, notwithstanding the lack of difference in academic outcomes. Thus, there is a research focus on student experience of learning with videos, independent of objective learning, which may encourage a virtuous cycle of engagement (especially with technical or so-called tedious topics) that can lead to learning independently of the educational quality of one particular video (Choe et al., 2019; Liu & Elms, 2019).

Furthermore, some research has found that videos can improve learning. Moravec et al. (2010) (biology concepts—nuclear structure/ribosomes) developed two narrated PowerPoint videos for students to use prior to lecture. There was a significant increase in performance on both comprehension- and application-level questions between the video and text-equivalent group, and 80% of respondents indicated that the "learn before lecture" videos were helpful in learning course material. Another study by Pickering (2017) (muscular anatomy) used a single procedural video and compared its effectiveness with a printed matter group. The video group showed a significant increase in learning gain compared to the printed matter group. Finally, Salina et al. (2012) developed a single procedural video (technique refresher). The video group showed a significant improvement in learning compared with the text-only group. This evidence suggests that, for learning both procedures and concepts, the video was either equal to, or superior to, the equivalent text/image-based material. This does not necessarily mean that video cannot be used with associated text/images, but it does suggest that video is an effective learning tool in its own right.

7.3.2 Student Viewpoint

It appears that students appreciate the availability and convenience of supplementary videos as a means of providing quick-fix answers to issues that arise during their core learning. The use of this particular medium may have been facilitated by familiarity with videos as part of the social media experience (Dzara et al., 2020).

In the past a student who needed clarification of a concept would have most probably approached an academic or tutor for assistance and/or delved into relevant textbooks. Anecdotally the willingness of students to approach a staff member (or use their textbook) appears to have dropped over the last few years, so they must be turning to other options. O'Malley et al. (2019) investigated both the extent to which and the reasons why medical students in Ireland used different resources to support their learning. In this study, half of the students surveyed said that if they had a physiology question, they would refrain from meeting face-to-face with their instructor to discuss difficult physiology concepts. Furthermore, 70% of these students reported that they would also refrain from seeking help from their physiology instructor via e-mail.

The question is: What resources do they use if they are not contacting the instructor? O'Malley et al. (2019) found that 97–99% of respondents agreed that they used web-based resources expressly to locate physiology information. YouTube (see Sect. 7.4.1) was undoubtedly the most common resource (76% agreed), 10% used a search engine, 5% used medical/health websites, 5% used textbooks, and less than 1% used the DVDs that come with some textbooks.

The problem with this type of student behavior is that they may be using educationally unsound material, which either may not answer their question or lead to an ongoing misconception. Educators should take note and provide resources where the students are already looking, and for this reason, we suggest that short supplementary videos may be an effective way of providing assistance for students regarding difficult topics (see Sect. 7.9).

For many students, supplementary videos have already become a substitute for one-on-one tutelage from an instructor. If you can't beat them, join them. Instead of denouncing the video medium for learning, some brave instructors have embraced this challenge so that their teaching experience can be brought to bear in video instruction. With the aid of helpful graphics as well (see Sect. 7.5.10), the finished product can then be made available for students both locally and globally.

7.4 Platforms Where Students Can Access Supplementary Videos

There are essentially three types of platform where students can access supplementary videos. The first is via the free-to-all world wide web, particularly by using popular video websites like YouTube or the Khan Academy. The second platform type is via commercial providers who use the Internet and a paywall to provide instruction and self-test material to subscribers; this includes UWorld, Pharmacology World, Pathoma, OnlineMedEd, Osmosis, etc. (Burk-Rafel et al., 2017; Tackett et al., 2018). The third type of platform is an institutional learning management system (or equivalent) which, unlike YouTube, is regulated by a particular academic institution. However, it should be pointed out that, in some cases, institutional supplementary videos have been uploaded to specific YouTube channels. The difference between YouTube videos accessed by a general search or by access to an institutional channel is clarified below.

Firstly, we will consider where students access videos outside of their institutional online structures, with our main focus on the YouTube site. The main questions here are the popularity of such sites, the types of video that can be accessed, and finally whether or not resources of this type are educationally effective.

7.4.1 YouTube and Related Platforms

Curran et al. (2020) recently carried out a literature scoping review of the utility, effectiveness, and validity of YouTube video resources in medical education. Of 113 articles identified by database searching, only 31 papers evaluated the usefulness (either learning or satisfaction) of YouTube for delivering content related to medical education. They found that although YouTube is a growing educational source of material for medical students, there was a lack of evidence-based regulation. For example, in the cardiovascular version of a series of studies by Azer (2014), they searched for YouTube videos focused on covering mechanisms or pathogenesis of the cardiovascular diseases in adults and compared the quality of the educational content (based on standardized criteria) to best-selling textbooks and emedicine.com. Overall, only a small proportion of YouTube videos are educationally relevant. Table 7.2 shows that the proportion is particularly low when the percentages are based on video number post preliminary selection.

7.4.2 Pervasiveness and Learning Benefits of YouTube Videos

Access to YouTube videos can either be "general" (i.e., enter a search term) or "specific" (i.e., access a specific named channel and search within that channel). Some channels cannot be accessed by a general search and can only be accessed when one is provided with a specific URL. In 2020 YouTube offered its own open, English-language channel, YouTube Learning, which claimed to be relevant for "studying for a test, want to learn a new skill or are just curious about the world around you," although it is not specifically aimed at university students. There are

Table 7.2 Scope and distribution of YouTube videos by discipline and topic

Author	Topic	Number of videos after preliminary selection	Proportion of videos of educational standard (%)
Azer (2012)	Surface anatomy	57	27
Azer et al. (2012)	Nervous system	129	47
Azer (2014)	Cardiovascular	29	55
Raikos and Waidyasekara (2014)	Heart	294	26
Van den Eynde et al. (2019	Gene rearrangement	70	24
Van den Eynde et al. (2019)	Antigen presentation	82	43

many other user-created channels that have English-language biomedical and life sciences content. For example, the introductory level channel "Crash Course" (2015) contains 47 anatomy and physiology videos (at the time of writing). The "Khan Academy" also contains physiology-related video content. Although these videos have been accessed several million times, to our knowledge there is as yet no review of their overall quality or effectiveness, although efforts to understand this have been made in some fields of surgical training (Celentano et al., 2017; Fischer et al., 2013), as well as in mathematics education (Kelly & Rutherford, 2017).

Although the sites mentioned receive a large number of views, this does not provide us with information about whether or not the videos provided useful information. However, some studies have been carried out under more regulated conditions and provide us with data concerning the proportion of a student cohort that accesses YouTube and the extent to which they find the content useful for learning. Although the number of studies is limited (Table 7.3), two things are apparent. Firstly, a relatively large proportion of students being studied accessed YouTube video material, and secondly, they generally find the content useful for learning. Some of the studies referred to in Table 7.2 use specific channels where the video content has been curated, whereas other studies (general) refer to material of (mostly) unknown origin. Thus, with the studies by Barry and Burns (Barry et al.,

Table 7.3 Prevalence of students accessing YouTube videos

Author	Topic	Students who accessed YouTube videos (%)	YouTube videos found to be useful (%)	Comments
Jaffar (2012)	Anatomy students	86	92	Agreed that it helped anatomy learning (Specific human anatomy channel)
Mesko et al. (2015)	Medical students	74		(General YouTube access)
Barry et al. (2016)	Medical and radiation therapy students	78	78	Agreed it was chief source of anatomy videos (General YouTube access)
Burns et al. (2020)	Dental students		95	Agreed it was a useful learning tool (General YouTube Access)
Richards-Babb et al. (2014)	Specialized chemistry channel	77 (viewed all or most)	53	Useful prior to exam (problem-based videos) (Specific channel)
O'Malley et al. (2019)	Medical students		83–86	Agreed that "generally useful for understanding physiological concepts" (General YouTube access) 41% preferred Khan Academy

2016; Burns et al., 2020), students appear to be under the impression that the (general) YouTube material they access is useful for learning. This is of concern simply because reference to Table 7.2 shows that only a small proportion of YouTube videos are educationally sound. In addition a study by O'Malley et al. (2019) investigated student use of social media. As part of this they were asked "If you have used online video clips to help you with your understanding of physiological concepts, where do you mostly source them from?". The student feedback indicated that 41% used Khan Academy, 36% did not indicate a preference, and 9% used the Crash Course Anatomy and Physiology Channel. Thus, it is apparent, in this case, that at least 50% of students using online videos were using non-refereed videos for their study.

7.4.3 Curation, Regulation, and Reporting: Promoting YouTube Videos That Are Credible

The question is: How can one minimize student dependence on what may turn out to be unreliable video information? There are essentially two approaches to this issue. The first is the free speech/*caveat emptor* approach we essentially have now; as part of a self-regulatory strategy, students should be taught digital literacy to assess the quality of the bioscience information that they encounter online (Mesko et al., 2015). The second is a curated approach based on either labeling (e.g., certification) or regulation, in which individual videos, or video producers, have to meet certain standards or criteria. The simplest form of labeling would be for individual bioscience teachers to provide links to the YouTube material that they regard as appropriate for the learning support of their own students; teachers already do this for textbooks (i.e., Recommended Reading lists), and teachers could do this for YouTube videos, although it would take additional time for each individual instructor to find this material. Another curation approach would be for trustworthy channel owners to be scrupulous about what appears on their platforms; the problem with channel owners being entrusted with this responsibility is that currently YouTube, and the web more generally, is monetized based on engagement, so channel-owners, even genuinely knowledgeable and able instructors like Salman Khan, are incentivized both financially and for the sake of branding to stray well outside their area of expertise (Braun & Eklund, 2019; Pottinger, 2018).

A third curation approach would be expert-led and independently reviewed, in which a learned society or scientific organization, such as the American Physiological Society, bestows a certification or label that video producers could apply for, much like the labels for organic food. Some fields in the biomedical sciences (e.g., robotic surgery) have looked into this strategy using a Delphi method, leading to a consensus statement (Celentano et al., 2020). The obvious problem with such certification is that there are different levels of students who need different levels of knowledge, and lower-level information invariably has to be shortened and simplified. There would need to be certifications for different levels, and experts would have to agree upon what each level of knowledge was.

For example, there is a place for the rapid-fire, humorous videos of Crash Course (Schmidt, 2015; Crash Course, 2015), but we cannot imagine those videos being of particular use to our university students because the bioscience information in those videos is rapidly diluted out by entertainment and scene-setting; there is too much cognitive load associated with irrelevant information. If Crash Course is genuinely focusing on communicating difficult conceptual matter, the breakneck speed of the videos is counterproductive. Yet, approximately 10% of the medical students in O'Malley et al. (2019) use Crash Course as their first choice. This highlights a problem with the unregulated *caveat emptor* approach for uncurated videos: the illusion of understanding.

The illusion of understanding is the proposition that a viewer may overestimate their understanding of material taught by video (especially animation) because the video is so effortless to watch (Daly et al., 2016; Paik, 2010). This effortlessness is associated with perceptual fluency, which is well-known to inflate judgments of learning (Bjork et al., 2013). The illusion of understanding is related to the Dunning-Kruger effect (the cognitive bias that people with comparatively low ability at a task overestimate their ability because they have no idea how little they know (Kruger & Dunning, 1999)); it also is related to the illusion of explanatory depth (the belief that we understand more about a phenomenon, such as gravity or a zipper, than we actually do because visual actions seem trivial and familiar when the mechanism is not inspected (Rozenblit & Keil, 2002)). In all these illusions and biases, the overconfidence arises from a visual familiarity combined with a lack of cognitive effort. Animations can create a visual familiarity with a process plus a cognitive laziness because the pictures advance automatically without requesting comprehension and assent, so animations can lead to similar overconfidence.

Thus, animations can potentially have two opposite effects, one contributing to learning and one undermining it. (1) The *enabling function* creates previously unimagined thoughts (thought expansion): compared with static images, animations can enlarge the set of possible cognitive processes, so viewers can perform more cognitive processing. (2) Animations can have a *facilitating function*: they can trigger simplistic thoughts (e.g., mental schemas of moving objects) that crowd out scrutiny and deeper thinking (i.e., thought capturing or retrieval fluency) (Schnotz & Rasch, 2005). When learning materials facilitate effortless engagement with the material, students may learn less precisely because the materials require less cognitive effort.

Diminishing relevant cognitive effort short-circuits active learning, whereas videos that enable new thoughts—which allow students to imagine what they could not previously conceive of—diminish irrelevant cognitive effort (attempting to visualize beyond one's ability). Paik and Schraw (2013) showed quite elegantly that reducing irrelevant cognitive effort aided learning, whereas reducing relevant cognitive effort did not; they compared the use of directive animation (pointing at the relevant element) improved learning, whereas representational animations (demonstrating how things moved) decreased learning. The animations promoted more optimistic meta-comprehension in low-proficiency learners but more pessimistic meta-comprehension in high-proficiency learners: the expertise reversal effect (Kalyuga, 2007). The poor insight that students arrive at during enjoyable and effortless learning is not unique to videos and animations; it has been demonstrated in classroom situations with live instructors as well (Deslauriers et al., 2019).

7.4.4 *Institutionally Based Supplementary Videos*

One of the advantages of supplementary videos developed within an educational institution is that, unlike many YouTube videos, the design and content can be regulated. Another advantage is that, as videos are often uploaded to learning management systems, student access can (in many cases) be tracked and then related to learning outcomes.

It should be noted that educational institutions can upload supplementary videos to a designated YouTube channel. In this case the quality of the material can be regulated. However, this type of material may (often) be accessed by a specific link (provided by the institution) and would not be accessed by a general YouTube search. It is difficult to establish the extent of this type of mechanism due to the limited extent of research material available.

Before considering the relevant research on supplementary videos that are used in a learning context, we first focus on what is essentially the "gold standard" in video design. This standard is a work in progress derived from a mixture of expert-led judgment (Celentano et al., 2020) and experimental studies (often in the educational psychology area) of defined student populations, and the current situation is addressed below. The findings of this research can be equally applied to videos used within core components of a course, e.g., within traditional teaching or within blended or flipped courses.

7.5 Best Practices for Production

7.5.1 *Elements of a High-Quality Supplementary Video*

The following section describes the results of several experimental studies (mainly from an educational psychology perspective) that attempt to establish the optimum configuration of educational videos. These results are applicable to videos in general, but we have included them to provide guidance for the development of supplementary video. These studies generally make comparisons between videos with different learning approaches (i.e., video versus video) or, in some cases, between video and a text equivalent. However, the video content is generally not specifically related to a student's core course of study. This approach is taken to try and equalize pre-knowledge of the participants prior to video exposure. This disadvantage is that students may be motivated to undertake the study but not in the same way that they would be if the material were presented as part of their studies. Thus, further testing in the learning environment is required to fully establish the efficacy of the various findings below.

In addition to the educational psychology material referred to above, some studies have been included that are derived from work within learning environments to underline areas not fully covered by experimental studies.

7.5.2 Viewer Engagement

The general idea is that the video structure optimizes student engagement and thus encouraging more and deeper thinking; if a student gets hooked, it leads to a virtuous cycle of further rewarding study. Engagement can stimulate deeper cognitive processes such as reflection, thus facilitating incorporation of video material with the viewer's long-term memory. The following approaches (Table 7.4, see also Brame 2016) should be considered when designing a supplementary video to achieve maximum student engagement (and hopefully learning). Some aspects (e.g., how often should an educational video be updated) remain controversial.

7.5.3 Topic, Aims, and Remit: Introducing What the Video Is About

A relatively simple, but often overlooked, approach is to explain to the viewer at the beginning of the video what the video is about and what they are required to do during the video. Guy et al. (2014) used the first 2 min of their videos to provide this

Table 7.4 Criteria for video design

1. Topic, aims, and remit
2. Timing and student control
3. Information overload and how to avoid it
4. Prompts for the viewer to answer questions or take other actions
5. Segmentation
6. Cueing
7. Dynamic or static images
8. Animations
9. Camera angle (procedural video)
10. Instructor visibility
11. Humor
12. Narration
13. Organization and combining design principles

explanation with the intention of stimulating initial engagement and ongoing interest in the following material. This "contract with the viewer" is also a matter of courtesy and fairness to the viewer, as many YouTube video producers provide titles (e.g., "best explanation") that hype, over-promise, or mislead in order to maximize monetization (Braun & Eklund, 2019; Pottinger, 2018). A similar effect can be obtained by providing a similar resource outside of the video, such as in the description or metadata accompanying each YouTube video. As content-producers on YouTube receive an income based on advertising to their viewers, it is only fair to let the users know what is in the video before they view it, and thus they can plan their approach and locate information more simply (Cojean & Jamet, 2018; Zhang et al., 2006).

7.5.4 Timing and Student Control

The duration of supplementary educational videos is an important consideration and extremely controversial. Some bioscience educators have claimed that videos longer than 9 min are a waste of effort (Brame, 2016) because longer videos result in high levels of viewer attrition (Guo et al., 2014). In Fig. 7.2a the duration of all the procedural videos ("in the wild" and experimental) referred to in this chapter is shown (where values are given) although two longer duration videos are not included (see Fig. 7.2a legend). In Fig. 7.2b the duration of all the conceptual videos ("in the wild" and experimental) are shown although two longer duration videos are not included (see Fig. 7.2b legend).

As can be seen from Fig. 7.2a, range of values has been trialed both in videos used by students as part of their studies ("in the wild") and within experimental situations (one video design versus a different video design) or video compared with equivalent static text and images (where the material covered is not directly related to core content). Among experimental studies using procedural content, relatively few listed the videos' durations. However, *in the wild* studies showed a range from approximately 3 to 20 min (Fig. 7.2a). With respect to conceptual videos, the majority (for *in the wild* and experimental) were found within the range of 3–17 min, with no major differences between the two groups (Fig. 7.2b).

Some aspects of how learning depends on timing relate to the learning outcomes (see Sect. 7.7), particularly if the information is complex or extensive. The pacing of how fast information is presented to students, as well as the student's control over sequencing, has long been known to determine how much students learn and whether they are potentially bored or overwhelmed (Milheim, 1990). Surprisingly, students sometimes do not spontaneously use their ability to pause the video when they would benefit from thinking independently or reviewing their own knowledge; instructors can provide pre-segmented videos to prompt students to stop and think (Biard et al., 2018) or introduce questions inside the video and ask students to pause and answer. Many researchers stress the importance of giving the learner elements to control in the video, as this discourages passivity and may encourage cognitive

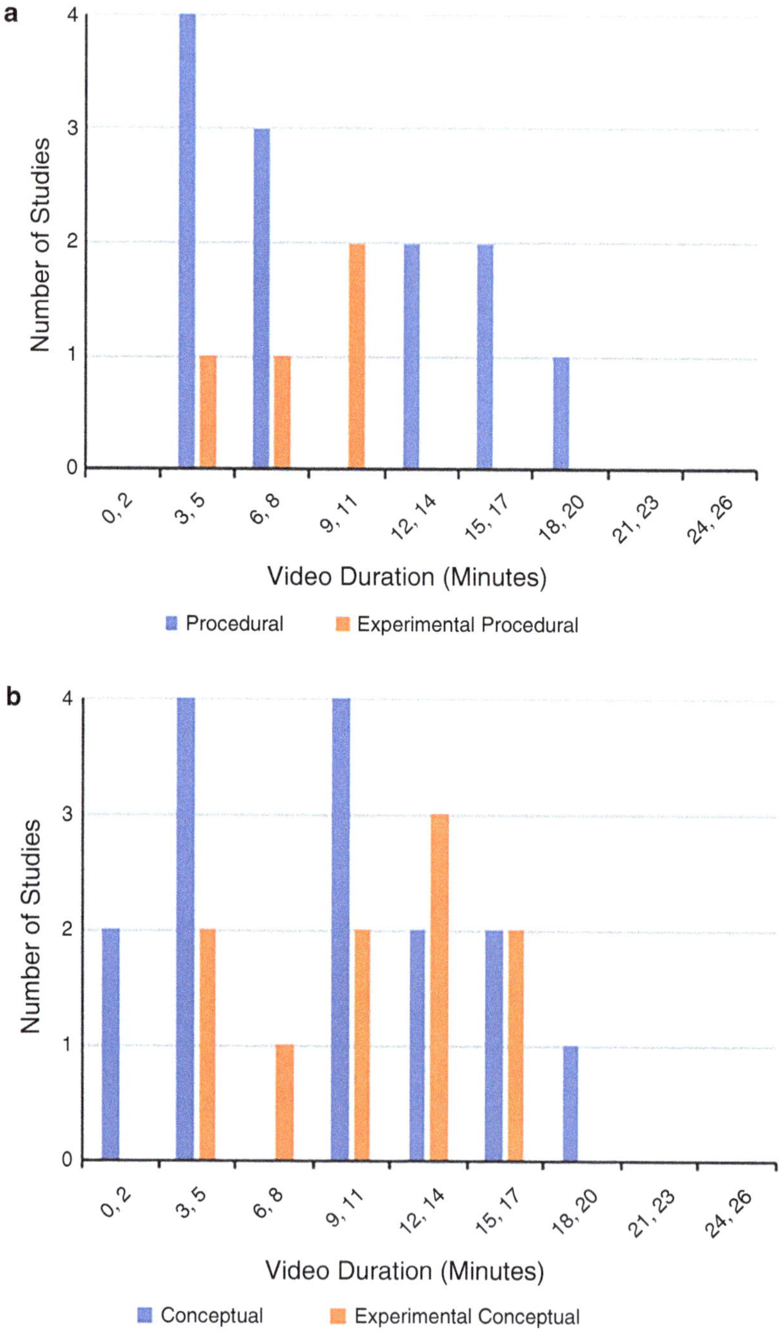

Fig. 7.2 Video durations from different studies. Where a single study developed multiple videos, a mean or median value has been used. Where a single study used different types of video, the data is shown separately. (**a**) Video durations from procedural and experimental procedural studies. Two studies are not shown with one procedural video in the range 30–32 min and another in the range

engagement. In some cases videos are presented in class as a way to launch discussion with immediacy and emotional engagement; in these cases there may be a necessary trade-off between community collaboration and control of pacing. Pacing and information rate is only one aspect of cognitive overload in video education.

7.5.5 Information Overload and How to Avoid It

One of the most important considerations is how complex the video is and whether or not it overloads the student viewer (Mayer & Fiorella, 2014). According to cognitive load theory (Paas & Sweller, 2014), there are three categories of cognitive load in a given lesson:

- Germane load, which is beneficial. It is mental effort that is both useful and necessary for learning the knowledge. These are mental manipulations (e.g., comparisons, analyses) to create schema that can enter long-term memory. For example, a student who is learning about compliance of the airways could compare this knowledge to previously learned information such as compliance in blood vessels.
- Intrinsic load, which is not directly beneficial. This is mental effort that is needed to approach the topic but does not positively contribute to building long-term memories. An example of intrinsic load would be novel vocabulary such as the word "propranolol"; for a student who is unfamiliar with pharmacology, this word is a series of nonsense syllables, but it is essential for mastering the topic of beta-blockade. Intrinsic load needs to be carefully managed.
- Extraneous load, which is detrimental. This is mental processing that can interfere with the learning outcome sought. A teacher-derived example is seductive details (e.g., the ghoulish story of how insulin overdosing was used to murder victims), but anxiety and imposter syndrome arising in the student can also create extraneous load. This is the type of mental activity educators seek to avoid in cases of cognitive overload.

A lack of matching between viewer and video will likely diminish engagement and learning outcomes. Cognitive load theory underlines the restricted cognitive capability of the learner as an important factor for the effective use of any learning material, including supplementary videos (Brame, 2016; Paas & Sweller, 2014; Sweller & Chandler, 1994). It is important to reduce any unnecessary factors that might contribute to cognitive load and at the same time strengthen the linking of new material (in short-term memory) to existing material (in long term memory). One

Fig. 7.2 (continued) 34–41 min. (**b**) Video durations from conceptual and experimental conceptual studies. Two studies are not shown, one for a conceptual video in the 27–29-min range and another for an experimental conceptual study with a video in the 30–32-min range

major source of potentially unnecessary cognitive load is "seductive details" (Mayer et al., 2001, 2020); these are facts or information that makes the knowledge seem more interesting, although the seductive detail is usually extraneous to the learning. If there is too big a gap between the learner's ability level and the video (as with a novice), the cognitive load may be too great to permit learning to occur. On the other hand, if the learner's ability is higher than the information provided, then the video may be ineffective (Kalyuga, 2007).

7.5.6 *Prompts for the Viewer to Answer Questions or Take Other Actions*

Prompts are instructions for the viewer to do something (e.g., answer a question) that is intended to encourage learning approaches that the viewer could do but might otherwise not (Renkl & Scheiter, 2015; Senchina, 2011). The role of prompts varies, for example, in some cases they may stimulate cognitive processes such as predicting the next step in a sequence or stimulate external action such as labeling, providing partial solutions to problems, or requesting self-explanations. Guy (2012) asked the students to take action while they viewed the video, e.g., generate a concept map, make a summary, or make a drawing. These various actions are meant to tap into active learning, as videos are often considered to be passive experiences. Further, Haagsman et al. (2020) generated videos with pop-up questions, and these significantly improved student scores compared to videos without questions. However, this effect was indirect as answers to specific conceptual problems were not improved. External approaches may assist learning or reduce learning depending upon the difficulty level of the prompt (Bodemer & Faust, 2006). It may be possible to offset the cognitive load induced by prompts by using complementary load reducing strategies. For example, De Koning et al. (2011) found that cueing of an animation offset the load produced by asking students to provide self-explanations.

Prompting in the form of tasks can also be added before or after the video. For example, in a study by Fiorella et al. (2020) (concerning dynamic versus static images within the video), additional studies were carried out where students carried out one of three tasks following the video presentation. They either explained what they had just seen, generated a relevant drawing, or just reviewed part of the video. It was found that the combination of the dynamic drawing video together with the student explaining post-video generated a significant learning improvement with respect to transfer (compared to the other post-video options). It is interesting that the dynamic drawing video was no better than the static image video by itself but only in combination with a post-video activity. This leads one to consider the best ways in which one can utilize supplementary videos in terms of their presentation context. For example, after a video about professional ethics in relationships between experimenters and participants during exercise physiology research, Senchina (2011) reinforced the video learning with interactive question sheets, small- and large-group discussion, and individual student reflections. Plainly, only a small part of the learning happens while the student is watching the video, but instead it happens inside the student's head during their processing of the information after watching.

7.5.7 *Segmentation*

Traditionally, video players come with basic control functions, e.g., play, pause, and rewind. These controls can assist in regulating information flow (pacing) according to particular learning needs (Delen et al., 2014; Mayer & Chandler, 2001; Merkt et al., 2018). However, more guidance may be required for beginners as they may not effectively use the controls. Breaking up the video with predetermined pacing may assist such learners, as this has proved to be a useful approach in computer-assisted learning (Mayer & Chandler, 2001). An extension of this approach can be found with the idea of segmentation. Rather than just adjusting the flow, one can break up the video at appropriate locations. For example, a video with a number of steps (e.g., dissection or laboratory preparation) can be divided in this way. The breaks provide pauses that allow the student time to reflect on the preceding information. In a study of second-year occupational therapy students, Biard et al. (2018) used a segmented video. During the pauses the video darkened and then continued to the next step. Improved learning outcomes were found for procedural learning, but this did not translate into significant differences for a recall test. Ibrahim et al. (2012) divided a 32-min entomology video into five conceptually separate video sections. In addition, each section started with a title and list of core concepts, and unnecessary content was removed from the video. When compared with a similar video without segmentation and the other modifications, the segmented video group performed significantly better on tests of acquisition and transfer.

7.5.8 *Cueing*

Inexperienced learners may have difficulty in working out what they should be looking at in the video. This may in turn increase cognitive load and impede learning. Thus, one can use cued instruction where salient points are highlighted in some way. For example, text, color, an arrow, or a symbol can be used to signal an important element. In turn the cue can occur in synchrony with the narration (De Koning et al., 2009; Mayer & Johnson, 2008). In addition to these approaches, certain words in the narration can also be stressed to draw attention to the point of interest.

7.5.9 Dynamic Versus Static Visuals

Video is intrinsically a method for capturing visual changes, but the extent of dynamic visualization is variable. For example, it may include a sequence of static PPT slides, static slides with dynamic highlighting (see cueing), sequential static problem-solving slides, instructor drawn images or problem-solving steps, ongoing procedural movement (e.g., dissection), an acted out clinical situation, or a "talking head" with ongoing movement. In addition, reasonably fast movement can be found when an animation is embedded within a video, and this special case is discussed separately (see below).

It certainly appears that video is more effective than equivalent static images/text (see above); however, the video may only contain a sequence of static images. Apart from these studies, there has been some research into aspects of dynamic visualization. Fiorella et al. (2020) found that there was no significant difference in student retention and transfer problems when comparing instructor-drawn visuals (instructor not visible) and static visuals (human kidney—structure and function). However, in an earlier study (Fiorella et al., 2019), instructor-drawn images in the video generated better scores than a video with static images (instructor not visible).

7.5.10 Animations

Sequential events such as biological processes (e.g., heart failure) are not concrete, so these concepts may be difficult to learn because they are impossible to mentally visualize as a single image in time. Thus, dynamic visualizations such as animations may offer a useful pathway to assist conceptual learning (Wouters et al., 2008). For example, it may be difficult for a student to visualize events occurring within a neuron or within skeletal muscle due to their abstract nature, e.g., the tiny fluxes of ions resulting in voltage changes during the action potential. Embedding a narrated animation into a short video makes it much more accessible in terms of platforms, and the video controls can substitute for the original animation controls. Neurophysiology animations developed by Guy (2012) to address student misconceptions were incorporated into narrated video form and successfully used in physiology teaching (RG unpublished results). Similarly Daly et al. (2016) developed an animated video (vascular neuroeffector transmission) for third-year physiology and pharmacology degree students, and this was compared with an equivalent video using static images. However, they found no significant difference in learning elicited by the animated video versus the video with static images.

One problem with dynamic visualizations is that they may significantly increase cognitive load (Brame, 2016). The student not only has to track changes in the image but also integrate the narration at the same time. It has been suggested that it is better to exclude dynamic visualizations and replace them with a series of static images (possibly with arrows indicating movement (Lowe et al., 2011). However, if there is no reasonable alternative to using a sequence of static images, then it may be possible to cue the dynamic visualization. For example, Boucheix and Lowe (2010) successfully used color cues that guided the student through the animation.

7.5.11 Camera Angle

With procedural video, there are approaches that can optimize the student experience. For example, Boucheix et al. (2018) compared the effect of different camera angles in a video of a manual procedure. They found that exchanging "over the shoulder" (e.g., as the instructor would view the process) and "face-to-face" views was better for learning than either view alone (or the control). This is worth considering when making a dissection video or a physiology laboratory preparation video.

7.5.12 Instructor Visibility

In a more conceptual style presentation (e.g., PowerPoint slides or "chalk and talk" equivalents), is it important to incorporate an image of the instructor within the video? Van Wermeskerken et al. (2018) produced a video with or without the presence of the instructor (eye contact—waist up view). No significant difference in learning results was found between the two situations. Although one may suppose that instructor presence might engage the student viewer, other studies have also shown learning may not be improved in this type of situation (Fiorella & Mayer, 2018). However, another study did find improved learning outcomes when the instructor appeared to establish eye contact with the viewer (Fiorella et al., 2019).

In a study by Wang and Antonenko (2017), videos were generated with the topics "similar triangles" (easy) or "trigonometric functions" (difficult). The videos were produced with or without the presence of the instructor (head and shoulders shown). In general, no significant change in learning was found between the two situations (instructor present/not present). However, recall was improved when the instructor was present in the easy video. Interestingly, the appearance of the instructor improved participants' estimation of their learning, increased student satisfaction, and reduced (perceived) mental work. Finally, Snelgrove et al. (2016) developed a "talking head" video where a patient is speaking about their sickness. The majority of the first-year student nurses (63%) agreed that, "The Talking Head video clip helped me to understand the nature of chronic illness." Thus, a "head only" video can be a useful approach. Guy et al. (2014) also utilized a "talking head" at the beginning of their videos.

7.5.13 Humor

Finally, it is worth mentioning something that is generally not often mentioned in the literature but certainly has played a part in lectures given by one of the authors (RG) and most probably by many physiology educators—humor. This approach can certainly be effective in engaging students, although, anecdotally, it seems educators are more reluctant to use this approach simply because their face-to-face presentations are being recorded. However, Guy et al. (2014) used a humorous, moving background behind the "talking head" video mentioned above (e.g., swimming fish, roller coaster, etc.). This was an attempt to increase student engagement with the videos. A large number of students made positive comments about this aspect of the short videos (RG unpublished results). Crowther et al. (2020) developed songs designed to assist students with problematic mathematical aspects of physiology (nine courses in two universities). Each song was associated with a PowerPoint slide that included the lyrics and the mathematical concept and converted to a video. Nearly half of the respondents suggested that the songs acted as useful mnemonic devices. Of the ten songs, student preference was for shorter ones ("jingles"). Apart from being a helpful approach, songs of this type may also serve a general function of engaging students, and there is no reason why they should not be incorporated as part of a longer video (see putting it together). Finally, Carnegie (2012) encouraged students to generate and review limericks related to the information-dense area of functional human anatomy. The project was well-received by the students, and a positive association was found between level of student participation in the limerick project and final course marks. Thus, it is worth considering the use of visual or text/auditory approaches that provide somewhat of a "break" from more traditional approaches to learning. We suggest that the most powerful use of such devices might be, as shown by Guy et al. (2014), as a short section of a longer video.

7.5.14 Narration

The narration for the videos generated by Guy et al. (2014) was conversational in order to engage the students. This type of approach has been shown to improve students' learning (Mayer, 2008). Some researchers have suggested that students prefer enthusiastic narration that includes both vocal enthusiasm (fast pace and tonal variation) and verbal enthusiasm including evaluations and emotions ("I love the next part") that improve social engagement with videos and thus learning (Brame, 2016; Liew et al., 2020).

7.5.15 Organization and Combining Design Principles

A range of different design principles can be used to enhance learning outcomes in short videos, but what happens when one combines two or more of the principles in the same video? One example can be found in a study by Guy et al. (2014) where different approaches were sequentially combined within a single video (Table 7.5). Although the video is relatively short, each section is designed to support understanding of the concepts and also to engage the viewer, particularly through the interactive elements. Thus, one can consider different ways of combining the design principles described above when generating a supplementary video.

7.6 Student Responses to Institutionally Provided Supplementary Videos

For the sake of clarity, we have divided the videos into procedural and conceptual categories. We have adopted the (somewhat artificial) view that procedural videos describe a series of visible steps necessary to reach a goal. On the other hand, conceptual videos tend to deal with abstract constructs. The reader is referred to work by Rittle-Johnson et al. (2015) for discussion on relationships between procedural and conceptual perspectives.

7.6.1 Procedural Videos

Procedural videos can clearly show the required actions, and the student can see and mentally follow the manual steps or the interactions occurring between human agents (behavioral modeling). However, there may be some degree of overlap between the two modeling situations (e.g., dissection videos), where both manual skills (what to do) are coupled with cognitive processes (what is the best way to progress). In many ways procedural videos are relatively straightforward, both from a development perspective and a student perspective, as one is dealing with an observable situation.

Several studies have used procedural videos to prepare students for dissection laboratories. Choi-Lundberg et al. (2016) developed dissection videos and found that most of the students agreed with the statements: "I felt more prepared for the dissection sessions" (91%); "I felt I learned more from the dissection sessions (90%)" when they had used the supplementary videos prior to their dissection laboratory. Students also felt that their dissections were better and were more confident in dissection sessions, particularly recognizing structures and the relationships between structures.

Table 7.5 Instructional process in an example conceptual video

Timing	Content	Intervention type	Purpose
Introduction 1–2 min	Talking head (instructor) with moving background (e.g., swimming fish)	See "instructor visibility and narration" sections above	Video content described and tasks student should complete during the content section are described e.g., writing a summary list, generating a concept map etc.
Content section (2–9 min),	Two to three simplified PPT slides with narration	See "Information overload and how to avoid it" section above See " Prompts" for the Viewer to Answer Questions or Take Other Actions	Students view simplified narrated images where concepts explained. Specified tasks undertaken
Final assessment/ feedback section (1–3 min)	A single multiple choice question (MCQ) (students pause) followed by class response to the question (from "clicker" data)(students pause). Finally, the most appropriate answer is given with reasons for inappropriate answers.	See " Prompts" for the Viewer to Answer Questions or Take Other Actions	Question encourages reflection re: concepts. Class response prompts reflection re : peer response. Instructor feedback encourages consolidation of concepts.

Saxena et al. (2008) also developed dissection videos, and student feedback indicated that 70% used the dissection videos for laboratory preparation and 21% agreed that the videos were useful with respect to understanding material. In a study by Langfield et al. (2018), 77% of the students accessed the dissection videos and reported that they used videos for both class preparation and exam revision. Topping

(2014) similarly found that student feedback indicated that dissection videos were useful in preparation for laboratory work and revision for examinations. They also rated them highly with respect to overall satisfaction.

Overall one can conclude that short dissection videos were popular with students for laboratory preparation and, in some cases, assisted with general understanding and exam revision. Although one might assume that higher levels of use would be beneficial, Langfield et al. (2018) reported that the 11% of students who viewed all the videos (Colthorpe et al., 2018) also had a significantly lower self-efficacy score. Self-efficacy is the belief that one is able to accomplish a task or reach a future goal (Brady-Amoon & Fuertes, 2011), so the lack of self-belief in the high-use video watchers suggests that the videos were often used as crutches by lost students rather than as a focused information gathering by self-regulated learners.

Procedural videos are not only used for dissection preparation. For example, Narula et al. (2012) generated ten videos (5 min each, e.g., atrial fibrillation) to support final-year medical students before their meeting patients. The students agreed (81%) that the videos were their preferred resource prior to patient contact. They also agreed (97%) that the videos were an effective way to learn about the symptom/disease. They also found that the videos significantly reduced the time needed to prepare prior to patient assessment (self-reported).

7.6.2 Conceptually Based Videos

In a study by Abdulla and O'Sullivan (2019), students were provided with videos to assist their understanding of difficult physiological concepts. Difficult topics were determined on the basis of student feedback. Students indicated that videos supported their learning and helped them to recollect knowledge (41%). They also felt that the visual nature of the tool assisted learning (27%) and provided a general review resource (18%). Similarly Evans (2011) found that 87% of students agreed the videos were of high quality and useful to their learning. Guy et al. (2018) developed 42 anatomy and physiology interactive videos, and these were accessed by ~50% of the cohort. Student feedback indicated that clips were engaging, assisted understanding of course content, and provided lecture support. Among the users, 25.7% visited a single video, 74% visited more than one video, and 49% of students showed repeat visits to one or more video. In addition Green et al., (2012) developed two types of video. For the homework videos 92% of respondents agreed with the statement, "I have a deeper understanding of the material discussed in lecture because of the homework solution videos." Students indicated that exam study tool (89%) and study supplement (76%) were their main reasons for use of homework videos. In terms of how they interacted with the videos, 33% watched the entire video from start to finish, and 26% re-watched certain segments based on homework responses.

With respect to the second type of video (concept-based), one group reported that the (concept-based) videos assisted them with achieving an increased level of understanding. In another study Morris and Chikwa (2014) developed two videos, and those who used the videos agreed that this resource had helped them to comprehend learning material. They appreciated the option to view material a number of times and also to make additional notes. Further, Kay and Kletskin (2012) developed problem-based videos where the background of the problem was explained and related to prior knowledge. Out of the total cohort, 68% used the videos and of those 45% agreed that they assisted integration of knowledge, review, and problem-solving. In another study Newsom et al. (2019) reported that students showed a positive response to the videos. Most students indicated that the videos improved course content understanding and improved their skill in applying this to patient cases. Finally, Daly et al. (2016) found that students preferred an animated video over a static control. However, no significant learning difference was found between the two groups, notwithstanding positive student feedback.

It is apparent that for both procedural and conceptual videos, there is a significant level of positive student feedback. This is very much in line with the tendency of students to utilize YouTube videos (see previous section) and to respond very positively to this type of learning resource.

7.6.3 Why Students Choose Not to Use Supplementary Videos

Some studies have examined why students chose not to access supplementary videos. The main reason was that students felt that the material had already been covered in the lecture and, to a lesser extent, time limitations (Abdulla & O'Sullivan, 2019; Green et al., 2012; Kay & Kletskin, 2012; Morris & Chikwa, 2014). Two studies found a small proportion of students were unaware of the videos (Kay & Kletskin, 2012; Morris & Chikwa, 2014), and another found some students simply forgot or had technical issues (Green et al., 2012). Interestingly the comments made by those students who did not think they needed the additional help actually performed significantly better than those who gave a different reason for non-use (Green et al., 2012). This highlights the importance of self-efficacy and fits with an anxiety model of engagement (see Chaps. 11 and 12 on attendance in this volume).

7.7 Improvements in Learning Outcomes: Strong and Weak Data

Learning gains have been measured in different research scenarios, which all suffer from one of two problems: failing to compare like-for-like or making measurements that are not ecologically valid to educators. Laboratory studies between video and text learning are the most meaningful test of the video form, but such studies are decontextualized from classroom study and examinations. In studies in the classroom, it would be unethical to randomize students so that some could access a learning resource, while others were not allowed. Comparing students who chose to access supplementary videos to those who did not access the videos may not be comparing like-for-like; these studies can compare student performance on the final exam (or selected questions), but they are open to confounding between student engagement and student lack of knowledge. Pre-post testing is when the same students are tested (using a test specific to the research study) before watching the video and then again after watching it; such studies measure knowledge gained but do not represent how students learn longer term for their course.

Eighteen studies were found where the effect of supplementary video availability on learning was tested. In all cases an improvement was found although the methodology varied from historical comparisons (nine studies), pre-post testing (one study), comparison of video with equivalent text (two studies), and comparison of video access with a no access control group (two studies). Two studies only examined correlations between video access and assessment performance, and a final two studies compared assessment scores between those students who did or did not access the supplementary videos.

Six of the studies using historical comparison controls all showed an improvement in assessment performance after introduction of supplementary videos (Evans, 2011; He et al., 2012; Moravec et al., 2010; Newsom et al., 2019; Roberts et al., 2018; Topping, 2014). These studies included a range of different areas and video types. One study used a "learn before lecture" approach by transferring material from biology lectures into short videos (Moravec et al., 2010), and another produced videos that covered some core concepts in physiology (Roberts et al., 2018). One study was classified as modal as it used anatomy preparation videos (Topping, 2014). Three studies covered difficult concepts in chemistry, embryology, and pharmacokinetics (He et al., 2012; Evans, 2011; Newsom et al., 2019). Another study, using pre-post assessment, which covered difficult physiology concepts, also showed higher scores following video availability (Abdulla & O'Sullivan, 2019).

Three further studies using historical controls (all anatomy preparation) did not find any improvement in exam score following introduction of the videos (Choi-Lundberg et al., 2016; Langfield et al., 2018; Saxena et al., 2008). However, these studies and two further engineering studies (Green et al., 2012; Pinder-Grover et al., 2011) did find significant correlations between use of the video resources and assessment outcomes.

Studies where final exam scores of users are compared with non-users (i.e., where there is no control group) do not provide conclusive evidence of learning gain. In the study by Guy et al. (2018), a relationship was found between video access and deep learning approaches, and thus it was possible that the learning gain was due to confounding that reflected engagement of those students who were more self-regulated and who made use of available resources. Similarly, correlations between video access and learning gain may also reflect higher levels of access by more able students. Thus, one cannot use this approach as hard evidence of video effectiveness. In addition, the single study using a pre-test post-test comparison did not have a control group, so this also limits the degree of confidence that one can have in video

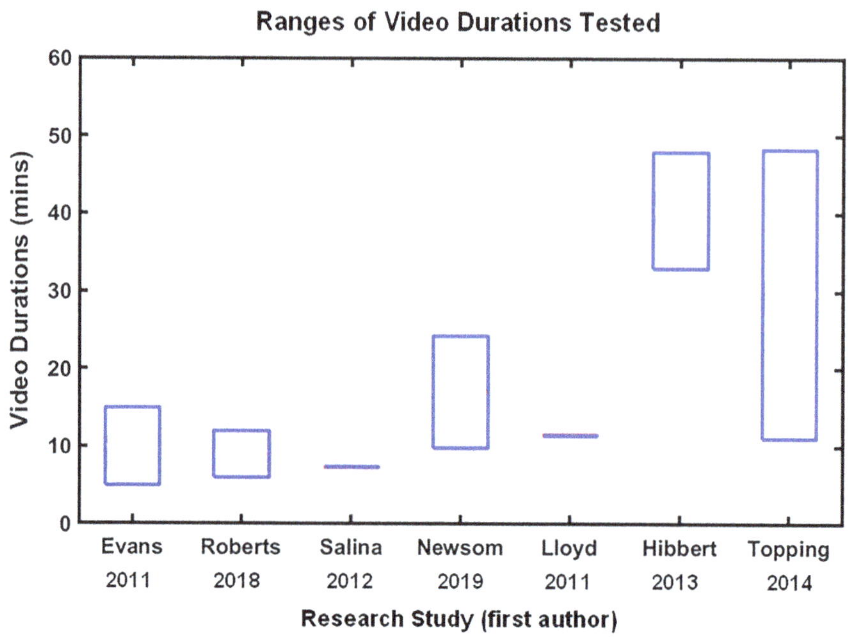

Fig. 7.3 Ranges of the video durations for the seven studies where learning improvement was demonstrated. Studies where individual video durations are given are indicated by a solid line

effectiveness. Further, in the "learn before lecture" study, the video resource did show a learning effect, but this effect was no greater than that achieved by equivalent text-based material. Additionally, one study did not provide a statistical basis for the learning improvement and is thus not considered further. Finally in the study by Morris and Chikwa (2014), a group with access to videos was compared with a group with no access. Although a significant learning gain was found for the video group, the comparison was made between those students who viewed the videos (rather than the whole test group). Thus, the study is biased, particularly as the viewing students only represented 56% of the group. It may be that the students who decided to access the videos were deeper learners as found in the study by Guy et al. (2018) and thus would be expected to perform at a higher level. For this reason, this study is not included among those that more reliably indicate learning gain.

Thus, out of the 18 studies, only 7 provide reasonable indicators of video effectiveness. Among these studies, three can be regarded as procedural (modal), and these include anatomy laboratory preparation, nursing technique, and clinical technique. The remaining four can be regarded as conceptual, with one study dealing with core concepts (renal physiology) and the final three studies dealing with conceptually difficult areas, embryology, statistics, and pharmacokinetics. The durations of the videos, for each study, are shown in Fig. 7.3.

Roberts et al. (2018) developed 13 videos with 2 to 4 covering a renal physiology core concept. During the first test year, 21/32 renal questions (MCQs) were the same as those used in the previous control year. For these questions (test year 1), the results were significantly higher than the control (72.7% vs. 64.4%, $P < 0.001$). During the second test year, 16/27 renal questions were the same as questions used in previous years. For these questions (test year 2), the results were again significantly higher than those in the control year (76.1% vs. 71.0%, $P < 0.001$). In another study Newsom et al. (2019) developed four concept-based videos where pharmacokinetic calculations were recorded as they were written together with simultaneous narration. The videos were considered to be at a conceptually high level as they covered additional and more complex situations than that found in core course material. The videos ranged from approximately 10–24 min (Fig. 7.3). Individual video effectiveness was not established, but there was significant learning improvement following introduction of the videos, with the final exam score in year 1 of 85.8% vs. 77.6% (intervention vs. control, $p < 0.001$) and in year 2 of 85.1% vs. 77.6% ($p < 0.001$). However, there was no improvement in year 3, possibly due to lower student access to the videos and a more difficult exam. Another study utilizing concept-based videos was that of Evans (2011). Five videos related to developmental processes (the first 10 min long and the remainder 5 min long) and a single review quiz video covered the embryology module (15 min) (Fig. 7.3). The first five videos incorporated PowerPoint animations, the PowerPoint slides were structured to maximize understanding, and synchronous narration was provided. Student results were compared to those of the previous control year and found to be significantly higher (78% vs. 66%, $P < 0.05$). In another study, a single high-level concept-based video was developed by Lloyd and Robertson (2011) to demonstrate various aspects of SPSS use (Fig. 7.3). The control group was given written material from an SPSS guide with equivalent static images of material shown in the video. The video group scored better than the control group on a set of related problems ($p < 0.001$). However, neither group demonstrated an optimal strategy for solving a statistical problem, and there was no difference between the groups in this respect.

With respect to procedural videos, Topping (2014) developed 13 anatomy videos (laboratory preparation) of the trunk region. It is interesting that the videos ranged in duration from just over 10–50 min (Fig. 7.3). The individual impact of the various videos was not established, but the average final laboratory examination score was 4% higher in the experimental group (82.9% vs. 78.9%, $P < 0.039$) than the historical control. The material that was covered in the videos was previously given in a teaching station scenario, where the instructor guided small groups through the same material given on the video. In the study by Salina et al. (2012), a single procedural video was developed to explain the method of moving a non-compliant patient from one position to another (Fig. 7.3). Students were randomized into the video or control group. The control group was given written information with identical words and sequence. The video group scored significantly better in the assessment ($p < 0.05$). Finally Hibbert et al. (2013) developed three procedural videos that covered clinical endocrinology techniques. Two dealt with diabetes and one with thyroid disease. Students who viewed the two diabetes videos performed significantly better in assessments related to each compared to students who did not view the videos (and who undertook standard revision methods) (diabetes history, 83.3% vs. 20.0%, $p = 0.007$; diabetes lower limb, 91.7% vs. 40%, $p = 0.024$). However, with respect to the thyroid video, no differences were found between the video and control groups.

The conclusion from this analysis is that, although at face value, there is support for the learning effectiveness of procedural and conceptual supplementary videos, the reality is not particularly convincing. This is evident when one considers the different types of supplementary video and the limited number of studies that provide support for the learning effectiveness of each. These ambiguities highlight the problems of evaluating an entire medium as opposed to a lesson. If it is possible to make a bad video, then concordant results from multiple studies would depend on certifying the quality of each tested video; likewise, if it is possible for students to respond to the novelty effect, then external oversight or careful baselining would be needed for each cohort tested. We suggest that further studies should be undertaken, with adequate controls, to provide a more substantial foundation for supporting the use of supplementary video across a wide range of disciplines.

7.8 Measuring Supplementary Video Usage

There are a variety of different metrics that have been used as indicators of student access (use) of supplementary videos. From an educators' point of view, it is important to have some idea of the proportion of students who access videos, the extent to which individual videos are accessed, and when (timing related to particular learning objectives) videos are accessed.

Some studies have provided data on the average number of times a single student accesses a single video. For example, Evans (2011) provided values of 2.8 (average of five conceptual videos) and 2.2 (quiz review), while Morris and Chikwa (2014) found a value of 2.7 (average of two conceptual videos). In addition, Langfield et al. (2018) found a value of 3.1 for 16 procedural (anatomy) videos. These studies demonstrate repeat visits by a student to a particular video and thus provide some evidence that the material is being used (as opposed to a student simply accessing a video and then not viewing). A slightly different approach was adopted by Guy et al. (2018), where 49% of the students viewed a video more than once, thus suggesting use rather than simply access.

Another metric that can be used is the average proportion of videos viewed. Saxena et al. found that on average 65% of anatomy videos were accessed (five videos) and 49% of radiology videos were accessed (three videos). In addition Tripodi (2018) found that on average 75% of 59 procedural videos (osteopathy) were accessed (semester 1 data), while 76% of 76 videos were accessed (semester 2 data). Finally, Choi-Lundberg et al. (2016) found an average of 73% of three procedural videos were accessed (anatomy year 2) and 68% of six procedural videos were accessed by year 3 students.

Another metric that has been used is to express the percentage of students who accessed at least one of the videos available in the study. In Fig. 7.4 the percentage of

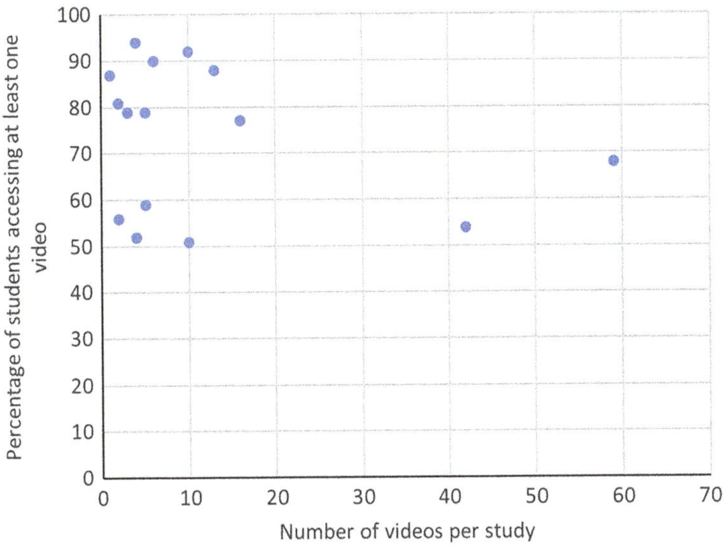

Fig. 7.4 Percentage of students accessing at least one of the available videos in a study. The data includes both conceptual and procedural videos. In some cases a single study used videos of different types, and these have been graphed separately

students is graphed against the number of videos in a particular study. Although there is no relationship between the two measures, there are two observations that arise from this graph. Firstly, the proportion of students viewing at least one video is above 50%, with a tendency to much higher values. Secondly, 85% of the studies (where access data was provided) developed ten or less supplementary videos.

Finally, some studies have measured the number of accesses to the supplementary videos on a day-to-day or week-by-week basis. For example, Langfield et al. (2018) plotted views per day for four sets of four videos (pre-laboratory anatomy procedural) for the first 8 weeks of the semester. The data indicated some access peaks before laboratory sessions and also a larger peak prior to the mid-semester practical examination. In another study Choi-Lundberg et al. (2016) developed procedural videos specifically to assist students in their preparation for dissection laboratories. However, only 28% of students accessed a specific video before the related dissection. On the other hand, approximately half of the students accessed all available videos by the end of semester. Thus, ongoing monitoring can be useful in determining the potential effectiveness of supplementary video by relating access to specific learning objectives.

It would be useful to have some standardized metrics that are used by any study using supplementary videos. The varied approaches for testing learning benefits have to balance the relevance to the classroom with the risks of confounding and challenges with fairness in running an experiment with students whose marks will be affected (Table 7.6). This and the difficulty in extracting relevant information from different studies make measuring learning benefit a thorny issue for researchers.

Table 7.6 Methodologies for testing learning benefits

Method	Ethically simple[a]	Free of confounding[b]	Ecological to classroom[c]
Results-based comparisons			
Pre-post (same students)	++	+++	0
Historical (different cohorts)	+	+	++
Students self-select	++	0	+++
Randomization (control group)	+/–	+++	0
No intervention	0	+	+
Delayed intervention	++	+++	++
Alternative intervention	0	++	0
Process-based evaluation			
Judgment of contents	+++	?	+++
Engagement/attendance	++	0	+/–
Student satisfaction	+++	0	+++

[a]0 means that there will be significant difficulty of remaining fair and ethically robust, especially in a classroom situation where summative tests affect the student. +/– means it depends on whether it occurs in a laboratory or classroom. + means that obtaining ethical approval, particularly for all students, will take preparation. ++ means many students (but not all) will agree. +++ means that there are no significant barriers to ethical approval and fairness

[b]0 means that there will be significant risk of confounding of the actual learning of students with other variables, such as student engagement or student preference for ease. ? means that the confounding will depend on the quality of theoretical model used. + means there is some risk of confounding due to varying student characteristics. ++ means that confounding is unlikely. +++ means that the comparison will be like-for-like

[c]0 means that the experiment is really more suited to a laboratory than a classroom context. +/– means that with the correct technology, it is possible to measure attendance or online engagement but that in many cases taking attendance intrudes upon the operation of a large classroom. + means that with significant encouragement students will take part in a useful research process while pursuing their studies. ++ means that with explanation students will take part in a useful research process while pursuing their students. +++ means that students will accept this research as not interfering with their studies

Thus, it would be useful for potential researchers to clearly state what metrics they are using and what these contribute toward and understanding of student supplementary video access (and hopefully use).

It is particularly important to consider these points, as technology is starting to enable more detailed analysis of student access to videos by providing information about what proportion of the video was played and which parts of the video were accessed. For example, Choi-Lundberg et al. (2016) determined that the median aggregate access time (for the semester) was close to the aggregate length of all of the available videos. They also found that the 75% percentile of accessors spent double the time on the videos, which suggests complete and repeat viewing of the videos.

7.9 Using Videos to Untangle Difficult Topics and Misconceptions

Certain aspects of learning may consistently prove problematic for students (Badenhorst et al., 2016). Although there have been several studies that attempt to address difficulty and misconceptions, there are relatively few where supplementary videos have been used to provide clear explanations (Guy, 2012). Because of the nature of complex concepts, the student is often required to construct a mental model of abstract processes (Chi, 2005). This is because the process being described is generally emergent rather than visible and visualization techniques need to be implemented (see animations and dynamic visualization above).

When one uses physiology as an example, there are a significant number of publications that address difficult topics and misconceptions (Cheng & Durairajanayagam, 2012; Colthorpe et al., 2018; Hintze & Zhao, 1999; Kay, 2008; Michael, 1998; Reinoso Tapia et al., 2019; Versteeg et al., 2019, 2020). However, it may be difficult to form a link between the solutions proposed by these various authors and the students who have the misconceptions. We suggest that a supplementary video may be a potentially effective vehicle to bridge research into difficult topics and misconceptions. Thus, the strategies now become embedded in a tool that is easily distributed to large numbers of students with the potential for remediating the misunderstandings that exist. Some studies used animations or simulations to assist student understanding of difficult concepts, and it may be that embedding these into a narrated video might improve access and student understanding (see Sect. 7.5.10).

The context of how videos are watched, particularly the mental tasks that students are expected to do after watching the video, is often overlooked. In addition to using a video to launch group discussions (Senchina, 2011), videos can demand that students perform homework, making the video part of a learning strategy rather than a passive source of information. For example, iBiology (2016) provides skills videos for early career biologists ranging from giving talks and experimental design to getting mentoring. Plainly these videos have a context of an entire scientific career, which is only meaningful if the videos are followed by action.

7.10 Conclusions

We have come to the following conclusions as a result of this study and (where appropriate) make suggestions with respect to further work.

1. Research on supplementary videos should, where possible, be separated into procedural and conceptual categories, despite the fact that these categories are artificial, and assessment of video-enhanced learning (even in mixed videos)

should distinguish between questions that elicit primarily rote factual, procedural, or conceptual knowledge.

Procedural videos are relatively straightforward as they demonstrate manual technique, person-to-person interaction, or individual narrative events. Most procedural videos present objects and relationships that can be directly visualized. They are generally short with most groups of studies in the 3- to 8- and 12- to 20-min durations. However, in those cases where there is evidence for learning gain, the video duration ranges from 8 to 50 min. Longer videos may thus approach the length of LCRs, and it is possible that the lower cognitive demands of this type of video might allow for effective student viewing. Experimental studies for this type of video are limited but have video durations within the range of 3–11 min.

The majority of studies related to concept-based videos have developed videos with durations of less than 17 min, with peak numbers in the 3- to 5- and 9- to 11-min ranges. Where learning gain has been demonstrated, the video durations range from 5 min to about 24 min. There are a significant number of experimental studies for this type of video with a reasonable correspondence between experimental and *in the wild* studies with respect to video length. There is significant evidence relating to the importance of providing concept-based videos of shorter duration. The simple reason is that they are much more cognitively demanding than procedural videos and, by their nature, tend to deal with much more abstract ideas, many of which are fundamentally emergent (Chi, 2005). As a result, students most probably can only maintain attention for relatively short periods of time.

2. Much more research needs to be carried out to establish learning gain in supplementary videos.

   In our opinion, the opportunity for the development of institutional supplementary videos, particularly in the conceptual area, has not been fully taken advantage of. It is clear that students with questions about course content (particularly difficult concepts) will quickly move to online resources, including commercial video platforms like YouTube. Although there is evidence that some institutions have uploaded supplementary videos into specific YouTube channels, there is also significant data suggesting that students will utilize general searches for answers to their questions. If the institution has not provided a reasonable range of supplementary videos to deal with student questions, then it is likely that students, whom we have taught to be self-directed learners, may end up using uncurated and sometimes incorrect material with YouTube or similar Internet sites.

   We conclude that educators need to develop supplementary videos (particularly in the conceptual area) in significant numbers or direct students to reliable online video material. Most of the studies covered in this chapter have demonstrated that simple concept-based videos can be relatively easily made using fairly simple techniques. The videos, to some extent, provide a technological equivalent to what may occur when students ask their questions directly to an educator (a situation that appears to be less common than in the past). Once developed, videos of this type can obviously be used for many years without further involvement of the educator, and the collection can be gradually built up with time. It is almost certain that if this is not done, students may well be gaining inappropriate conceptual perspectives by using non-curated material.

3. Educators should encompass the development of supplementary videos as a normal part of their work. Although some effort is required initially, the benefits both in time and student learning may be substantial.

   When educators start to develop their own supplementary videos, they should take note of the findings of experimental studies (as described in this chapter). Experimental studies generally do not test students with material related to their core learning and, as such, lack an important motivational aspect. However, there are many sensible approaches that can be incorporated into a video that will most likely assist in making it as effective as possible by, for example, lowering cognitive load or by segmenting approaches. Our analysis of published work demonstrates that very few studies make reference to the gold standard approaches that have been developed (mainly) in the educational psychology area. Thus, it is important for educators to develop videos and for researchers to ensure that developed videos meet an appropriate standard in terms of their potential for assisting learning. Additionally, researchers should provide clear metrics related to student access of the videos made available to them. In this way we may see an improvement in clarity relating to learning outcomes and student use of supplementary video.

4. We conclude that the findings from a significant body of experimental research are not effectively being utilized during the development of supplementary videos. We suggest that incorporation of these "gold standard" approaches may improve the learning effectiveness of these video resources. In addition, researchers should provide clear metrics regarding student use of video resources.

   It seems reasonable that the most important supplementary videos to develop are those that deal with the most difficult concepts and with student misconceptions. Although there is significant research into misconceptions (at least from a physiology perspective), it appears to us that a supplementary video may be an effective vehicle for incorporating suggested strategies to deal with these issues. Animations (and simulations) represent one approach that has been used to demonstrate abstract concepts; however, they are mostly used directly within a required IT framework. It is quite rare to find studies where animations or simulations are placed within supplementary video together with narration. In fact, there is evidence that this approach may be more effective than direct student use of these resources. The benefit of placing this material within a video is that it is much easier to distribute to a wide audience.

5. We suggest that supplementary video should be used to convey ideas relating to difficult concepts or student misconceptions. This might well include the use of narrated animations (or simulations) to facilitate understanding of abstract ideas.

   We suggest that the outlook is positive for the use of supplementary educational videos. As the name suggests, supplementary videos supplement in-class activities, especially when a student has forgotten, is reviewing, or has never learned the concept. This is the opposite role of videos in flipped classroom, which cover material that is considered required, and may even be tested in the next classroom meeting. On the basis of published work, supplementary video is a very much underutilized resource by faculty, although students use it all the time. The penalty for institutions not keeping up with the trend of students using unregulated media is a potential increase of inaccurate understanding kindled by unregulated sources (Hornsey et al., 2020) and the possibility of promoting student misconceptions. Supplementary video from any source should incorporate the strategies arising from educational research and should be effectively curated to ensure the optimum learning environment for students.

6. It is important to provide effective curation of supplementary videos to ensure that students only access high-quality learning materials. Scientific organizations and learned societies have several roles to play in making sure that this curation is expert-led. These include academic-level guidance, production advice, and certification of both quality and level. These functions should not be entrusted exclusively to third-party and for-profit education providers.

References

Abdulla, M. H., & O'Sullivan, E. (2019). The impact of supplementing powerpoint with detailed notes and explanatory videos on student attendance and performance in a physiology module in medicine. *Medical Science Educator, 29*(4), 959–968. https://doi.org/10.1007/s40670-019-00780-8

Azer, S. A. (2012). Can "YouTube" help students in learning surface anatomy? *Surgical Radiologic Anatomy, 34*(5), 465–468. https://doi.org/10.1007/s00276-012-0935-x

Azer, S. A. (2014). Mechanisms in cardiovascular diseases: How useful are medical textbooks, eMedicine, and YouTube? *Advances in Physiology Education, 38*(2), 124–134. https://doi.org/10.1152/advan.00149.2013

Azer, S. A., AlEshaiwi, S. M., AlGrain, H. A., & AlKhelaif, R. A. (2012). Nervous system examination on YouTube. *BMC Medical Education, 12*(126), 1–8.

Badenhorst, E., Hartman, N., & Mamede, S. (2016). How biomedical misconceptions may arise and affect medical students′ learning: A review of theoretical perspectives and empirical evidence. *Health Professions Education, 2*(1), 10–17.

Banerjee, S. (2021). To capture the research landscape of lecture capture in university education. *Computers & Education, 160*, 104032. https://doi.org/10.1016/j.compedu.2020.104032

Barry, D. S., Marzouk, F., Chulak-Oglu, K., Bennett, D., Tierney, P., & O'Keeffe, G. W. (2016). Anatomy education for the YouTube generation. *Anatomical Sciences Education, 9*(1), 90–96. https://doi.org/10.1002/ase.1550

Bassili, J. N. (2008). Media richness and social norms in the choice to attend lectures or to watch them online. *Journal of Educational Multimedia and Hypermedia, 17*(4), 453–475.

Biard, N., Cojean, S., & Jamet, E. (2018). Effects of segmentation and pacing on procedural learning by video. *Computers in Human Behavior, 89*, 411–417. https://doi.org/10.1016/j.chb.2017.12.002

Bjork, R. A., Dunlosky, J., & Kornell, N. (2013). Self-regulated learning: Beliefs, techniques, and illusions. *Annual Review of Psychology, 64*, 417–444.

Bodemer, D., & Faust, U. (2006). External and mental referencing of multiple representations. *Computers in Human Behavior, 22*(1), 27–42. https://doi.org/10.1016/j.chb.2005.01.005

Boucheix, J.-M., Gauthier, P., Fontaine, J.-B., & Jaffeux, S. (2018). Mixed camera viewpoints improve learning medical hand procedure from video in nurse training? *Computers in Human Behavior, 89*, 418–429. https://doi.org/10.1016/j.chb.2018.01.017

Boucheix, J.-M., & Lowe, R. K. (2010). An eye tracking comparison of external pointing cues and internal continuous cues in learning with complex animations. *Learning and Instruction, 20*(2), 123–135. https://doi.org/10.1016/j.learninstruc.2009.02.015

Brady-Amoon, P., & Fuertes, J. N. (2011). Self-Efficacy, Self-Rated abilities, adjustment, and academic performance. *Journal of Counseling & Development, 89*(4), 431–438.

Brame, C. J. (2016). Effective educational videos: Principles and guidelines for maximizing student learning from video content. *CBE—Life Sciences Education, 15*(4), es6.

Braun, J. A., & Eklund, J. L. (2019). Fake news, real money: Ad tech platforms, profit-driven hoaxes, and the business of journalism. *Digital Journalism, 7*(1), 1–21. https://doi.org/10.1080/21670811.2018.1556314

Burk-Rafel, J., Santen, S. A., & Purkiss, J. (2017). Study behaviors and USMLE Step 1 performance: Implications of a student self-directed parallel curriculum. *Academic Medicine, 92*(11S), S67–S74.

Burns, L. E., Abbassi, E., Qian, X., Mecham, A., Simeteys, P., & Mays, K. A. (2020). YouTube use among dental students for learning clinical procedures: A multi-institutional study. *Journal of Dental Education*. https://doi.org/10.1002/jdd.12240

Carnegie, J. A. (2012). The use of limericks to engage student interest and promote active learning in an undergraduate course in functional anatomy. *Anatomical Sciences Education, 5*(2), 90–97. https://doi.org/10.1002/ase.1264

Celentano, V., Browning, M., Hitchins, C., Giglio, M. C., & Coleman, M. G. (2017). Training value of laparoscopic colorectal videos on the World Wide Web: A pilot study on the educational quality of laparoscopic right hemicolectomy videos. *Surgical Endoscopy, 31*(11), 4496–4504.

Celentano, V., Smart, N., McGrath, J., Cahill, R. A., Spinelli, A., Challacombe, B., Belyansky, I., Hasegawa, H., Munikrishnan, V., Pellino, G., Ahmed, J., Muysoms, F., Saklani, A., Khan, J., Popowich, D., Ballecer, C., & Coleman, M. G. (2020). How to report educational videos in robotic surgery: An international multidisciplinary consensus statement. *Updates in Surgery*, 1–7. https://doi.org/10.1007/s13304-020-00734-5

Cheng, H. M., & Durairajanayagam, D. (2012). Misconceptions highlighted among medical students in the annual International Intermedical School Physiology Quiz. *Advances in Physiology Education, 36*(3), 229–232. https://doi.org/10.1152/advan.00089.2011

Chi, M. T. (2005). Commonsense conceptions of emergent processes: Why some misconceptions are robust. *The Journal of the Learning Sciences, 14*(2), 161–199.

Choe, R. C., Scuric, Z., Eshkol, E., Cruser, S., Arndt, A., Cox, R., Toma, S. P., Shapiro, C., Levis-Fitzgerald, M., Barnes, G., & Crosbie, R. H. (2019). Student satisfaction and learning outcomes in asynchronous online lecture videos. *CBE—Life Sciences Education, 18*(4), ar55.

Choi-Lundberg, D. L., Cuellar, W. A., & Williams, A. M. (2016). Online dissection audio-visual resources for human anatomy: Undergraduate medical students' usage and learning outcomes. *Anatomical Sciences Education, 9*(6), 545–554. https://doi.org/10.1002/ase.1607

Cojean, S., & Jamet, E. (2018). The role of scaffolding in improving information seeking in videos. *Journal of Computer Assisted Learning, 34*(6), 960–969. https://doi.org/10.1111/jcal.12303

Colthorpe, K. L., Abe, H., & Ainscough, L. (2018). How do students deal with difficult physiological knowledge? *Advances in Physiology Education, 42*(4), 555–564. https://doi.org/10.1152/advan.00102.2018

CrashCourse. (2015). The nervous system, part 2—Action! Potential!: Crash course A & P #9. Accessed 2021-03-01. https://youtu.be/OZG8M_ldA1M

Croker, K., Andersson, H., Lush, D., Prince, R., & Gomez, S. (2010). Enhancing the student experience of laboratory practicals through digital video guides. *Bioscience Education, 16*(1), 1–13.

Crowther, G. J., Wessels, J., Lesser, L. M., & Breckler, J. L. (2020). Is memorization the name of the game? Undergraduates' perceptions of the usefulness of physiology songs. *Advances in Physiology Education, 44*(1), 104–112. https://doi.org/10.1152/advan.00112.2019

Curran, V., Simmons, K., Matthews, L., Fleet, L., Gustafson, D. L., Fairbridge, N. A., & Xu, X. (2020). YouTube as an educational resource in medical education: A scoping review. *Medical Science Educator.* https://doi.org/10.1007/s40670-020-01016-w

Daly, C. J., Bulloch, J. M., Ma, M., & Aidulis, D. (2016). A comparison of animated versus static images in an instructional multimedia presentation. *Advances in Physiology Education, 40*(2), 201–205. https://doi.org/10.1152/advan.00053.2015

De Koning, B. B., Tabbers, H. K., Rikers, R. M. J. P., & Paas, F. (2009). Towards a framework for attention cueing in instructional animations: Guidelines for research and design. *Educational Psychology Review, 21*(2), 113–140. https://doi.org/10.1007/s10648-009-9098-7

De Koning, B. B., Tabbers, H. K., Rikers, R. M. J. P., & Paas, F. (2011). Improved effectiveness of cueing by self-explanations when learning from a complex animation. *Applied Cognitive Psychology, 25*(2), 183–194. https://doi.org/10.1002/acp.1661

Delen, E., Liew, J., & Willson, V. (2014). Effects of interactivity and instructional scaffolding on learning: Self-regulation in online video-based environments. *Computers & Education, 78*, 312–320. https://doi.org/10.1016/j.compedu.2014.06.018

Deslauriers, L., McCarty, L. S., Miller, K., Callaghan, K., & Kestin, G. (2019). Measuring actual learning versus feeling of learning in response to being actively engaged in the classroom. *Proceedings of the National Academy of Sciences of the United States of America, 116*(39), 19251–19257.

Dzara, K., Chen, D. T., Haidet, P., Murray, H., Tackett, S., & Chisolm, M. S. (2020). The effective use of videos in medical education. *Academic Medicine, 95*(6), 970.

Emanuel, E. J. (2020). The inevitable reimagining of medical education. *Journal of American Medical Association, 323*(12), 1127–1128.

Evans, D. J. (2011). Using embryology screencasts: A useful addition to the student learning experience? *Anatomical Sciences Education, 4*(2), 57–63. https://doi.org/10.1002/ase.209

Fiorella, L., & Mayer, R. E. (2018). What works and doesn't work with instructional video. *Computers in Human Behavior, 89*, 465–470. https://doi.org/10.1016/j.chb.2018.07.015

Fiorella, L., Stull, A. T., Kuhlmann, S., & Mayer, R. E. (2019). Instructor presence in video lectures: The role of dynamic drawings, eye contact, and instructor visibility. *Journal of Educational Psychology, 111*(7), 1162–1171. https://doi.org/10.1037/edu0000325

Fiorella, L., Stull, A. T., Kuhlmann, S., & Mayer, R. E. (2020). Fostering generative learning from video lessons: Benefits of instructor-generated drawings and learner-generated explanations. *Journal of Educational Psychology, 112*(5), 895–906. https://doi.org/10.1037/edu0000408

Fischer, J., Geurts, J., Valderrabano, V., & Hügle, T. (2013). Educational quality of YouTube videos on knee arthrocentesis. *Journal of Clinical Rheumatology, 19*(7), 373–376.

Gorwa, R., Binns, R., & Katzenbach, C. (2020). Algorithmic content moderation: Technical and political challenges in the automation of platform governance. *Big Data & Society, 7*(1). https://doi.org/10.1177/2053951719897945

Green, K. R., Pinder-Grover, T., & Millunchick, J. M. (2012). Impact of screencast technology: Connecting the perception of usefulness and the reality of performance. *Journal of Engineering Education, 101*(4), 717–737.

Guo, P. J., Kim, J., & Rubin, R. (2014). How video production affects student engagement: An empirical study of MOOC videos. In *Proceedings of the first ACM conference on Learning@ scale conference* (pp. 41–50).

Guy, R. (2012). Overcoming misconceptions in neurophysiology learning: An approach using color-coded animations. *Advances in Physiology Education, 36*(3), 226–228. https://doi.org/10.1152/advan.00047.2012

Guy, R., Byrne, B., & Dobos, M. (2018). Optional anatomy and physiology e-learning resources: Student access, learning approaches, and academic outcomes. *Advances in Physiology Education, 42*(1), 43–49. https://doi.org/10.1152/advan.00007.2017

Guy, R., Byrne, B., & Rich, P. (2014). Supporting physiology learning: The development of interactive concept-based video clips. *Advances in Physiology Education, 38*(1), 96–98. https://doi.org/10.1152/advan.00096.2013

Haagsman, M. E., Scager, K., Boonstra, J., & Koster, M. C. (2020). Pop-up questions within educational videos: Effects on students' learning. *Journal of Science Education and Technology., 29*, 713–724. https://doi.org/10.1007/s10956-020-09847-3

He, Y., Swenson, S., & Lents, N. (2012). Online video tutorials increase learning of difficult concepts in an undergraduate analytical chemistry course. *Journal of Chemical Education, 89*(9), 1128–1132. https://doi.org/10.1021/ed200685p

Hibbert, E. J., Lambert, T., Carter, J. N., Learoyd, D. L., Twigg, S., & Clarke, S. (2013). A randomized controlled pilot trial comparing the impact of access to clinical endocrinology video demonstrations with access to usual revision resources on medical student performance of clinical endocrinology skills. *BMC Medical Education, 13*, Article 135.

Hintze, T. H., & Zhao, G. (1999). Common misconceptions that arise in the first-year medical physiology curriculum concerning heart failure. *Advances in Physiology Education, 22*(1), S260–S267.

Hornsey, M. J., Finlayson, M., Chatwood, G., & Begeny, C. T. (2020). Donald Trump and vaccination: The effect of political identity, conspiracist ideation and presidential tweets on vaccine hesitancy. *Journal of Experimental Social Psychology, 88*, 103947.

iBiology. (2016). *iBioEducation: iBiology courses*. Retrieved from https://www.ibiology.org/online-biology-courses/

Ibrahim, M., Antonenko, P. D., Greenwood, C. M., & Wheeler, D. (2012). Effects of segmenting, signalling, and weeding on learning from educational video. *Learning, Media and Technology, 37*(3), 220–235.

Jaffar, A. A. (2012). YouTube: An emerging tool in anatomy education. *Anatomical Sciences Education, 5*(3), 158–164. https://doi.org/10.1002/ase.1268

Kalyuga, S. (2007). Expertise reversal effect and its implications for learner-tailored instruction. *Educational Psychology Review, 19*(4), 509–539. https://doi.org/10.1007/s10648-007-9054-3

Kay, B. (2008). Common misconceptions perpetuated. *Advances in Physiology Education, 32*(3), 244–245.; author reply 246–247. https://doi.org/10.1152/advan.90157.2008

Kay, R., & Kletskin, I. (2012). Evaluating the use of problem-based video podcasts to teach mathematics in higher education. *Computers & Education, 59*(2), 619–627. https://doi.org/10.1016/j.compedu.2012.03.007

Kay, R. H. (2012). Exploring the use of video podcasts in education: A comprehensive review of the literature. *Computers in Human Behavior, 28*(3), 820–831. https://doi.org/10.1016/j.chb.2012.01.011

Kelly, D. P., & Rutherford, T. (2017). Khan Academy as supplemental instruction: A controlled study of a computer-based mathematics intervention. *The International Review of Research in Open and Distributed Learning, 18*(4), 70–77.

Kruger, J., & Dunning, D. (1999). Unskilled and unaware of it: How difficulties in recognizing one's own incompetence lead to inflated self-assessments. *Journal of Personality and Social Psychology, 77*(6), 1121.

Langfield, T., Colthorpe, K., & Ainscough, L. (2018). Online instructional anatomy videos: Student usage, self-efficacy, and performance in upper limb regional anatomy assessment. *Anatomical Sciences Education, 11*(5), 461–470. https://doi.org/10.1002/ase.1756

Liew, T. W., Tan, S. M., Tan, T. M., & Kew, S. N. (2020). Does speaker's voice enthusiasm affect social cue, cognitive load and transfer in multimedia learning? *Information and Learning Sciences, 121*(3/4), 117–135. https://doi.org/10.1108/ILS-11-2019-0124

Liu, C., & Elms, P. (2019). Animating student engagement: The impacts of cartoon instructional videos on learning experience. *Research in Learning Technology, 27*, 2124. https://doi.org/10.25304/rlt.v27.2124

Lloyd, S. A., & Robertson, C. L. (2011). Screencast tutorials enhance student learning of statistics. *Teaching of Psychology, 39*(1), 67–71. https://doi.org/10.1177/0098628311430640

Lowe, R., Schnotz, W., & Rasch, T. (2011). Aligning affordances of graphics with learning task requirements. *Applied Cognitive Psychology, 25*(3), 452–459. https://doi.org/10.1002/acp.1712

Mayer, R. E. (2008). Applying the science of learning: Evidence-based principles for the design of multimedia instruction. *American Psychologist, 63*(8), 760–769.

Mayer, R. E., & Chandler, P. (2001). When learning is just a click away: Does simple user interaction foster deeper understanding of multimedia messages? *Journal of Educational Psychology, 93*(2), 390–397. https://doi.org/10.1037//0022-0663.93.2.390

Mayer, R. E., & Fiorella, L. (2014). Principles for reducing extraneous processing in multimedia learning: Coherence, signaling, redundancy, spatial contiguity, and temporal contiguity principles. In *The Cambridge handbook of multimedia learning* (pp. 279–315).

Mayer, R. E., Fiorella, L., & Stull, A. (2020). Five ways to increase the effectiveness of instructional video. *Educational Technology Research and Development, 68*(3), 837–852.

Mayer, R. E., Heiser, J., & Lonn, S. (2001). Cognitive constraints on multimedia learning: When presenting more material results in less understanding. *Journal of Educational Psychology, 93*(1), 187–198.

Mayer, R. E., & Johnson, C. I. (2008). Revising the redundancy principle in multimedia learning. *Journal of Educational Psychology, 100*(2), 380–386.

Merkt, M., Ballmann, A., Felfeli, J., & Schwan, S. (2018). Pauses in educational videos: Testing the transience explanation against the structuring explanation. *Computers in Human Behavior, 89*, 399–410. https://doi.org/10.1016/j.chb.2018.01.013

Mesko, B., Gyorffy, Z., & Kollar, J. (2015). Digital literacy in the medical curriculum: A course with social media tools and gamification. *JMIR Medical Education, 1*(2), e6. https://doi.org/10.2196/mededu.4411

Michael, J. A. (1998). Students' misconceptions about perceived physiological responses. *Advances in Physiology Education, 19*(1), S90–S98.

Milheim, W. D. (1990). The effects of pacing and sequence control in an interactive video lesson. *Educational & Training Technology International, 27*(1), 7–19.

Moravec, M., Williams, A., Aguilar-Roca, N., O'Dowd, D. K., & Wakimoto, B. (2010). Learn before Lecture: A strategy that improves learning outcomes in a large introductory biology class. CBE—Life Sciences. *Education, 9*(4), 473–481. https://doi.org/10.1187/cbe.10-04-0063

Morris, C., & Chikwa, G. (2014). Screencasts: How effective are they and how do students engage with them? *Active Learning in Higher Education, 15*(1), 25–37. https://doi.org/10.1177/1469787413514654

Narula, N., Ahmed, L., & Rudkowski, J. (2012). An evaluation of the '5 Minute Medicine' video podcast series compared to conventional medical resources for the internal medicine clerkship. *Medical Teacher, 34*(11), e751–e755. https://doi.org/10.3109/0142159X.2012.689446

Newsom, L., Proctor, R., Marshall, L. L., & Liao, T. V. (2019). Implementation and evaluation of problem-based video podcasts in an introductory pharmacokinetics course. *Currents in Pharmacy Teaching and Learning, 11*(12), 1213–1220. https://doi.org/10.1016/j.cptl.2019.09.003

O'Callaghan, F. V., Neumann, D. L., Jones, L., & Creed, P. A. (2017). The use of lecture recordings in higher education: A review of institutional, student, and lecturer issues. *Education and Information Technologies, 22*(1), 399–415. https://doi.org/10.1007/s10639-015-9451-z

O'Malley, D., Barry, D. S., & Rae, M. G. (2019). How much do preclinical medical students utilize the internet to study physiology? *Advances in Physiology Education, 43*(3), 383–391. https://doi.org/10.1152/advan.00070.2019

Paas, F., & Sweller, J. (2014). Implications of cognitive load theory for multimedia learning. In *The Cambridge handbook of multimedia learning* (2nd edn., pp. 27–42).

Paik, E. S. (2010). *Learning with animation and illusions of understanding*. Thesis. University of Nevada at Las Vegas. https://doi.org/10.34917/1884997

Paik, E. S., & Schraw, G. (2013). Learning with animation and illusions of understanding. *Journal of Educational Psychology, 105*(2), 278–289. https://doi.org/10.1037/a0030281

Pickering, J. D. (2017). Measuring learning gain: Comparing anatomy drawing screencasts and paper-based resources. *Anatomical Sciences Education, 10*(4), 307–316. https://doi.org/10.1002/ase.1666

Pinder-Grover, T., Green, K. R., & Millunchick, J. M. (2011). The efficacy of screencasts to address the diverse academic needs of students in a large lecture course. *Advances in Engineering Education, 2*(3), 1–28.

Pottinger, N. E. (2018). Don't forget to subscribe: Regulation of online advertising evaluated through YouTube's monetization problem. *Kentucky Law Journal, 107*(3), 515–546.

Raikos, A., & Waidyasekara, P. (2014). How useful is YouTube in learning heart anatomy? *Anatomical Sciences Education, 7*(1), 12–18. https://doi.org/10.1002/ase.1361

Ramos-Rincón, J. M., Belinchón-Romero, I., Sánchez-Ferrer, F., Torre, G. M., Harris, M., & Sánchez-Fernández, J. (2017). The reach of Spanish-language YouTube videos on physical examinations made by undergraduate medical students. *Journal of Educational Evaluation for Health Professions, 14*, 31. https://doi.org/10.3352/jeehp.2017.14.31

Reinoso Tapia, R., Delgado-Iglesias, J., & Fernández, I. (2019). Learning difficulties, alternative conceptions and misconceptions of student teachers about respiratory physiology. *International Journal of Science Education, 41*(18), 2602–2625. https://doi.org/10.1080/09500693.2019.1690177

Renkl, A., & Scheiter, K. (2015). Studying visual displays: How to instructionally support learning. *Educational Psychology Review, 29*(3), 599–621. https://doi.org/10.1007/s10648-015-9340-4

Richards-Babb, M., Curtis, R., Smith, V. J., & Xu, M. (2014). Problem solving videos for general chemistry review: Students' perceptions and use patterns. *Journal of Chemical Education, 91*(11), 1796–1803. https://doi.org/10.1021/ed500280b

Rittle-Johnson, B., Schneider, M., & Star, J. R. (2015). Not a one-way street: Bidirectional relations between procedural and conceptual knowledge of mathematics. *Educational Psychology Review, 27*(4), 587–597. https://doi.org/10.1007/s10648-015-9302-x

Roberts, J. K., Chudgar, S. M., Engle, D., McClain, E. K., Jakoi, E., Berkoben, M., & Lehrich, R. W. (2018). Digital chalk-talk videos improve knowledge and satisfaction in renal physiology. *Advances in Physiology Education, 42*(1), 146–151. https://doi.org/10.1152/advan.00131.2017

Rozenblit, L., & Keil, F. (2002). The misunderstood limits of folk science: An illusion of explanatory depth. *Cognitive Science, 26*(5), 521–562.

Salina, L., Ruffinengo, C., Garrino, L., Massariello, P., Charrier, L., Martin, B., Santina Favale, M., & Dimonte, V. (2012). Effectiveness of an educational video as an instrument to refresh and reinforce the learning of a nursing technique: A randomized controlled trial. *Perspectives in Medical Education, 1*(2), 67–75. https://doi.org/10.1007/s40037-012-0013-4

Saxena, V., Natarajan, P., O'Sullivan, P. S., & Jain, S. (2008). Effect of the use of instructional anatomy videos on student performance. *Anatomical Sciences Education, 1*(4), 159–165. https://doi.org/10.1002/ase.38

Schmidt, J. T. (2015). "Crash course": Unleashing YouTube on the US survey. *The Journal of the Gilded Age and Progressive Era, 14*(2), 284–286.

Schnotz, W., & Rasch, T. (2005). Enabling, facilitating, and inhibiting effects of animation in multimedia learning: Why reduction of cognitive load can have negative results on learning. *Educational Technology Research and Development, 53*, 47–58.

Schwartz, M. (2013). Khan academy: The illusion of understanding. *Online Learning Journal, 17*(4), 183760. https://www.learntechlib.org/p/183760/

Senchina, D. S. (2011). Video laboratories for the teaching and learning of professional ethics in exercise physiology curricula. *Advances in Physiology Education, 35*(3), 264–269.

Snelgrove, S., Tait, D. J. R., & Tait, M. (2016). Teaching psychology to student nurses: The use of 'Talking Head' videos. *Research in Learning Technology, 24*(1), Article 30891. https://doi.org/10.3402/rlt.v24.30891

Stockwell, B. R., Stockwell, M. S., Cennamo, M., & Jiang, E. (2015). Blended Learning Improves Science Education. *Cell, 162*(5), 933–936. https://doi.org/10.1016/j.cell.2015.08.009

Sweller, J., & Chandler, P. (1994). Why some material is difficult to learn. *Cognition and Instruction, 12*(3), 185–233.

Tackett, S., Slinn, K., Marshall, T., Gaglani, S., Waldman, V., & Desai, R. (2018). Medical education videos for the world: An analysis of viewing patterns for a YouTube channel. *Academic Medicine, 93*(8), 1150–1156.

Topping, D. B. (2014). Gross anatomy videos: Student satisfaction, usage, and effect on student performance in a condensed curriculum. *Anatomical Sciences Education, 7*(4), 273–279. https://doi.org/10.1002/ase.1405

Tripodi, N. (2018). First-year osteopathic students' use and perceptions of complementary video-based learning. *International Journal of Osteopathic Medicine, 30*, 35–43. https://doi.org/10.1016/j.ijosm.2018.09.004

Van den Eynde, J., Crauwels, A., Demaerel, P. G., Van Eycken, L., Bullens, D., Schrijvers, R., & Toelen, J. (2019). YouTube videos as a source of information about immunology for medical students: Cross-sectional study. *JMIR Medical Education, 5*(1), e12605. https://doi.org/10.2196/12605

van Wermeskerken, M., Ravensbergen, S., & van Gog, T. (2018). Effects of instructor presence in video modeling examples on attention and learning. *Computers in Human Behavior, 89*, 430–438. https://doi.org/10.1016/j.chb.2017.11.038

Versteeg, M., van Loon, M. H., Wijnen-Meijer, M., & Steendijk, P. (2020). Refuting misconceptions in medical physiology. *BMC Medical Education, 20*(1), Article 250. https://doi.org/10.1186/s12909-020-02166-6

Versteeg, M., Wijnen-Meijer, M., & Steendijk, P. (2019). Informing the uninformed: A multitier approach to uncover students' misconceptions on cardiovascular physiology. *Advances in Physiology Education, 43*(1), 7–14. https://doi.org/10.1152/advan.00130.2018

Wang, J., & Antonenko, P. D. (2017). Instructor presence in instructional video: Effects on visual attention, recall, and perceived learning. *Computers in Human Behavior, 71*, 79–89. https://doi.org/10.1016/j.chb.2017.01.049

Witchel, H. J., Thompson, G. A., Jones, C. I., Westling, C. E., Romero, J., Nicotra, A., Maag, B., & Critchley, H. D. (2020). Spelling errors and shouting capitalization lead to additive penalties to trustworthiness of online health information: Randomized experiment with laypersons. *Journal of Medical Internet Research, 22*(6), e15171.

Wouters, P., Paas, F., & van Merriënboer, J. J. G. (2008). How to optimize learning from animated models: A review of guidelines based on cognitive load. *Review of Educational Research, 78*(3), 645–675. https://doi.org/10.3102/0034654308320320

Zhang, D., Zhou, L., Briggs, R. O., & Nunamaker, J. F. (2006). Instructional video in e-learning: Assessing the impact of interactive video on learning effectiveness. *Information & Management, 43*(1), 15–27. https://doi.org/10.1016/j.im.2005.01.004

Chapter 8
Aligning Assessment Goals with the Current and Future Technologies Needed to Achieve Them

Melanie M. Cooper and Michael W. Klymkowsky

Abstract The issue of assessment involves two interdependent drivers: the purpose (s) of the assessment and how such assessments can be applied at scale, that is, in large and, in some cases, remote settings. The simplest assessment goal, to sort students by what content they know or can recognize as correct, often involves a variety of "forced-choice" or fill in the blank questions that are readily analyzed by computers. Higher-level assessments that evaluate the extent to which students can access and apply their knowledge to new situations (as opposed to remembering previously presented examples), and can be used to develop students' working knowledge, demand more sophisticated Socratic approaches aimed at making student presumptions explicit, together with their relevance and implications. Progress along these lines involves the automated analysis and response to drawn responses (graphs and such), as in the beSocratic™ system. Future extensions will require an iterative feedback system that can analyze students' textual responses "on the fly" and pose disciplinarily relevant and clarifying Socratic questions. We consider the current state of affairs in achieving this goal.

8.1 Introduction

Knowing what students know is a difficult proposition (National Research Council, 2001). While it is relatively easy to ascertain whether students are able to remember facts, apply rules or heuristics, or solve algorithmic problems, understanding how students' reason about more complex ideas is much more difficult. Assessment of simple recall and algorithmic problems can readily be achieved by multiple-choice

M. M. Cooper
Present Address: Department of Chemistry, Michigan State University, East Lansing, MI, USA
e-mail: mmc@msu.edu

M. W. Klymkowsky (✉)
Department of Molecular, Cellular Developmental Biology, University of Colorado Boulder, Boulder, CO, USA
e-mail: michael.klymkowsky@colorado.edu; klym@colorado.edu

H. J. Witchel, M. W. Lee (eds.), *Technologies in Biomedical and Life Sciences Education*, Methods in Physiology, https://doi.org/10.1007/978-3-030-95633-2_8

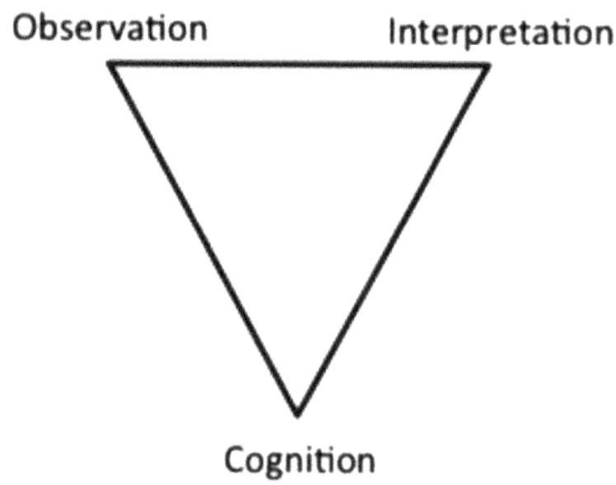

Fig. 8.1 The assessment triangle

tests that are easily administered and machine graded. Such approaches have become commonplace in many areas of higher education, particularly where large numbers of students are involved. However, we know that the ways that students are assessed may impact the choices that students, and instructors, make about teaching and learning. We can assume that students are rational actors; if they see that their grades will be determined solely by whether they can remember facts and perform rote calculations, then that is what they will learn to do. Indeed, the adage "If you don't assess what is important, then what's assessed becomes important" captures this reality (Miller & Parlett, 1974). Such recall and algorithmic (heuristic)-based assessments can lead to fragmentation and trivialization not only of what students know but of how curricula are designed and delivered so that outcomes can be assessed in this way. That being said, it is unlikely that most faculty would place memorization and the rote application of mathematical formulae as their goals, which leads us to the central problem—how to assess student reasoning on more complex tasks, at scale, without increasing costs in terms of time and personnel. In this chapter we discuss ways in which technology can be harnessed to address this problem.

8.1.1 Designing Assessments That Elicit Evidence of Learning

We start by considering theoretical aspects of assessment design. Perhaps the most influential approach was described in the National Academy of Sciences publication *Knowing What Students Know* (National Research Council, 2001). The authors of this consensus report proposed that assessments be considered as evidence-driven arguments. That is, assessments should elicit evidence that can be used to support an argument about what students know and are able to do. A visual representation of this process is known as the assessment triangle (Fig. 8.1), where the three vertices represent (1) the model of cognition that supports assessment design; (2) the observations that the assessment design is expected to elicit; and (3) how those observations will be interpreted. For example, if we want to know what students can memorize or recognize, we would design very different test items than if we are interested in whether students are able to use their knowledge and apply it to new situations.

8.1.2 Models of Cognition

Such differences stem from the model of cognition that guides assessment development. Over the years, many models of learning have been proposed (National Academies of Sciences, Engineering, and Medicine, 2018; National Research Council, 1999) and with them concomitant approaches to designing learning environments and assessments. It is beyond our scope here to review them all, but in general and over time, there has been a paradigm shift away from behaviorist learning models (Gagne, 1965; Skinner, 1954) such as direct instruction and associated measures of learning based on what we will call traditional tests, tests often based on right/wrong and forced-choice answers, factual recall, and the application of procedures and equations. Over time, most researchers have moved to more constructivist perspectives on learning and assessment. That is, instead of viewing knowledge as something that is transferred from instructor to student, learning is viewed as an act of construction by the student, through a process aided by the instructor and shaped by the design of the learning materials. Interestingly, as pointed out by Shepard, the shift from a behaviorist to a constructivist learning model is often *not* accompanied by a shift in assessment design and use (Shepard, 2000). Too often constructivist pedagogical approaches, based on active engagement of students with specific terms and concepts, and their application, are not accompanied by constructivist rethinking of course design or assessments. For example, the highly cited report on the efficacy of "active learning" strategies, as opposed to "passive" lectures (Freeman et al., 2014), relies almost exclusively on multiple-choice assessment instruments known as concept tests or concept inventories that, as we will discuss, are often incompatible with constructivist learning theories.

Even within constructivist learning models, there are an array of variants as to just how knowledge is constructed, that is, how learning occurs, and how students' prior knowledge is integrated (and modified) in the course of their development of a coherent understanding of disciplinary ideas and their application. At the same time, there has been a shift from models that focus on replacing students' incorrect ideas (misconceptions) with correct ideas (Posner et al., 1982) to viewing student ideas as resources that may or may not be useful, or even be recognized as relevant in a particular situation. Through iterative assessment and associated feedback, commonly termed formative assessment, students' ideas can be reshaped and woven together to construct a disciplinarily coherent and applicable framework (Hammer, 2000). Different models of, and approaches to, learning require (or are satisfied by) different assessment strategies.

8.1.3 Assessment: From Simple to Complex

Here we consider how various assessments align with different cognitive models (or not) and the evidence of learning that they have the potential to elicit. If we are simply interested in whether a student has memorized a fact or is able to use an equation to calculate a numerical answer, assessment design can be quite simple. Typically, such questions are written so that there is a single unambiguously correct response. Consider the question "in what molecule is genetic information stored in organisms?" The answer is DNA. A multiple-choice or fill in the blank response will suffice to determine whether a student "knows" or can recognize the correct answer. As we discuss later, answering such questions does not provide us with any evidence that the student has a coherent knowledge framework that incorporates an understanding of where genetic information comes from, how it is stored, how it is used, or what features of DNA, and the cellular systems involved in its replication, repair, and access make it a uniquely suitable molecular system for information storage. We should not expect that students who are assessed through memorization/recognition-type questions will have developed the disciplinary expertise required. Such assessments are, however, easy to write and grade and can be administered online. Indeed, there has been a proliferation of publisher developed, textbook-associated test banks (as witnessed by a simple Google search). By their very nature, such test bank questions address surface-level knowledge. While fears of cheating (see Chap. 10 in this volume) can drive various anti-cheating strategies (such as randomizing question order), they do not address the underlying nature of the assessment instrument, what we might term its "resolution"—the level of understanding its use can reveal. Such exams rely on simplistic models about how people learn and can be counterproductive: supporting an overconfidence (a type of Dunning-Kruger effect) among both students and instructors that students' knowledge and instructional efficacy is much more robust than it actually is.[1] It is unlikely that students who are routinely tested in this manner will develop disciplinary expertise, that is, the ability to apply their knowledge to new and/or more ambiguous situations. For example, students who scored in the 75th percentile of the national average in the USA (using a multiple-choice test) have been shown to have considerable misunderstandings about important ideas central to chemistry and biology (Williams et al., 2015). In fact, there is evidence that such forced-choice instruments tend to overestimate what students know (Lee et al., 2011) and that they tell us very little about what students can do with their knowledge. It should also be noted that such assessment items do not align with modern theories of cognition.

There are a few types of assessment instruments that are based on a model of cognition, have a research-based development protocol, and can be administered online and machine graded. These are what are now known as concept tests or inventories. Pioneered by Treagust (1988) and Hestenes et al. (1992), these multiple-choice tests are designed not to determine what students know, but rather to identify

[1] Reverse Dunning-Kruger effects and science education

what "misconceptions" students may hold, that is, what ideas and presumption students incorrectly apply and that need to be addressed through instruction. There are now hundreds of such inventories, across many fields of study (Klymkowsky & Garvin-Doxas, 2020; National Research Council, 2012).

The model of cognition underlying concept tests and inventories, while acknowledging that students construct knowledge, implies that if we can identify misconceptions, they can be replaced or overwritten by instruction. Concept tests have had a major impact in physics, where students' understanding of the macroscopic world conflicts with established physical laws. Understanding why we are not projected into space by the Earth's rotation requires principles not readily deduced from everyday experiences.

In early work Strike and Posner (Posner et al., 1982) proposed that once a misconception is identified, instruction can induce cognitive dissonance that will lead students to become dissatisfied with their understanding—allowing the correct idea to replace the incorrect one. While this approach has been modified over the years, there is, in fact, little evidence that this is how learning actually works for the typical student or that this "replacement" approach brings about conceptual change, particularly for complex phenomena—which is the case for scientific processes, whether chemical, physical, or biological. Nevertheless, concept inventories are often used to monitor "learning outcomes" and to supply evidence for improved outcomes associated with specific educational strategies. In such cases, pre- and post-instruction test scores are used to calculate "learning gains" (Freeman et al., 2014; Hake, 1998), but what exactly has been learned is rarely, if ever, clearly stated, making such conclusions rather unconvincing.

Part of the problem lies with the multiple-choice format of concept test/inventories. While a student may choose the correct answer, we have no evidence about why that answer was chosen. Conversely, choosing an incorrect answer may not mean that the student has a misconception, but rather they may have misunderstood the question or are using reasoning and presumptions that the item writer did not intend. Certainly, if the distractors (wrong answers) are based on research on how students might answer that question, a student's choice of distractor can provide some insights into student thinking (Garvin-Doxas and Klymkowsky, 2008), but without further evidence, it is difficult to determine the cause of such problems. When distractors do not reflect actual student thinking (as is often the case when instructors construct the test items), students may more easily rule out implausible answers which can improve scores, while understanding may remain unaltered (Gierl et al., 2017). Another possibility is that such "lures" may actually implant ideas that students did not have previously (Butler et al., 2007; Butler & Roediger, 2008).

One final type of multiple-choice assessment that relies on extensive research on student thinking is the "ordered multiple choice" (OMC) instrument, which treats student learning as a progression of increasingly sophisticated ideas. These instruments were first introduced by Briggs and co-workers, as a way to provide effective and rapid diagnostic information about student learning. Because they are not a typical dichotomous (right/wrong) test, they provide more evidence about how students are thinking about a particular phenomenon (Briggs et al., 2006). OMC items are often based on learning progressions, which are "successively more complex ways of thinking about an idea that might reasonably follow one another in a student's learning" (Smith et al., 2006). In OMC instruments the distractors are developed from research on student thinking and are ordered in levels of increasing sophistication using item response theory (IRT) (Embretson & Reise, 2013). That is, the answers are not right or wrong, but rather they differ in sophistication. Such OMC assessment instruments require a great deal of prior research and psychometric analysis before they are considered reliable and valid.

8.2 A Constructivist Approach

8.2.1 Constructivist Assessments for Constructivist Pedagogies and Learning Models

Because we focus here on how technology can support the assessment of learning, we have so far considered the most common types of questions that lend themselves to automated delivery and grading on various instructional platforms. However, we hope that we have convinced the reader that multiple-choice, fill in the blank, and true/false questions generally fail to provide convincing evidence about what students understand and perhaps even more importantly what they can do with their knowledge. Furthermore, these types of questions typically test only fragmentary knowledge, which does not align with constructivist models of how students learn. While we do not expect our students to become experts after 1 or 2 (or even 4) years of learning in a subject, our goal as instructors is to support students toward the development of knowledge that is useful and not inert—that is, toward a more expert-like understanding.

It is clear that experts' knowledge is structured differently from that of novices (National Research Council, 1999). Experts' knowledge cannot be reduced to sets of isolated facts; it is connected, conditionalized depending on circumstances, and reflects a deep understanding of the subject matter—the ability to recognize what is likely to be relevant and irrelevant to a specific situation. To develop such expertise requires that students are encouraged to construct more expert-like knowledge frameworks, a process that often involves multiple attempts, informed feedback (coaching), and repeated revisions. If we want to evaluate whether students are moving toward expertise, we must learn how to distinguish and evaluate more expert-like thinking. To do that, we must provide students with learning materials and formative and summative assessments that help them develop more connected, contextualized, and useful knowledge frameworks.

In much of our current work, we have adopted a constructivist model of cognition and used a resources framework (Hammer, 2000) through which we attempt to help students develop, recall, and use appropriate resources (which may be facts, skills, concepts, over-arching principles, and practices) that can be connected to produce a

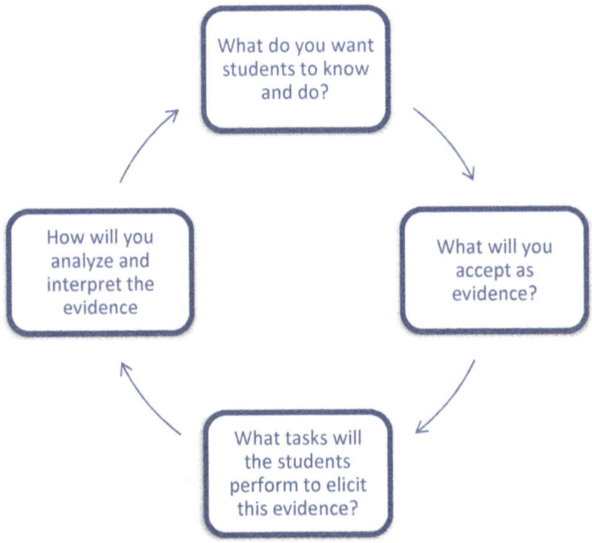

Fig. 8.2 Evidence-centered design

coherent and useful response to a prompt. In general, this requires that students are able to construct (rather than choose from a menu) their own responses, which consist of written explanations and arguments that may include or rely upon drawn diagrams, models, graphs, and pictures. In general, both modalities (written and drawn) are required to elicit evidence about what students know and can do. This, of course, poses a technological challenge; it is much more difficult to recognize and respond to students' constructed responses than it is to recognize the correct answer in a forced-choice instrument. However, if it is evidence that we are after, we must allow students to speak with their own words and drawings.

In our work we have used a modified version of evidence-centered design (ECD) (Mislevy & Riconscente, 2011) to design assessments that can elicit appropriate evidence. As shown in Fig. 8.2, ECD involves a number of steps:

1. Develop statements about what students should know and be able to do. This means explicitly acknowledging the model of cognition to be addressed.
2. Describe what evidence would convince us that students are able to use that knowledge in ways that we have specified.
3. Design tasks that have the potential to elicit such evidence.
4. Consider how the evidence elicited by the assessment will be analyzed and interpreted.

This is an iterative process, and, in practice, we find that it is typically necessary to repeat the cycle a number of times. Reframing of questions and prompts is also often needed to elicit informative student responses.

8.2.2 beSocratic: A Constructivist Teaching and Assessment System

With these ideas in mind, let us consider the ways in which technology can support the delivery and evaluation of well-designed, beyond forced-choice, assessment tasks. Here we describe our work on the development and use of the beSocratic system in the context of courses in general chemistry, organic chemistry, and molecular biology. These are assessment tasks that have been administered to tens of thousands of students and so reflect a particularly promising approach.

beSocratic is a web-based assessment platform that allows students to complete tasks by drawing diagrams, graphs, and reaction mechanisms, by writing explanations and arguments, with minimal constraints (Bryfczynski, 2012; Cooper et al., 2014). That is, the evidence elicited by the task prompt is not constrained by the way the student must respond as is the case with forced-choice assessment systems. The system is designed so that many (although not all) types of diagrams and graphs can be automatically analyzed and responded to, but perhaps more importantly instructors can target known issues with student drawings and graphs and provide contextualized feedback that is increasingly specific, as necessary. An obvious extension of such a system would be the automated analysis and input-driven feedback to written responses. While not yet implemented, there are a number of promising strategies applicable to this task which are briefly discussed at the end of this chapter.

8.2.3 Examples of beSocratic Activities

Here we describe a few beSocratic activities and the evidence they have provided about student learning. Non-covalent interactions or intermolecular forces (IMFs), that is, the forces that occur between molecules and between regions of larger molecules, are a fundamental idea in chemistry with significant implications for biological systems, relevant to molecular structure and properties and to how and why molecules interact (stick together) and come apart. Whether it be to explain phase changes, chemical reactions, enzyme-substrate binding, the effects of solvent and pH on the behavior of biomolecules and pharmaceuticals, or why geckos can walk up walls, the general principles are the same. These non-covalent interactions that go by a variety of names that include van der Waals interactions, London dispersion forces, and hydrogen bonding arise from the same phenomenon—nonuniform electron density distribution (whether permanent or induced) resulting in dipoles that lead to electrostatic attractions between molecules or between parts of large molecules.

In the first study, our goal was to determine whether students understood the nature of IMFs. The evidence we sought to elicit was simply "do students know what intermolecular forces are." We used beSocratic to record students responses to questions that asked them to describe their understanding of various types of

IMFs, to draw three molecules (of ethanol), and to show where IMFs were operating (Cooper et al., 2015). Perhaps not surprisingly, students were able to provide the kinds of descriptions of IMFs that are prevalent in conventional textbooks, but to our consternation, these descriptions were accompanied by drawings that in general tended to represent IMFs as forces operating within small molecules. This was true for all types of IMFs, and we have since replicated this study in many settings, with similar findings. Interestingly previous work, using multiple-choice instruments, failed to reveal such problematic responses (Schmidt et al., 2009). We also note that subsequent findings, in which students were enrolled in a transformed curriculum where the use of beSocratic was routine, showed marked improvements in student understanding of IMFs (Williams et al., 2015).

While such studies were revealing, what they did not elicit was any evidence about what factors (resources) students used to describe IMFs. Because IMFs are so central to understanding chemical and biological processes, we pushed further with activities designed to elicit more and better evidence about what resources students use and how they weave them together to provide explanations. Students were asked to write and draw explanations for what happens when two neutral atoms or molecules move toward each other; responses were captured in beSocratic and were coded by hand (Becker et al., 2016; Noyes & Cooper, 2019). The codes applied captured the type of reasoning that students used to answer the question and could be applied to both the written and drawn explanations. Student responses generally fell into three classifications of increasing sophistication. The lowest level corresponded to students who said the molecules are attracted but did not explain why. At level 2 students were able to leverage ideas about coulombic attractions—they were able to explain that the molecules were attracted because of opposite charges but did not provide a mechanism for how neutral molecules could have unequal charge distributions—which only occurred at level 3. We have replicated this study many times in many different institutions. Recently, working with the Automated Analysis of Constructed Responses (AACR) group (Urban-Lurain et al., 2015), we have been able to automate the coding of student responses (Noyes et al., 2020). This process is time-consuming and personnel intensive. Humans must code many responses to train the AACR system, but once trained the system is able to code student responses with the same accuracy as human coders. We return to the automated analysis of student responses shortly.

We describe one more example of a constructivist task designed to elicit the resources that students use as they reason about chemical phenomena to illustrate the importance of the prompt. In this study, we were interested in how students reason about acid-base reactions (Cooper et al., 2016). The methodology was similar to that described earlier: that is, student responses were characterized by the type of reasoning and the resources they used to address the prompt. Initially, we found that students' responses were at a rather low level. When students were asked to explain what was happening in a particular acid-base reaction, they provided descriptive responses about what acid-base reactions are and how acid-base reactions occur. By changing the prompt (Fig. 8.3) to make our expectations more explicit, much more sophisticated responses could be elicited from a similar cohort

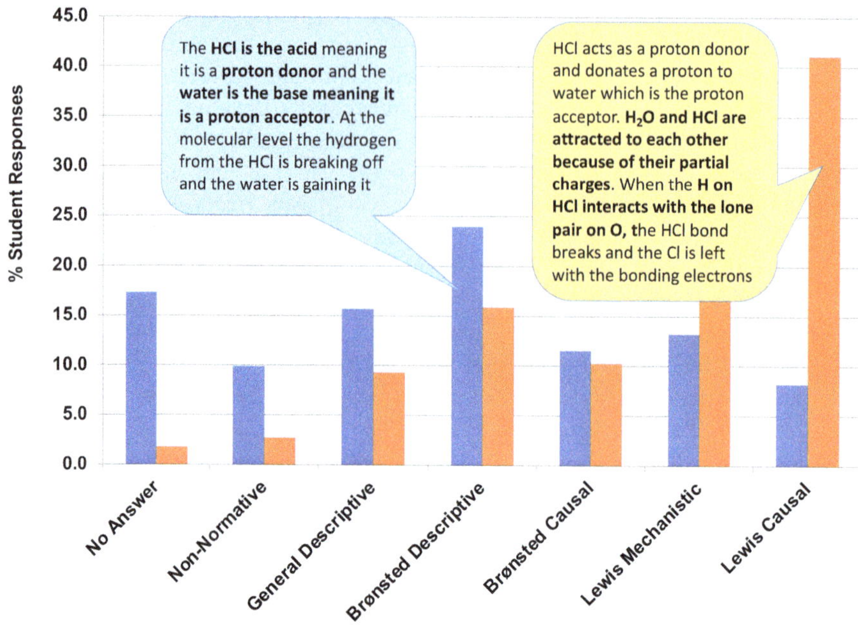

Fig. 8.3 The change in student responses by changing the scaffolding in the prompt

of students. Indeed, it has been our experience that it takes several attempts to design task prompts that elicit informative student responses and so provide clearer evidence of student thinking (Cooper et al., 2016). That is, we have had to go through the ECD assessment design cycle a number of times before we alight on a prompt that "works" to elicit a revealing response from the student.

One, perhaps not obvious, lesson has been that asking students to "explain" typically elicits disappointing results because students are not given sufficient guidance about what the instructor is looking for in the response, what exactly are they expected to explain. On the other hand, if the prompt is too scaffolded, the student is led to the appropriate answer, often without understanding why, a result that leads to an overestimate of the students' capability, in a similar way to multiple-choice items. Identifying a "goldilocks" prompt, not too prescriptive, not too vague, often requires a multiple trial-and-error process.

8.3 Feedback

8.3.1 The Role of Feedback

Up to now, we have not addressed the role of feedback in assessment. Assessments can generally be divided into formative and summative. In the past, particularly at the university level, most assessment has been summative where the purpose of the assessment is to sort students, assign a grade, or allow them to pass through a gateway to further study. However, if we are to teach, that is, to introduce students to and help them develop more expert-like thinking that involves more than right/wrong, it becomes important to use formative assessment tasks to provide feedback to both student and instructor. Feedback to the student can, if done appropriately, drive their learning. Models of such interaction include the Socratic dialogue and the peer-review and author revision process, where students (authors) are asked to articulate their assumptions and explain and justify their reasoning. Feedback to the instructor is just as important: it can provide evidence about where students "are" so that instruction can be tailored to address areas of confusion or omission, what students do not know and perhaps have not been taught, as well as what they need to know and consider as relevant in order to construct an adequate answer. It can also be an impetus for course and curriculum redesign. Formative assessment is a stepping-stone to subject mastery and improved instruction.

For formative assessment to be effective, we know that students must receive timely and appropriate feedback (Hattie & Timperley, 2007). However, it has also been shown that, for various reasons, many students do not actually use the feedback they receive (Winstone et al., 2017) (see Chap. 9 in this volume). Nicol and Mcfarlane-Dick (2006) point out that while higher education has moved toward constructivist pedagogies, a parallel shift in formative assessment and feedback has generally not occurred. That is, formative assessment and feedback are seen as the responsibility of the instructor, where the feedback is a transmission process from instructor to student. Providing information about whether students responses are right or wrong, and perhaps even their strengths and weaknesses, may not bring about the kinds of changes we imagine or hope for. For example, some feedback is complex and not easily decoded and translated by the student into effective action. Responding to, and learning from feedback requires the same kinds of opportunities to actively construct and test the validity of the student's understanding. For feedback to be effective, both the giver and the receiver of the feedback must understand the purpose and must engage with the feedback in a meaningful way. This often involves an extended dialogue—which of course means more time and effort on the part of both instructor and student: we will discuss approaches to this later.

Winstone et al. (2017) interviewed students and found that there were a number of barriers that deterred the use of feedback. For example, some feedback was too technical and could not readily be implemented by the student. The most common problems with feedback, however, had to do with grades. Students tend to ignore feedback if they have received a good (enough) grade; conversely if their grade was poor, many students did not use the feedback provided. Prior unrealistic expectations about the time and effort it would take to improve led to a kind of helplessness and the feeling that using feedback would be pointless. Often instructors can contribute in negative ways by conveying their eagerness (need) to move on, to "cover" the course content—reinforcing the implicit message that individual student learning is of secondary importance. It is suggested that for students to develop the agency and self-regulation to effectively use feedback, it should not be provided in the context of a graded assignment. Additionally, students should have the opportunity and time needed to engage with the feedback and consider (by themselves, or with others) its meaning and how it can be used.

8.3.2 Instructional Technology, Formative Assessment, and Feedback

Most learning management systems provide the ability for multiple-choice assessments to provide feedback about right and wrong answers, as well as feedback of various types for incorrect responses. More sophisticated versions are found in "cognitive tutors" (D'mello & Graesser, 2013) but tend to be focused on mathematics, usually algebra (Pane et al., 2014), and the development of automaticity for rote cognitive skills or for training of instrument use. Such adaptive approaches have also found use to help medical students interpret diagnostic tests (Romito et al., 2016). However, if we want to support students, and provide feedback for more complex (constructivist) tasks, the technological solution must provide the capability to respond to students drawing and writing.

As noted earlier, beSocratic allows instructors to provide contextual responses to some kinds of simple diagrams, drawings, and graphs. For example, we might ask a student to draw an energy profile for a reaction, draw curved arrows to indicate an organic reaction mechanism, indicate the direction of energy transfer in a system, or predict the behavior over time of a gene expression network. beSocratic uses a system of rules inputted by the author/instructor, and feedback can be designed based on the context. Figure 8.4a shows a sample activity in which a student is asked to provide evidence that they know what kind of reaction is being discussed and what the energy change during the reaction is.

The student draws a response, which in this case is incorrect (there should be only one maximum), and receives feedback that they must respond to and then attempts to redraw the diagram. The feedback can be in increasingly directed tiers, if desired. When the student has successfully drawn the initial diagram (Fig. 8.4b), it can be transferred to the next task, where the student must respond to a new question (Fig. 8.4c).

This example shows some of the ways in which beSocratic can be used, by responding to graphical input, bringing forward prior drawings for comparison to use in a new task, and how written and drawn responses may be combined to provide a fuller response and more evidence about what the student knows and can do. We have developed over 100 beSocratic activities with these capabilities that are in regular use in the context of reimagined general and organic chemistry and biology courses (Cooper et al., 2019; Cooper & Klymkowsky, 2013; Klymkowsky et al., 2016). We have published a range of studies where we have compared students from transformed courses that use beSocratic activities as formative assessments to

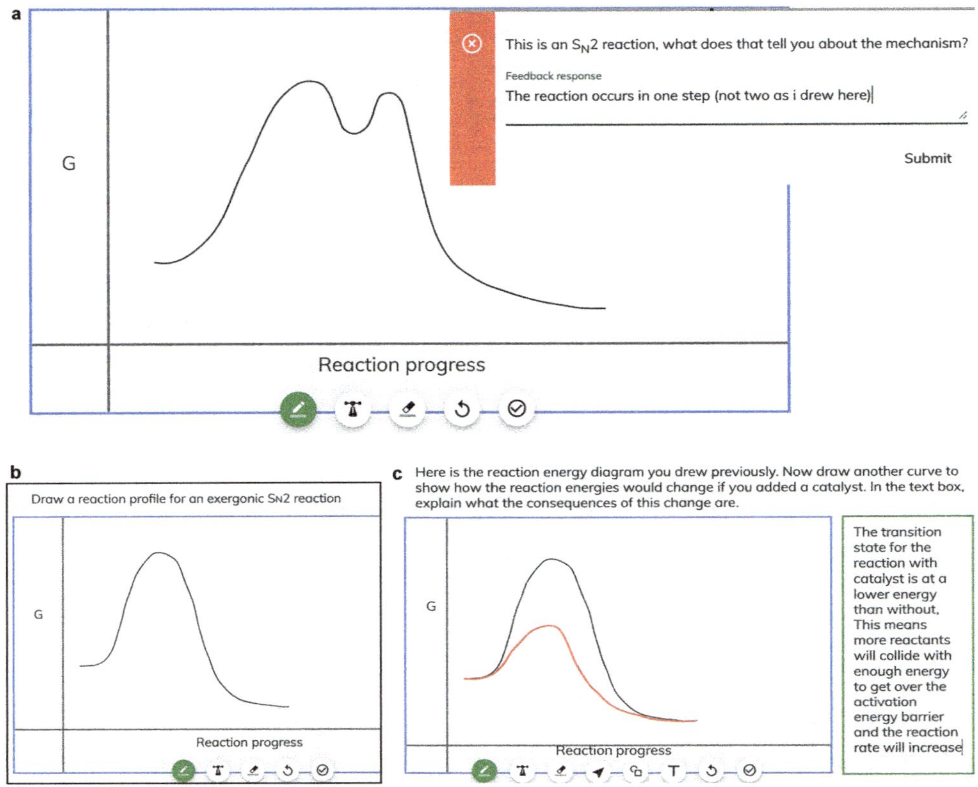

Fig. 8.4 (**a**) Feedback provided to students after drawing an incorrect energy diagram. (**b**) The correct diagram. (**c**) A new task showing how the student draws a new curve relative to the original, correct energy diagram

comparable cohorts of students who are enrolled in more traditional courses that use more conventional assessment systems. These studies reveal that students in transformed courses are better able to draw Lewis structures (Cooper et al., 2012), predict properties (Underwood et al., 2015), explain acid-base reactions (Cooper et al., 2016; Crandell et al., 2019), and nucleophilic substitutions (Crandell et al., 2020), among others.

For us, the next step is to integrate feedback into the written components of student explanations. We have already shown that we can train a machine grader to recognize differing levels of sophistication in student responses (Noyes et al., 2020), but what we cannot do yet is provide students with appropriate, useful discipline-specific feedback in a timely manner. Our goal over the next few years is to integrate beSocratic with natural language processing and machine learning systems to provide such feedback (Foltz et al., 2000).

We would be remiss if we did not acknowledge that there are a large range of adaptive technologies available for instruction and assessment, and these are discussed in detail in Chap. 5 in this volume.

8.4 The Future of Technology-Based Assessment

By now we hope to have convinced you that the future of assessment lies not in more complex types of forced-choice instruments (Klymkowsky & Garvin-Doxas, 2020), but rather in well-designed open response tasks that require both written and drawn elements that can be automatically recognized, responded to, and graded (if necessary). Indeed, many of the elements necessary to make this a reality are already in place; we can recognize and respond to a subset of student drawn elements, and we can train systems to identify different levels of written responses, based on original human coding. Systems involving chatbots and the more sophisticated "smart conversational AI" are commonplace, often taking the place of human customer service. Certainly, these systems can be frustrating to deal with on occasion (Powell, 2019), but it is not a stretch to imagine that in the future students will interact with AI systems by drawing, writing, and speaking, with the AI providing accurate and relevant contextual feedback, scaffolded support, and challenging (Socratic) questions or grading students' responses. We envision a future in which examinations are no longer necessary, where concerns about cheating can be minimized (see Chap. 10 in this volume), where continuous formative tasks provide students with the opportunities needed to develop more expert-like levels of thinking, and where the evidence elicited by them is convincing and robust.

References

Becker, N. M., Noyes, K., & Cooper, M. M. (2016). Characterizing students' mechanistic reasoning about London dispersion forces. *Journal of Chemical Education*. https://doi.org/10.1021/acs.jchemed.6b00298

Briggs, D. C., Alonzo, A. C., Schwab, C., & Wilson, M. (2006). Diagnostic assessment with ordered multiple-choice items. *Educational Assessment, 11*(1), 33–63.

Bryfczynski, S. P. (2012). *BeSocratic: An intelligent tutoring system for the recognition, evaluation, and analysis of free-form student input* (UMI No. 3550201). Doctoral dissertation. Clemson University.

Butler, A. C., Karpicke, J. D., & Roediger, H. L., III. (2007). The effect of type and timing of feedback on learning from multiple-choice tests. *Journal of Experimental Psychology: Applied, 13*(4), 273.

Butler, A. C., & Roediger, H. L. (2008). Feedback enhances the positive effects and reduces the negative effects of multiple-choice testing. *Memory & Cognition, 36*(3), 604–616.

Cooper, M. M., & Klymkowsky, M. W. (2013). Chemistry, life, the universe and everything: A new approach to general chemistry, and a model for curriculum reform. *Journal of Chemical Education, 90*, 1116–1122. https://doi.org/10.1020/ed300456y

Cooper, M. M., Kouyoumdjian, H., & Underwood, S. M. (2016). Investigating students' reasoning about acid–base reactions. *Journal of Chemical Education, 93*(10), 1703–1712. https://doi.org/10.1021/acs.jchemed.6b00417

Cooper, M. M., Stowe, R. L., Crandell, O. M., & Klymkowsky, M. W. (2019). Organic chemistry, life, the universe and everything (OCLUE): A transformed organic chemistry curriculum. *Journal of Chemical Education*. https://doi.org/10.1021/acs.jchemed.9b00401

Cooper, M. M., Underwood, S. M., Bryfczynski, S. P., & Klymkowsky, M. W. (2014). A short history of the use of technology to model and analyze student data for teaching and research. In R. S. Cole & D. Bunce (Eds.), *Tools of chemistry education research* (pp. 219–239). American Chemical Society.

Cooper, M. M., Underwood, S. M., Hilley, C. Z., & Klymkowsky, M. W. (2012). Development and assessment of a molecular structure and properties learning progression. *Journal of Chemical Education, 89*(11), 1351–1357. https://doi.org/10.1021/ed300083a

Cooper, M. M., Williams, L. C., & Underwood, S. M. (2015). Student understanding of intermolecular forces: A multimodal study. *Journal of Chemical Education, 92*(8), 1288–1298. https://doi.org/10.1021/acs.jchemed.5b00169

Crandell, O. M., Kouyoumdjian, H., Underwood, S. M., & Cooper, M. M. (2019). Reasoning about reactions in organic chemistry: Starting it in general chemistry. *Journal of Chemical Education, 96*(2), 213–226. https://doi.org/10.1021/acs.jchemed.8b00784

Crandell, O. M., Lockhart, M. A., & Cooper, M. M. (2020). Arrows on the page are not a good gauge: Evidence for the importance of causal mechanistic explanations about nucleophilic substitution in organic chemistry. *Journal of Chemical Education, 97*(2), 313–327. https://doi.org/10.1021/acs.jchemed.9b00815

D'mello, S., & Graesser, A. (2013). AutoTutor and affective AutoTutor: Learning by talking with cognitively and emotionally intelligent computers that talk back. *ACM Transactions on Interactive Intelligent Systems (TiiS), 2*(4), 1–39.

Embretson, S. E., & Reise, S. P. (2013). *Item response theory*. Psychology Press.

Foltz, P. W., Gilliam, S., & Kendall, S. (2000). Supporting content-based feedback in on-line writing evaluation with LSA. *Interactive Learning Environments, 8*(2), 111–127.

Freeman, S., Eddy, S. L., McDonough, M., Smith, M. K., Okoroafor, N., Jordt, H., & Wenderoth, M. P. (2014). Active learning increases student performance in science, engineering, and mathematics. *Proceedings of the National Academy of Sciences of the United States of America, 111*, 8410–8415. https://doi.org/10.1073/pnas.1319030111

Gagne, R. M. (1965). *The conditions of learning*. Holt Rinehart & Winston.

Gierl, M. J., Bulut, O., Guo, Q., & Zhang, X. (2017). Developing, analyzing, and using distractors for multiple-choice tests in education: A comprehensive review. *Review of Educational Research, 87*(6), 1082–1116. https://doi.org/10.3102/0034654317726529

Hake, R. R. (1998). Interactive-engagement versus traditional methods: A six-thousand-student survey of mechanics test data for introductory physics courses. *American Journal of Physics, 66*(1), 64–74.

Hammer, D. (2000). Student resources for learning introductory physics. *American Journal of Physics, 68*, S52–S59.

Hattie, J., & Timperley, H. (2007). The power of feedback. *Review of Educational Research, 77*, 81–112.

Hestenes, D., Wells, M., & Swackhamer, G. (1992). Force concept inventory. *The Physics Teacher, 30*(3), 141–158. https://doi.org/10.1119/1.2343497

Klymkowsky, M. W., & Garvin-Doxas, K. (2020). Concept inventories: Design, application, uses, limitations, and next steps. In *Active learning in college science* (pp. 775–790). Springer.

Klymkowsky, M. W., Rentsch, J. D., Begovic, E., & Cooper, M. M. (2016). The design and transformation of biofundamentals: A nonsurvey introductory evolutionary and molecular biology course. *CBE Life Sciences Education, 15*(4). https://doi.org/10.1187/cbe.16-03-0142

Lee, H.-S., Liu, O. L., & Linn, M. C. (2011). Validating measurement of knowledge integration in science using multiple-choice and explanation items. *Applied Measurement in Education, 24*(2), 115–136. https://doi.org/10.1080/08957347.2011.554604

Miller, C. M., & Parlett, M. (1974). *Up to the mark: A study of the examination game*. Society for Research into Higher Education.

Mislevy, R. J., & Riconscente, M. M. (2011). Evidence-centered assessment design. In S. Downing & T. Haladyna (Eds.), *Handbook of test development*. Erlbaum.

National Academies of Sciences, Engineering, and Medicine. (2018). *How people learn II: Learners, contexts, and cultures*. https://doi.org/10.17226/24783.

National Research Council. (1999). *How people learn: Brain, mind, experience, and school.* National Academies Press.

National Research Council. (2001). In J. W. Pellegrino, N. Chudowsky, & R. Glaser (Eds.), *Knowing what students know: The science and design of educational assessment.* National Academies Press.

National Research Council. (2012). In S. R. Singer, N. R. Nielson, & H. A. Schweingruber (Eds.), *Discipline-based education research: Understanding and improving learning in undergraduate science and engineering*. National Academies Press.

Nicol, D. J., & Macfarlane-Dick, D. (2006). Formative assessment and self-regulated learning: A model and seven principles of good feedback practice. *Studies in Higher Education, 31*(2), 199–218. https://doi.org/10.1080/03075070600572090

Noyes, K., & Cooper, M. M. (2019). Investigating student understanding of london dispersion forces: A longitudinal study. *Journal of Chemical Education.* https://doi.org/10.1021/acs.jchemed.9b00455

Noyes, K., McKay, R. L., Neumann, M., Haudek, K. C., & Cooper, M. M. (2020). Developing computer resources to automate analysis of students' explanations of London dispersion forces. *Journal of Chemical Education, 97*(11), 3923–3936. https://doi.org/10.1021/acs.jchemed.0c00445

Pane, J. F., Griffin, B. A., McCaffrey, D. F., & Karam, R. (2014). Effectiveness of cognitive tutor algebra I at scale. *Educational Evaluation and Policy Analysis, 36*(2), 127–144. https://doi.org/10.3102/0162373713507480

Posner, G. J., Strike, K. A., Hewson, P. W., & Gertzog, W. A. (1982). Accommodation of a scientific conception: Toward a theory of conceptual change. *Science Education, 66*(2), 211–227.

Powell, J. (2019). Trust me, I'm a chatbot: How artificial intelligence in health care fails the Turing test. *Journal of Medical Internet Research, 21*(10), e16222.

Romito, B. T., Krasne, S., Kellman, P. J., & Dhillon, A. (2016). The impact of a perceptual and adaptive learning module on transoesophageal echocardiography interpretation by anaesthesiology residents. *BJA: British Journal of Anaesthesia, 117*(4), 477–481.

Schmidt, H.-J., Kaufmann, B., & Treagust, D. F. (2009). Students' understanding of boiling points and intermolecular forces. *Chemistry Education Research and Practice, 10*, 265–272. https://doi.org/10.1039/B920829C

Shepard, L. A. (2000). The role of assessment in a learning culture. *Educational Researcher, 29*(7), 4–14. https://doi.org/10.3102/0013189X029007004

Skinner, B. F. (1954). The science of learning and the art of teaching. *Cambridge, Mass, USA, 99*, 113.

Smith, C. L., Wiser, M., Anderson, C. W., & Krajcik, J. (2006). Focus article: Implications of research on children's learning for standards and assessment: A proposed learning progression for matter and the atomic-molecular theory. *Measurement: Interdisciplinary Research & Perspective, 4*(1–2), 1–98.

Treagust, D. F. (1988). Development and use of diagnostic tests to evaluate students' misconceptions in science. *International Journal of Science Education, 10*, 159–169. https://doi.org/10.1080/0950069880100204

Underwood, S. M., Reyes-Gastelum, D., & Cooper, M. M. (2015). Answering the questions of whether and when student learning occurs: Using discrete-time survival analysis to investigate how college chemistry students' understanding of structure-property relationships evolves. *Science Education, 99*(6), 1055–1072. https://doi.org/10.1002/sce.21183

Urban-Lurain, M., Cooper, M. M., Haudek, K. C., Kaplan, J. J., Knight, J. K., Lemons, P. P., Lira, C. T., Merrill, J. E., Nehm, R. H., Prevost, L. B., Smith, M. K., & Sydlik, M. (2015). Expanding a national network for Automated Analysis of Constructed Response assessments to reveal student thinking in STEM. *Computers in Education Journal, 6*(1), 65–81.

Williams, L. C., Underwood, S. M., Klymkowsky, M. W., & Cooper, M. M. (2015). Are noncovalent interactions an Achilles heel in chemistry education? A comparison of instructional approaches. *Journal of Chemical Education, 92*, 1979–1987. https://doi.org/10.1021/acs.jchemed.5b00619

Winstone, N. E., Nash, R. A., Rowntree, J., & Parker, M. (2017). 'It'd be useful, but I wouldn't use it': Barriers to university students' feedback seeking and recipience. *Studies in Higher Education, 42*(11), 2026–2041. https://doi.org/10.1080/03075079.2015.1130032

Chapter 9
The Use of Video, Audio, and E-Portfolios to Provide Feedback

Sabine G. Uijl and Renée M. Filius

Abstract Feedback is considered one of the most important factors for learning, especially for deep learning, which is important in higher education. However, providing and processing good qualitative feedback is quite complex and will not always result in the pursued learning effect. After a theoretical basis of deep learning and feedback, we look into the use of feedback to best promote deep learning without increasing instructor load. This is important not only for online education, where scalability is an option, but also for on-campus education. The use of video and audio feedback as well as the use of peer feedback is discussed as easy-to-implement didactic tools with impact. Furthermore, e-portfolios are discussed in the context of gathering feedback for student progress and development, whether or not to be used for assessment purposes.

9.1 Deep Learning

Before we dive into feedback as a concept in all its forms and shapes, we would like to start with the purpose of feedback as a didactic tool in higher education. Feedback is and has been considered more and more important over the years, so it attracts more attention in the literature as well as in education development itself. In studies and programs involving more practical skills, such as medicine or any profession in which communication skills are important, feedback is a commonly applied didactic tool. In the more theoretical studies, feedback is still deemed important, but traditionally these are more focused on written feedback on essays after they are finished. Yet, times are changing, and feedback is increasingly incorporated throughout the process of learning. This is related to the change in the conception of teaching in higher education, where the main emphasis is slowly moving from the instructor-focused and content-oriented conception of teaching to a student-oriented conception focused on the student learning (Entwistle, 2000). In that shift, feedback

S. G. Uijl (✉) · R. M. Filius
Utrecht University, Utrecht, The Netherlands
e-mail: s.g.uijl@uu.nl

H. J. Witchel, M. W. Lee (eds.), *Technologies in Biomedical and Life Sciences Education*, Methods in Physiology, https://doi.org/10.1007/978-3-030-95633-2_9

becomes an important tool both to help students progress and to measure student achievements (e.g., feedback on certain levels achieved in a rubric) (Entwistle, 2000; van der Vleuten, 2016).

The purpose of feedback in higher education is mostly for students to accomplish deep learning. Deep learning has been described by different authors (Biggs, 1987; Entwistle et al., 1974). It is generally understood as learning to understand. The antonym of deep learning is described as surface learning. When deep learning is learning to understand, surface learning is described as learning to reproduce. For example, in an exam where a lot of facts need to be memorized, surface learning may be the best approach, while for understanding a certain principle and being able to apply this is a new situation, deep learning is essential (Entwistle, 2000). This example immediately shows that deep and surface learning are not opposites, but different approaches to learning. Moreover, there is no strict division between a deep learning and surface learning approach; that is, a student does not consciously choose either for deep or surface learning. Nor is it the case that a person is only capable of one way of learning. Deep and surface learning are both the extremes of a continuum. In relation to that, a third approach often mentioned in relation to surface and deep learning is strategic or achievement learning, which has a place on this continuum. Strategic or achievement learning is aimed at getting the best results in assessment (Entwistle, 2000). It depends on the personality of the student and on factors related to the learning context—e.g., content, didactic tools, timing, and interest of the students—in which part of the continuum the learning takes place (Biggs, 1978; Marton, 1983).

For higher education, we strive for deep learning. The times when higher education was aimed at transferring knowledge from the professors' heads into the students' heads are long gone. With the speed of new knowledge being created, in many academic fields, knowledge now has an expiration date. For that reason, transfer of knowledge from one generation to the next is not only outdated, but has become the definition of lagging behind. The amount of knowledge produced, in combination with the availability of knowledge to all, makes it impossible to be able to know all the latest knowledge in a certain field, let alone in surrounding fields. Therefore, in higher education, we want to teach the students to deal critically with the abundance of information around them, and we want them to integrate this with what they already know. Furthermore, we want them to use this to create themselves. We even take it to the next level: students need to understand the content of what they are taught in a certain context, and they need to be able to apply the knowledge in different situations. This goal of education can only be reached when deep learning takes place. Based on the known literature, we define deep learning as the following (Filius et al., 2018a):

- The initiation of critical thinking
- The integration of what the student is learning with what he or she already knows
- Creating new connections between different concepts

In this era of increasing globalization, rapid innovations, and an exponential growth of scientific evidence in all fields, the need for deep learning in higher education is even more prominent (Blouin et al., 2009). Our students need to be prepared to be flexible and adaptive to a continuously changing world. They will need to develop an open and critical attitude toward the abundance of available information, being able to filter, interpret, and process the different information sources. Moreover, jobs will change, and even if they stay the same, the requirements and context of these professions will change.

The recent COVID-19 pandemic is an example of the complex issues the world faces and will continue to face. The virus and its measures had an impact on every part of our societies. It also impacted many disciplines, from healthcare to politics, to economics, to technical solution-seeking, to ethical issues. Only by aiming for deep learning can we train engineers who are equipped to design the ventilation systems needed to avoid the spreading of viruses, since they will need to communicate with clinical physicists to understand the dynamics of transmission. Only by aiming for deep learning, we can train biomedical scientists who can contribute to the prevention and cure of not only COVID-19 but also future viruses. Only by aiming for deep learning can we expect our students to be able to contribute to the multidimensional solutions needed for the current and future problems of this world.

9.2 Community of Inquiry

It is clear that the dominant learning approach in our higher education system needs to be organized in a way that promotes deep learning. This can be accomplished through learning goals, teaching methods, curriculum, and assessment (Biggs, 1987; Filius et al., 2018b; Hall et al., 2004), which should be aligned (Biggs & Tang, 2011). In learning, the student is part of an ecosystem, a learning community. When aiming for deep learning, a theoretical framework can help in including this whole ecosystem or learning community. Given the incorporation of online and blended aspects in higher education, the framework we choose is the Community of Inquiry as described by Anderson and Garrison (1995) as the didactic framework because it shows the different components in such a learning community. Taking such a framework into account can help in promoting deep learning since it includes the actors and dynamics in learning.

The Community of Inquiry framework describes the components of a critical collaborative learning community, aimed at engaging students in deep and meaningful learning (Anderson & Garrison, 1995; Garrison et al., 2000). It is called Community of *Inquiry* since inquiry relates to the content of a problem or question, which includes a critical discourse and is frequently multidisciplinary. Inquiry furthermore incorporates research methods like information gathering and synthesis of ideas. Finally, it typically has a small-group feature. The importance of community in Community of Inquiry relates to the need for social and communal aspects of critical reflection in learning. Social cognitive theory shows that social interaction, between instructors and learners and among learners, significantly affects student learning and improves student motivation. Social and peer acceptability are powerful

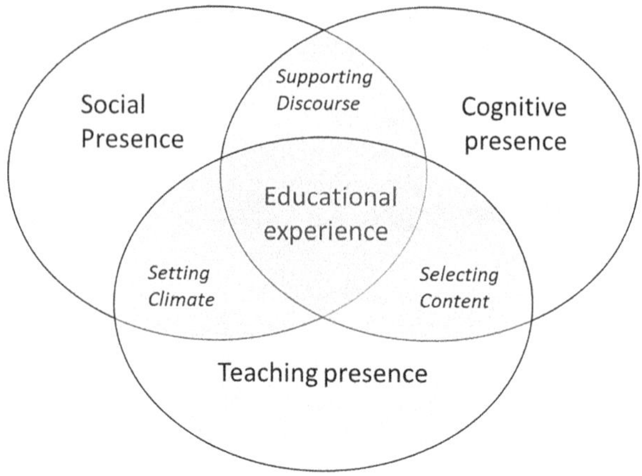

Fig. 9.1 Community of inquiry, after the Anderson and Garrison model (Anderson & Garrison, 1995; Garrison et al., 2000)

in education and increase the likelihood of adopting effective learning behaviors (Bandura, 1999). The Community of Inquiry framework is shown in Fig. 9.1.

A Community of Inquiry has three elements: social presence, cognitive presence, and teaching presence. As visualized in the figure, the educational experience is the part where all three elements are present and therefore overlap. Social presence relates to the degree to which participants of a Community of Inquiry feel socially and emotionally connected. This involves open communication, group cohesion, interactivity, and personal and affective indicators. Cognitive presence is related to exploration, construction, resolution, and confirmation of understanding. This involves collaboration and reflection. Teaching presence relates to the design, facilitation, and direction of cognitive and social processes. The purpose of teaching presence is facilitating personally meaningful and educationally worthwhile learning outcomes in the Community of Inquiry. It involves elements like instructional design and organization, facilitation of discourse, and direct instruction.

Although the Community of Inquiry framework is constructed for online education, in the current developments in higher education, we believe the framework is broadly applicable. We are moving toward including technology in all education: online, blended, and on-campus education. Moreover, we would like to add that it is likely that the Community of Inquiry also applies to hybrid education. Where on-campus education is the traditional classroom education without online aspects and online education does not include any physical encounters, blended education is used when online elements are incorporated in the on-campus education. With hybrid education, we mean the combination of on-campus education and online education during the same learning activity, i.e., part of the students engage in the learning activity on-campus, and the other part of the students engage online in the same learning activity at the same time. The use of online education has been increasing rapidly over the last few decades, with a climax in the lockdown situations of the COVID-19 pandemic in 2019–2021, when all education was largely moved online instantly. However, many of the theoretical frameworks of our didactic concepts and tools are based on traditional on-campus education. The translation of on-campus education to online education requires a different didactic framework. The Community of Inquiry framework can help design education aiming for deep learning in these different forms of education. The balance of the elements of the Community of Inquiry feeds into the design and learning environment for the education to be designed. We sincerely believe that post COVID-19, more elements of online education will be permanently included in our regular higher education programs (Mozelius, 2020). For the use of different didactic elements in your education, the Community of Inquiry can serve as a checkpoint: which of the elements do I want to influence; which elements are influenced by the didactic tools I use in relation to the learning context I am using these tools in; and what does that mean for the balance between the elements of the Community of Inquiry. To make this more concrete—and to come back to the topic of this chapter—we will use the example of feedback. For an instructor to implement feedback in the education as a didactic tool to reach deep learning, the provider can choose the form of feedback, i.e., oral, written, audio, or video. One can imagine that oral feedback in an on-campus setting has a different impact than written feedback on a paper sent to a participant of online education through email. Written feedback is possible in all settings, while oral feedback in an online setting can be mimicked by video calling in an online setting. Different facilitation is necessary in an online setting compared to an on-campus setting. In this chapter, we will focus on feedback elements that are applicable to online education, since these also apply to blended or hybrid education. Furthermore, we (the authors) believe that on-campus education without being open to implementing online tools for specific purposes is a very poor form of education, making all education at least blended.

In conclusion, our learning context is defined by the Community of Inquiry framework, and the purpose of the learning context is to reach deep learning. In this context, feedback is a very powerful tool (Hattie & Timperley, 2007). But there are conditions that need to be met for feedback to be powerful in learning. Only a well-constructed concept and execution of feedback, in an optimally matched learning environment, will help education to reach its aims. This is what the following sections will be about.

9.3 Feedback

Feedback is a powerful didactic tool to promote students learning, as described by Hattie and Timperley (2007). Feedback is meant to be able to reduce the discrepancies between the students' current understanding and performance of the learning task or outcome aimed for and the understanding or performance that the education at that point in time aims for (Ramaprasad, 1983; Sadler, 2010). Feedback consists of information on aspects of one's understanding or performance, which can be provided by any source. This means that feedback is not necessarily provided by an instructor or peer, but can also be provided by a parent or by a particular experience. Moreover, a book or even the student themselves can provide feedback on an understanding or performance. Askew and Lodge (2000) define feedback as all dialogues that facilitate and promote learning. This dialogue can take place in informal as well as formal settings. This can be referred to as relational feedback, which can be contrasted with transactional feedback, meaning the act of an instructor providing information on a presented performance to a student. This difference is described as feedback models Mark 1 and Mark 2 by Boud and Molloy (2013a), where Mark 1 relates to transactional feedback with the instructor as the actor in the feedback and Mark 2 relates to relational feedback, introducing an active role of the learner. We deem the latter an important factor in feedback and follow Duijnhouwer's definition of feedback, which is "information provided by an external agent regarding some aspect(s) of the learner's task performance, intended to modify the learner's cognition, motivation and/ or behavior for the purpose of improving performance" (Duijnhouwer, 2010).

Boud (2000) and Sadler (2010) indicate the impact of feedback beyond the educational setting. We already mentioned the abundance of information being produced in our continuously changing world. Feedback can help to equip students to become self-initiating seekers and users of information, a skill they will need to be able to continue to learn and adapt in their future careers (Boud, 2000). Feedback leads to competencies beyond knowledge and skills; it helps students develop independent judgment, problem-solving capabilities, self-evaluation, and reflection (Sadler, 2010). But feedback in itself is not enough. The aspects needed for feedback to work are the following:

- Quality of the feedback
- Feedback culture
- Feedback literacy

In the following sections, we will discuss these aspects.

9.3.1 *Quality of Feedback*

In order for feedback to reach its goal, the quality of the feedback is important. In this context, we aim for feedback that promotes the deep learning approach of the students. In relation to the quality of feedback, Yang and Carless (2013) described three dimensions, which are overlapping with the Community of Inquiry, namely, as a cognitive dimension (related to the cognitive presence), a social affective dimension (related to the social presence), and a structural dimension (related to the teaching presence). All three dimensions are both important for the quality of feedback and for deep learning. The cognitive aspect relates to the discipline or domain-specific aspects of the problem at hand. This dimension should inform and guide the students in determining or understanding the difference between their own performance on, and the discipline- or domain-specific aspects of, the problem.

The social affective dimension relates to the power imbalance between an instructor and a student in a feedback situation. The power of an instructor as an assessor may influence the safety of a feedback situation from the point of view of the student. Recognition of this imbalance is a first step to tackle this problem and to upgrade the quality of the feedback. An instructor can do this by being attentive to the regulation of emotions and empathy. Otherwise, the feedback may not reach its goal.

The third dimension is the structural dimension, meaning the timing, order, and form in which the feedback is provided (Yang & Carless, 2013). For feedback to reach its goal, it should be delivered in a timely fashion and in a form that suits the learning activity (Henderson et al., 2019). For example, for a writing assignment, written feedback can be appropriate. The timing of the feedback is most suitable during the writing process rather than after the essay is finished (Wiggins, 2012). Our own teaching experience confirms that students who have already moved on to other academic activities do not have the motivation to take the time to dive into feedback on finished assignments or courses, with the goal to incorporate the feedback into their current and future learning. In contrast, feedback during the writing process can be used for the learning process and can open a dialogue with the feedback provider (see also in the next Sect. 9.3.2). For a performance task, like clinical or laboratory work, oral feedback directly after the demonstration of the task will have a greater impact on the student learning. Here a dialogue on the reason a student performed the task in a certain order, or in a specific way, may also help the student to reflect and integrate the feedback in their learning.

9.3.2 Feedback Culture

A strong feedback culture can be distinguished by a number of aspects, as described by London and Smither (2002). Although this originates from the human resources literature, it is highly applicable to the educational environment (Henderson et al., 2019). As described above, the quality of feedback is part of a feedback culture, as is the emphasis of the organization on the importance of feedback. Facilitating and providing support for feedback are essential in creating a feedback culture. Providers of feedback should be trained on how to provide useful feedback of good quality. In the context of higher education, instruction to provide feedback to promote deep learning is an important aspect. Furthermore, a shared focus of the feedback can help in creating a feedback culture. A shared focus can be directed to clear standards, like the learning goals, deep learning aspects, and certain valued behaviors.

The feedback culture consists of more than providing high-quality feedback. Feedback also needs follow-up. The follow-up of feedback can be stimulated by ensuring that performance improvements following feedback are recognized. There are two different forms of follow-up after feedback we like to distinguish: feed forward and a feedback dialogue. Feed forward is an individual action, when the content of the feedback is used to improve the performance in the future (Moore & Wallace, 2012). In a feedback dialogue, students get the opportunity to ask elaborating questions that can help them relate the feedback on their performance to improvements they will be able to implement in the future. Moreover, the dialogue about the feedback aims to elaborate on the arguments in the feedback, leading to deeper thinking about and more understanding of the feedback, thus helping in achieving deep learning (Filius et al., 2018c).

A powerful aspect of a strong feedback culture is that of role models. When instructors actively seek feedback and show how they receive and incorporate this feedback, students will feel less vulnerable and more part of a feedback-friendly environment (Parsell & Bligh, 2001). This aspect is an important tool to reach a feedback culture on a smaller scale. If an instructor is keen on incorporating feedback in education, despite acting in an organization where the culture does not naturally support this, the instructor can create a feedback culture within the class or course by actively seeking feedback and enabling a feedback seeking setting for the students. As a start, it may help to make giving feedback and responding to it a mandatory part of the course or curriculum. This can be followed by a reflection session where the instructor can discuss with the classroom how the feedback conversation is conducted. Similarly, in a higher education setting lacking a feedback culture, a specific curriculum can still be designed to support a feedback culture for the instructors and students involved. Instructors and curriculum designers may even use this to showcase the benefits of incorporating feedback in learning and start a movement in their organization, potentially leading to future change.

9.3.3 Feedback Literacy

Even with high-quality feedback in a strong feedback culture, feedback does not necessarily reach its goal. It all depends on the interpretation of the receiver of the feedback and the perception of the receiver toward the feedback (Boud & Molloy, 2013b). This is defined as feedback literacy by Sutton (2012). Feedback literacy can be seen as a set of generic practices, skills, and attributes acquired by a series of situated learning practices. Feedback literacy is related to the educational identity of learners and has three dimensions: knowing, being, and acting.

Knowing is an epistemological dimension, related to feedback *on* and feedback *for* knowledge. Feedback *on* knowledge is more instrumental and provides the receiver with information on the quantity and quality of knowledge they presented. Feedback *for* knowing includes guidance on how an academic performance can be improved. Therefore, knowing mainly impacts the receiver of feedback.

Being is an ontological dimension, referring to the educational being of the learner. Feedback can have an impact on the confidence of the receiver; it can have either a confidence-boosting or a confidence-crushing effect. Feedback therefore influences the relationship of the receiver to their (educational) environment and influences the being of the learner. Learners need to develop confidence in their own academic ability. For some learners, this can be an anxiety-provoking experience, whereas others simply learn and present the learnings.

Acting is the third dimension of feedback literacy, which is a practical dimension. Feedback literacy requires learners to act upon the given feedback. This acting is also called feed forward. This requires specific skills, i.e., reading, interpreting, and using feedback. These skills can be developed by practice. Students need to appreciate the feedback they receive. Based on the feedback, they need to make appropriate evaluative judgments, and they need to actively engage with the provider of the feedback, leading to actions for further learning. Leenknecht et al. (2019) investigated the dimensions of a feedback-seeking behavior and concluded that students aiming to learn from the activity will use more deep learning strategies and will seek more feedback.

9.4 Feedback in Blended and Online Education

In traditional on-campus education, oral feedback is commonly used, especially as informal feedback during a course or in individual student trajectories. Also, with written feedback, instructors often elucidate on the feedback given during class meetings. This way, students are more actively involved so that they can better incorporate the feedback. In online education, this involvement with feedback is more difficult to achieve, especially when interaction in online education is mainly asynchronous, which means that the response can be given at a much later time than immediately after the comment or question is placed. Feedback is usually given in a written form, and the interaction after the feedback, the feedback dialogue, is difficult to organize and therefore often forgotten (Filius et al., 2018a).

Sound online education requires a different didactic design than the didactic design of on-campus education. Learning methods used in on-campus education are not always directly transferable to online education, partly because the dynamics and pace vary. Yet, during the instant move of all higher education to online as a result of the measures against COVID-19, there was often no time to rethink the design, and many instructors had to turn to an online replica of their on-campus education. For example, a lecture became a video recording, and a discussion in working groups became an online discussion in groups. Student evaluations in the Netherlands showed that students reasonably accepted this rapid conversion to online education during the first few months. They were sympathetic about the workload of instructors. But as time passed, students became more frustrated and increasingly demanded a better quality of online education.

The use of feedback is typically a learning method that suits the didactic design in online education. The way the education is organized and the didactic tools used should be adjusted to the learning goals. When we discuss the different forms of feedback, and their applicability in education, we are not restricted in thinking in either online or on-campus education. The tool that suits the purpose of the feedback best should be incorporated. Also in on-campus education, it is relatively easy and more and more common to incorporate digital tools, making this a realistic starting point for educational design, turning on-campus education into blended learning.

Dialogues are an important aspect of feedback (Sargeant et al., 2009). In an online setting with mostly asynchronous education, this may be more difficult to organize. However, it is still possible. In our own studies, we saw that the chat function in small-scale online education was used for social interactions as well as content-related interaction. This broad use of the chat function incorporated the social presence of the Community of Inquiry framework and led to a safe environment for group discussion and a feedback-friendly culture. During these courses, general feedback was provided by the instructor in the assignment chats, resulting in a dialogue on content in an open and friendly way (Uijl et al., 2017). Moreover, if in online education the learning goal should require it, one-on-one meetings can be organized for individual oral feedback through video calling or by telephone.

The latter is, of course, time-consuming and less scalable, something that is associated with online education. Therefore, we studied what feedback interventions could promote deep learning in scalable online education (Filius et al., 2018b). Feedback management was deemed important, as well as peer feedback and the possibilities of automated feedback. Feedback management refers to four different possible interventions. One is to collect student information in advance, i.e., by asking students to introduce themselves online. This relational approach creates a feeling of personal commitment and adheres to the social presence and the feedback culture. The second intervention on feedback management is the use of a dashboard to monitor progress. We will come back to this in Sect. 9.9.1. The third intervention for feedback management is restricting the feedback in scalable online education to essentials, especially for instructor feedback. Students can be asked to prepare a feedback request from instructors, making it clear for the instructor what the specifics of the feedback request are. Finally, as will come back in the next section, an early feedback management intervention is training the students how to provide feedback. This can include discussing the feedback: not just the content or aim of the feedback but also the quality and process of the feedback. Also, rating feedback in a peer feedback situation helps in training students to provide feedback. Next to rating, the possibility in an online environment to "like" feedback that you have received with a thumbs up button also helps in creating a strong feedback culture and the use of feedback for deep learning.

9.5 Peer Feedback

In peer feedback, a student is feedback provider as well as receiver. This comes with some changes in the traditional instructor-student dynamic. There are many positive sides to incorporating peer feedback in education (Filius et al., 2018c; Liu & Carless, 2006; van Popta et al., 2017). In peer feedback, there is no hierarchy between feedback receiver and provider, eliminating the power issue in the social affective dimension we described when we discussed the quality of feedback. The equal relation between students adheres to the creation of a safe setting for an open dialogue (Nicol & Macfarlane-Dick, 2006; Prins et al., 2006). The actual process of working together also resonates in the Community of Inquiry framework, initiating learning involvement and promoting deep learning.

We distinguish three main advantages of peer feedback, all three of them enhancing deep learning:

- The added role of the student as feedback provider
- The feedback provider not being an expert
- The scalability option

The student role of feedback provider initiates active involvement in learning, and with that, it initiates deep cognitive processing (Lundstrom & Baker, 2009). To be able to provide feedback, the student needs to relate the peer's work to the learning goals. This requires a different level of understanding and will influence not just their assessment to their peers work, but also of their own work.

Providing useful feedback requires a different set of skills, and feedback training for students before they engage in peer feedback is required (Filius et al., 2018c). Instructions are essential to guide students to give honest and useful feedback and to avoid the tendency of peers to fall back on bland, superficial, inaccurate, and/ or inconsistent feedback (Nilson, 2003). Instruction can also help explain to students the added value of peer feedback. Students are quite often inclined to think that peer feedback can add little because it is not given by an expert. In a previous study (Filius et al., 2018c), we found that for that very reason, peer feedback can encourage deep learning; after all, this is why students look more critically at peer feedback than at feedback given by an instructor (which they more likely blindly consider "true").

Next to instructions, the creation of a dialogue also enhances the quality of peer feedback as well as the possibility for the receiver to improve based on the peer feedback. In our previous study, we found that when students had to inform their feedback provider to which extent the feedback they received helped them to learn deeply, feedback providers tended to try harder for quality feedback beyond friendly but uncritical remarks. In our study, students had to evaluate the feedback they received and indicate the extent to which it helped them achieve deep learning. As a result, feedback providers went to greater efforts to formulate good feedback; after all, they knew they would be assessed on it, even if the assessment was formative. It also turned out to be a solution to the problem that students often hardly read the feedback they received. The feedback receivers, in fact, read the peer feedback they received more carefully, so they could give a substantiated evaluation to the feedback providers (Filius et al., 2018c). By organizing peer feedback this way, you not only decrease the instructor's workload, but you also teach the students valuable skills beyond the topic of the course. The skills required for providing feedback are essential future skills, since in professional life feedback from peers is common ground, especially in a strong feedback culture (Liu & Carless, 2006).

In peer feedback, the provider is a peer and not an expert. Students themselves usually prefer feedback from an instructor over peer feedback (Ertmer et al., 2007). They see the instructor as the expert in the field, and therefore the feedback is accepted. The viewpoint of students to their education can also be considered traditional related to this. The students seek knowledge or "the truth," which the instructor as expert presumably will provide. Students usually express some doubts about the content when they receive peer feedback. However, from our study on dialogic peer feedback in online education, we learnt that this more skeptical approach to the provided feedback by peers also has its positive sides on learning (Filius et al., 2018c). It stimulates students to enter into a dialogue about the feedback and therefore about their own work and the possibilities to improve. With instructor-provided feedback, a student more easily accepts the feedback as it is and does not take an active approach of decomposing the feedback. In line with this, peer feedback urges students to look up the theoretical backgrounds to be sure that the feedback makes sense. In contrast, if specific theoretical chapters are mentioned by the feedback provider (either instructor or peer), the urge to look this up is virtually nonexistent. This all indicates that deep learning is promoted by using peer feedback.

The third advantage of peer feedback is the option to upscale. From the organizational point of view, peer feedback has the advantage of scalability. One can ask students to review more peers, with the result that a student will receive feedback on one assignment from different reviewers, without increasing the instructor load. As mentioned when discussing feedback management interventions to promote deep learning, feedback training is essential for peer feedback to reach its goals. Feedback providers should know what they are doing and how they are doing this. In a previous study, we asked the students to rate the feedback that was provided by their peers. This way, the provider received feedback on the provided feedback, again initiating the dialogue and increased learning, creating a feedback loop. The use of a like button also helps to get instant feedback on provided feedback and creates a positive attitude toward feedback-providing (Filius et al., 2018c).

9.6 Forms of Feedback

Feedback is usually given orally or in writing. More recently—in line with blended and online education—audio and video feedback have been added. These feedback forms have the opportunity to include narrative information in a natural way. We argued earlier for a close connection between the understanding or performance of the task and the form and timing of the feedback. The desired learning outcomes or the task being performed, will usually suggest the optimal choice between oral, video, audio, and written feedback. The use of the voice, visuals, or written language forms the basis for the formats of feedback that are discussed in this chapter. In Table 9.1, we show the different feedback types and how these variants differ from each other.

The most commonly used feedback type, combining visual- and voice-based feedback, is oral feedback. With oral feedback, we mean verbally provided feedback with the provider and receiver being in the same room. Oral feedback is, therefore, usually most suitable in two situations. The first is in a practice situation, where the

Table 9.1 Feedback types

Feedback type	Real-time	Personal connection	Detailed or interactive	Scalable
Visual and voice-based:				
Oral	Yes	Yes	Yes	No
Video (one-take)	No	Yes	Partly	Yes[a]
Video call	Yes	Yes	Yes	No
Voice-based:				
Audio (one-take)	No	Partly	Partly	Yes[a]
Audio (record in pieces)	No	Yes	Yes	No
Phonecall	Yes	Partly	Yes	No
Written-based:				
Written	No	No	Yes	No

[a]See Sect. 9.7 on the ways to efficiently provide these feedback forms

student shows a skill or performs a task. The instructor or supervisor can deliver oral feedback on the aspects of that task. Timing is critical in order for the student to be able to process the oral feedback; preferably, the feedback is given directly after the task, and immediately following this with a feedback dialogue can enhance the integration of the feedback in the learning, leading to improvement of the task based on the feedback. When it is not possible to deliver the feedback immediately after the task, videotaping the act where the skill or task is shown is an alternative. However, the moment of timely feedback is then lost. For the feedback to still be received and incorporated, the feedback provider and the student can watch the video together in a feedback session. This enhances the feedback dialogue because of the opportunity to look back at and discuss the performance.

Visual-based feedback usually involves the use of video. In the next section, we describe formats that video feedback can be taken, namely, as a talking head and as a screencast. The main difference between voice-based feedback and visual-based feedback is the influence on the degree of social presence, which is the degree to which "the other" in a communication appears to be a "real" person (Kreijns et al., 2011). The social presence seems greater with visual-based feedback because of the ability to see the feedback provider and pick up social cues. However, the possibility of interaction between the feedback provider and the feedback receiver also affects social presence. This possibility is available in both visual-based feedback and voice-based feedback—especially when this is done in real-time.

With deep learning in mind, written feedback has proven less effective compared to oral feedback. The natural dialogue in oral feedback plays a role here (Agricola et al., 2020; Carless et al., 2011; Merry & Orsmond, 2008). Also, students themselves indicate a larger appreciation for oral feedback over written feedback (Agricola et al., 2020). Still, feedback on essays or written assignments may well be provided in a written form, especially when details of the written text are important. But an oral dialogue would enhance the incorporation of the feedback. In higher education practice, thesis reports and final essays are often graded by the instructor in a written form, and the students receive feedback on the work they handed in together with the grade.

When using written feedback, there are options to promote deep learning. One can ask students to come up with specific questions, or best practices can be used to give a context to general written feedback. When a dialogue is added, it can also enrich the instructor-student relation.

As shown in the "real-time" column in Fig. 9.1, feedback can be given either synchronously (real-time) or asynchronously. Asynchronously given feedback means that the feedback giver gives the feedback at a random time and in such a way that the feedback receiver does not have to perceive it at the same time. Asynchronous feedback can be given in, for example, a text file or a pre-recorded audio or video file. The advantage of asynchronous feedback is that it can be reviewed more often by the receiver. This can help students better understand and process the feedback. Repeated playback can certainly also help in situations where students tend to react emotionally to very critical feedback first and only later understand the meaning of the feedback.

With synchronous feedback, a dialogue can flow more smoothly because the other person can respond immediately, and it is easy to prompt and probe. With asynchronous feedback, there is more time between the action and the response. Our research showed that this makes follow-up questions more difficult (Filius et al., 2018a).

The question remains how the form and structure of the feedback can be adjusted to promote both learning from the interaction and individual reflection. To be more precise: the challenge is to organize feedback in a way that fosters the cognitive, the social, and the structural dimensions of high-quality feedback. We hope this overview will help educational designers and instructors in making those decisions.

Now that we have given a background in the feedback types, we will address more specifically the use of and video feedback and audio feedback, as well as the use of e-portfolios for feedback.

9.7 Video Feedback

Video feedback is less studied, despite its potential for delivering feedback, especially in a blended or online situation. In video, facial expressions and social cues can be seen, and that may stimulate the feeling of personal commitment. Feedback is a form of complex and—sometimes—difficult communication. Considering that this kind of communication is best delivered through rich media, including both verbal and nonverbal communication, it is surprising that video feedback is not used or studied more often (Borup et al., 2012). When looking for an alternative to oral feedback, one would say that video feedback, with the possibilities of including facial expressions and body language, would be the most communicative choice; however, in instances where oral feedback is not possible, educators much more commonly provide written feedback. This may be related to technical challenges in video editing. However, the technical issues in recording and sharing short videos with a mobile phone are possibly not the most challenging anymore. We expect the natural vanity of the feedback provider and the unacquaintedness with seeing yourself on film and sharing this for professional purposes to be more of a hurdle in the relative underuse of video feedback.

In a recent review, Mahoney et al. (2019) assessed the use of video feedback in higher education. In addition to video feedback in a general way, we will follow the distinction of two specific types of video feedback by Mahoney et al., namely, a talking head and a screencast.

9.7.1 Talking Head

Talking head video feedback is literally videotaping yourself while you talk to the camera (e.g., to a smartphone) as if you are talking to the feedback receiver. Like every recording of yourself, this is always awkward at the start. When we asked our students to deliver video presentations and upload them online, students tended to retake the recordings numerous times, to make sure there were no long silences or hitches. This took so much time that it was at the expense of the actual learning goals. Considering that making beautiful videos was not one of the learning goals, we had to tell students not to retake any recording more than twice. This is also a good starting point for providing video feedback.

Delivering talking head video feedback has been studied by Lamey (2015). Interestingly, he describes his own experience with video feedback as a philosophy lecturer from the feedback provider perspective:

> Prior to using video, my comments on student essays were evenly divided between the intellectual substance of the paper and mechanical issues, such as grammar and spelling. With video the ratio has shifted decisively in favor of substantive comments. Rather than draw attention to every instance of bad writing, I now single out one or two as representative, which I draw to the student's attention by reading them aloud in my feedback. (Lamey, 2015)

His feedback content became more personal and richer, since it mainly focused on intellectual substance. Because of the effort to make a video for a student, the video feedback included more positive feedback and included feed forward. These were aspects that he included much less in the written feedback he gave on student essays. Another upside was the time spent on providing feedback. After gaining some experience, he made feedback videos of around 4 minutes per student, in one take, without taking elaborate notes beforehand.

Most students in Lamey's study preferred video feedback over written feedback on essays; they found the video feedback more informative, easier to understand, and richer because it includes the tone of voice. They also indicated it fostered a greater connection with the feedback provider since seeing the feedback provider giving feedback creates a connection between provider and receiver via the feedback content. Furthermore, students indicated the video offered constructive criticism regarding how to improve future performance. The criticism by students consisted of the lack of focus in the video feedback on individual passages in a paper. This would require notes when reading the paper, which is more associated with written feedback. This student concern could be addressed efficiently by making some annotations onto the student's assignment when initially reading it and later when recording the video using those as speaking notes.

Finally, students mentioned an awkwardness in receiving videos by their instructor (Lamey, 2015). Videos were associated by students as informal communication, and they had to get used to receiving videos from an instructor. This issue is related to the visual aspects of this type of feedback and will therefore probably have less influence on the perception of audio feedback, which we will discuss later in this chapter. This was, however, in 2015, and we expect with the global implementation of video in education—especially since 2020—that this issue by now has become nearly extinct.

Similar results were found in other studies on talking head instructor video feedback on written essays (Crook et al., 2012; Henderson & Phillips, 2015). Students were also very enthusiastic about this way of receiving feedback, with similar reasoning. Furthermore, instructors gave similar argumentation why they were positive about video feedback. Henderson and Phillips (2015) included two different instructors and their videos in this study. The structures of the 5-minute feedback videos were analyzed, and these were similar for both instructors, suggesting a helpful way of structuring your video feedback as presented in Box 9.1:

Box 9.1: How to Structure Video Feedback (Henderson & Phillips, 2015)

Salutation
Relational work (recognition and valuing of the student)
Evaluative summary
Textual issues
Commenting on the substance of the assignment with an emphasis on feed forward
Engaging with the conclusions, arguments, logic, justification, and literature included in the assignment
Valediction and invitation (round off and invitation to proceed)

Considering the huge benefits for both instructors and students in implementing video feedback, we find it remarkable that written feedback on essays is still the first choice. Of course, the feedback form fits the assignment form when a written essay is provided with written feedback. Instructors may also feel hurdles before starting with video feedback. Besides the mentioned awkwardness, technological issues may play a role, and the surroundings for videotaping need to be controlled—a quiet setting with little distraction is necessary. However, these perceived hurdles may not be as high as they seem when you look at the practicalities of video feedback. It is important to realize and communicate that it is still about the content of the feedback and not about professional video-making. Every instructor can be expected to have access to a smartphone with a video function. When reading an essay, speaking notes can be added to the paper, and after the paper is read, it is advised to directly videotape the feedback, using the structure provided in the text box.

In the (post) corona-era we are living in, we are more used to working with video, and hurdles that may have stopped instructors from trying video feedback before 2020 are now not that valid anymore. Especially in these days of overburdened workloads for instructors, one would expect that a method, which not only saves time but also is experienced by students as a better way to receive feedback and feed forward, would have found its way into the educational practice.

This is even more true in higher education, where aiming for deep learning can take place with the instructor as guide or coach, rather than as the all-knowing expert, telling in detail what is wrong with the work the students handed in. Here, it is logical to highlight the social presence, as described in the Community of Inquiry. The social presence in online education increases when there is an interaction, especially when this interaction included a face, a tone of voice, and both verbal and nonverbal communication. For students this indeed triggers the feed forward: to see the face of the instructor telling you he or she expects a certain improvement at a certain point in time is a stronger call to action than a written note with new deadlines (Borup et al., 2012).

9.7.2 Screencast Video Feedback

With screencast feedback, the recording of a computer screen is combined with an audio recording of the feedback provider. Although it includes video images, it does not include the physical image of the feedback provider. This has an impact on both the benefits and pitfalls of video feedback. In a screencast, the feedback provider can better pinpoint parts of the assignment that need improvement, and the relation of the individual feedback statements to specific parts of the assignment is more clear for students. On the other hand, the visual clues in the interaction are lost when the provider is not seen and only heard. A combination screencast is also possible, in which a small talking head of the feedback provider is concurrently shown in the corner of the screen (Klappa, 2015). This has the potential of the advantages of the talking head, although nuances may be lost when a small image of the feedback provider is included.

9.8 Audio Feedback

As you can imagine, audio feedback has some of the same advantages as video feedback. Moreover, audio feedback can be recorded in fragments, while the instructor is reading the essay or paper. This diminishes the two-step approach we presented in video feedback: using annotations as speaking notes for the recording.

Studies on the student perception of audio feedback on written assignments show similar positive outcomes as we mentioned for video feedback: it is felt to be more personal, richer, more in-depth and more focused on feed forward, compared to written feedback, as is shown in the following examples:

> the majority of respondents feel very positive towards the audio feedback they received as 85% agreed or strongly agreed that the feedback helped them see what they had missed out from their coursework, and 88% agreed or strongly agreed that the feedback would help them improve their coursework. The students' comments also confirm the point about quality and detail, with 75% of students saying that they felt the audio feedback was more detailed than written comments. (Lunt & Curran, 2010)

> A sizeable majority of students found audio feedback to be beneficial and reported that it contributed to their learning. Some of the comments from the focus group include: it was 'easy to understand what they (lecturers) meant'. 'I found audio feedback very useful. The feedback I received was clear and understandable. It was like having a 'mini tutorial' and 'good for distance learning'.' As students in the course come from a wide geographical area, this is an important finding. (Gould & Day, 2013)

> Typical comments from the audio feedback group include 'This was so much more explanatory [than written feedback], it helped me understand it a lot more than just having a short sentence about it'. (Voelkel & Mello, 2014, a comparison study of audio and written feedback)

> Overall, student perceptions of the audio feedback were positive. In terms of the feedback being clear and easy to follow, 95% of students strongly agreed, agreed or somewhat agreed. Audio feedback was also successful in identifying areas where students 'did well', with 98% of students strongly agreed, agreed or somewhat agreed. For identifying areas where students 'were wrong', 94% of students strongly agreed, agreed or somewhat agreed. Audio feedback was considered more personal than written feedback, with 92% of students strongly agreed, agreed or somewhat agreed. The majority of students were satisfied with the amount of feedback received with 94% of students strongly agreed, agreed or somewhat agreed. (Parkes & Fletcher, 2017)

> In an email received about three weeks after audio commenting was first used a student offered the following: We've had written comments twice and verbal comments twice now. Let me guess—this is someone's research project right? Let me just save you some time. The verbal feedback is much, much, much better than the written. I said the same thing when I talked to you on-campus last month. So can you just send me the voice comments from here on out, say there is no comparison between the two at all and nix the written stuff? That's probably not going to happen, but I thought it was worth a shot! (Ice et al., 2019)

In our own studies on online teaching experiences—years before the corona pandemic—students also expressed the upsides of audio feedback, but also addressed the technical hurdles. We include some quotes of our own students:

> I think the principle of providing audio/video feedback is tone of voice—with typing constructive criticism can sound harsh, however I think typing is more practical because SoundCloud and other recording websites are difficult to use and quite often you can end up wasting a lot of time trying to use them.

> I think there a bigger threshold to providing audio and video feedback than typed feedback, but I also believe the result is of better quality. First of all, it takes more effort to record and second of all, it's much more personal and much less anonymous. This is also the reason why I believe people will prepare better for it (psychological aspect of approval from peers).

Voelkel and Mello additionally studied student results after audio feedback compared to written feedback and found no differences in outcome. Ice et al. did find increased retention of content after audio feedback compared with written feedback (Ice et al., 2019; Voelkel & Mello, 2014).

In their assessment of audio feedback in online education, Oomen-Early et al. (2008) found mixed results from the student perspective. The upsides of social presence and personal relation were definitely there, but the approval was not as overwhelming as in the earlier mentioned studies and the studies on video feedback. In their studies, about half of the students indicated they preferred written feedback, and over 80% of students would prefer a blended approach with a combination of audio and written feedback. This would of course completely eliminate the upside of the gained time for instructors in giving audio (or video) feedback instead of written feedback.

In both video feedback and audio feedback, the instructor was the feedback provider and the student the feedback receiver. However, we indicated the potential benefits of peer feedback earlier in the chapter. Next to benefits for deep learning of peer feedback receivers, the student will learn a lot—maybe even more—from providing feedback to fellow students. Imagine the benefits in our reach with video or audio peer feedback. The authors studied the effect of audio peer feedback on deep learning (Filius et al., 2019). As expected, we assessed the results both for feedback providers and for feedback receivers. As seen in video and audio instructor feedback, the feeling of personal commitment was important. In the peer feedback, this effect was seen both for feedback providers and for feedback receivers. Personal commitment is an important aspect of the Community of Inquiry and is directly related to deep learning. We found a stronger relationship of the mechanisms that trigger deep learning in feedback providers than in feedback receivers, again indicating that peer feedback is an interesting approach to deep learning.

Students reported that video and audio feedback more resemble the classroom experience, but in peer feedback, they were a bit hesitant to actually audio tape their own feedback. This made them feel vulnerable and sometimes even uncomfortable. A safe and feedback-friendly environment is essential here. And providing the students with examples (both good and bad) of peer feedback may help in overcoming an initial awkwardness (Filius et al., 2018c), and it may even help in bonding and increase the social presence. A change in culture and/or attitude is needed to fully incorporate both video and audio recordings as feedback tools in higher education. In our view, this is worth striving for, since it can offer an efficient and effective way of providing feedback to learners, suiting the current demands of instructors (less allocated time for more students).

9.9 Feedback in E-Portfolios

In the previous sections, feedback was related to a single assignment or a separate part of the students' work. Feedback is often seen as formative assessment: it gives the student insight into discrepancies between the current understanding or performance of the learning task and the desired learning outcome. Usually the students' final exam and/or essay grade is then the summative assessment. In higher education, there is a movement away from this traditional view on high-stakes assessment. The idea behind this is that there has been a tendency to overassess. Students have become straight A seekers (achievement/ strategic learning approach), and the grades seem to matter more than the content of the education or the development of the student (deep learning approach). A new concept of assessing has emerged, called programmatic assessment (van der Vleuten, 2016; see also Chap. 8 in this volume).

In a programmatic approach to assessment, assessment is no longer the result of one test; instead, the individual assessments can be seen as data points. Moreover, assessment no longer consists of a single evaluation presented as a verdict; but instead, the aim of assessment has become a way to provide feedback to the learner. The form of the feedback can differ, but the idea is that when the complexity of the task increases, the feedback becomes richer in content. This way, learning value is increased. The combination of data points gives the student insight into his or her own learning and personal development. This way, the progress in a program is no longer a box-ticking exercise of exams for the separate parts of that program, but an overview of the individual development of the student, which can be related to study parts, but also to skills and (personal) learning goals. No individual test or assessment point is utterly reliable, but the complete overview can approach a robust appraisal of the student's abilities or performance (van der Vleuten, 2016). A portfolio is useful to display a student's qualitative efforts as well as collecting all the data points—and to register all the non-numerical feedback—in order for students or gain a nuanced overview of their own progress and learning.

9.9.1 *E-Portfolios*

E-portfolios are digital tools to register the progress of students. The development of technology and the increased availability of web-based platforms to document and present work in multimedia formats have led to widespread use of e-portfolios (Clark & Eynon, 2009; Duncan-Pitt & Sutherland, 2006). Over time, e-portfolios have moved beyond the purpose of assessment in its traditional form to one that more deeply facilitates and enhances student learning. This is well captured in the following definition of e-portfolios by Duncan-Pitt and Sutherland (2006):

> A system that belongs to the learner, not the institution; populated by the learner not their examiner; primarily concerned with supporting learning not assessment; for life-long and life-wide learning not a single episode or a single course; that allows learners to present multiple stories of learning rather than just a simple aggregation of competencies; and, importantly, where access to them is controlled by the learner who is able to invite feedback to support personal growth and understanding. (Duncan-Pitt & Sutherland, 2006)

This definition fits the programmatic learning approach as described earlier, with the difference that it does support this different approach of assessment, defined as multiple data points with feedback giving insight into the student learning and progress.

With this in mind, the e-portfolio is a very valuable instrument, with its strengths especially in the electronic format, because of the following reasons:

- Digital formats are easy to rearrange and edit.
- It allows for inclusion of diverse multimedia formats.
- Hyperlinks can make a connection to documents and link to external sources/references.
- The electronic format is portable and mobile (Stefani et al., 2007).

We strongly believe that the person owning the e-portfolio should see the purpose of that e-portfolio. If an e-portfolio is just something students need to perform on top of their education, there will be a lack of intrinsic motivation. In line with the Duncan-Pitt definition, the implementation of the e-portfolio should start when the students enter a program; students need to be taught what the e-portfolio is and what purpose it has in their personal development. If you want to implement an e-portfolio to showcase progress and development, reflection is an important factor. Students then should be taught what (self-)reflection is and how they can use this to set goals, make decisions, and work on personal development. This way, e-portfolios tie in with feedback. Any feedback can show progress, and feed forward can lead to further development. Therefore, feedback, from any provider (even the student themselves) and in any form (text, pictures, video, audio, links to websites, et cetera), can have a place in an e-portfolio. This is in line with Roberts et al. (2016), who also suggest making room for the students to decide what content should be part of the e-portfolio.

E-portfolio tools can also be used for feedback only, without any connection to assessment, like the reflection app, studied by Könings et al. (2016). This app for medical resident students gave freedom in the form of feedback students sought during their residency. Students' preferences between different forms of feedback by may differ, and this app had different options to capture learning moments for reflection. The most used form of reflection was text (used by all students), followed by pictures. Only a few students uploaded an audio or video recording, which again points out the feeling of awkwardness and vulnerability in audio- or videotaping yourself for professional (educational) purposes.

9.9.2 E-Portfolios for Assessment

When feedback and assessment come together, like in programmatic assessment, we talk about an assessment *as* learning (Dochy & Segers, 2018; Segers et al., 2003). The predecessors of assessment as learning are assessment *of* learning (the traditional tests assessing whether the knowledge and skills of the student are at the required level) and assessment *for* learning, which is ongoing assessment that allows instructors to monitor students on a day-to-day basis and modify their teaching based on what the students need to be successful. Assessment *as* learning provides students with the timely, specific feedback that they need to make adjustments to their learning.

In assessment as learning, complex problem-solving offers rich information on how a learner is progressing and what his current level of proficiency is. The assessment is as much situated as the learning itself is and can be done by the community, by peers, or by instructors. Moreover, in assessment as learning, assessment of group learning is as important as the learning of the individual (Dochy & Segers, 2018). In assessment as learning, students engage in peer and self-assessment; they learn to make sense of information, relate it to prior knowledge, and use it for new learning. In this view of assessment, deep learning, feedback, the use of e-portfolios, and assessment come together. Continuous constructive feedback from different stakeholders gives continuous information to the student, who can gather all that he or she deems important in the e-portfolio. The e-portfolio becomes an overview of the learning of the student and evidence of acquired competencies, skills, and learning goals achieved. The students, as owners of their e-portfolio, can choose what material they will use to showcase specific competencies and knowledge in a programmatic assessment approach. That way, the student remains the owner of the material, and collecting feedback can be incorporated without the risk that every piece of feedback gathered turns into a summative assessment. This ties in with the definitions and applications of feedback as described in this chapter. Boud and Molloy (2013a) define this as feedback Mark 2, including the active role of learners as constructors of their own understanding, leading to a sustainable model of feedback:

> The framework proposed is premised on students committed to becoming effective practitioners in the domain of their study. Such commitment places them as active learners who seek to do whatever they need to understand what is required of them, what constitutes good work within the context of their study and whether their efforts at producing good work meet the appropriate standards and criteria within the knowledge domain. Learners rarely enter courses prepared for this, so there is a need to help develop their capacity, and disposition, to operate effectively to seek and utilise feedback. (Boud & Molloy, 2013a)

9.10 Future Vision and Research Needed

In the previous sections, we looked at feedback aiming for deep learning. We stated that the importance of feedback to promote deep learning in students is evident. The rapid transition to online education due to COVID-19 has resulted in a large group of instructors and students gaining experience with new online education applications. These experiences also provide opportunities for the future (Mozelius, 2020). This is especially relevant for the future on-campus education, as it offers the opportunity to take a fresh look at educational design and critically examine which part of education is best on-campus and which part is best online and remote. We suggest that educational designers make evidence-based choices, both about the form of education (online, on-campus, blended, or hybrid) and about the didactic methods that are used. As described in this chapter, there are many ways to implement feedback to promote deep learning. Instructors need to decide what will best serve the learning purposes: peer feedback, instructor feedback, or both and whether to use written or face-to-face oral feedback, or rather video or audio feedback.

In feedback choices, not only the learning aims are important. Another factor is the workload in higher education. Even before COVID-19, the increasing academic workload was worrying (Gregory & Lodge, 2015). The pandemic only worsened these trends. We provided options to implement feedback in a way that could immediately reduce the workload of instructors, while it is at the same time increasing deep learning of students: using peers as feedback providers and choosing video and audio as the main feedback form. Wide implementation of these didactic forms may help in coping with the issues at hand regarding workload, by creating more effective and more in-depth use of time of both instructors and students.

9.10.1 Peer Feedback

The strength of peer feedback to promote deep learning by giving the students the role of feedback provider, and by stimulating critical thinking in the feedback receiver of peer feedback, is evident. Moreover, peer feedback is an easily implemented tool that relieves the instructor of providing all feedback. It therefore immediately results in a reduced workload, and—on top of that—the learning effects for students increase with peer feedback, as discussed earlier. There is however one condition: students need to be motivated and trained how to provide deep, constructive, and critically valid feedback aiming for deep learning. This feedback training is not necessarily dependent on the course or program that needs to introduce peer feedback. For reasons of efficiency, the organization can choose one online training focused on how to provide feedback that stimulates deep learning. Students can follow the training, or parts of an online training can be available for students at or before the start of a course with peer feedback.

9.10.2 Video Feedback

A wider implementation of video in feedback could also immediately decrease the instructor load while promoting deep learning in students. Again, there are no technical barriers for instructors to implement video feedback in their education; they can just record the feedback with their smartphone and share it with the student. When deep learning is the goal, the suggested 5-minute videos have many advantages over written detailed feedback on assignments and essays.

Moreover, video feedback is now mostly used in online learning situations, but this way of providing feedback is, of course, also applicable in on-campus, blended, or hybrid education. The added value is substantial, and in a higher education setting, it promotes deep learning more than written feedback.

9.10.3 Qualitative Feedback for Programmatic Assessment

Since feedback is essential to learning, incorporating the arc of a student's feedback as an assessment could be used to make a "portfolio" of the student's development akin to programmatic assessment. If students gather feedback throughout a course, or even throughout a program, this feedback can be used for summative assessment, with the potential of including richer information of the progress and development of the student compared to a final exam at the end of a course or program (van der Vleuten, 2016). This demands active student learning and student ownership of the gathered feedback, including the freedom to choose the material used as data points in assessment. Moreover, with the increased attention to individual learning paths and flexibility in education, this way of looking at student progress and assessment could provide a very valuable solution. This way of using feedback for assessment requires a large-scale culture change and a completely different way of looking at assessment in general. The assessment is about the progress of the student, and the student is leading the process. For implementation of programmatic assessment at a curriculum level, a curriculum revision, facilitating this form of assessment, is necessary (van der Vleuten, 2016).

9.10.4 Artificial Intelligent Assistants

The future of feedback can be significantly influenced using artificial intelligence assistants/ edubots to provide feedback. Many commercial websites use chatbots to answer the most commonly occurring questions from customers. In education, one can imagine that the use of artificial educational assistants—edubots—can take over certain instructor tasks (see also the beSocratic system described in Chap. 8 in this volume). The result can be more instructor time to engage in in-depth discussion, aiming for deep learning. Edubots can provide solutions for content delivery and can provide basic and administrative feedback and supervision. Goel and Polepeddi (2016) show that artificial intelligence assistants are already appearing as a disruptive alternative to traditional teaching assistants, by using an episodic memory of questions and answers from previous course editions to interact with students. Students in their MOOC, unaware of this teaching assistant being non-human, preferred this assistant over their human colleagues (Goel & Polepeddi, 2016). Currently, technology is not yet ready to use artificial intelligence assistants to offer feedback to promote deep learning, but it is technically possible to program these edubots to prompt questions that require deeper thinking of students and therefore promote deep learning. The use of artificial intelligence in higher education is a relatively new and quickly developing research field (Popenici & Kerr, 2017). Future research can be directed toward cutting-edge technical possibilities in combination with the needs in higher education: personalization, flexibility, and scalability.

References

Agricola, B. T., Prins, F. J., & Sluijsmans, D. M. A. (2020). Impact of feedback request forms and verbal feedback on higher education students' feedback perception, self-efficacy, and motivation. *Assessment in Education: Principles, Policy & Practice, 27(1)*, 6–25.

Anderson, T. D., & Garrison, D. R. (1995). Critical thinking in distance education: Developing critical communities in an audio teleconference context. *Higher Education, 29*(2), 183–199.

Askew, S., & Lodge, C. (2000). Gifts, ping-pong and loops: Linking feedback and learning. In S. Askew (Ed.), *Feedback for learning* (pp. 1–17). Routledge.

Bandura, A. (1999). A social cognitive theory of personality. In L. Pervin & O. John (Eds.), *Handbook of personality* (2nd ed., pp. 154–196). Guilford.

Biggs, J. B. (1978). Individual and group differences in study processes. *British Journal of Educational Psychology, 48(3)*, 266–279.

Biggs, J. B. (1987). *Student approaches to learning and studying. Research monograph.* Australian Council for Educational Research. https://files.eric.ed.gov/fulltext/ED308201.pdf.

Biggs, J. B., & Tang, C. (2011). *Teaching for quality learning at university: What the student does* (3rd ed.). McGraw-Hill Education.

Blouin, R. A., Riffee, W. H., Robinson, E. T., Beck, D. E., Green, C., Joyner, P. U., Persky, A. M., & Pollack, G. M. (2009). Roles of innovation in education delivery. *American Journal of Pharmaceutical Education, 73*(8), 154.

Borup, J., West, R. E., & Graham, C. R. (2012). Improving online social presence through asynchronous video. *The Internet and Higher Education, 15*(3), 195–203. https://doi.org/10.1016/j.iheduc.2011.11.001

Boud, D. (2000). Sustainable assessment: rethinking assessment for the learning society. *Studies in Continuing Education, 22*(2), 151–167. https://doi.org/10.1080/713695728

Boud, D., & Molloy, E. (2013a). Rethinking models of feedback for learning: The challenge of design. *Assessment & Evaluation in Higher Education, 38(6)*, 698–712.

Boud, D., & Molloy, E. (2013b). *Feedback in higher and professional education: Understanding it and doing it well.* Routledge.

Carless, D., Salter, D., Yang, M., & Lam, J. (2011). Developing sustainable feedback practices. *Studies in Higher Education, 36(4)*, 395–407.

Clark, J. E., & Eynon, B. (2009). E-portfolios at 2.0-surveying the field. *Peer Review, 11(1)*, 18–23.

Crook, A., Mauchline, A., Mawc, S., Lawson, C., Drinkwater, R., Lundqvist, K., Orsmond, P., Gomez, S., & Park, J. (2012). The use of video technology for providing feedback to students: Can it enhance the feedback experience for staff and students? *Computers & Education, 58(1)*, 386–396.

Dochy, F., & Segers, M. (2018). *Creating impact through future learning: The high impact learning that lasts model*. Routledge.

Duijnhouwer, H. (2010). *Feedback effects on students' writing motivation, process, and performance*. Doctoral dissertation, Utrecht University. Utrecht University Repository.

Duncan-Pitt, L., & Sutherland, S. (2006). An introduction to the use of eportfolios in professional practice. *Journal of Radiotherapy in Practice, 5(2)*, 69–75.

Entwistle, N. J. (2000). *Promoting deep learning through teaching and assessment: Conceptual frameworks and educational contexts*. Paper presented at the first annual conference of the Teaching and Learning Research Programme. Leicester. Retrieved from http://www.etl.tla.ed.ac.uk/publications.html

Entwistle, N. J., Thompson, J., & Wilson, J. (1974). Motivation and study habits. *Higher Education, 3(4)*, 379–396.

Ertmer, P. A., Richardson, J. C., Belland, B., Camin, D., Connolly, P., Coulthard, G., Lei, K., & Mong, C. (2007). Using peer feedback to enhance the quality of student online postings: An exploratory study. *Journal of Computer-Mediated Communication, 12(2)*, 412–433.

Filius, R. M., De Kleijn, R. A. M., Uijl, S. G., Prins, F. J., Van Rijen, H. V. M., & Grobbee, D. E. (2018a). Challenges concerning deep learning in SPOCs. *International Journal of Technology Enhanced Learning, 10(1&2)*, 111–127.

Filius, R. M., De Kleijn, R. A. M., Uijl, S. G., Prins, F. J., Van Rijen, H. V. M., & Grobbee, D. E. (2018b). Promoting deep learning through online feedback in SPOCs. *Frontline Learning Research, 6(2)*, 92–112.

Filius, R. M., De Kleijn, R. A. M., Uijl, S. G., Prins, F. J., Van Rijen, H. V. M., & Grobbee, D. E. (2018c). Strengthening dialogic peer feedback aiming for deep learning in SPOCs. *Computers and Education, 125(10)*, 86–100.

Filius, R. M., De Kleijn, R. A. M., Uijl, S. G., Prins, F. J., Van Rijen, H. V. M., & Grobbee, D. E. (2019). Audio peer feedback to promote deep learning in online education. *Journal of Computer Assisted Learning, 35*, 607–619.

Garrison, D. R., Anderson, T., & Archer, W. (2000). Critical inquiry in a text-based environment: Computer conferencing in higher education. *The Internet and Higher Education, 2(2&3)*, 1–19.

Goel, A. K., & Polepeddi, L. (2016). *Jill Watson: A virtual teaching assistant for online education.* Georgia Institute of Technology. http://hdl.handle.net/1853/59104

Gould, J., & Day, P. (2013). Hearing you loud and clear: Student perspectives of audio feedback in higher education. *Assessment and Evaluation in Higher Education, 38(5)*, 554–566.

Gregory, M. S., & Lodge, J. M. (2015). Academic workload: The silent barrier to the implementation of technology-enhanced learning strategies in higher education. *Distance Education, 36(2)*, 210–230.

Hall, M., Ramsay, A., & Raven, J. (2004). Changing the learning environment to promote deep learning approaches in first-year accounting students. *Accounting Education, 13(4)*, 489–505.

Hattie, J., & Timperley, H. (2007). The Power of Feedback. *Review of Educational Research, 77(1)*, 81–112.

Henderson, M., & Phillips, M. (2015). Video-based feedback on student assessment: Scarily personal. *Australasian Journal of Educational Technology, 31(1)*, 51–66.

Henderson, M., Phillips, M., Ryan, T., Boud, D., Dawson, P., Molloy, E., & Mahoney, P. (2019). Conditions that enable effective feedback. *Higher Education Research & Development, 38*(7), 1401–1416.

Ice, P., Curtis, R., Phillips, P., & Wells, J. (2019). Using asynchronous audio feedback to enhance teaching presence and students' sense of community. *Online Learning, 11(2)*, 3–25.

Klappa, P. (2015). *Innovative pedagogies series: Videos for learning and teaching.* Higher Education Academy. https://www.heacademy.ac.uk/system/files/peter_klappa_final2.pdf

Könings, K. D., van Berlo, J., Koopmans, R., Hoogland, H., Spanjers, I., Ten Haaf, J., & Van Merriënboer, J. J. G. (2016). Using a smartphone app and coaching group sessions to promote residents' reflection in the workplace. *Academic Medicine, 91*, 365–370.

Kreijns, K., Kirschner, P. A., Jochems, W., & Van Buuren, H. (2011). Measuring perceived social presence in distributed learning groups. *Education and Information Technology, 16*, 365–381.

Lamey, A. (2015). Video feedback in philosophy. *Metaphilosophy, 46(4&5)*, 691–702.

Leenknecht, M., Hompus, P., & Van der Schaaf, M. (2019). Feedback seeking behaviour in higher education: The association with students' goal orientation and deep learning approach. *Assessment & Evaluation in Higher Education, 44(7)*, 1069–1078.

Liu, N. F., & Carless, D. (2006). Peer feedback: the learning element of peer assessment. *Teaching in Higher Education, 11*(3), 279–290. https://doi.org/10.1080/13562510600680582

London, M., & Smither, J. W. (2002). Feedback orientation, feedback culture, and the longitudinal performance management process. *Human Resource Management Review, 12*(1), 81–100.

Lundstrom, K., & Baker, W. (2009). To give is better than to receive: The benefits of peer review to the reviewer's own writing. *Journal of Second Language Writing, 18(1)*, 30–43.

Lunt, T., & Curran, J. (2010). 'Are you listening please?' The advantages of electronic audio feedback compared to written feedback. *Assessment and Evaluation in Higher Education, 35(7)*, 759–769.

Mahoney, P., Macfarlane, S., & Ajjawi, R. (2019). A qualitative synthesis of video feedback in higher education. *Teaching in Higher Education, 24*(2), 157–179.

Marton, F. (1983). Beyond individual differences. *Educational Psychology, 3(3&4)*, 289–303.

Merry, S., & Orsmond, P. (2008). Students' attitudes to and usage of academic feedback provided via audio files. *Bioscience Education, 11(1)*, 1–11.

Moore, C., & Wallace, I. P. (2012). Personalizing feedback for feed-forward opportunities utilizing audio feedback technologies for online students. *International Journal of e-Education, e-Business, e-Management and e-Learning, 2(1)*, 6–10.

Mozelius, P. (2020). Post Corona adapted blended learning in higher education. Mid Sweden University, Faculty of Science, Technology and Media, Department of Computer and System Science. In D. Remenyi, K. A. Grant, & S. Singh (Eds.), *Responding to Covid-19: The university of the future*. ACIL.

Nicol, D., & Macfarlane-Dick, D. (2006). Formative assessment and higher self-regulated learning: A model and seven principles of good feedback practice. *Studies in Higher Education, 31*(2), 199–218.

Nilson, L. B. (2003). Improving student peer feedback. *College Teaching, 51(1)*, 34–38.

Oomen-Early, J., Bold, M., Wiginton, K. L., Gallien, T. L., & Anderson, N. (2008). Using asynchronous audio communication (AAC) in the online classroom: A comparative study. *MERLOT Journal of Online Learning and Teaching, 4(3)*, 267–276.

Parkes, M., & Fletcher, P. (2017). A longitudinal, quantitative study of student attitudes towards audio feedback for assessment. *Assessment & Evaluation in Higher Education, 42(7)*, 1046–1053.

Parsell, G., & Bligh, J. (2001). Recent perspectives on clinical teaching. *Medical Education, 35*, 409–414.

Popenici, S. A. D., & Kerr, S. (2017). Exploring the impact of artificial intelligence on teaching and learning in higher education. *Research and Practice in Technology Enhanced Learning, 12*(22).

Prins, F. J., Sluijsmans, D. M. A., & Kirschner, P. A. (2006). Feedback for general practitioners in training: Quality styles and preferences. *Advances in Health Sciences Education, 11(3)*, 289–303.

Ramaprasad, A. (1983). On the definition of feedback. *Behavioral Science, 28*, 4–13.

Roberts, P., Maor, D., & Herrington, J. (2016). ePortfolio-based learning environments: Recommendations for effective scaffolding of reflective thinking in higher education. *Journal of Educational Technology & Society, 19*(4), 22–33.

Sadler, D. R. (2010). Beyond feedback: Developing student capability in complex appraisal. *Assessment & Evaluation in Higher Education, 35*(5), 535–550.

Sargeant, J. M., Mann, K. V., Van der Vleuten, C. P. M., & Metsemakers, J. F. (2009). Reflection: A link between receiving and using assessment feedback. *Advances in Health Sciences Education: Theory and Practice, 14(3)*, 399–410.
Segers, M., Dochy, F., & Cascallar, E. (2003). *Optimising new modes of assessment: In search of qualities and standards*. Kluwer Academic.
Stefani, L., Mason, R., & Pegler, C. (2007). *The educational potential of e-portfolios*. Routledge.
Sutton, P. (2012). Conceptualizing feedback literacy: Knowing, being, and acting. *Journal Innovations in Education and Teaching International, 49*, 31–40.
Uijl, S., Filius, R., & Ten Cate, O. (2017). Student interaction in small private online courses. *Medical Science Educator, 27(2)*, 237–242.
van der Vleuten, C. P. M. (2016). Revisiting 'Assessing professional competence: From methods to programmes'. *Medical Education, 50(9)*, 885–888.
van Popta, E., Kral, M., Camp, G., Martens, R. L., & Simons, P. R. J. (2017). Exploring the value of peer feedback in online learning for the provider. *Educational Research Review, 20*, 24–34.
Voelkel, S., & Mello, L. V. (2014). Audio feedback—Better feedback? *Bioscience Education, 22(1)*, 16–30.
Wiggins, G. (2012). Seven keys to effective feedback. *Educational Leadership, 70(1)*, 10–16.
Yang, M., & Carless, D. (2013). The feedback triangle and the enhancement of dialogic feedback processes. *Teaching in Higher Education, 18(3)*, 1–13.

Chapter 10
Academic Cheating: How Can We Detect and Discourage It?

John C. McLachlan and Izabella Penier

Abstract We define cheating as the intentional use of prohibited *materials*, *processes* or *assistance* to undermine the validity of the assessment and gain an unfair advantage. We explore various ways in which cheating can occur and classify these under the three terms of the definition. Improper use of materials includes plagiarism, access to prohibited materials before the test and access to prohibited material during the test. Improper processes include denial of material to others and attempts to influence the performance of others by social means. Improper assistance includes impersonation, ghost writing and collusion. Addressing these issues involves questions of deterrence, detection and sanction. We consider technical means by which cheating can be detected, with an emphasis on electronic and online assessment. We then consider cultural, contextual, socio-demographic and personal factors which may have impact on cheating. Finally, we explore ways by which assessments can be designed to reduce the likelihood and impact of cheating. Design of a rewarding educational environment and attention to the cultural and personal factors which might promote cheating are important. We also recommend that assessments be delivered electronically where possible, to aid the detection of cheating, and that proprietary programmes which calculate various relevant statistics are used. We suggest that online solutions which permit proctoring are preferable to non-proctored approaches and that assessments are best designed as rating rather than ranking exercises.

10.1 Definition and Significance of Academic Cheating

'Cheating' is variously defined. Here we adopt a definition, modified from Cavanagh (2014), that cheating is the intentional use of prohibited *materials*, *processes* or *assistance* to undermine the validity of the assessment and gain an unfair advantage. While validity is itself variously defined operationally (Downing, 2003), we believe

J. C. McLachlan (✉) · I. Penier
University of Central Lancashire, Preston, UK
e-mail: jcmclachlan1@uclan.ac.uk; ipenier@uclan.ac.uk

H. J. Witchel, M. W. Lee (eds.), *Technologies in Biomedical and Life Sciences Education*, Methods in Physiology, https://doi.org/10.1007/978-3-030-95633-2_10

we should be most concerned with predictive validity: the ability of the test to predict later performance in some setting since this is of most practical value to those analysing the outcome of the assessment. Related terms such as 'unfair means' and 'academic misconduct' are often encountered in academic settings.

In this article, we address cheating carried out by students. Of course, faculty may engage in 'cheating' as well, by revealing assessment materials to candidates. This may occur as a result on individual favouritism, bribery, or where institutional rankings are at stake but is outside the scope of this chapter.

The property under test (i.e. the knowledge, skill or ability that the assessment is meant to measure) is generally known as the construct, and various aspects of the test may be construct relevant or irrelevant (Haladyna & Downing, 2004). Construct irrelevance relates to systematic sources of error which can obscure the property you are trying to test. For example, speededness (the possibility that a candidate may run out of time to record material that they do indeed know) is generally considered construct irrelevant. Unseen closed book exams are sometimes considered as merely a test of memory, which might be viewed as irrelevant to the construct of understanding, but in our view, recall under exam conditions requires the advance construction of mental schemas, which are themselves markers for understanding.

A recent general overview of assessment may be found in Yudkowsky et al. (2019). The context of the assessment is an important determinant of the kinds of cheating that might take place. Written assessments may have selected response or constructed response formats, with the former being represented by multiple choice items in various formats, and the latter by short answer questions (including very short answer, one- or two-word styles), essays or other written formats, which may be expensive to mark. The assessment may be delivered in paper or electronic versions, with greater or lesser degrees of invigilation or proctoring, and electronic delivery may be via a physical test centre or remotely through online methods. Practical assessments, where candidates are observed performing a task, are perhaps less susceptible (but not immune) to cheating. Assessments may be in 'closed book' format, where candidates are forbidden from using support materials during the assessment itself, or 'open book' where defined or undefined support materials are permitted or indeed required. Online exams may have closed or open-web equivalents. 'Bring Your Own Device' approaches, where the candidate sits an online assessment in a proctored environment but brings their own laptop or computer, pose their own particular challenges (Sindre & Vegendla, 2015). In the United Kingdom in 2020, under the influence of COVID-19, a number of summative assessments were switched to unproctored online versions.

Harper et al. (2021) indicate that students self-report that they have cheated on non-oral exams, including selected response formats (e.g. multiple choice tests) almost as often as in written assignments such as essays (26.6% vs. 36.7%). Conversely, far fewer faculty report having *detected* cheating on multiple choice quizzes compared to written assignments (12.5% vs. 70.7%), probably due to the availability of plagiarism detection software (see Fig. 10.1). In general, a number of forms of cheating are more easily detected in electronically delivered exams, in keeping with the theme of this book. Because of the imbalances described by Harper et al., we

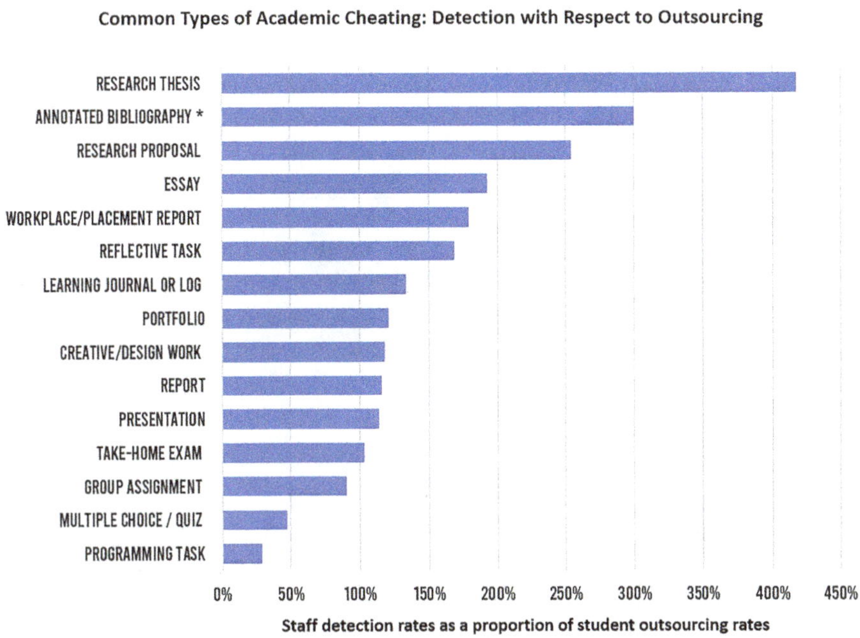

Fig. 10.1 Staff detection of cheating methods with respect to student reports of outsourcing to others. *Also includes literature reviews. With permission adapted from the original data in Harper et al. (2020). Higher Education Research & Development, 40(2), 263–278

devote consideration to electronic means of detection of cheating on exams, including in selected response formats.

Cheating is important not just for the validity of candidate attempts to pass the assessment but also as a marker for future behaviour. Unprofessional behaviour by medical students during training is a predictor of later poor clinical performance (Papadakis et al., 2008; Sawdon & McLachlan, 2020). Studies that have investigated the relation between academic integrity and business ethics show that student decisions to cheat might be related to decisions to engage in unethical behaviour in the workplace (Nonis & Swift, 2001; Ogilby, 1995; Sims, 1993). Plagiarism may indicate deeper flaws about lack of integrity (Bretag, 2013). Furthermore, some studies have found a relation between the level of college cheating and a country's corruption index (Magnus et al., 2002).

10.2 What Forms of Cheating Are Commonly Employed by Students?

We will classify these under the headings of materials, processes and assistance. A brief overview is provided in Table 10.1.

Table 10.1 A classification of cheating methods

Improper use of materials	Plagiarism
	Accessing test material before the test
	Accessing test material during the test
Improper processes	Denial of materials to others, e.g. by destruction
	Social exclusion, e.g. from self-study groups
Prohibited assistance	Impersonation
	Ghost writing
	Collusion during the test

10.2.1 Improper Use of Materials

10.2.1.1 Plagiarism

Plagiarism, originally from a Greek root indicating 'kidnapping', is the presentation of another's work as one's own. It may refer to words, computer code or, indeed, images or ideas, but here we will focus on written texts submitted for assessment purposes. As we will discuss further in a later section, it is important to distinguish between intentional plagiarism, with the intention of gaining an unfair advantage, and unintentional plagiarism, caused by a misunderstanding of the rules or requirements of academic writing. A variety of methods may be employed by candidates to disguise plagiarism, including translation from foreign languages and replacement of key words with synonyms (so-called Rogetising or Thesaurusing. Perhaps apocryphally, this process has led to 'Ulysses S. Grant' becoming 'Ulysses S. Award'). There are even paraphrasing tools available online for students, such as GoParaphrase (www.goparaphrase.com) and Paraphrasing Online (www.paraphrasingonline.com) (Cited from Bretag et al., 2019).

10.2.1.2 Access to Prohibited Materials Before the Test

Candidates may gain access to the test content before the point of test administration, enabling them to memorise answers, rather than address the items in the test from their own understanding. This access may be in the form of electronic hacking of administrative computer systems, physical theft of hard-copy materials, copying of items from a previous closed book exam for use in later exams or even bribery of test designers or administrators.

Since most exam items are drawn from a limited assessment bank, copying of items from one test administration may be of use to future candidates in future exams. For selected response tests where candidate knowledge is 'sampled' from a wider pool and where there are high stakes depending on successful completion of the test, some commercial organisations will offer to pay candidates to memorise or record items, so they can be sold as learning experiences, or candidates may spontaneously create a shared pool of such 'stolen' items for the benefit of themselves or others.

10.2.1.3 Access to Prohibited Materials During the Test

Candidates may access prohibited support materials during the test itself. These may be in the form of illegally introduced physical materials, electronic means such as the use of mobile phones or access to the Internet or other learning materials in settings where such uses have been forbidden.

10.2.2 Improper Processes

10.2.2.1 Denial of Materials to Others

In competitive assessments, where the outcomes of value are limited to a subset of candidates, one observed form of cheating is the denial of necessary learning materials to others. Previously, this took the form of physical destruction or theft of resources such as library books and journal articles. More recently this may take the form of interference with electronic records, such as altering online resources such as Wikipedia to be incorrect when they are a commonly used source of information.

10.2.3 Prohibited Assistance

10.2.3.1 Impersonation

Candidates may substitute an individual believed to be more expert than themselves in the assessment, a practice which may be slightly easier in digital or remote assessments. In our experience, the impersonation may be quite complex, involving, for instance, use of false passports and other documentation.

10.2.3.2 Ghost Writing

Where submitted constructed response work is required, this represents the hiring of a substitute to prepare or write the relevant submission.

10.2.4 Collusion

Candidates may arrange collusion with other candidates at live-presence or online assessments or non-candidates at online assessments.

10.3 How Can Cheating Be Reduced by Electronic and Other Means?

We will distinguish between *detecting*, *deterring* and *sanctioning* cheating and consider these under the same three headings as previously in this chapter.

A significant method of detecting cheating applies to all cheating methods, however. This is whistleblowing, when a candidate not involved in the cheating process comes forward to indicate that cheating has taken place. In our experience, this is generally retrospective, and the cheating suspects are rarely named. Instead, a candidate who is concerned that cheaters are gaining a relative advantage may come forward to describe the method of cheating and ask for modifications to the assessment process so that it cannot happen again.

A common issue with sanctions for cheating is the degree of confidence with which cheating can be identified. Since a sanction for cheating can have serious consequences for the individual, accusations may result in civil claims against the organisation, as well as posing ethical challenges in making them (Haney & Clarke, 2007; Tippins, 2009). Since, as we will see, some methods give a statistical probability of cheating rather than a certainty, imposing sanctions is therefore problematic.

One resolution to this statistical uncertainty is to require candidates to be prepared to undertake the assessment again, under circumstances which preclude the suspected form of cheating; sometimes this is done in an ad hoc fashion orally by faculty members. It is essential to make clear to candidates in advance that this may happen, and that, since it is based on statistical possibilities, is not necessarily in itself an accusation of cheating. It should be regarded as a confirmatory test of the candidate's ability.

10.3.1 Improper Materials

10.3.1.1 Plagiarism

Plagiarism detection may be divided into extrinsic and intrinsic forms. Extrinsic plagiarism detection compares the target text with materials published elsewhere. Intrinsic plagiarism detection focuses solely on the text itself, seeking to address the question, 'Did a single author write this text?' and is discussed under 'Ghost Writing'.

There are a variety of extrinsic detection tools (Kanjirangat & Gupta, 2016), but the authors' experience largely lies with Turnitin,[1] which is widely used in academic circles. The advantage of plagiarism detection software is that cheating may be identified with a high degree of certainty, enabling both detection and sanction to take place. In our experience, the great majority of cheating sanctions occur around plagiarism issues and may involve calculations of frequency and severity. The visible presence of sanctions may act as a deterrent.

10.3.1.2 Access to Prohibited Materials Before the Test

Candidates may have item preknowledge due to a variety of different means. These may include physical or electronic theft of items or the capture of items administered in previous tests. This form of cheating may seem difficult to detect, but there are consistent patterns of outcomes which offer a fingerprint for the detection of such preknowledge. These may include similarities between candidates' specific responses, and the issue is discussed under Collusion.

Here we focus on the items themselves rather than the candidates.

The application of item response theory (IRT) offers one way of detecting item preknowledge (Zimmermann et al., 2016). Under IRT the absolute difficulty of an item can be calculated along with its relationship to the difficulty of other items in the test for a particular candidate cohort. Consistent shifts in item difficulty, item discrimination and local dependence (showing that responses to items on the test are related to each other) may signal that a group of candidates have answered similarly because they had item preknowledge. A combination of these methods may therefore indicate which items were released to some candidates before the formal test administration. However, detecting shifts in item difficulty or discrimination over time requires that tests be equated with each other, and if the pool of items is already compromised, this may not be straightforward. By contrast, showing local dependence does not depend on test equating.

Such approaches are not successful if a large proportion of items are available in advance of the test and a large proportion of candidates have access to them. Similarly, if a small pool of items is released in advance and only a small group of candidates has access to them, this is difficult to detect. However, in these cases similarities of candidate responses might be successful in detecting item preknowledge.

---

[1] https://www.turnitin.com/regions/uk

10.3.1.3 Access to Prohibited Materials During the Test

This category we are viewing as distinct from collusion between individuals during the course of the assessment. It would include the smuggling of materials into a proctored environment. Interestingly, this is one of the earliest recorded forms of cheating, and candidates in the Chinese examinations were searched before entering the exam and sequestered during the course of it (Brickman, 1961 cited in Whitley, 1998). More modern versions include the use of digital communications such as mobile phones during the course of an invigilated examination or access to prohibited materials during the course of an online examination.

Setting aside the self-evident need for conscientious invigilation during proctored examinations to detect the use of smuggled materials, a variety of electronic safeguards may be employed. Computer systems may have automatic locking to prevent access to forbidden areas of the Internet or communications, and many assessment platforms permit keystroke analysis which could detect access to non-permitted materials.

Where constructed response items are involved, plagiarism detection software recognises large-scale copying from external sources.

Constructed response ‘short answer’ questions in online timed assessments may be particularly prone to copying of answers from elsewhere, since they combine the challenge of a potentially speeded assessment with the need to recall and rewrite large amounts of information. Here the answer may be to convert them into assessments which may be undertaken over a longer time scale: in other words, to convert them into an assignment. For assignments the rules are clearer to candidates, and they are more likely to be wary of plagiarism, preferring instead to reframe the information in the way generally seen as desirable.

10.3.2 Improper Processes

10.3.2.1 Denial of Materials to Others

Destruction or sequestration of material is frequently identified by other candidates in the preparation part of the assessment. It may be alleviated by the provision of electronic resources accessed from reliable sites. Attempts may be made to influence the outcomes of other candidates by a variety of means which can include denigration of character, social exclusion, deliberate misleading and other methods.

Reducing these approaches is again dependent on other candidates coming forward to share information. Such methods are most commonly encountered in assessments intended for ranking rather than rating. Switching to pass/fail methodologies and criterion referencing is therefore preferable to norm referencing in many situations, and the purpose of the assessment must be carefully considered.

10.3.3 Prohibited Assistance

10.3.3.1 Impersonation

Dealing with impersonation is similar in both proctored online and face-to-face test administration. Candidates may be required to bring photo ID such as passports to the venue or show these to the camera where online proctoring is in place. It is also possible to use biometric data such as voice or face recognition, fingerprints or even retinal scans (Rosmansyah et al., 2020). Identity is harder to confirm in online unproctored tests, but an intermediate step may be to use videoconferencing software alongside unproctored exam software. In this approach, candidates may be required to run videoconferencing software alongside the exam software and to have their computer cameras switched on. Invigilators can review any candidate at any point in the exam and record or screenshot their faces and ID documents.

In the absence of such facilities, detection of online impersonation becomes progressively more difficult. Checking the IP address used during the exam may identify instances where this is different from that used during the course of normal teaching.

10.3.3.2 Ghost Writing

Ghost writing refers to candidates engaging another individual to produce written materials or assignment on their behalf. Where there is a financial arrangement, this may be known as contract cheating (Rundle et al., 2019). While ghost writing or contract cheating may seem very difficult to detect, intrinsic plagiarism detection employs tools to identify 'writing style' (Kuznetsov et al., 2016). If a document appears to have more than one author, then this may emerge during the analysis. Alternatively, a style very different from that of the candidate in other circumstances may be observed. Either of these circumstances may be indicative of cheating. Extensive consideration of contract cheating in Australia can be found on the website: https://cheatingandassessment.edu.au

10.3.4 Collusion

Collusion in invigilated in-person assessments is best prevented (rather tautologically) by vigilance on the part of the invigilator. Online assessments may have a proctoring process, by which candidates are required to work in defined environments, such as being alone in the room, having a clear desk and being required at intervals to rotate the camera.

There are a variety of approaches to detect collusion having taken place subsequent to the administration of the test. These are particularly suitable for electronic and online delivery methods. For remote exams, IP addresses may identify candidates in physical proximity to each other. Timestamps can identify candidates looking at the same item at the same time, and activity within the application may be logged, to trace keystrokes and user interactions.

Subsequent to the assessment, patterns of similar answers can be identified, particularly when common incorrect answers are given. One example is the Harpp-Hogan Index (Harpp & Hogan, 1993). This calculates 'errors in common' (EIC) between candidates and 'exact errors in common' (EEIC). If the ratio EEIC/EIC exceeds 0.75, this is taken as an indication that collusion *may* have taken place. A calculation of the probability of obtaining such results by chance can be added (Nath & Lovaglia, 2009). If a seating plan record is kept, and if the probability of cheating between two nearby candidates is less than 0.001, then further action, such as interviewing the candidates, may be taken. Items with no correct answers may be inserted into the test to increase the chances of detection.

In addition to detecting similarities of wrong answers, it is also possible to consider the similarity of correct answers. In one approach based on Angoff's method (McManus et al., 2014), a computer programme called Acinonyx was deployed on Royal College examinations. This approach was successful in identifying rare cases of possible collusion with a statistical probability that could lead to licensing body sanction.

Knowledge that such systems are in place, and that random seating plans may be used for in-person exams, may act as a deterrent, but sanctions may be challenging to implement on a purely statistical basis. Nonetheless, recording of seating plans is a valuable addition to security methods. If two candidates with high similarity in responses are found to have been seated in close proximity, greater confidence can be placed in cheating having occurred.

Other deterrents and ways of alleviating the problem may include use of alternative questions, such that candidates do not receive exactly the same test, though this then raises issues of test reliability and is only feasible where a large bank of tested items with psychometric data is available, such as when national examinations are regularly provided by computer (Cantillon et al., 2004).

Where multiple choice items are used, cloning and automatic item generation may be used to construct very similar but different versions of a given item. This means that candidates are not sitting exactly the same version of the test. In cloning approaches, non-essential details of the item such as the age and gender of subjects described in the stem may be varied, so that the item becomes harder to recognise. In automatic item generation, essential aspects may be changed so that the correct answer varies between versions. A combination approach may involve the use of 'testlets'. These are blocks of items which may be used in various combinations to construct different tests. The advantage of this is that the psychometric properties of the testlets may be characterised in advance, so that test difficulty may be compared and corrected between test versions.

10.4 What Are the Factors Promoting Cheating?

10.4.1 Cultural Factors

Cheating is not viewed in the same way in every part of the world. Most often, but not always, it is considered a serious form of academic misconduct, but even then, it is not punished with the same severity. This implies that attitudes towards cheating vary and are culturally and contextually determined. Cross-cultural studies on the occurrence and acceptance of academic dishonesty show that students' propensity to engage in dishonest practices can be linked to social norms, ethical values, and divergent socialising practices, all of which are culturally specific. In the words of Chudzicka-Czupała et al. (2016 p. 643): 'Culture frames norms for students, perceptions of how difficult/easy it may be to get away with dishonest behavior . . ., and the sense of how moral/amoral cheating is'.

In the era of steadily growing student mobility, this has a significant ramification for institutions of higher education worldwide. The number of international students in OECD countries has doubled since 2000 (OECD Publishing, 2015). As universities tackle more and more multinational and culturally diverse cohorts of students, it is important for the university leadership and lecturers to have an understanding of divergent perceptions on academic dishonesty and how native cultures can influence students' academic conduct. Such an understanding is crucial in preventing dishonesty among students and promoting ethical codes of academic conduct.

What have we learned from cross-cultural studies on academic cheating?

There are profound differences in views on academic integrity between Western countries on the one hand and post-socialist and developing countries on the other.

A number of studies researching different attitudes to academic dishonesty between Western European/American students and Eastern European students found out that in the so-called transition countries (i.e. countries that have transitioned from communist to market economy), academic misconduct was endemic—its prevalence was far greater than in developed countries. Students from central and Eastern Europe, such as Russia, Ukraine, Hungary, Croatia or Poland have been found to be more likely to cheat than Western European or American students. In post-socialist/transition countries, not only incidences of academic cheating were higher; there were also fewer malpractices reported by students, and the likelihood of students being angered by whistle-blowers was greater (see Chudzicka-Czupała et al., 2013, 2016; Farkas & Orosz, 2012; Grimes, 2004; Hrabak et al., 2004; Lupton & Chapman, 2002; Lupton et al., 2000; Magnus et al., 2002; Poltorak, 1995). Similar comparative cross-cultural studies have been conducted on students from Asia (Curtis et al., 2012; Rawwas et al., 2004), Africa (Saana et al., 2016), Israel (Magnus et al., 2002) and the Middle East (McCabe et al., 2008); all of them reported significant differences in the perception of academic misconduct.

There are theoretical explanations for the cultural predictors, which can shed light on student misconduct. These theories on national culture differences turned out to be very useful for managing international teams in the wake of growing globalisation of business, but they have been also successfully contextualised in relation to student performance. The cultural theories that explain the variance in students' intention to cheat is Hall's theory of high and low-context national cultures (1976), Hofstede's (1980) theory of organisations and cultures and Trompenaars and Hampden-Turner's model of national culture differences (1993, 1997).

High-context cultures, such as Japan, China, the Middle East, South America, Africa, Eastern and Southern European countries, are more relational and group-oriented, i.e. they have strong collectivist character. That means that members of these cultures value long-term interpersonal relationships and seriously take into account views of colleagues, friends and family. In the high-context communitarian countries, people see themselves first and foremost as members of a group that has clearly outlined boundaries differentiating between insiders and outsiders. Decisions and activities focus on interpersonal relationships and often involve a central person who has authority. High-context cultures rely on internalised understandings of the group dynamics and need less verbally explicit or formal communication.

On the other hand, low-context countries, such as Australia, Norway, Denmark, Sweden, Finland, Canada or the United States, put more emphasis on individualism and diversity as well as transactional interactions. Members of the low-context individualist countries regard themselves as independently thinking and acting individuals. They often form loose and wide networks to achieve particular goals. Their relationships have shorter duration and are governed by external rules that are codified. Members are also more task-oriented, and achieving goals is more important than maintaining relationships. Low-context cultures are less exclusive, more accessible for outsiders and easier to participate in because rules are public, and relationships are easier to form, especially if they are related to accomplishing tasks.

The differences between low-context cultures and high-context cultures can have powerful ramifications for students' conduct. Students from low-context cultures characterised by high individualism scores are more self-reliant and have little concern about opinions of others, (Chudzicka-Czupała et al. 2016; Hampden-Turner & Trompenaars, 1993, 1997; Hofstede, 1991). They are more likely to follow their own mind, rather than give in to any form of social coercion. Conversely, for students from high-context cultures, norms of the collectivist society and the perception of peers' behaviour play a much bigger role in the ethical decision-making process (McCabe et al., 2008). For example, a 2008 study by McCabe et al. compared opinions and experiences of students from three Lebanese universities and seven American universities and found strong evidence that Lebanese students were more likely to engage in cheating because they saw others cheat and because they knew other students would approve of their dishonest behaviour.

When it comes to academic assessment, which is individualised and highly competitive, students from high-context cultures may underperform and so exhibit a bigger intention to cheat. High-context cultures value teamwork, which is why students from high-context cultures are more likely to collaborate to gain advantage. Moreover, in group-oriented societies, like Lebanon, helping weaker students, even on individual assignments, may be seen not only as a socially acceptable practice but also as an act of moral obligation. Another study by Saana et al. (2016) conducted among Ghanaian undergraduate students had similar findings. The study examined students' self-reported understanding of academic misconduct and their involvement in specific dishonest behaviours, as well as their motivation factors. The authors reported that the most frequently occurring malpractice was allowing colleagues to copy one's work, followed by copying colleagues' work with their permission. Saana et al. (2016) argued that a great majority of students consented to cheating: 'they would copy their friends' work (with permission) and reciprocate this by helping friends to also cheat' (p. 7). It is clear that in collectivist cultures, students' lenience towards academic dishonesty, especially in the form of collusion, springs from subjective perception of cheating as a form of solidarity and self-help, which is at variance with how cheating is viewed in more individualistic societies. Similarly, the earlier example of Lebanese students, if 'viewed through a collectivistic lens', shows that 'the Lebanese students are behaving exactly the way they were raised to behave—working together to navigate a difficult task' (McCabe et al., 2008 p. 464).

The dichotomy between low- and high-context cultures was also addressed by Trompenaars's model of national culture differences. An important component of that framework is the difference between universalism and particularism (Hampden-Turner & Trompenaars, 1997; Luthans & Doh, 2012). Universalism implies that social practices should be applied in all situations without modification. Particularism, on the other hand, is based on belief that circumstances dictate how practices should be applied and reality is seen as far more subjective. Particularism prioritises relationships over rules. In universalistic cultures, such as Western cultures, where rules are placed above relationships, cheating is considered unacceptable. In particularistic cultures, such as post-socialist, Middle Eastern or African countries, on the contrary, cheating behaviours may be regarded with greater degree of moral relativism (Parsons, 1937/1968, 1951/1991).

Geert Hofstede's (1980) theory of organisations and cultures can be also usefully applied to the educational system. In particular, his concept of power distance can explicate differences in conduct in international cohorts of students. People in countries with large power distance accept and respect power and authority. In countries with small power distance, people are reluctant to accept power and authority merely because of their own lower position in the social hierarchy. They have relatively little respect for authority or expertise, and in some countries with the lowest power distance scores, like the United States or Australia, members of the society may have very little trust in scholars, experts or government officials. Instead, they will have confidence in 'impersonal "truth" which can in principle be obtained from any competent person' (Hofstede, 1986 p. 313) or even arrived at by themselves.

It is not difficult to see why power distance can explain the relationship between the country of origin and students' decision to engage in academic dishonesty. For example, in India with a high Power Distance Index (PDI), young students are penalised if they do not copy-paste (Chaurasia, 2016). Likewise, Saana et al. (2016) observed that in Ghana, which scores very high on power distance dimension, learning by heart from a textbook or lecture notes and then repeating the information verbatim during exams 'is accepted and even encouraged' (p. 12). Ghana's high PDI means that Ghanaian students accept a hierarchical order in which, to quote Hofstede out of context, '"wisdom" is transferred in the relationship with a particular teacher (guru)' (Hofstede, 1986 p. 313) who 'outline[s] paths [for student] to follow' (Hofstede, 1986 p. 313).

Unlike small power distance societies in the West, where students are expected to 'find their own paths' (Hofstede, 1986, p. 313), Indian and Ghanaian students may not see development of their own critical thinking as their main educational goal. Instead, their upbringing and their educational system may imply that their own ideas do not matter, and it is important to know what experts think (Baxter Magolda, 1999). Saana et al. (2016) do not draw on the power distance theory to explain why Ghanaian students are more likely to engage in such form of cheating as plagiarism; but the theory makes it is easy to see why students who do not feel that they can generate valid knowledge might exhibit behaviours that would be considered dishonest and unacceptable in the West.

However, plagiarism should not be considered as isolated to developing countries or failed individuals in the West, as was made apparent in two cases of thesis plagiarism by major German political figures, German Minister of Defence, Karl-Theodor zu Guttenberg, as well as Annette Schavan, Federal Minister of Education and Research. Even Ursula von der Leyen, the current President of the European Commission, had her thesis investigated for plagiarism long after her graduation from the Medical University of Hannover. The thesis was found to have 'obvious flaws, particularly in the introduction', but the investigation concluded, 'the pattern of the plagiarism does not indicate a fraudulent intent' (BBC News, 2016).

Moreover, differences between national cultures are not as absolute as they may seem, and there are also inconsistencies between Hofstede's and Trompenaars's research and classifications that placed some countries, such as Mexico, Germany, China, France, Japan or Singapore in different places on the individualist-collectivist spectrum (Luthans & Doh, 2012). These discrepancies show the dynamic and complex nature of national cultures. Consequently, it would be interesting to see if the behaviours of students from the transition countries of the former Soviet Union and central and Eastern Europe have changed as a result of 20 years of transformation and acculturation with Western values and educational models.

Finally, as we will argue in the section on contextual factors, academic integrity can be enhanced by a healthy learning environment where students are educated about what constitutes academic misconduct, and all forms of fraudulent behaviour are condemned and prosecuted. For example, a study on academic cheating in Austria, Portugal, Romania and Spain (Teixeira & Rocha, 2006), which compared prevalence of cheating between Eastern and Western European students of Economics and Management (Austria, Portugal, Romania and Spain) found that the East/West divide can be reversed by robust anti-cheating measures, which made cheating difficult and introduced repercussions for students who had been caught. In the three countries with greater sanctions for cheating—Austria, Portugal and Romania—the prevalence of cheating was lower. In Spanish universities, where campus cultures were more 'copying-favourable' (Teixeira & Rocha, 2006, p. 198), students were much more inclined to cheat.

10.4.2 Contextual Factors

Contextual factors related to campus cultures, such as codes of honour, disciplinary rules and sanctions, can have a significant impact students' perception about the commonality and acceptability of cheating. A learning environment that is healthy will discourage misconduct. According to Chudzicka-Czupała et al. (2016), who draws on the research by Crown and Spiller (1998), Whitley (1998), Jordan (2001), McCabe et al. (2002), Volpe et al. (2008) and Naghdipour and Emeagwali (2013), 'a healthy learning environment and campus culture are among the factors often cited as important in the prevention of cheating and the creation of an academic integrity climate' (p. 642). For example, if the higher education system is corrupt and allows such malpractices as bribery, nepotism, research misconduct, discrimination, etc. (as it often was the case in the transition countries), the perception and awareness of the general malfunctioning of the educational system may lead to higher predisposition to cheat among students (Chudzicka-Czupała et al. 2016; Osipian, 2008).

McCabe et al. (2008) suggest that in settings where lenience towards academic dishonesty prevails, students may feel they have no choice but to cheat to keep a level playing field. Błachnio and Weremko (2011) and Chapman et al. (2004) contend that seeing others cheat is also important not only because individuals' model their behaviour on peer but also because they may judge their own dishonest behaviour as 'less morally reprehensible' (O'Rourke et al., 2010, p. 47). On the other hand, campus culture that is intellectually stimulating, highly motivating, inspirational and performance-driven is more likely encourage students to excel and report dishonest practices of 'free riders', that is, students who get higher grades by unfair means.

Even though academic cheating takes place in both developed and developing countries, in poorer countries it can be exacerbated by poor English literacy (Bretag, 2013), lack of access to resources and insufficient support from tutors (Saana et al., 2016). According to McCabe and Treviño (1996), cheating is more common in institutions where the ratio of teachers to students is low, i.e. lecturers have a high number of students to supervise and a high number of assignments and exams to check. This factor, in conjunction with cultural factors, such as large power distance, may explain provenance of cheating in developing countries.

Another factor increasing the likelihood of students engaging in cheating is lack of awareness of what constitutes cheating (e.g. insufficient anti-plagiarism training) and what are institutional regulations on cheating (Henning et al., 2014; Ryan et al., 2009). In such cases, students may engage in unethical practices inadvertently. On the other hand, as Henning et al. (2014) and Saana et al. (2016) have shown, awareness of the policies, regulations and penalties associated with cheating is not a sufficient deterrent if students do not have a proper understanding of how these rules and penalties work. In Saana et al.'s study (2016), including 131 Ghanaian students, who underwent freshman orientation programmes, 93% of students thought they were familiar with the regulations, but only 31% knew practical means of avoiding plagiarism and understood the full implications of being caught (p. 10–11).

If there are no training programmes about different forms of academic dishonesty and how to avoid them, the occurrence of cheating is likely to be high (Saana et al., 2016, p. 10). This is particularly true about guidelines for information referencing, which, if misunderstood, will fail to prevent plagiarism. According to Jordan (2001), Mahmud and Bretag (2013) and Okoroafor et al. (2016), students who have a true understanding of guidelines, bolstered by robust trainings, will cheat less. Other studies (Barrett & Malcolm, 2006; Chanock, 2008; Devlin, 2002) also emphasised the importance of continuous and repeated training at both undergraduate and postgraduate levels.

Finally, Bretag (2013) argued that not only students but other 'stakeholder groups' should be given 'much more induction, information, training, and support to ensure that they have the necessary understanding and skills to fulfil their academic responsibilities'. These stakeholders include 'educational policy makers, senior managers, teaching academics and advisors, students at all levels, researchers, funding bodies, editors, and reviewers' all of whom are part of an education system that unfortunately gives preference to 'tangible awards' (diplomas, titles, promotions, grants) over 'the intrinsic value of learning and knowledge creation'.

10.4.3 Personal and Socio-demographic Factors

Students give different personal excuses for engaging in academic dishonesty. According to Henning et al. (2014), students acting in bad faith are most likely to justify their cheating behaviours by claiming that they did not consider academic dishonesty as serious, that they wanted to 'save time' or that they were 'overloaded with demands' at school and elsewhere. Other excuses included 'lack of educational value of the assignment', peer and family pressure, teachers' leniency, institutional laxity or a desire to 'raise the grade'. Interestingly this last factor can affect even high-achieving students who do not have to engage in misconduct to pass. In a large-scale study of students in a Canadian university who submitted written tests for a regrade, Ottaway et al. (2017) found that the students who were most likely to alter their resubmitted test for a higher mark were the better performing students in the class (and female). This kind of cheating is of marginal value to the student, yet it suggests how varied and complex are the factors that impact ethical choices made by students.

Earlier studies from different regions examined psychological and socio-demographic factors that may make individuals more likely to cheat in an effort to create profiles of cheaters. These profiles included such predictors as introversion and extroversion, narcissism, locus of control, competitive achievement and motivation (Chudzicka-Czupała et al., 2016; Farkas & Orosz, 2012; Kerkvliet & Sigmund, 1999; Whitley, 1998). Some of them also tried to link motives underlying academic misconduct with personal value systems (inclination towards idealism or relativism, opportunism, level of tolerance or individual moral philosophies (Rawwas et al., 2004). Other studies investigated relationships between gender and tendency to cheat (Hilbert, 1987; McCabe, 2009; Theart & Smit, 2012) or age and gender and tendency to cheat (Hendershott et al., 1999; Kerkvliet & Sigmund, 1999).

Unfortunately, cheater profiles turned out to be inconsistent and contradictory. For instance, some studies found out that female students are the least likely to cheat (Davis et al., 1992; Michaels & Miethe, 1989; Newstead et al., 1996), whereas other studies (Fisher, 1970; Stevens & Stevens, 1987; Vitro & Schoer, 1972) found no such correlation. Likewise, some studies (Bowers, 1964; Haines et al., 1986; Newstead et al., 1996) reported that the likelihood of getting involved in cheating decreases with age, whereas Tang and Zuo (1997) found that older students cheat more.

These discrepancies in findings may suggest that personal and socio-demographic factors, such as gender, age or personal value system, may also be culturally sensitive. In other words, students may have psychological or socio-demographic characteristics that predispose them to cheat that can be further reinforced by their cultural backgrounds, for example, being raised in a society that values collectivism. But what really matters is how universities promote academic integrity, especially among students from academically disadvantaged backgrounds. A campus culture that is aware of cultural differences among students and their different attitudes towards cheating can more effectively mitigate against the risk of students engaging in unethical behaviour. As we have argued earlier, we believe that cheating can be reduced by proper training that engages with the worldviews and values systems inherited by students.

10.4.4 Rational Choice Theory

Rational choice theory (Cornish & Clarke, 1987) suggests that individuals operate on what they perceive to be a rational basis. Within this, the factors that influence whether or not they are likely to cheat depend on three factors: probability of detection, cost (severity of sanction) and the speed at which detection occurs. On this basis, inhibiting cheating would rely on increasing the likelihood of rapid detection and the severity of the resulting punishment.

However, as we have already noted, this individual-level analysis is intermixed with social, personal and cultural factors. Even on its own terms, it may be problematic. We have noted that some systems for the detection of cheating give a probabilistic indication, rather than a definitive one. In this environment severe sanctions may be problematic. But in addition, the general theory of humans as rational actors can be challenged. As noted above, Ottaway et al. (2017) noted candidates cheating when the advantage was marginal. We ourselves have observed instances of cheating in formative assessments, where there is no gain to the individual—indeed there may be harms, arising from incorrect feedback on performance. Perhaps there is an element of 'self-esteem' measurement in assessment, in which just receiving a higher formative score is seen as a benefit.

10.5 How Can Cheating Be Inhibited?

Academic dishonesty will probably never be eradicated; but a more thorough understanding of the cultural and contextual factors that underpin it could help university administrators to curtail this malpractice.

Evidence shows that universities that actively engage in promoting a culture of academic integrity and accountability have a lower prevalence of academic misconduct. Better promotion of ethical norms regarding academic conduct, consistency and enforcement of sanctions, ethical codes of conduct as well as investment in anti-plagiarism software or video monitoring can all contribute to creating a learning environment in which cheating is extremely difficult and dangerous for students.

International students, in particular, need to be inducted into the educational system of the host country as early as possible so they can get an in-depth understanding of ethical standards and sanctions associated with violating them. Mandatory training should be offered not only for freshmen but should continue throughout the course, and repercussions of cheating should be regularly discussed with students in class. Infringement of rules should be consistently penalised as fear of getting caught and punished might be the most powerful deterrent of cheating (Chudzicka-Czupała et al., 2016)

As McCabe et al. (2008) pointed out, one way of preventing academic misconduct in collectivist cultures is 'developing collectivistic-appropriate teaching strategies that emphasize and take advantage of the power of collaborative work' (p. 464). In the words of McCabe et al. (2008):

> Colleges and universities should not, in our view, implement policies that may be fundamentally at odds with societal norms since student transgressions of such norms are almost predictable. Of course this does not preclude the possibility that a school might desire to hold its students to a higher standard. But in that case, it is imperative that the school properly inform and orient its students concerning these norms and the consequences for violators. Perhaps the most important policy implication is that 'one size does not fit all' when it comes to academic integrity. (p. 464–465)

Finally getting students involved in designing and implementing anti-cheating regulations, i.e. having senior students instruct junior students on the standards of academic conduct, may be a useful intervention in preventing academic dishonesty and promoting professionalism. Saana et al. (2016) found that students rightly perceive that their lecturers have a better understanding of institutional rules concerning cheating, which can be explained by the fact that these rules are prepared by the faculty members with little or no input from students. Therefore, we believe that getting students involved in revising and relaying these regulations to their peers might be an effective way of disseminating and enforcing the rules among students.

10.6 How to Modify Assessment to Reduce Cheating and Encourage Cooperation

In addition to the modifications to the educational cultural environments described in the preceding sections, certain assessment design features are likely to reduce the probability of cheating for any given assessment method. These have previously been described but are summarised again here and in Table 10.2.

There are some preliminary remarks that can be made, however. Foremost is the purpose of the assessment, whether it serves to *rank* candidates with respect to each other or *rate* their ability against an absolute standard. This is the distinction between norm referencing and criterion referencing in standard setting (McLachlan et al., 2012). Ranking candidates against each other turns the process into a competitive one, in which improper competitive processes are likely to occur. Rating against an absolute standard, which every candidate can pass, is preferable.

However, candidates may then co-operate to pass. Co-operation can be built into the assessment programme, with candidates being required to form co-operative groups, though this then creates problems of assessing individual contributions to group work and rooting out free riders and bad faith actors.

We can also say that in general, electronic delivery is preferable to hard-copy written assessments, since cheating software can then readily be used. In terms of identification of candidates to avoid impersonation, stringent methods such as requiring photo ID, photographing candidates and employing biometric data are worth consideration. And proctor training and continual vigilance is essential both for in-person and online examinations.

If the assessment consists of selected response items, then electronic delivery allows collusive methods and item preknowledge to be detected, as described above. Use of automatic item generation, item cloning and use of testlets ensure that candidates do not all sit exactly the same test.

For constructed response items, such as short answer questions, timed online exams are problematic, since they create a speededness challenge. It is better to replace these with assignments which can be completed in the candidates' own time against a more realistic timescale.

In-person attendance is probably more secure than online participation. In-person attendance should be accompanied by allocated seating plans, not known to candidates in advance, and retained after the exam for subsequent analysis. If online engagement is required, then live proctoring or videoconferencing adds to the security.

Table 10.2 Kinds of assessment: cheating hazards and remedies

In all circumstances				Use computer aided where possible Ensure vigilant proctoring, both online and in person Check and confirm candidate identities		
Delivery of exams	Assessment mode	Nature of response	Assessment type	Main cheating hazard	Detection	Mitigation
Exam conditions	Written	Selected response	Multiple-choice questions and extended matching items	Item preknowledge	Item response analysis	Item cloning Use of testlets
				Collusion during test	Response similarity (Haarp-Hogan, Acinonyx) Seating plan comparison	Seating plan employed
				Prohibited access during test	Proctor vigilance	Proctor vigilance
				Non-candidate completion	Identity checking	Identity Confirmation
		Constructed Response	Essays, short answer questions and data interpretation and problem-solving questions	Item preknowledge	Response similarity	Computer delivery Candidate awareness
				Collusion during test	Seating plan Comparison	Seating plan employed
				Prohibited access during test	Proctor vigilance	Proctor vigilance
				Non-candidate completion	Identity recording	Identity confirmation
Not under exam conditions	Written	Constructed response	Essays, reports, portfolios, theses	Plagiarism	Extrinsic and intrinsic plagiarism software	Candidate awareness
				Impersonation	Identity recording	Identity confirmation
Under exam conditions	Observed	Structured response	Objective structured clinical exams, multiple mini interviews	Item preknowledge	Statistical analysis of outcomes	Candidate awareness
				Collusion during course of assessment	Proctor vigilance	Sequestering students who have completed parts
				Impersonation	Identity recording	Identity confirmation
		Free form	Oral exams, vivas, performances	Impersonation	Identity recording	Identity confirmation

Table 10.3 Education integrity platforms

Platform	Functionalities	Internet domain
Turnitin (including iThenticate, Originality, Similarity, Feedback Studio, Gradescope)	Plagiarism detection, remote assessment delivery and proctoring, student feedback delivery	www.turnitin.com
Proctorio	Remote assessment delivery and proctoring, plagiarism detection	www.proctorio.com
ExamSoft	Remote assessment delivery and proctoring	www.examsoft.com
Google Assignments	Remote marking, originality assessment	https://edu.google.com/assignments/
Respondus	Remote assessment delivery and proctoring	https://web.respondus.com/
Maxinity	Exam management Online healthcare learning	www.maxinity.co.uk/
proctorU	Remote assessment delivery and proctoring	www.proctoru.com/

It is a powerful advantage to deploy the assessment via a commercial platform such as ExamSoft, Maxinity MaxExam and Speedwell, which have built in cheating detection and/or proctoring capabilities.

If the assessment consists of constructed response items, then plagiarism detection approaches such as Turnitin are essential. A list of well-known education integrity platforms is shown in Table 10.3; a complete review of their features and privacy issues is not currently available but would be a welcome addition to the literature.

Preventing 'access to prohibited materials' also raises the question of why the materials were prohibited in the first place. It is arguable that giving students access to books, notes or the Internet creates a situation which is closer to workplace practice than preventing such access, and it has been proposed that this should be the norm even for high stakes assessments in healthcare (Teodorczuk et al., 2018).

Publicising the existence of methods which increase the likelihood of detection is itself a deterrent, as is the severity of the possible consequences. For instance, a healthcare student detected in cheating may well face fitness to practice sanctions, which could terminate their career. However, since many of the detection systems give a probabilistic statement rather than an absolute certainty, an alternative of requiring suspected candidates to re-take the assessment under carefully monitored conditions may be a valuable alternative.

Finally, the cultural and personal factors described in Sect. 10.4 should be taken into consideration when designing assessments. It is essential that assessment design takes into account, and is responsive to, the nature of the candidate pool.

References

Barrett, R., & Malcolm, J. (2006). Embedding plagiarism education in the assessment process. *International Journal for Educational Integrity, 2*, 38–45.

Baxter Magolda, M. B. (1999). *Creating contexts for learning and self-authorship*. Vanderbilt University Press.

BBC News. (2016-03-09). German Defence Minister Von der Leyen cleared of plagiarism. https://www.bbc.co.uk/news/world-europe-35769001. Accessed 2021-01-15.

Bowers, W. J. (1964). *Student dishonesty and its control in college*. Bureau of Applied Social Research, Columbia University.

Bretag, T. (2013). Challenges in addressing plagiarism in education. *PLoS Medicine, 10*(12), e1001574. https://doi.org/10.1371/journal.pmed.1001574

Bretag, T., Harper, R., Burton, M., Ellis, C., Newton, P., Rozenberg, P., Saddiqui, S., & van Haeringen, K. (2019). Contract cheating: A survey of Australian university students. *Studies in Higher Education, 44*, 1837–1856.

Błachnio, A., & Weremko, M. (2011). Academic cheating is contagious: The influence of the presence of others on honesty. A study report. *International Journal of Applied Psychology, 1* (1), 14–19.

Brickman, W. W. (1961). Ethics, examinations, and education. *School and Society, 89*, 412–415.

Cantillon, P., Irish, B., & Sales, D. (2004). Using computers for assessment in medicine. *BMJ, 329*, 606–609.

Cavanagh, T. (2014). Cheating on online assessment tests: Prevalence and impact on validity. *Collected Faculty and Staff Scholarship, 174*. https://scholar.dominican.edu/all-faculty/174

Chanock, K. (2008). When students reference plagiarised material—What can we learn (and what can we do) about their understanding of attribution? *International Journal for Educational Integrity, 4*, 3–16.

Chapman, K., Davis, R., Toy, D., & Wright, L. (2004). Academic integrity in the business school environment: I'll get by with a little help from my friends. *Journal of Marketing Education, 26*, 236–249. https://doi.org/10.1177/0273475304268779

Chaurasia, A. (2016). Stop teaching Indians to copy and paste. *Nature, 534*(7609), 59.

Chudzicka-Czupała, A., Grabowski, D., Mello, A. L., Kuntz, J., Zaharia, D. V., Hapon, N., Lupina-Wegener, A., & Börü, B. (2016). Application of the theory of planned behavior in academic cheating research—Cross-cultural comparison. *Ethics and Behavior, 26*, 638–659. https://doi.org/10.1080/10508422.2015.1112745

Chudzicka-Czupała, A., Lupina-Wegener, A., Borter, S., & Hapon, N. (2013). Students' attitude toward cheating in Switzerland, Ukraine and Poland. *The New Educational Review, 32*, 66–76.

Cornish, D. B., & Clarke, R. V. (1987). Understanding crime displacement: An application of rational choice theory. *Criminology, 25*, 933–948.

Crown, D. F., & Spiller, M. S. (1998). Learning from the literature on collegiate cheating: A review of empirical research. *Journal of Business Ethics, 17*, 683–700.

Curtis, M. B., Conover, T. L., & Chui, L. C. (2012). A cross-cultural study of the influence of country of origin, justice, power distance, and gender on ethical decision making. *Journal of International Accounting Research, 11*, 5–34. https://doi.org/10.2308/jiar-10213

Davis, S. F., Grover, C. A., Becker, A. H., & McGregor, L. N. (1992). Academic dishonesty: Prevalence, determinants, techniques, and punishments. *Teaching of Psychology, 19*, 16–20.

Devlin, M. (2002). *Minimising plagiarism*. Centre for Studies in Higher Education, University of Melbourne. http://www.cshe.unimelb.edu.au/assessinglearning/docs/PlagMain.pdf

Downing, S. (2003). Validity: On the meaningful interpretation of assessment data. *Medical Education, 37*, 830–837.

Farkas, D., & Orosz, G. (2012). Why Hungarian high school students cheat? *Practice and Theory in Systems of Education, 7*, 279–292.

Fisher, C. T. (1970). Levels of cheating under conditions of informative appeal to honesty, public affirmation of value, and threats of punishment. *Journal of Educational Research, 64*, 12–16.

Grimes, P. W. (2004). Dishonesty in academics and business: A cross-cultural evaluation of student attitudes. *Journal of Business Ethics, 49*, 273–290. https://doi.org/10.1023/B:BUSI.0000017969.29461.30

Haines, V. J., Diekhoff, G. M., LaBeff, E. E., & Clark, R. E. (1986). College cheating: Immaturity, lack of commitment, and the neutralizing attitude. *Research in Higher Education, 25*, 342–354.

Haladyna, T. M., & Downing, S. M. (2004). Construct-irrelevant variance in high-stakes testing. *Educational Measurement: Issues and Practice, 23*, 17–27.

Hall, E. T. (1976). *Beyond culture*. Anchor Books/Doubleday.

Hampden-Turner, C., & Trompenaars, A. (1993). *The seven cultures of capitalism: Value, systems for creating wealth in the United States, Britain, Japan, Germany, France, Sweden, and the Netherlands*. Piatkus.

Hampden-Turner, C., & Trompenaars, A. (1997). *Riding the waves of culture: Understanding cultural diversity in business*. Nicholas Brealey.

Haney, W. M., & Clarke, M. J. (2007). Cheating on tests: Prevalence, detection, and implications for online testing. In E. M. Anderman & T. B. Murdock (Eds.), *Psychology of academic cheating* (pp. 255–288). Academic Press.

Harper, R., Bretag, T., & Rundle, K. (2021). Detecting contract cheating: Examining the role of assessment type. *Higher Education Research and Development, 40*(2), 263–278. https://doi.org/10.1080/07294360.2020.1724899

Harpp, D. N., & Hogan, J. J. (1993). Crime in the classroom: Detection and prevention of cheating on multiple choice exams. *Journal of Chemical Education, 73*, 306–311.

Hendershott, A., Drinan, P. F., & Cross, M. (1999). Gender and academic integrity. *Journal of College Student Development, 40*, 345–354.

Henning, M. A., Ram, S., Malpas, P., Sisley, R., Thompson, A., & Hawken, S. J. (2014). Reasons for academic honesty and dishonesty with solutions: A study of pharmacy and medical students in New Zealand. *Journal of Medical Ethics, 40*(10), 702–709.

Hilbert, G. A. (1987). Academic fraud: Prevalence, practices, and reasons. *Journal of Professional Nursing, 3*(1), 39–45.

Hofstede, G. (1980). *Culture's consequences: International differences in work-related values*. Sage.

Hofstede, G. (1986). Cultural differences in teaching and learning. *International Journal of Intercultural Relations, 10*, 301–320. https://doi.org/10.1016/0147-1767(86)90015-5

Hofstede, G. (1991). *Cultures and organizations: Software of the mind*. McGraw-Hill.

Hrabak, M., Vujaklia, A., Vodopedivec, I., Hren, D., Marušić, M., & Marušić, A. (2004). Academic misconduct among medical students in a transition country. *Medical Education, 38*, 276–285. https://doi.org/10.1111/j.1365-2923.2004.01766.x

Jordan, A. (2001). College student cheating: The role of motivation, perceived norms, attitudes and knowledge of institutional policy. *Ethics and Behavior, 11*, 233–247. https://doi.org/10.1207/S15327019EB1103_3

Kanjirangat, V., & Gupta, D. (2016). Study on extrinsic text plagiarism detection techniques and tools. *The Journal of Engineering Science and Technology Review, 9*, 9–23.

Kerkvliet, J., & Sigmund, C. L. (1999). Can we control cheating in the classroom? *The Journal of Economic Education, 30*, 331–343. https://doi.org/10.1080/00220489909596090

Kuznetsov, M. P., Motrenko, A., Kuznetsova, R., & Strijov, V. V. (2016). Methods for intrinsic plagiarism detection and author diarization. In *CLEF (Working notes)* (pp. 912–919).

Lupton, R. A., Chapman, K. J., & WEISS, John E. (2000). International perspective: A crossnational exploration of business students' attitudes, perceptions, and tendencies toward academic dishonesty. *Journal of Education for Business, 75*(4), 231–235.

Lupton, R. A., & Chapman, K. J. (2002). Russian and American college students' attitudes, perceptions, and tendencies towards cheating. *Educational Research, 44*, 17–27. https://doi.org/10.1080/00131880110081080

Luthans, F., & Doh, J. P. (2012). *International management: Culture, strategy, and behavior* (8th ed., pp. 123–125). McGraw-Hill. ISBN 978-0-07-811257-7.

Magnus, J. R., Polterovich, V. M., Danilov, D. L., & Savvateev, A. V. (2002). Tolerance of cheating: An analysis across countries. *The Journal of Economic Education, 33*, 125–136. https://doi.org/10.1080/00220480209596462

Mahmud, S., & Bretag, T. (2013). Postgraduate research students and academic integrity: 'It's about good research training'. *Journal of Higher Education Policy and Management, 35*, 432–443.

McCabe, D. L. (2009). Academic dishonesty in nursing schools: An empirical investigation. *The Journal of Nursing Education, 48*, 614–623.

McCabe, D. L., Feghali, T., & Abdallah, H. (2008). Academic dishonesty in the Middle East: Individual and contextual factors. *Research in Higher Education, 49*, 451–467. https://doi.org/10.1007/s11162-008-9092-9

McCabe, D. L., & Treviño, L. K. (1996). What we know about cheating in college longitudinal trends and recent developments change. *The Magazine of Higher Learning, 28*, 28–33.

McCabe, D. L., Treviño, L. K., & Butterfield, K. (2002). Honor codes and other contextual influences on academic integrity: A replication and extension of modified honor code settings. *Research in Higher Education, 43*, 357–378. https://doi.org/10.1023/A:1014893102151

McLachlan, J. C., Illing, J., Rothwell, C., Margetts, J. K., Archer, J., & Shrewsbury, D. (2012). *Developing an evidence base for the Professional and Linguistic Assessments Board Test*. Commissioned by the General Medical Council. https://www.gmc-uk.org/-/media/documents/gmc-plab-revision_pdf-57909436.pdf

McManus, I. C., Chis, L., Fox, R., Waller, D., & Tang, P. (2014). Implementing statistical equating for MRCP (UK) parts 1 and 2. *BMC Medical Education, 14*, 204. https://doi.org/10.1186/1472-6920-14-204

Michaels, J. W., & Miethe, T. D. (1989). Applying theories of deviance to academic cheating. *Social Science Quarterly, 70*, 870–885.

Naghdipour, B., & Emeagwali, O. L. (2013). Students' justifications for academic dishonesty: Call for action. 2nd World Conference on Educational Technology Researches—WCETR2012. *Procedia—Social and Behavioral Sciences, 83*, 261–265. https://doi.org/10.1016/j.sbspro.2013.06.051

Nath, L., & Lovaglia, M. (2009). Cheating on multiple choice exams: Monitoring, assessment, and an optional assignment. *College Teaching, 57*, 3–8.

Newstead, S. E., Franklyn-Stokes, A., & Armstead, P. (1996). Individual differences in student cheating. *Journal of Educational Psychology, 88*, 229–241.

Nonis, S., & Swift, C. O. (2001). An examination of the relation between academic dishonesty and workplace dishonesty: A multicampus investigation. *The Journal of Education for Business, 77*, 69–78.

OECD Publishing, International Migration Outlook. (2015). Foreign students worldwide and in OECD countries, 2000-12: Millions. https://doi.org/10.1787/migr_outlook-2015-graph13-en

Ogilby, S. M. (1995). The ethics of academic behavior: Will it affect professional behavior? *Journal of Education for Business, 71*, 92–97.

Okoroafor, A. U., Henning, M. A., Chibuike, O. M., & Rajput, V. (2016). Disclosing academic dishonesty: Perspectives from Nigerian and New Zealand health professional students. *Ethics & Behavior, 26*(5), 431–447. https://doi.org/10.1080/10508422.2015.1055494

O'Rourke, J., Barnes, J., Deaton, A., Fulks, K., Ryan, K., & Rettinger, D. A. (2010). Imitation is the sincerest form of cheating: The influence of direct knowledge and attitudes on academic dishonesty. *Ethics & Behavior, 20*, 47–64. https://doi.org/10.1080/10508420903482616

Osipian, A. L. (2008). Political graft and education corruption in Ukraine: Compliance, collusion, and control. *Demokratizatsiya: The Journal of Post-Soviet Democratization, 16*, 323–344. https://doi.org/10.3200/DEMO.16.4.323-344

Ottaway, K., Murrant, C., & Ritchie, K. (2017). Cheating after the test: Who does it and how often? *Advances in Physiology Education, 41*(3), 368–374.

Papadakis, M. A., Arnold, G. K., Blank, L. L., Holmboe, E. S., & Lipner, R. S. (2008). Performance during internal medicine residency training and subsequent disciplinary action by state licensing boards. *Annals of Internal Medicine, 148*, 869–876.

Parsons, T. (1937/1968). *The structure of social action*. Free Press.

Parsons, T. (1951/1991). *The social system*. Routledge.

Poltorak, Y. (1995). Cheating behavior among students of four Moscow institutes. *Higher Education, 30*, 225–246. https://doi.org/10.1007/BF01384098

Rawwas, M. Y. A., Al-Khatib, J. A., & Vitell, S. J. (2004). Academic dishonesty: A cross-cultural comparison of U.S. and Chinese marketing students. *Journal of Marketing Education, 26*, 89–100. https://doi.org/10.1177/0273475303262354

Rosmansyah, Y., Hendarto, I., & Pratama, D. (2020). Impersonation attack-defense tree. *International Journal of Emerging Technologies in Learning (iJET), 15*, 239–246.

Rundle, K., Curtis, G. J., & Clare, J. (2019). Why students do not engage in contract cheating. *Frontiers in Psychology, 10*, 2229.

Ryan, G., Bonanno, H., Krass, I., Scouller, K., & Smith, L. (2009). Undergraduate and postgraduate pharmacy students' perceptions of plagiarism and academic honesty. *American Journal of Pharmaceutical Education, 73*, 1.

Saana, B. M. E., Ablordeppey, N. J. M., & Karikari, T. (2016). Academic dishonesty in higher education: Students' perceptions and involvement in an African institution. *BMC Research Notes, 9*, 234. https://doi.org/10.1186/s13104-016-2044-0

Sawdon, M., & McLachlan, J. C. (2020). '10% of your medical students will cause 90% of your problems': A prospective correlational study. *BMJ Open, 10*, e038472. https://doi.org/10.1136/bmjopen-2020-038472

Sims, R. L. (1993). The relationship between academic dishonesty and unethical business practices. *The Journal of Education for Business, 68*, 207–212.

Sindre, G., & Vegendla, A. (2015). E-exams versus paper exams: A comparative analysis of cheating-related security threats and countermeasures. *NISK Journal*, 34–45.

Stevens, G. E., & Stevens, F. W. (1987). Ethical inclinations of tomorrow's managers revisited: How and why students cheat. *Journal of Education for Business, 63*, 24–29.

Tang, S., & Zuo, J. (1997). Profile of college examination cheaters. *College Student Journal, 31*, 340–346.

Teixeira, A., & Rocha, M. (2006). A comparative analysis. *Research in Comparative and International Education, 1*, 198–209. https://doi.org/10.2304/rcie.2006.1.3.198

Teodorczuk, A., Fraser, J., & Rogers, G. D. (2018). Open book exams: A potential solution to the 'full curriculum'? *Medical Teacher, 40*, 529–530.

Theart, C. J., & Smit, I. (2012). The status of academic integrity amongst nursing students at a nursing education institution in the Western Cape. *Curationis, 35*, E1–E8.

Tippins, N. T. (2009). Where is the unproctored Internet testing train headed now? *Industrial and Organizational Psychology, 2*, 69–76.

Vitro, F. T., & Schoer, L. A. (1972). The effects of probability of test success, test importance, and risk of detection on the incidence of cheating. *Journal of School Psychology, 10*, 269–277.

Volpe, R., Davidson, L., & Bell, M. C. (2008). Faculty attitudes and behaviors concerning student cheating. *College Student Journal, 42*, 164–175.

Whitley, B. E. (1998). Factors associated with cheating among college students: A review. *Research in Higher Education, 39*, 235–274. https://doi.org/10.1023/A:1018724900565

Yudkowsky, R., Park, Y. S., & Downing, S. M. (Eds.). (2019). *Assessment in health professions education*. Routledge.

Zimmermann, S., Klusmann, D., & Hampe, W. (2016). Are exam questions known in advance? Using local dependence to detect cheating. *PLoS One, 11*(12), e0167545.

Part III
How Educational Technologies Transcend the Classroom

Chapter 11
Attendance Debate Part 1. Attendance and Performance: A New Landscape in the Era of Online Teaching

Louise Robson and Christine A. Kauffman

Abstract Attendance in class is essential for student success! How many times have we said this to our students, with the belief that we were right? However, is this really a valid statement? Is there actually evidence to support this, and is attendance at class the only marker for engagement that is at our disposal as educators? This chapter reviews what is known about attendance and student engagement. It argues that the central dogma of attendance being critical for student success is not always valid and that access to additional resources such as lecture capture can also be a predictor of success. We discuss the impact of lecture capture availability on attendance and performance and how this can provide a more inclusive learning environment, allowing students who cannot attend class the potential to succeed. The key message is that the value of in-class attendance or lecture capture as markers of engagement really depends on the type of course being delivered, the learning outcomes for that course, and how these are being assessed. This means that, as educators, it is critical that we evaluate the literature to make informed decisions on what are valid engagement markers for our own students. The chapter finishes with a consideration of what engagement might look like in the new normal of the COVID-19 pandemic.

This chapter and Chap. 12 in this volume are meant to be read together as part of a debate on the topic of whether lecture attendance is a determinant of student performance and learning in the era of lecture capture recording.

L. Robson (✉)
School of Biosciences, University of Sheffield, Sheffield, UK
e-mail: l.robson@sheffield.ac.uk

C. A. Kauffman
College of Medicine, University of Central Florida, Orlando, FL, USA

H. J. Witchel, M. W. Lee (eds.), *Technologies in Biomedical and Life Sciences Education*, Methods in Physiology, https://doi.org/10.1007/978-3-030-95633-2_11

11.1 Attendance in Class as a Predictor of Success

We all want to be able to know how well our actions will help us to achieve our goals, to look at an action or circumstance and say "if this is done that will happen" and know it for a certainty. Educators are no different. We want to be able to look at our students and know which ones will do well and which ones will do poorly. We want to define a series of traits and actions that will predict success. The students are in the same position. They want that same answer, the magic bullet that will give them the keys to success. Educational psychologists and sociologists have been working in learning labs to better define how we learn and retain new knowledge for the long term. Practicing educators have been working to study these same aspects in the real ecology of the active classroom. These researchers have confirmed for us things that we already believed to be true and surprised us with unsuspected new knowledge. The one thing that they have not been able to give us is that perfect pattern of behavior that will predict who will succeed and who will fail. We can look at a population of students and predict what proportion of them will succeed or fail. For example, college and graduate program admissions exams are fantastic at doing this. However, these types of metrics tell us things about populations and not necessarily individual students. No magic bullet yet.

Historically, both through research, anecdotal observation, and a pure belief in what we know is right, educators have been confident in one thing that we can tell our students that will allow them to succeed: attend class. Being present in the classroom was thought to be so vital for learning that it has become dogma. Attendance was considered a key aspect for a student to succeed. Our students have not always agreed with us, as evidenced by declining attendance in any session that did not require it. So, who is correct? The faculty member with years of experience in the content area that they teach as well as the education of students? Or the student who ignores advice to go their own path and yet somehow succeeds anyway? There is a lot of literature that would agree with the proverbial faculty member on this. However, there is also a growing body of evidence to suggest that the original dogma may no longer be as universally true as we have always thought.

Colleges, universities, and medical schools generally do not require attendance in lecture-based classes. This allows the young adult and adult learners in these settings to make their own decisions on their learning needs. Despite giving them this choice, educators have always made a strong recommendation for attendance. There is literature to back them up. Credé et al. (2010) published a large meta-analysis on attendance and performance that included over 50 published articles from 1927 to 2009 and over 20,000 students. Their findings suggested a strong correlation between attendance and performance ($r = 0.44$). While we all know that correlation does not equal causation, their finding supports this recommendation to students. They further subdivided their data to look at science versus nonscience-based classes. While the number of students in the science-based classes was smaller (just over 8500), it is still quite impressive and showed a slightly higher correlation with attendance and performance ($r = 0.49$) than the group as a whole (Credé et al.,

2010). Society has changed greatly from 1927 to 2009, most notably for education with the introduction of the Internet. In the study, they attempted to take this into account by looking at the size of the correlation between attendance and performance and the year that the study was published. They found no significant correlation. Their suggestion from this data was that the introduction of the Internet did not remove the positive association that attendance has on performance (Credé et al., 2010).

What was it about attendance that made it a marker of performance? The literature does not truly enlighten us on why. There is correlational evidence on the characteristics of those who choose to attend. Credé et al. (2010) included some of this in their meta-analysis. They found a weak positive correlation of attendance with attributes like conscientiousness ($r = 0.22$), high school GPA ($r = 0.16$) and academic motivation ($r = 0.10$), and a negative correlation with year in college ($r = -0.011$) in their general undergraduate population (Credé et al., 2010). If you look at students in the medical school population, Stegers-Jager et al. (2012) looked at year 1 and 2 medical students ($n = 600$) in the Netherlands using the Motivational Strategies for Learning Questionnaire (MSLQ) and self-reported attendance pattern. They found small positive correlations with lecture attendance and attributes like task value ($r = 0.26$), organization ($r = 0.25$), time management ($r = 0.27$), and effort regulation ($r = 0.31$) (Stegers-Jager et al., 2012). Also using the MSLQ, prospectively collected attendance was positively correlated with organization ($r = 0.34$), peer learning ($r = 0.31$), and help seeking ($r = 0.47$) and negatively with self-efficacy ($r = -0.33$) and effort regulation ($r = -0.33$) in second-year medical students in the USA (Kauffman et al., 2018). Boulton et al. (2019) surveyed first- and second-year university students in the UK and found that attendance correlated with social motivations ($r = 0.26$) to attend college (e.g., to socialize with friends). While these low-level correlations begin to give the picture of the person who chooses to attend, they do not necessarily tell us why the attendance itself is thought to actually lead to improved performance. One major thought is that attendance is a marker of the ever-elusive engagement.

11.1.1 What Is Engagement?

Student engagement is something that all educators are attempting to attain, but it is less easy to define and measure (see Chap. 4 in this volume on engagement). The most robust research on engagement seems to be in the primary and secondary school literature; however, even their definition of engagement is still at odds. In the *Handbook of Research on Student Engagement*, the editors summarized it well by saying "Although researchers have reached consensus that student engagement is multidimensional, agreement on the multidimensionality differs from agreement on the number and types of engagement dimensions" (Christenson et al., 2012). These dimensions included the relationships that students form with their teachers and peers as well as behaviors such as attendance, participation, and effort displayed in the classroom (Christenson et al., 2012). Their goal with this research is to get the students to graduate from high school and move on to postsecondary education. In higher education, our goals are obviously different, but the definition of engagement in this population is equally not well defined. Kahu (2013) reviewed the four major perspectives that are used to both define and research engagement at the college and university level: behavioral, psychological, sociocultural, and holistic. These different perspectives give wide-ranging ways to look at engagement in the higher education student from observable behaviors to self-reported cognitive and affective aspects, including participation. What these different perspectives that Kahu describes have in common is that they refer to global engagement in the university community as a whole and not just to engagement in the individual class. This global look at engagement is also seen in the medical education sphere as well. Peters et al. (2019) give us an article with a goal of giving us a practical approach to engagement rather than a theoretical one. The article gives twelve tips on how to improve engagement as outlined by an international group of medical education institutions who have been recognized for their strong student engagement. This more practical approach gives some more concrete examples for expanding engagement in the program as a whole but still does not define for us what engagement in the classroom is. While their focus was again on enhancing a student's global engagement, they also acknowledge that participation is a part of that engagement (Peters et al., 2019). Given that many of these constructs include "participation," it is no wonder that faculty members would perceive in-person class attendance to be the first step to demonstrating this engagement.

11.1.2 Is Our Understanding of Engagement Based on Our Own Experience as Students?

Attendance at in-class sessions may link into our own perception of the engaged learner. As educators, do our own learning experiences shape our approach to how current students learn and what engagement looks like? Do we think that because this was how we engaged that it is the only way for our current students to do so? Or as educators do we want to share the learning experience of our students in the classroom, exploring knowledge and understanding together? Whatever the reasons are, given the changing face of how we deliver teaching, is it valid to insist on classroom attendance for all? We would argue that a change is needed in our understanding of what an engaged learner looks like, moving away from the attendance-in-class centric viewpoint, to one that draws on a range of resources. The big question is, of course, which resources should be used as indicators for student engagement?

11.2 Attendance and Performance in Medical Education in the Current Era

Context is everything. The Credé et al. (2010) meta-analysis data was very compelling and suggested that faculty were correct all along to recommend attendance as a key to success. However, these students were primarily general undergraduate college students mostly working before the era of the Internet. While the authors attempted to take this into account, their data was collected from articles published more than 10 years ago now. While the Internet was in full swing from 2000 to 2010, things have changed even further since then. As described later, the expansion of lecture capture in higher education is one of the facets that no longer make attendance a strong predictor of performance.

One group where this is true is in medical education. In the USA, it is a common practice to use lecture capture in the preclinical curriculum. We start with an example from the research from one of the authors in this chapter (Kauffman) who looked at attendance and performance in second-year preclinical medical students at the University of Central Florida College of Medicine during a gastrointestinal and renal systems course (Kauffman et al., 2018). In this program, attendance at lecture-based sessions was voluntary, and the sessions were all recorded and available to the students afterward. The course did have a subset of mandatory sessions that were either team-based learning or small-group case-based, but these were a significant minority of the available sessions. Attendance was prospectively collected, and performance was assessed based on the score on the MCQ final examination. There was no correlation between attendance and performance ($r = -0.20$; $p = 0.09$) (Kauffman et al., 2018). In addition, attendance was quite low with the mean attendance being only 21.9% of the sessions, and one-third of the participants did not attend a single session that they weren't required to (zero attendance). This was initially a surprising finding; when the group with zero attendance ($n = 26$) was compared to the remainder of the class that had any level of attendance ($n = 52$), the zero attendance group performed better (mean 84.7% v 82.9%; $p = 0.02$). The greater than zero attendance group of students had lower MCAT scores (mean percentile 78.7 v 84.2, $p = 0.05$) suggesting that attendance did not hurt them per se (Kauffman et al., 2018). This same group was followed after the course was over and their attendance pattern was compared with their performance on Step 1 of the United States Medical Licensing Examination (USMLE). Step 1 is a multiple-choice knowledge-based assessment which covers all of the basic science content from the first 2 preclinical years and was taken approximately 4 months after the course was complete. There was no correlation with attendance and performance on Step 1 ($r = -0.15$; $p = 0.21$), though there was a very prominent correlation with course performance and Step 1 ($r = 0.813$; $p < 0.001$) (Kauffman et al., 2019).

Eisen et al. (2015) made similar findings in a required dermatology course at the University of California—Davis School of Medicine. This was a lecture-based course where all of the materials, including recordings of the sessions, were available to the students online. Attendance was collected prospectively and compared to the final MCQ examination score. Class attendance was similarly low with an average of only 38.7%. They found no correlation between attendance and performance in a univariate analysis for the whole group ($n = 101$, $p = 0.77$) and in a multivariate analysis that controlled for age, gender, MCAT score, number of children, and reason for attending class ($n = 62$, $p = 0.54$). The mean performance on the exam for the whole group was 86%, and the authors state that this is similar to the other courses at their program (Eisen et al., 2015). These two courses had very similar contexts despite being at different institutions. The students had access to the material regardless of attendance, and the assessments were knowledge-based MCQ examinations.

What if the context is changed? Instead of the materials being available, what if they were only available if the student attended class? Instead of knowledge-based MCQ assessments, what if performance-based assessments are used instead? Deane and Murphy (2013) published a study with just such a context. They looked at fourth-year medical students on an obstetrics and gynecology clinical rotation in Dublin, Ireland. The course included both clinical duties and didactic tutorials. Attendance was not required but was strongly encouraged and over all very high (89% overall, 89% clinical, and 90% tutorial). The didactic materials were only available as hard copies if you attended the session. Assessments were multifaceted, and the study looked at the combined performance on a MCQ examination, an Observed Structured Clinical Examination (OSCE), short answer, and an oral examination. Overall attendance positively correlated with performance on the combined examination ($r = 0.48$, $p < 0.001$) even when adjusted for age, sex, country of origin, previous examination failure, and the timing of the rotation within the academic year. The same held true for both clinical attendance ($r = 0.44$, $p < 0.001$) and tutorial attendance ($r = 0.42$, $p < 0.001$) when adjusted for the same factors (Deane & Murphy, 2013). Schillerstrom and Lutz (2013) had a similar context. They looked at third-year psychiatry clerkship students over 2 academic years at the University of Texas Health Science Center at San Antonio. The course was made up of primarily clinical experience with supplementary weekly didactic sessions. The course rule was that students could have up to 3 excused absences, and of the 433 students in the study, only 1 student had more than 3 absences leaving overall attendance very high. They found higher performance on the National Board of Medical Examiners Subject Examination in those with no absences vs those with any number (80 versus 77.8; $p = 0.006$) (Schillerstrom & Lutz, 2013).

Again, context is important. The things that seem to differentiate these two groups of studies is whether or not the materials were available to the students without attending the sessions and if performance-based assessments were a part of the construct. The studies in this example that have a purer knowledge-based assessment as the final goal do not show that attendance is necessary for performance. The construct being assessed in these MCQ assessments is knowledge, and the method of dissemination of that knowledge is through lectures and reading assignments. Knowledge is an important construct to assess, and well-developed MCQ examinations are very successful at doing this in a reproducible and standardized way. However, the students do not need to consume that content in the presence of a faculty member if all they are going to be asked to do is regurgitate that knowledge in an MCQ examination. Another part of this context is that we are talking about medical students. These are highly motivated and high-performing subset of students who are much more likely to maintain a study schedule without attending lecture-based classes in person.

For contrast, let's look at another study that did show a correlation of attendance and performance. Al Khaja et al. (2019) looked at attendance at lecture-based resource sessions during a PBL-based pharmacology course at the College of Medicine and Medical Sciences at Arabian Gulf University. This included over 1400 medical students across years 2 through 4 of the preclinical curriculum. They found a range of moderate correlations depending on which portion of the curriculum they looked at ($r = 0.264–0.493$, $p < 0.001$). However, they also showed significantly diminishing attendance, beginning at 78.7% at the start of 2nd year and ending at 22.1% by the end of the 4th year (Al Khaja et al., 2019). While these findings would clearly be recognized as statistically significant, with only 22% attendance how much of a truly meaningful difference can there be? One of the most striking findings in the studies that did not find a correlation with attendance and performance was the overall very low levels of attendance (21.9% and 38.7%) with overall good performance of the cohort as a whole (Eisen et al., 2015; Kauffman et al., 2018). Even with the concession that the study could have missed a real effect by chance due to the size of the study populations, these studies showed overall good performance of the cohorts and the students passing the course. The vast majority of these students are not attending the lecture-based sessions in person, yet there is not a rash of failures or significant decrease in total class performance.

In years past, medical students needed to attend the sessions to get the content. Admittedly, this is a bit of an overstatement. The content was available in the form of large reference books that could be difficult to assimilate on their own. The lecture-based curriculum available to the students framed the level of content needed for the course. Lecture attendance was higher, and it would make sense that attendance would correlate with performance. This is no longer really the case. The lecture content can be available at home through lecture capture and streaming technology, and the availability of digestible third-party resources has exploded.

The content needed for medical education is expansive, but it is reasonably well defined. In the USA, no matter what program that a student attends, they all need to prepare for the USMLE at the end of the preclinical years of the curriculum. This exam is difficult and has been used to rank the students for the highly competitive next stage in physician training (NRMP, 2018). While the individual universities may have their own goals and objectives for their students that influence their curriculum, they cannot escape the fact that they need to help prepare the students to have the medical knowledge to do well on this exam. The consistency of the content for all medical students in the USA and the need to obtain a good score on the national licensing exam have led to an explosion of third-party resources for this purpose. That means that these third-party resources marketed to prepare students for the USMLE can also be useful for their course studying as well. These resources come in many formats including question banks, spaced repetition flashcard programs, pictorial mnemonics, and pre-recorded lectures to name a few. Studies have shown that the use of these resources during the academic year is extensive. Burk-Rafel et al. (2017) surveyed two classes of preclinical students at the University of Michigan about their study habits in preparation for the USMLE Step 1. Of the 274 respondents, 77% began studying during their second year before the dedicated study period for the exam began. Greater than 99% of students used the *First Aid* Review Book and *UWorld* question bank with 90% using the *Pathoma* lecture series. The use of the practice questions was impressive with students reporting completing an average of 3597 (SD = 1611) questions (Burk-Rafel et al., 2017). Kauffman et al. (2019) showed that >80% of their study population at the University of Central Florida used the same four resources during the second-year preclinical course: *UWorld*, *Pathoma*, *First Aid*, and *Sketchy*.

One of the arguments for attendance in a lecture-based curriculum is the need to guide the students on how to apply the content rather than just regurgitate the information back. Much effort has been placed on developing second- and third-order multiple-choice questions that assess this application of knowledge. However, as Burk-Rafel et al. (2017) pointed out, these students are completing thousands of unique MCQs as a part of their study practice. How does completing this massive number of practice MCQs affect how the students interact with ones we use for assessment? Can they change what was meant to be an application question to just a recognition question? There are only so many ways to describe a scenario for a student in an MCQ. We will use heart failure as an example. Whether I want to ask a question about physiology, pathophysiology, or clinical treatment, I still need to succinctly describe the patient in the vignette. If the students have gone through enough of these practice questions, then they may have become intimately aware of all of those ways to describe heart failure in an MCQ. Would my MCQ then become less of a piecing together of clues in the vignette requiring reasoning to figure out and more of a quick recognition question? While this question is outside the scope of this chapter (but see Chap. 5 in this volume), it is food for thought on how these outside resources have changed what is going on in our classroom.

11.3 Lecture Capture: A Marker of Engagement?

One of the supplementary resources that has increased in use over the last few years is lecture capture, delivered by third-party providers such as Echo360 and Panopto. Lecture captures are recordings of live in-class lectures and typically record the presentation slides and audio from a session, although some also record a video of the person delivering the session. While lecture captures have been around for several years, their use has massively expanded over the last few years, with more and more universities adopting an institutional approach to providing lecture captures for their students. At least some of this increase has been driven by student demand, particularly around supporting students with additional learning support needs.

An additional use of lecture captures is in flipped approaches, where students are provided with lecture captures in advance of an in-class session that supports understanding and provides an opportunity to problem solve (see Chap. 6 in this volume). This flipped approach has been widely utilized during the COVID-19 pandemic, with captures replacing in-class didactic teaching due to social distancing impacting on the ability to deliver sessions in attendance to large groups. This is something we will come back to at the end of the chapter, when we reflect on the current situation in terms of the value of didactic lecture delivery and lecture capture moving forward.

11.3.1 The Problem of Researching the Effects of Lecture Captures

Lecture capture certainly generates fierce debate within the higher education sector, with educators holding strong views on the value (or not) of captures being available (Dommett et al., 2020; Morris et al., 2019; Nordmann et al., 2019; Owston et al., 2011). Looking at the literature, what is clear is that different studies find differing impacts on attendance and performance. One of the reasons for this is that it is impossible to dissect out the impact of a single factor on attendance and performance. As scientists, we would want to design our pedagogical research in a similar way to our scientific research, with matched control and test groups, with large sample sizes, and having the availability of lecture capture as the only variable. The reality is somewhat different. Ethically and morally, we cannot place students into different groups, give them access to different resources, and then observe the impact on summative assessments. One way round this would be to investigate the impact of lecture capture on formative aspects of assessment. However, engagement with formative assessments varies across students, and this would tend to skew any findings. In addition, matching groups is almost impossible. Recording attendance at lectures when you normally do not do so can also skew your findings, with students who would normally not attend now attending and influencing the findings. There is therefore a real challenge when considering the outcome of any pedagogical research, and it is essential as scholars that we recognize this in our work. These challenges are some of the reasons why the literature on the impact of lecture capture is very mixed. Other reasons include the different cohorts that are studied, e.g., discipline and year of study. For example, Nordmann et al. (2019) shows that there are differences in terms of the impact of lecture capture depending on the year of study of undergraduate students, indicating that years of study in higher education is an important consideration when evaluating lecture capture research. The study by Morris et al. (2019) showed different patterns of use of captures across different disciplines. As we evaluate the published literature on the impact of lecture capture, it is critical that we understand the limitations of these studies and recognize that there is no consensus on whether lecture capture is valuable to student learning (or not), because there is always going to be the caveat that it depends on the specific situation.

11.3.2 Lecture Capture and Attendance

One of the common arguments put forward against the use of lecture capture is the impact on student attendance at lectures (Biggs et al., 2001; Leadbeater et al., 2013; Morris et al., 2019; Nordmann et al., 2019; Traphagan et al., 2010; Wiese & Newton, 2013). Educators often raise concerns about student attendance falling, and, as already discussed, implicit in this is a worry about student engagement. There are also some anecdotal concerns that the social experience of the live lecture will be different if many students choose not to attend. What is clear from the literature is that there is a mixed picture in terms of the impact on attendance. One of the largest studies investigating the impact of lecture capture was carried out by Morris et al. (2019). Using data from across the institution, they showed that there was an impact on attendance, with median attendance significantly lower in lectures with a capture than those without a capture (81% versus 85.7%, with and without lecture captures, respectively). This difference was observed across the majority of the different disciplines, including the medical, healthcare, and STEM programs (science, technology, engineering mathematics). Interestingly, the data looking at different years groups in this study reinforces the findings from Nordmann et al. (2019), with year 1 undergraduate students showing no impact on attendance from lecture capture, while later undergraduate years and also postgraduate level students showing a drop in attendance when captures were available. Leadbeater et al. (2013) observed a 15% drop in attendance of students in year 2 of a medical science program, while in the USA there was a 9% drop in attendance at geology lectures (Traphagan et al., 2010). Another study in Canada (Wiese & Newton, 2013) reviewed the impact on attendance in a third-year undergraduate biological science cohort. They asked students to indicate their perception of the impact of lecture capture availability on attendance and observed that the majority of students (84%) reported no change. Only 2% and 1% of students indicated a drop in attendance, of 25% and 50%, respectively. However, in this study they also evaluated whether different student groups within a cohort were more likely to not attend. Using the revised two-factor study process questionnaire 2 approach (Biggs et al., 2001), they divided the students into groups that took either a deep or a surface approach to learning. Those students who were identified as surface learners from the questionnaire were more likely to state they would attend fewer lectures if lecture captures were available. This study raises another important point when considering the impact of lecture capture: a cohort of students on the same course in the same year will act differently, and it is important that this is recognized in our research evaluation. In contrast to these studies, others show no or limited impact on attendance when lecture capture is available. Two different reviews of the published literature found no evidence for a universal reduction in attendance (Nordmann & McGeorge, 2018; Pursel & Fang, 2012). In the Pursel and Fang paper, only 20% out of 47 studies actually showed a decrease in attendance. Overall, the published literature suggests there is only a small impact on attendance.

The educational climate has changed dramatically over the last 20 years, and there are now many factors that may impact on students' ability to attend class (Dommett et al., 2020; MacKay, 2019). In the Dommett et al. (2020) study, key factors impacting on student attendance at lectures were the availability of online lecture resources (including lecture capture) and difficulty of attendance (cost of commuting, scheduling). Similar reasons were reported by MacKay (2019). In the UK, for example, more students work to support their studies than previously, reducing the flexibility a student has. If they have a shift at work that clashes with a timetabled class and that work is needed for them to pay their bills, etc., then it is understandable that they would choose to miss class and use a lecture capture to catch up afterward. Financial considerations are also going to play a role in how a student behaves. For example, a student who commutes to university may choose to miss class if this means they are travelling at peak times, which is more expensive. Similarly, a student may opt to miss class if they commute and only have a single class in a day. While students report several reasons why they might not attend class, a survey of educators highlighted that they felt the availability of online resources was key in determining attendance (Dommett et al., 2019). There is therefore something of a mismatch between why students do not attend and why educators think they do not.

As educators it is time that we recognize the additional demands on students and that there are genuine reasons why a student may miss class. Rather than insist on attendance at everything, we need to evaluate what aspects absolutely require in-class attendance and accept that that a student may not be able to attend everything.

11.3.3 Impact of Lecture Capture Use on Performance

Another factor that is raised relates to captures impacting negatively on student learning and performance (Edwards & Clinton, 2019; Owston et al., 2011). Some educators suggest that captures lead to a surface approach to learning and that this can have a negative impact on student performance in assessments. An additional concern is a lack of opportunity to ask questions at the time the lecture is delivered, which may lead to misunderstanding of the capture content and a false perception of mastery. However, this is only an issue with lecture capture if the student would have asked a question in the live lecture and the vast majority of students don't. It is also important that we recognize that students develop a range of different skills in their programs, and while making lecture captures of some teaching sessions might be appropriate, it is not a valid replacement for attendance at more practical-based sessions, e.g., clinical and practical skills development sessions.

Students themselves certainly feel that having access to captures enhances or strongly contributes to their learning. Typically, studies have shown between 53 and 95% of students, when asked, indicated they thought captures had a positive impact (Danielson et al., 2014; Robson, 2020; Traphagan et al., 2010; von Konsky et al., 2009). Thematic analyses of student comments also show that students are very positive about lecture capture (Dommett et al., 2019; MacKay, 2019). However, is student perception of the impact on their learning actually translated into improved performance? As we consider the evidence around this area, we need to remember that this type of research is complicated by the fact that there is always more than one variable within a given study. Learning is not limited to in-class attendance nor watching lecture captures. There are many other resources that students can draw on, e.g., textbooks, research papers, and peer-learning activities, complicating our interpretation of studies looking at a single factor on performance.

Research looking at the impact of lecture captures on student performance is, like the attendance research, a mixed picture, with some data showing no impact and others both positive and negative impacts. This is likely to reflect the fact that the studies have been carried out on a range of different disciplines and cohorts and that there will be variation within a cohort. In addition, just like the evaluation of the value of in-class attendance on student engagement and success, it is also important to consider both the type of teaching being captured and the assessment methodology, i.e., testing of knowledge, understanding versus application and critical analysis, problem-solving, and clinical and practical skills.

There are a range of studies showing no correlation between lecture capture use and performance. The performance of medical laboratory science degree program students in multiple-choice questions (MCQs) and short answer questions (SAQs) was not correlated with attendance or lecture capture availability (Doggrell, 2020). In the study by von Konsky et al. (2009), no significant difference was observed in lecture capture use between failing and passing third-year engineering students. It is unclear in this study what type of examination assessment was being used; therefore, it is not possible to draw conclusions based on whether it was knowledge or higher-level skills that were being tested. Interestingly, a greater variation in the passing students use of captures was observed in comparison to the failing students, i.e., more failing students had low use, whereas in the passing groups, some students had high use, while others low. This aligns with the findings from Robson (2019), where some high-achieving students didn't view many captures, while in the fail student group, almost all students had low numbers of capture views. Postgraduate medical school studies also show no correlation with lecture capture views and performance in the MCQ-based final assessment, nor with the United States Medical Licensing Examination Step 1 (Kauffman et al., 2019).

In contrast, there was some correlation in a study by Nordmann et al. (2019). When all 4 years of the undergraduate cohort were combined as one group, there was also no correlation between lecture capture use and assessment performance. However, when the data were divided into different year groups, then some correlation was observed. In the first year, a three-way positive correlation was observed between lecture attendance, recording use, and grade point average (GPA) from

MCQ (recall)-based assessment. In later years, no correlation with recordings use and performance was observed. These later assessments were essays that tested more than just knowledge/recall but also assessed skills around critical analysis and constructing coherent arguments. This suggests that higher-level skills are not positively influenced by the availability of lecture captures, but importantly they also do not appear to be negatively impacted. An uplift in performance on an MCQ-assessed course in Canada was also observed after the introduction of lecture capture, with mean cohort grades increasing by 5% (Wiese & Newton, 2013). These data are similar to findings from Robson (2019), where actual views of lecture captures taken from analytics data were positively correlated with academic performance in an academic year long, year 2 bioscience module. The more lecture captures a student had watched, the more likely they were to achieve a higher grade in the examination assessments. In the Robson study, correlations were observed with both MCQ and also SAQ /essay-based assessments. Having said that, there were a small number of high-achieving students who did not use the lecture captures, suggesting that these students were confident enough in their skills and ability to study that they didn't need the captures to succeed. Of importance, this study actually mapped the longitudinal use of lecture captures across the academic year. Those students who accessed lecture captures early in the academic year were more likely to have higher views, space their learning across the teaching weeks, and do well in their assessments. In contrast, those who had few views early in the academic year tended to have lower views overall and show sharp increases in views just before the two examination periods, consistent with a massed learning approach. It is therefore unsurprising that low views were correlated with low performance, given that a massed learning approach is known to be ineffective. In contrast, lecture capture views in a third (final) year course were not correlated with academic performance in an open book, data handling, and interpretation assessment, which required students to apply their knowledge and understanding to solve a novel problem. Interestingly, in another final-year module, an invigilated, examination-based essay and data interpretation assessment was associated with an increase in course performance, when mapped across several cohorts. Comparing 2 years before the introduction of lecture capture and 2 years afterward (with minimal changes in topics taught, lecturers and assessment method), there was a significant increase in the number of students achieving the two top awards (firsts and upper second). Before the introduction of lecture capture, 50% of students attained the two highest classifications, while after lecture capture was introduced this rose to 65% of students. The assessment required students to demonstrate knowledge and understanding of topics, show an ability to analyze experimental data, and demonstrate critical analysis and synthesis, i.e., it was not simply recall of information. Assessments were set by the same two individuals, who had years of experience, and were reviewed by external examiners to align difficulty levels as much as possible. While we cannot rule out other factors that have contributed, from the teaching perspective, the big change was the introduction of lecture capture.

In contrast, Owston et al. (2011) showed that the student-declared rate of access to lecture captures was inversely associated with grades achieved in assessments in freshman classes. They found that students who declared that they accessed the recordings less often (once a month or less) achieved higher grades than those who accessed the recordings four or more times a week, or higher. The student grades in the study were drawn from MCQ examinations, midterm assessments (type not stated), and assignments (coursework). However, the multiple types of assessment may mask a correlation to knowledge-based assessments such as MCQs. It is also important to note that the use of lecture capture data were also self-declared, and it is possible that higher achieving students may have picked up on cues around lecture captures having a negative impact, and this influenced their declaration. A negative impact on exam performance was also reported by Edwards and Clinton (2019), although they also observed that term time views were positively correlated with coursework grades. Selvig et al. (2015) also saw that students who watched lecture captures were less likely to do well in a histology assessment.

These studies reinforce the point that the type of learning outcomes and assessment method is important when considering the impact of lecture capture. Overall, it is not possible to say that lecture capture does, or does not, impact on student assessment performance. There are multiple factors that need to be taken into consideration, including the type of assessment, the year of study of the students, if the course is a knowledge-based one or more application, and whether there are practical skills aspects that need to be assessed. It is also important to consider that a student cohort on a course is diverse, and they will not use the captures in the same way, on the same course, in the same year. It is clear that strong students may not need to access lecture captures to succeed but also that weaker students often use captures in an inappropriate way (e.g., binge-watching before examinations). Therefore, we would strongly advise educators to ensure that more guided support on how to use captures is provided, particularly (but not exclusively) to support these weaker students. In addition, while there is minimal evidence for a positive impact of lecture captures on performance, we would argue that the benefit of lecture capture in terms of leveling the playing field for those students who have specific learning support needs means that it is something that should be used whenever it is appropriate to do so.

11.3.4 Lecture Capture: A Marker for Engagement?

So, is watching lecture captures a marker for engagement? We would argue that yes, student use of lecture capture can be a marker for one kind of student engagement. Students who are using lecture captures are engaging with a resource that is available, just like attending lectures, accessing presentation slides, completing formative quizzes, and participating in online discussion forums. This engagement has been measured in multiple ways including watch time, rewatches, and click-through rates. These metrics allow us to know that the student is accessing the materials, just like attendance in the live lecture allows us to know that the student is exposed to that experience. However, none of these metrics, including attendance at the live lecture, actually tell us what the student does with the content.

On balance, the evidence in the literature, and from our own experiences, has illustrated that for the majority of students, having the lecture captures available is certainly of benefit to how they feel. Students value the ability to check something that was said in class, they state that they are able to produce more detailed sets of notes around the aspects that were covered, and they have reduced anxiety levels. There are significant benefits to those students who have learning support needs such as dyslexia, with lecture capture contributing to an inclusive campus where all students have the opportunity to succeed. While having captures available is positive from a student perspective, it is less clear whether the captures have a positive impact on performance. There is minimal evidence in the literature that there is a negative impact, however, and given the positive impact on campus inclusivity, we would argue that for the majority of students having lecture captures is important.

However, we also recognize that the value of lecture captures is not "one size fits all," both in terms of type of course and for some learners. In courses that have a high clinical or practical component, attendance at sessions is essential, and lecture captures cannot replace these live sessions. In this context, captures (of targeted pre-recorded material) can be beneficial to help prepare students for practical/skills classes, i.e., using a flipped approach to ensure students get maximum benefit from the live class session. In addition, no matter how much guidance we provide on how to space their learning and use captures effectively, a small minority of students will choose to ignore this advice. Remember, undergraduate and postgraduate level study requires students to take responsibility for their learning strategies and time management. We can provide the information they need, but some will always choose to take their own path.

Overall, we would argue that for some lectures (and in some cases active learning using online tools), the availability of lecture capture has a positive impact, and capture views can be used as a measure of engagement. For sessions teaching more practical or clinical skills, we would suggest lecture captures alone are not going to be sufficient to allow students to meet the learning outcomes, and in-class attendance is required.

11.4 How Do Students Use Lecture Captures?

Lecture captures allow students to revisit teaching sessions, particularly to catch up on something they may have missed, or didn't understand (Dommett et al., 2019). The captures reduce the pressure on students to make sure they write everything down that was key; if they miss something, they can always go back and relisten to the lecture. Of course, this must be balanced against developing the skill of taking notes on key points, something that many graduates will be doing in their future careers, and something educators have also raised (MacKay, 2019). The aspect of

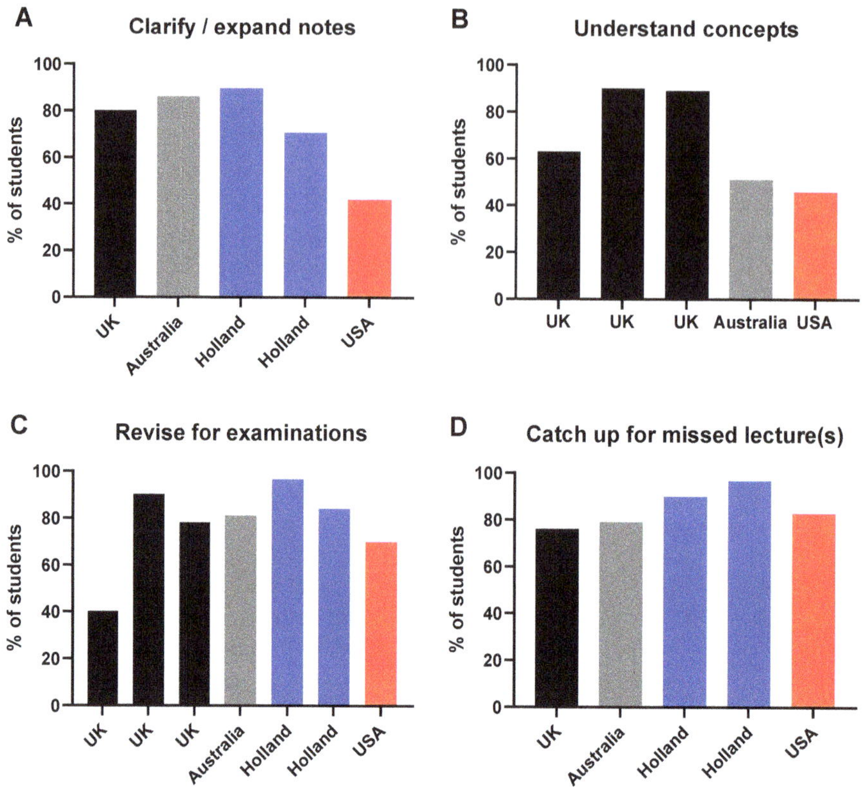

Fig. 11.1 Student-declared use of lecture captures. % of students from a number of individual studies selecting the option: (**a**) Clarifying/expanding their notes. (**b**) Using the lecture captures to understand concepts. (**c**) Using the lecture captures to revise for examinations. (**d**) Using the lecture captures to catch up after missing a lecture(s). The color coding indicates studies carried out in different countries. Data were extracted from Doggrell (2020); Dommett et al. (2020); Gorissen et al. (2012); Leadbeater et al. (2013); Robson (2016); Traphagan et al. (2010)

noting down the key points from a lecture is also part of the learning process itself (Kiewra, 1989), highlighting the importance of students developing this skill.

A number of studies have asked students to report how they use lecture captures to support their learning, and the findings across these studies are actually very similar, suggesting that students across a range of global institutions tend to use the captures in the same way, Fig. 11.1. The main reasons were to (a) clarify or expand their notes from the lecture, (b) help them understand concepts, (c) revise for the examinations, and (d) catch up on a missed lecture. It is not surprising that a and b are key reasons for using lecture captures, as it is well-known that note-taking in class is a skill that needs to be developed over time (van der Meer, 2012). In addition, using captures to support reviewing material for examinations is also not surprising, although when the captures are used for preparation for examinations is important. Finally, if a student misses a lecture, then using the lecture capture is, of course, an easier way of catching up than trying to use peer lecture notes, as they can see the context of the slide content from the audio track.

One of the interesting findings around the availability of lecture capture is the impact on student anxiety. A number of studies have shown a reduction in anxiety when captures are available (Dommett et al., 2019; Robson, 2016; Traphagan et al., 2010). This is presumably because, if a student misses something in the lecture, or the lecture itself, they have an opportunity to "go back in time" to rewatch the section or lecture they missed. The importance of this cannot be overestimated, because there are a number of reasons that may mean a student cannot attend a lecture, e.g., student illness. If a student has an ongoing issue that impacts on their ability to come to class, then having the lecture captures means they are going to be less disadvantaged in terms of their learning, and importantly captures therefore make university a more inclusive place to learn:

> I suffer from .... and had a serious flare and missed some lectures. In the past I've suffered from anxiety over missed lectures or forced myself to walk to lectures making my condition worse. Having lecture capture meant I could stay at home and recover properly without missing out on information and it's been a major relief—undergraduate student.

The positive impact on inclusivity is never more evident when considering the critical role lecture captures play in student support. Some students with specific learning disabilities struggle to assimilate and write down information in a live didactic lecture (Mortimore & Crozier, 2006). Studies looking at the impact of lecture capture have also raised this, with students with dyslexia more likely to report issues with note-taking. A study by Nightingale et al. (2019), looking at biological science students, observed that 74% of the dyslexic participants agreed that they struggled to take notes, compared with only 12% of the non-dyslexic group. These students are therefore at a disadvantage in their studies compared to their peers. Having access to recordings of the lectures means they can go back to relisten and rewatch the parts they missed, or didn't understand, and this provides key support for their learning. The value of such support should not be underestimated. Anecdotally, staff who provide support for students with learning differences have indicated that without lecture capture some of their students would have dropped out of university. In the Nightingale et al. (2019) study, they reported that there was no significant difference in the performance of the students with versus without specific learning differences, after use of either textbooks or lecture captures as supplementary learning material. However, it is not clear from this study whether any differences had been observed previously, in the absence of lecture captures, i.e., had the availability of the lecture captures made a difference. There was a "no revision" group, but the number of dyslexic participants was too small to provide meaningful information. In this study, it was clear that dyslexic students reported more issues with lectures and taking notes and that they appeared to be using more online resources to get round these issues. Therefore, while there is no direct evidence from this study that lecture capture has a positive impact on the performance of dyslexic students, it is clear that it forms part of a suite of resources that can provide additional support:

> I have dyslexia and chronic fatigue, if it wasn't for lecture capture, I would have had to drop out of university—undergraduate student.

It should also be recognized that there are wider benefits to lecture captures being available to everyone, not just those with a specific learning disability. For example, not all students who have specific learning differences are diagnosed before they attend university. Many are only subsequently diagnosed once they arrive and sometimes not even until their final year. In addition, many students who do not have specific learning disabilities also benefit. In a large institutional wide study in the UK, 58% of students indicated that they had some form of note-taking difficulties (Dommett et al., 2020). These students can therefore also benefit from being able to revisit lectures through captures. As mentioned earlier, students also often have many demands on their time. Some have to work, so that they can support their studies, and if this clashes with timetabled lectures, then there is often not the flexibility to allow attendance in class. Others have caring responsibilities or commute to study (incurring financial and time costs), all of which impact on a students' ability to be in class. A key point that we need to consider as educators is that not all non-attending students choose to miss class; many have to miss class for reasons that lie outside of their control. For these students having access to captures ensures they are not disadvantaged and as stated earlier highlights the role captures have to play in providing an inclusive learning environment for students.

Irrespective of why lecture captures are used by students, it is critical that they are supported in how to use captures effectively (Nordmann, Küepper-Tetzel, et al., 2020). Simply providing lecture captures and assuming students know how to use them is insufficient. We must also guide them on the way to use the captures to support their learning. Indeed, unless we support students with this, lecture captures could add to the workload for students and thereby impact negatively on their learning and achievements.

The key support messages for students when the captures are a supplement to in attendance, or are being used in a flipped approach, are actually very similar. For both the important aspect is for students to use the captures as part of their self-regulated learning activities.

For use as a supplement after an in attendance session students are advised to (Nordmann, Küepper-Tetzel, et al., 2020):

1. Attend classes.
2. Take effective notes in the lectures.
3. Afterward, only dip into those parts of the capture you missed, or didn't understand.
4. Watch those parts at normal speed, so you don't miss any key points.
5. Review the capture within a couple of days after the in-class session, after they have consulted their notes from class and identified what aspects need additional work.
6. Avoid watching the whole lecture capture again, unless you missed the live lecture.
7. Use the recordings to identify where you need additional help, and ask for it.

For flipped lecture captures, the advice is the same, except that we would advise students to use the captures like a live lecture, i.e., press start and run to the end without stopping, taking effective notes (point 2 above). Then go back into the parts that were missed or not understood (point 3 onward).

While many institutions host guidance for lecture capture use on their generic student support pages, this is probably insufficient as it stands to provide the level of support that is needed by students. Indeed, we would suggest that institutions need to take an overarching approach to supporting students, embedding support across all aspects of the student experience. Joining up the messages from academic departments, central support services, technical support teams, dyslexic and disability support, and student communications ensures students are getting the message about how to use lecture captures as they participate in a range of activities across campus. Such an approach was implemented at the University of Sheffield in the 2018/2019 academic year. The student-facing guides from Nordmann et al. (2018; Nordmann, Küepper-Tetzel, et al., 2020) formed the framework to deliver the key messages (as listed above) to students. The approach involved:

1. Providing resources to departments to use in handbooks and webpages.
2. Display of the key guidance on digital displays across the campus.
3. An engaging video reinforcing points, released on university social media (University of Sheffield, 2019).
4. Work with the generic study skills and dyslexia and disability services to embed guidance into training sessions and support services.
5. Guidance displayed the point of access of lecture captures on the LMS.
6. Guidance also embedded into the library digital literacy tools.

The initial plan was to evaluate the impact of this approach after 2 years of implementation, although the COVID-19 pandemic has meant that plans around an evaluation will have had to be altered, as students have been using lecture captures in a flipped way since March 2020.

11.4.1 Staff Perceptions and the Impact on the Type of Teaching

While lecture capture is valued by students, some educators are often wary, questioning its impact on a range of learning and teaching aspects. A study by Dommett et al. (2020) compared both student and educator views of lecture capture. 84% of students had positive views, compared to only 50% of educators. A number of studies, which span a range of institutions, have collated educator comments around lecture capture, and these typically show the same set of concerns (Dommett et al., 2019; MacKay, 2019; Morris et al., 2019). These main concerns are:

1. Student attendance will decrease.
2. Students will start to see lectures as simply for learning facts rather than developing higher-level understanding and an ability to apply knowledge to problem solve.
3. Student learning will become superficial, as students focus on captures post-lecture rather than complete broader reading around the topic.
4. Students will reproduce verbatim words spoken in lecture captures in assessments.
5. Being captured changes teaching styles.
6. Student note-taking skills will be compromised.
7. Students will binge-watch lectures before examinations.

Another concern raised by educators is around the impact of lecture capture on what they say in lectures. Some educators feel that being recorded will stifle their ability to use anecdotes, as they will worry about these being taken the wrong way (Dommett et al., 2019). Our thoughts on this are that if you are unclear whether something might be taken the wrong way on a lecture capture, and then perhaps it's not something that should be said in the live lecture either!

Educators also raise concerns about their own performance rights, whether lecture captures will be used to replace them, and if they can be used in disciplinary action. For these reasons, it is essential that institutions have a robust lecture capture policy in place. By having a detailed policy, available academics can be reassured about how lecture captures will be used by the institution, and students are aware of the possible reasons why lecture captures may not be available. A review of UK university policies can be found in Ibrahim et al. (2021).

Institutions have introduced a wide range of policies around lecture capture, including opt in (where educators have to act to have their lectures recorded) and opt out (where educators have to act to prevent the recording of their lectures, for valid reasons of course). Reasons why a lecture may not be recorded include:

1. Sensitive content
2. Unpublished research data
3. Guest speaker delivering a session
4. Type of session, e.g., one that has significant student participation that would be impacted by being recorded
5. Copyright issues

Whatever the policy approach that is implemented, it is important that the institution, departments, and educators ensure that student expectations are being met, i.e., by being very clear what is and is not being recorded. Students need to know in advance what is and is not being recorded, so that expectations are managed, and the students see lecture capture availability as a positive experience. Managing student expectations (and not just around lecture captures) is particularly important in countries that have, or have moved toward, a more commercial model of higher education, e.g., the introduction of higher student fees in the UK from 2012 onward. When fees are in place alongside student satisfaction surveys (e.g., the

National Student Survey—NSS, UK), the results from which are used to determine countrywide rankings, then an institution/program can find themselves moving down the rankings if student expectations are not met and managed.

Of course, while some educators raise concerns, others embrace lecture capture as being of benefit to their students. Interestingly, some educators who were initially against it can often change their minds once they have used it. This is actually the case for one of the authors (Robson), who initially was quite skeptical about the value, worrying about attendance and whether students would turn into superficial learners. However, having implemented it in a third-year lecture course, the benefits to the students became very clear, and there were also unexpected benefits for staff, as students stopped e-mailing basic questions (because they were revisiting the lecture captures to work the answers out for themselves) and instead started e-mailing more insightful, higher-level questions, often from their additional reading. The perception is that by providing the captures, students were able to grasp the basics more easily, and by their own learning activities, and this left more time for extended reading around the subject. This pattern of e-mail activity has continued with all courses where Dr Robson uses lecture captures. Other positive aspects that educators have raised in studies around lecture capture include:

1. They are a useful tool for students to go back to earlier material to help with current courses.
2. Lecture captures can enable independent learning in students.
3. Lecture captures provide an opportunity to add review/revision pointers.
4. Lecture captures open up new ways of teaching.

This final point (4) leads onto the final aspect to consider in this section the impact of lecture capture on teaching style. Educators have indicated that while lecture captures are valuable to record a didactic lecture delivering content, it is not appropriate to record more interactive sessions as it might stifle student participation (Dommett et al., 2019; Morris et al., 2019). In the Morris et al. (2019) study, 26% and 28% of educators in two different academic years indicated they had altered their teaching style, although the majority had not.

Most lecture captures are typically of didactic lectures, what we would call one-way teaching. This is where an educator talks to students, explaining facts, ideas, and concepts, but there is no active participation by students within class. Indeed, educators have indicated that they would prefer not to record active participation sessions, as this may have a negative impact on student participation. Students themselves are also more likely to watch lecture captures of didactic sessions. Danielson et al. (2014) showed that there was an inverse relationship between how likely students were to view a capture versus the interactivity of a session, but no correlation in terms of the type of content covered in a lecture, i.e., clinical, problem-solving, and applied and non-applied basic science. This focus on captures of didactic sessions links back to concerns raised by educators around the lecture and subsequent capture being seen by students as a way of obtaining information to pass the exam, stifling higher-level thinking around application of knowledge and understanding.

Interestingly, a study by Nordmann, Clark, et al. (2020) showed in a cross institutional study that educators who valued an active approach to teaching also were more likely to have a positive viewpoint about lecture capture. This might seem somewhat incongruous; however, our own anecdotal experience from talking to students is that recording sessions that incorporate active learning approaches, whether mixing these with didactic teaching or running standalone active sessions, is valued by students (Szpunar et al., 2013; see also Chap. 6 (flipped classrooms) in this volume). Given class sizes, these active sessions have mostly utilized anonymous online polling platforms, getting students to answer application and problem-solving questions in class after (unrecorded) peer discussions. These polling platforms provide opportunities for students to test their knowledge and understanding and form a key part of student self-regulated learning. Students can answer questions on an individual basis or after group activities where they bounce ideas off each other (peer learning). If questions are set as application or problem-solving, then this really helps students identify if they have understood a topic. This "active" approach to teaching has long been known to provide real benefits to student understanding and their ability to problem solve, even if students don't always recognize the value. Previously, posing questions in class would have involved the educator giving the question and asking for volunteers to verbally answer in class or picking out an individual to answer the question. This approach has a number of pitfalls. If volunteers are requested, then the same confident individuals may answer the questions each time, leaving the other students feeling ignored or unsupported. If students are selected by the educator, then this can introduce significant anxiety for students and may put them off attending class at all. Of course, we would anticipate that capturing a teaching session that required more oral whole class discussion could impact on the effectiveness of the session. More work is certainly needed to evaluate the impact of capturing active learning sessions on both how participants act in class and whether these captures have value to students after class.

11.5 Attendance Is Not Necessarily Required for Performance

We would argue that with the material available outside of the classroom in the form of lecture-capture and third-party resources, that attendance is not necessarily required for success on knowledge-based examinations in medical education. That being said, we are not preparing these students to just state facts back to us. We do want them to use their knowledge in real-world situations and in medical education specifically patient care. This requires many more competencies than just knowledge. We need to encourage professional identity formation, critical thinking/clinical reasoning, and procedural, teamwork, and communication skills. These types of constructs do not lend themselves to a lecture-based curriculum and will require interactions with the more experienced others, i.e., the faculty member. We do not argue that the students can learn to become doctors all by themselves. Nor do we argue that preclinical basic science content needed to prepare students for the clinical part of the curriculum should be done by the student entirely on their own. However, we do argue that it might be time to consider changing the focus of the time we do interact with our learners in the same space and change the focus of our assessments to prioritize other constructs in addition to just knowledge.

The world continues to change around us, e.g., changes to assessments created by regulatory bodies. These outside changes could have the potential to influence how students engage with our content, e.g., attendance at the live lecture or use of lecture captures. In the USA, the National Board of Medical Examiners (NBME) who organizes and runs the USLME announced that the Step 1 of the examination will be only reported as "pass or fail" beginning in 2022 (USMLE, 2019). This will eliminate the three-digit score that has been used to rank US medical students for residency applications. Since they will now only need to pass this exam in order to obtain a medical license, this will change their focus in the preclinical years of the program. Those of us who are working in this area see it as a game-changer in how the students will interact with the curriculum. In the UK, the General Medical Council has announced that they will begin their own national licensing examination, the Medical Licensing Assessment. This will be a must-pass exam that will be required of all medical students before they can join the medical register starting in the 2024–2025 academic year. Depending on the breadth and depth of what is covered in this examination and how it is reported, this could also have a significant effect on how those students perceive the medical curriculum. It could encourage the students to focus on what is presented in the classroom or further encourage them to rely on third-party resources. The end result is, once again, that context is everything. These regulatory changes will change the context of medical education in ways that we cannot yet fully predict.

When we look at the need for attendance, we need to look at the full context and not just make the blanket statement that attendance is needed for success. We need to consider the age and maturity level of the group of students in question and the purpose of the class (knowledge versus skills development). For some populations, the focus of our in-person sessions needs to move away from knowledge acquisition and move toward the skills that clearly need interactions with other human beings to develop.

11.6 The Future of Digital Technology in Learning and Teaching: Lessons Learned from COVID-19

We have all experienced the changes that have occurred as a result of the COVID-19 pandemic on our lives in general and in whatever area of education we work in. Changes that have told us that being physically present together in a space is potentially a dangerous activity. In March 2020, the way we deliver bioscience teaching had to make a dramatic move online. While some institutions already had experience with delivery of online distance learning programs, many institutions and educators did not. Yet all educators were asked to move their teaching online at short notice, in some cases within less than 3 days. This move to online required us to think about our teaching delivery in a different way and of course the use of lecture captures was at the forefront of some aspects of planning for this emergency pivot. Having lecture captures available from previous years meant that content-based lectures were delivered with relative ease, with students accessing these in an asynchronous way (i.e., in their own time). However, providing online content alone, as illustrated by the ten top rules review by Nordmann, Horlin, et al. (2020) is not sufficient. Support for students is required, and so educators need to put in place online synchronous sessions, e.g., using a variety of different platforms such as Blackboard Collaborate, Microsoft Teams, and Zoom. Overall, what we did was move from an in-attendance-at-lectures-based approach to an emergency-flipped approach.

Fast forward 6 months and as the world settled into the new normal, the measures we had taken in the emergency pivot were able to shape our significantly amended approaches to teaching. A structured flipped approach was taken, with new recorded (and chunked) lecture captures providing content, examples, and evidence, released on a weekly basis to students for them to access and use like the live lecture. These were supported with weekly synchronous, online active learning sessions, providing opportunities for students to use anonymous online polling platforms to answer questions and test their ability to apply knowledge and understanding. Recording these online sessions meant that students who could not attend online were also able to potentially benefit. Work by Robson (2020) illustrates the value students place on an asynchronous lecture/synchronous active learning approach. Around 80% of students strongly agreed that the asynchronous lecture captures supported their learning. 62% strongly agreed that the synchronous, online active learning sessions allowed them to apply their knowledge and understanding, while 49% strongly agreed that being able to revisit the recordings of the active sessions was valuable (even though the majority had attended the sessions).

So where does this leave us in terms of the future, when we are in a world where our actions are not dictated by the risk of COVID-19? We would like to speculate that teaching will not go back to the traditional, didactic approach of the past. Indeed, there has been speculation that the lecture as a didactic teaching session is redundant (MacKay, 2019). We are hopeful that educators will recognize the value of flipped approaches to student learning, providing videos and activities in advance of the class, with in-class sessions focusing more on supporting student development of skills around application of knowledge, ideas, and concepts (see Chap. 6 in this volume). Do we really want to go back to the lecture theatre and just stand and talk to a group of students? Or do we want to interact with them, facilitating their development as questioning and critical thinkers? Of course, we are not saying ditch all didactic sessions; rather mix things up and use a range of approaches. We could even keep some of the online, interactive sessions, given the value of putting students into virtual breakout rooms to discuss answers to questions.

What does the new normal mean in terms of engagement? Well, engaging with pre-class resources will be critical to student success, and this could mean viewing lecture captures or accessing other resources made available before class. For lecture captures, it will be essential to guide students on how to use these captures effectively, so that they are able to identify the key learning points and also know that spacing learning is essential. In addition, we would expect those students who participate in the active in-class sessions may be more likely to succeed, particularly in those extended, application-type questions. If this is the case, then attendance will still be of value in terms of measuring student engagement. As we move to implement the new normal, we should of course still seek to evaluate the effectiveness of our new approaches. The COVID-19 pandemic has had a huge impact on teaching approaches, in a time period of mere months. We don't think there has ever been a period where change and innovation were introduced in such a short time period, with faculty learning and implementing new and exciting ways to teach and engage with students. In this way, the pandemic has been instrumental in driving forward innovation in learning and teaching.

References

Al Khaja, K. A. J., Tayem, Y., James, H., Jaradat, A., & Sequeira, R. P. (2019). Pharmacology and therapeutics resource session attendance and academic performance of pre-clerkship medical students in problem-based learning curricula. *BMC Medical Education, 19*(1), 269. https://doi.org/10.1186/s12909-019-1699-3

Biggs, J., Kember, D., & Leung, D. Y. (2001). The revised two-factor Study Process Questionnaire: R-SPQ-2F. *The British Journal of Educational Psychology, 71*(Pt 1), 133–149. https://doi.org/10.1348/000709901158433

Boulton, C. A., Hughes, E., Kent, C., Smith, J. R., & Williams, H. T. P. (2019). Student engagement and wellbeing over time at a higher education institution. *PLoS One, 14*(11), e0225770. https://doi.org/10.1371/journal.pone.0225770

Burk-Rafel, J., Santen, S. A., & Purkiss, J. (2017). Study behaviors and USMLE step 1 performance: Implications of a student self-directed parallel curriculum. *Academic Medicine, 92* (11S Association of American Medical Colleges Learn Serve Lead: Proceedings of the 56th Annual Research in Medical Education Sessions), S67–S74. https://doi.org/10.1097/acm.0000000000001916

Christenson, S. L., Reschly, A. L., & Wylie, C. (2012). Preface. In *Handbook of research on student engagement*. Springer.

Credé, M., Roch, S. G., & Kieszczynka, U. M. (2010). Class attendance in college: A meta-analytic review of the relationship of class attendance with grades and student characteristics. *Review of Educational Research, 80*(2), 272–295. https://doi.org/10.3102/0034654310362998

Danielson, J., Preast, V., Bender, H., & Hassall, L. (2014). Is the effectiveness of lecture capture related to teaching approach or content type? *Computers & Education, 72*, 121–131. https://doi.org/10.1016/j.compedu.2013.10.016

Deane, R. P., & Murphy, D. J. (2013). Student attendance and academic performance in undergraduate obstetrics/gynecology clinical rotations. *JAMA, 310*(21), 2282–2288. https://doi.org/10.1001/jama.2013.282228

Doggrell, S. A. (2020). No apparent association between lecture attendance or accessing lecture recordings and academic outcomes in a medical laboratory science course. *BMC Medical Education, 20*(1), 207. https://doi.org/10.1186/s12909-020-02066-9

Dommett, E. J., Gardner, B., & van Tilburg, W. (2019). Staff and student views of lecture capture: A qualitative study. *International Journal of Educational Technology in Higher Education, 16*(1), 23. https://doi.org/10.1186/s41239-019-0153-2

Dommett, E. J., Gardner, B., & van Tilburg, W. (2020). Staff and students perception of lecture capture. *The Internet and Higher Education, 46*, 100732. https://doi.org/10.1016/j.iheduc.2020.100732

Edwards, M. R., & Clinton, M. E. (2019). A study exploring the impact of lecture capture availability and lecture capture usage on student attendance and attainment. *Higher Education, 77*(3), 403–421. https://doi.org/10.1007/s10734-018-0275-9

Eisen, D. B., Schupp, C. W., Isseroff, R. R., Ibrahimi, O. A., Ledo, L., & Armstrong, A. W. (2015). Does class attendance matter? Results from a second-year medical school dermatology cohort study. *International Journal of Dermatology, 54*(7), 807–816. https://doi.org/10.1111/ijd.12816

Gorissen, P., van Bruggen, J., & Jochems, W. (2012). Students and recorded lectures: Survey on current use and demands for higher education. *Research in Learning Technology, 20*(3), article 17299. https://doi.org/10.3402/rlt.v20i0.17299

Ibrahim, Y., Howarth, A., & Stone, I. (2021). Lecture capture policies: A survey of British universities. *Postdigital Science and Education, 3*, 144–161. https://doi.org/10.1007/s42438-020-00102-x

Kahu, E. R. (2013). Framing student engagement in higher education. *Studies in Higher Education, 38*(5), 758–773. https://doi.org/10.1080/03075079.2011.598505

Kauffman, C. A., Derazin, M., Asmar, A., & Kibble, J. D. (2018). Relationship between classroom attendance and examination performance in a second-year medical pathophysiology class. *Advances in Physiology Education, 42*(4), 593–598. https://doi.org/10.1152/advan.00123.2018

Kauffman, C. A., Derazin, M., Asmar, A., & Kibble, J. D. (2019). Patterns of medical student engagement in a second-year pathophysiology course: Relationship to USMLE Step 1 performance. *Advances in Physiology Education, 43*(4), 512–518. https://doi.org/10.1152/advan.00082.2019

Kiewra, K. A. (1989). A review of note-taking: The encoding-storage paradigm and beyond. *Educational Psychology Review, 1*(2), 147–172.

Leadbeater, W., Shuttleworth, T., Couperthwaite, J., & Nightingale, K. P. (2013). Evaluating the use and impact of lecture recording in undergraduates: Evidence for distinct approaches by different groups of students. *Computers & Education, 61*, 185–192. https://doi.org/10.1016/j.compedu.2012.09.011

MacKay, J. R. D. (2019). Show and 'Tool': How lecture recording transforms staff and student perspectives on lectures in higher education. *Computers & Education, 140*, article 103593. https://doi.org/10.1016/j.compedu.2019.05.019

Morris, N. P., Swinnerton, B., & Coop, T. (2019). Lecture recordings to support learning: A contested space between students and teachers. *Computers & Education, 140*, article 103604. https://doi.org/10.1016/j.compedu.2019.103604

Mortimore, T., & Crozier, W. R. (2006). Dyslexia and difficulties with study skills in higher education. *Studies in Higher Education, 31*(2), 235–251. https://doi.org/10.1080/03075070600572173

Nightingale, K. P., Anderson, V., Onens, S., Fazil, Q., & Davies, H. (2019). Developing the inclusive curriculum: Is supplementary lecture recording an effective approach in supporting students with Specific Learning Difficulties (SpLDs)? *Computers & Education, 130*, 13–25. https://doi.org/10.1016/j.compedu.2018.11.006

Nordmann, E., Calder, C., Bishop, P., Irwin, A., & Comber, D. (2019). Turn up, tune in, don't drop out: The relationship between lecture attendance, use of lecture recordings, and achievement at different levels of study. *Higher Education, 77*(6), 1065–1084. https://doi.org/10.1007/s10734-018-0320-8

Nordmann, E., Clark, A., Spaeth, E., & MacKay, J. (2020). Lights, camera, active!, Appreciation of active learning predicts positive attitudes towards lecture capture. *Higher Education (Dordrecht)*. https://doi.org/10.1007/s10734-020-00674-4. https://psyarxiv.com/2jgcs/

Nordmann, E., Horlin, C., Hutchison, J., Murray, J.-A., Robson, L., Seery, M. K., & MacKay, J. R. D. (2020). Ten simple rules for supporting a temporary online pivot in higher education. *PLoS Computational Biology, 16*(10), e1008242. https://doi.org/10.1371/journal.pcbi.1008242

Nordmann, E., Küepper-Tetzel, C. E., Robson, L., Phillipson, S., Lipan, G. I., & McGeorge, P. (2018). Lecture capture: Practical recommendations for students and lecturers (student support guide).

Nordmann, E., Küepper-Tetzel, C. E., Robson, L., Phillipson, S., Lipan, G. I., & McGeorge, P. (2020). Lecture capture: Practical recommendations for students and instructors. *Scholarship of Teaching and Learning in Psychology*. https://doi.org/10.1037/stl0000190

Nordmann, E., & McGeorge, P. (2018). Lecture capture in higher education: Time to learn from the learners. Retrieved from http://psyarxiv.com/ux29v.

NRMP NRMPDRaRC. (2018). *Results of the 2018 NRMP Program Director Survey*. National Resident Matching Program.

Owston, R., Lupshenyuk, D., & Wideman, H. (2011). Lecture capture in large undergraduate classes: Student perceptions and academic performance. *The Internet and Higher Education, 14*(4), 262–268. https://doi.org/10.1016/j.iheduc.2011.05.006

Peters, H., Zdravkovic, M., João Costa, M., Celenza, A., Ghias, K., Klamen, D., Mossop, L., Rieder, M., Devi Nadarajah, V., Wangsaturaka, D., Wohlin, M., & Weggemans, M. (2019). Twelve tips for enhancing student engagement. *Medical Teacher, 41*(6), 632–637. https://doi.org/10.1080/0142159x.2018.1459530

Pursel, B., & Fang, H. N. (2012). Lecture capture: Current research and future directions.

Robson, L. (2016). *Lecture capture—student perceptions and unexpected benefits*. Paper presented at the advance HE STEM conference.

Robson, L. (2019). *Effective use of lecture capture and inclusive education*. Paper presented at the advance HE teaching and learning conference.

Robson, L. (2020). *Active learning in action—breaking down the barriers to inclusive education*. Paper presented at the advance HE conference.

Schillerstrom, J. E., & Lutz, M. (2013). Academic performance in the context of a "three excused absences" psychiatry clerkship policy. *Academic Psychiatry, 37*(3), 171–174. https://doi.org/10.1176/appi.ap.12050091

Selvig, D., Holaday, L. W., Purkiss, J., & Hortsch, M. (2015). Correlating students' educational background, study habits, and resource usage with learning success in medical histology. *Anatomical Sciences Education, 8*(1), 1–11. https://doi.org/10.1002/ase.1449

Stegers-Jager, K. M., Cohen-Schotanus, J., & Themmen, A. P. (2012). Motivation, learning strategies, participation and medical school performance. *Medical Education, 46*(7), 678–688. https://doi.org/10.1111/j.1365-2923.2012.04284.x

Szpunar, K. K., Khan, N. Y., & Schacter, D. L. (2013). Interpolated memory tests reduce mind wandering and improve learning of online lectures. *Proceedings of the National Academy of Sciences of the United States of America, 110*(16), 6313–6317. https://doi.org/10.1073/pnas.1221764110

Traphagan, T., Kucsera, J. V., & Kishi, K. (2010). Impact of class lecture webcasting on attendance and learning. *Educational Technology Research and Development, 58*(1), 19–37. https://doi.org/10.1007/s11423-009-9128-7

University of Sheffield. (2019). Lecture capture—time travel for students. https://digitalmedia.sheffield.ac.uk/id/1_xwyj9gw4

USMLE. (2019). Change to pass/fail score reporting for step 1. https://www.usmle.org/incus/.

van der Meer, J. (2012). Students' note-taking challenges in the twenty-first century: Considerations for teachers and academic staff developers. *Teaching in Higher Education, 17*(1), 13–23. https://doi.org/10.1080/13562517.2011.590974

von Konsky, B. R., Ivins, J., & Gribble, S. J. (2009). Lecture attendance and web based lecture technologies: A comparison of student perceptions and usage patterns. *Australasian Journal of Educational Technology, 25*(4). https://doi.org/10.14742/ajet.1130

Wiese, C., & Newton, G. (2013). Use of lecture capture in undergraduate biological science education. *The Canadian Journal for the Scholarship of Teaching and Learning, 4*(2). https://doi.org/10.5206/cjsotl-rcacea.2013.2.4

Chapter 12
Attendance Debate Part 2. Lecture Capture, Attendance, and Exam Performance in the Biosciences: Exploring Rare Exceptions to the Link Between Attendance and Performance in the Era of Online Teaching

Harry J. Witchel, Richard Guy, Christopher Torrens, Kenneth Langlands, and Sheila A. Doggrell

Abstract We argue that there is extensive and overwhelming evidence that increased attendance at lectures typically improves university bioscience students' performance on assessments. Nevertheless, video lecture capture recording and other educational technologies have mushroomed, which may undermine this link because these technologies provide alternative routes for engagement. Furthermore, a few studies of postgraduate and advanced university students have demonstrated no correlation between attendance and performance in their biomedical cohorts. These discrepancies highlight that the correlation linking learning to classroom attendance of lectures must be addressed mechanistically. Over-extrapolating from most classrooms to all classrooms is erroneous. The link is complicated by other

This chapter and Chap. 11 in this volume are meant to be read together as part of a debate on the topic of whether lecture attendance is a determinant of student performance and learning in the era of lecture capture recording.

H. J. Witchel (✉)
Department of Neuroscience, Brighton and Sussex Medical School, Brighton, UK
e-mail: h.witchel@bsms.ac.uk

R. Guy
Victoria University, Melbourne, VIC, Australia

C. Torrens
Royal College of Surgeons, Dublin, Ireland
e-mail: christophertorrens@rcsi.ie

K. Langlands
University of Wolverhampton, Wolverhampton, UK

S. A. Doggrell
Queensland University of Technology, Brisbane, QLD, Australia
e-mail: sheila.doggrell@qut.edu.au

H. J. Witchel, M. W. Lee (eds.), *Technologies in Biomedical and Life Sciences Education*, Methods in Physiology, https://doi.org/10.1007/978-3-030-95633-2_12

educational factors and needs to be considered holistically for the specific cohort, classroom, and policy being investigated. We suggest that some cohorts' unusually high self-regulation and maturity of study skills may explain the discrepancies in the literature. Exceptionally, classroom-based skills learning, such as microscopy and histology, may not have the same potential for video-enhanced home study as other fact-based bioscience subjects. Mandatory attendance policies need to be investigated explicitly, as most previous investigations at the university level have considered such policies as an ancillary issue, and many found little evidence for performance benefit. A range of models are re-evaluated, including single-population models that suggest how lecture attendance explains or supports learning and two-population models that may suggest that the statistical link is partly a result of confounding. We propose specific requirements for reporting future research in this field to aid editors and referees; these considerations may also aid experimental designers.

12.1 Introduction

12.1.1 Has Technology Diluted the Link Between Classroom Attendance and Performance?

The debate over whether university students learning the biosciences should be required to attend lectures (Cutler et al., 2016; Lamb et al., 2020; St. Clair, 1999) has been sharpened by the ubiquity of lecture capture recording (LCR) videos (Feldman, 2013; Kanter, 2012; Subramaniam et al., 2013). Students certainly want classroom attendance at lectures to be non-compulsory (Daud et al., 2017; Zazulia & Goldhoff, 2014). This has rekindled the research question as to whether students benefit from attending lectures. Recently a range of carefully run studies for learning in the biosciences have shown that in their classrooms there was no statistical relationship between lecture attendance and exam performance (Chap. 11 in this volume; Azab et al., 2016; Doggrell, 2020b; Eisen et al., 2015; Kauffman et al., 2019). However, previously both meta-analyses and systematic reviews (Credé et al., 2010; Doggrell, 2020a) have suggested that classroom attendance is beneficial to test scores. The Credé et al. (2010) meta-analysis of 28,304 students (68 separate studies) among many undergraduate disciplines found a strong correlation between attendance and performance, and they concluded:

> class attendance [is] a better predictor of college grades than any other known predictor of academic performance, including scores on standardized admissions tests such as the SAT, high school GPA [grade point average], study habits, and study skills.

The discrepancy between this unequivocal conclusion and the recent studies showing no effect is striking and needs to be addressed. Note that in this chapter we are mostly focusing on lectures, but there are also similar questions to be asked about tutorials (Alexander & Hicks, 2016) and clinical/skills learning (Cruz et al.,

Table 12.1 The link between attendance and exam performance helps to assess

1. At-risk students
(a) How much attendance at classroom lectures relates to professionalism
2. The value and quality of lectures
(a) The value of lecturing as a teaching form
(b) Whether attendance at classrooms lectures should be mandatory
(c) The quality of teaching by an individual course or institution
3. If final grades should incorporate low-stakes coursework as well as the final exam

2019; Deane & Murphy, 2013). In some previous studies, attendance data from tutorials, lectures, seminars, and other learning events have been aggregated (Newman-Ford et al., 2008); we feel that this aggregation is useful for quality assessment at the institutional level, but our chapter here is focused on the mechanistic issues relating to whether lecture attendance is necessary—or even beneficial—to student learning.

Several rationales for carefully testing the link between classroom attendance and exam performance are listed in Table 12.1. First, as shown in part one of the table, this question is relevant to instructors who may want to use classroom attendance as a marker for at-risk students (Tower et al., 2015) or as a requirement for student progression (St. Clair, 1999). Proponents of requiring lecture attendance suggest that it is a marker of professionalism (Deane & Murphy, 2013; Smith, 2012), an incentive for engagement (An Bord Altrainais, 1994), as well as a guarantor of student enculturation into—and support of—a community of peers (Hassel & Lourey, 2005). Some students would prefer to minimize their attendance at all sessions, not just lectures. It is statistically well-established that many educational benefits accrue from engaging deeply with a community of peers (Astin, 1993).

Second, the benefits of attending a lecture are important to understand for educators because it relates to the value of the lectures. On a small scale, this issue is part of institutional quality assessment, but on a large scale this issue is a quality assessment question about the whole of academia—it questions the entire role of lecturing. Lectures have been a cornerstone in the practice of higher education for over 800 years (Friesen, 2011) and of knowledge dissemination since antiquity. Even today lectures form the mainstay of how academics communicate at scientific meetings. The presumed purpose of all this lecturing has been to benefit the listeners. It would be pretty ironic if the main beneficiary for lecturing was not the listeners—if they learned little—and was instead the speaker, "the sage on the stage"; this seems to violate the whole point of student-centered learning (King, 1993). As one surveyed student stated in (Zazulia & Goldhoff, 2014):

> I'm actually a little offended that someone, somewhere thinks I'm less of a student, less professional, or that I won't be as good a physician simply because I've found a more effective/efficient way for me to learn....Seems like professors would care more about us actually learning the material than just seeing us in the seats.

In answer to that kind of opinion, many education researchers, chasing after student-centered learning and engagement, are obsessed with replacing lectures with active learning or with social/team learning. These researchers presume that group work is better than lectures at developing student team dynamics. These researchers should ask the students themselves. Counterintuitively, passive lectures may be more valuable than group work for development of social and team dynamics in the students. In one study 67% of undergraduate medical students (especially in the earlier preclinical classes) still found lectures useful, and if they were replaced, the students would want clinical sessions, not group work or active learning (Mustafa et al., 2014); even in Kauffman et al.'s (2019) study of postgraduate medical students in Florida, there was no statistical difference in student ratings of live lectures from live group work (in person cases). Some students obviously like lectures (Mattick et al., 2007). One theory even proposes that the authority (sage) and the group attention (stage) contribute to the lecture's long held fascination in human history (Charlton, 2006). So it remains controversial as to who benefits more: student or lecturer. If it is possible that lecturing primarily benefits the vanity of the speaker or the institution promoting him (or more rarely her, see Chaps. 2 (inclusivity) and 3 (students as partners) in this volume), then sincere educators should be very interested in learning the truth of this research. Thus, this topic poses a grander question about whether sitting in live lectures is ineffective, or at least suboptimal, for information transfer and learning for students (Matheson, 2008; Schwartzstein & Roberts, 2017).

Third, this question also relates to whether low-stakes coursework should be able to engender a passing final grade when a student plainly fails the final exam. Many courses include ongoing assessments such as weekly quizzes or Individual Readiness Assurance Tests as ways to guarantee both studying and attendance; in essence, these low-stakes tests incentivize timeliness and compliance (Nieder et al., 2005). Similarly, some courses include course work (Doggrell, 2020c) and attendance (Popovic et al., 2018) as part of the final course grade. If the distribution of marks between these different elements versus the final exam is balanced incorrectly and if the calculation of a passing grade allows the light-touch ongoing assessments to compensate for low scores on the high stakes final exam, there is a risk of qualifying students who are incompetent (Doggrell, 2020c). That is, compliant students who learn little may be allowed to complete professional degree programs in allied healthcare professions, particularly in countries that do not require a national exam for being registered to practice.

12.1.2 The Approach of This Review

Some reviews of this research topic have divided up the previous research by year of publication (Credé et al., 2010; Doggrell, 2020a) because the historical development of educational technology has led to many changes in how students go about learning day to day. In particular, a lot of recent educational research has debated the benefits of video lecture capture recording. Students love it, but lecturers are divided over its benefits (Morris et al., 2019) because in some cases it seems to vastly lower student attendance (Albon et al., 2020). Perhaps students are voting with their feet, silently voicing their opinion that sitting in lectures does not help them learn (Ikonne et al., 2018). As early as 2002, Diana Laurillard in *Rethinking University Teaching* suggested that lectures had become an unreliable and outmoded learning method for transferring information from instructors to students, enfeebled by centuries of university tradition; she asked, "Why aren't lectures scrapped as a teaching method?" More recently, Ezekiel Emanuel (2020) suggested that regularly repeated live classroom lectures were "a waste of everyone's time" for preclinical service teaching of healthcare students; his argument was that video recording makes the cost of local classroom lectures untenable and their comparative shortcomings indefensible. Instead, he proposed that local lectures be replaced by video lectures of "a few outstanding professors" combined with casually employed tutors and markers who would support individual learners. Such speculative predictions may make many lecturers (and their educational institutions) nervous that they could be easily replaced by video lectures (St. Clair, 1999). If true, educational technology will help administrators and accountants inaugurate a new age of educational utopia, or dystopia, depending on your point of view (Harden, 2013).

So this chapter will outline the following consequential research question: whether classroom attendance during lectures in the biosciences is necessary, sufficient, contributory, or indicative for the ability of students to learn. For our purposes "learning" is defined as test performance, which is a conventional oversimplification for addressing this research question. Of course, simplifications (such as equating learning with test performance) are reductionist, but they are inevitable in order to gather quantitative evidence. So also, this chapter will consider and contextualize some of these simplifications in terms of their meaning.

In particular, making causal arguments via statistics has recently become even more controversial (Wasserstein & Lazar, 2016), and attempts to correlate lecture attendance with test performance require many caveats. The first is that statistics are subject to expert judgment (Brownstein et al., 2019). The second is that the psychological link between attendance and learning is neither simple nor linear. This link is multifactorial—lecture attendance is not the only player in the learning story. As such we want to understand why a few recent studies have found no link between attendance and performance although most previous research has. When we explain these discrepancies in the literature, we reject the trivial view that the discrepancies are due to widespread poor research on only one side. However, we will critically consider that the discrepancies arise because there is a small effect that can be drowned out in some contexts. In summary, we are searching for the potential sources of the previous discrepancies, which we believe are bona fide. These sources of discrepancies will be outlined in terms of the mechanism for learning, the research processes, and the statistical ambiguities. Given the complexity, this chapter does not aim to be the last word on the topic but a springboard for discussion. The goals of this chapter are to:

(A) Consider the most important topics that are relevant to this field from a mechanistic point of view in order to answer the questions, "Why do results in this field diverge? What are the differences that make the difference?"
(B) Explore the potential mechanisms of the discrepancies in order to distinguish statistical issues versus education factors.
(C) Identify the factors that affect the linkage between learning and attending lectures, in order to suggest approaches to further this research.

12.1.3 Statistical and Educational: Mechanistic Sources of Discrepancies

Because the benefits of lecture attendance are meant to accrue over the longer term, studies meant to clarify the link between attendance and exam performance are unlike laboratory-based educational psychology studies that allow for control of all the factors while altering just one variable (see the discussion of "grain size" of engagement in Chap. 4 (engagement) in this volume). Studies testing this link have many of the challenges inherent in sociology, where the link could be affected by extraneous changes relating to the time or the cohort involved. Nevertheless, a majority of studies do find a link between attendance and exam performance (Doggrell, 2020a). Some of these studies have found large effects (Colby, 2005; Moore, 2003) or medium-sized effects (Doggrell, 2018; Sade & Stroud, 1982), but more commonly the effects for bioscience lectures have been modest (Doggrell, 2019). When genuine but modest statistical links are found between an educational activity (e.g., attendance) and exam performance, there can be a spectrum of explanations. At one end, there is the possibility that when all students attend lectures, they all benefit a little bit. This is the single-population model, and the effects of the intervention (lecturing) can be described as "a rising tide lifts all boats." This would support a policy where lecture attendance is strongly encouraged or required for everyone, particularly if there are many students near the borderline of failing. At the opposite end of the spectrum, a modest link could result from a two-population model, where there is a mixture of two types of students in the class, (1) self-regulated and (2) poorly regulated, and only one group benefits from the intervention. At one extreme is a two-population model entitled "the rich get richer," where only self-regulated students benefit because they have the skills to consistently make use of the opportunity. At the other extreme is "the ceiling model," a situation where a few of the poorly regulated students who attend lectures benefit enormously from the experience, whereas most students seemingly fail to benefit at all (i.e., the students stuck on the ceiling are already studying at maximum effectiveness from home). An example of this might be where absenteeism is a marker for that individual student's generalized disengagement (see Chap. 4 in this volume). For instance, poorly regulated students may choose to skip classes if they misjudge their ability to self-regulate their learning, such as when first year undergraduate students residing at university discover beer. Because such students are extrinsically motivated, they view themselves as just having fun, but they literally do not possess or have not developed (either through formal guidance or via experience) the learner characteristics that allow them to focus and diligently accomplish tasks. This would support a policy where students as a rule should not be mandated to attend; instead, it suggests that educators should focus on student support policies that triage or select for the rare students who would benefit greatly from being mandated to attend lectures; this kind of tracking would be very unpopular in a medical school, as educational tracking has a history of stigmatization and encouraging inequality (Hanushek & Wößmann, 2006).

Educational causes of the previous research discrepancies would depend on the mechanisms for learning. In this era where a student has a choice of learning materials and ways to engage with learning any topic, students describe at least five mechanisms why attending a lecture—or any learning activity for that matter—might be uniquely useful to them (Westrick et al., 2009). (1) Attending a lecture could provide the student with an *explanation* (or information) that otherwise might be unavailable. This may be due to additional clarity, and it harks back to the lectures of yore, when the lecture might be the only readily available source on the topic with information at the correct level. (2) Attending a lecture could provide the student with a uniquely *engaging experience*. That is, the student might be motivated to spend more time on task, or even to pursue further reading and study. (3) Attending a lecture could provide the student with a particularly *memorable experience*, after which the presented material stays in the mind longer than after other exposures of equal duration to the same material. (4) Attending a lecture could provide the student with *emphasis* so that the student would know what studying to prioritize. In this case the instructor highlights the most important information for the student, who is allowed to deprioritize much of the information from indigestible textbooks and other vast Internet sources (Valverde & Schmidt, 1997). (5) Attending a lecture could provide the student with enjoyable or valued *social interactions*, chances to participate, opportunities to ask questions, and reinforcements of their relationships with the instructor and other students. Instructors would claim these social experiences are essential for building a community and professionalism.

This chapter does not propose to provide a final answer to whether exam performance is linked to attendance—because the answer is that it depends on the situation. Instead we aim to contextualize these questions—and the answers thus far provided by the literature. Our goal is to critically analyze and evaluate the important parameters that impinge on the links between attendance and exam performance. This consideration should provide insight for teaching practitioners, as well as helping researchers to design future studies that measure the relevant mechanistic factors, by providing clear recommendations for reporting such research. For example, there are studies on this topic that do not state whether LCR was available (Lukkarinen et al., 2016; Salamonson et al., 2009). A key goal of this chapter is to encourage referees and editors to demand that the relevant parameters and contextual factors are reported in manuscripts because we believe that the black box approach of simply reporting statistics correlating attendance and exam performance has run its course. A more mechanistic approach is now required.

12.2 Educational Differences

12.2.1 *Differences in the Student Cohorts*

12.2.1.1 Postgraduate Versus Undergraduate Student Cohorts: Age, Selection, and Self-Regulation

One possible reason for the discrepancies is that the association between lecture attendance and academic outcomes may depend on the student cohort. Student cohorts may differ in terms of prior academic outcomes, which are a major predictor of academic outcomes (Astin, 1993; McKenzie & Schweitzer, 2001). One type of cohort with systematically excellent previous academic accomplishments are post-graduate students studying medicine (allopathic and osteopathic), dentistry, and pharmacy. Whereas most medical students in the UK (and many other European and African countries) are undergraduates who typically start at age 18, students in medicine or dentistry in the USA (and other anglophone countries including Australia and Canada) are effectively required to complete another undergraduate degree before receiving their doctoral degree in medicine (e.g., the M.D.); thus, students in medical programs in the USA are postgraduates. Such postgraduate students (who have received a first undergraduate university degree but are working toward a professional degree) are sometimes called predoctoral students in order to distinguish them from students in the medical specialties who already have a postgraduate medical degree (such as an M.D.). These predoctoral postgraduate students are generally at least 4 years older than undergraduates (i.e., they typically start studying for their professional degree at age 22), and they are likely to have more mature and self-regulated study skills. When medical schools use selection for excellence to limit admissions to high-achieving university graduates, all admitted students must have evolved successful strategies to study independently at the university level during their first degree program. Postgraduate students who are self-regulated learners may have little need for the external guidance and discipline of lecture attendance. Among recent studies of postgraduate healthcare professional training programs in the USA, three have shown that there was clearly no association between classroom attendance and exam performance (see Table 12.2).

For example Kauffman et al. (2018, 2019) showed that students in their University of Central Florida postgraduate (i.e., predoctoral) medical degree program had no correlation between lecture attendance in a year 2 pathophysiology course and exam performance. The exams they monitored were both their local exam in pathophysiology as well as a national examination (USMLE Step 1). It appears that the predoctoral medical students in Kauffman et al.'s cohort (and possibly in the USA more generally) are focused on content for the national exam rather than on deeply understanding the local lectures, which are thus less important to attend. Their students used third-party resources instead (see below). By contrast, European undergraduate medical students (see Fig. 12.1, panel a) are focused on locally set exams, so the lectures are more relevant. The high marks in Kauffman et al. probably

Table 12.2 Recent studies of postgraduate degrees for healthcare professions, in which some have no link between attendance and performance

Author (date)	Link	Location	PG education	Course	Avg score %
Kauffman et al. (2018)	N	USA, Central Florida	Medicine	Pathophysiology	84–87
Eisen et al. (2015)	N	USA, UC Davis	Medicine	Dermatology	82–90
Azab et al. (2016)	N	USA, Tufts	Dentistry	Year 2	83–95
Shumway et al. (2018)	Y	USA, Louisville	Dentistry	Oral Pathology	75–78
Selvig et al. (2015)	Y	USA, Michigan	Medicine	Histology	86–91
Ikonne et al. (2018)	WP	USA, Arizona	Osteopathy	Year 1	81

Link, correlation between attendance and exam scores; N, no correlation; Y, significant correlation; WP, weak positive correlation that did not reach significance; Avg score %, average final test score

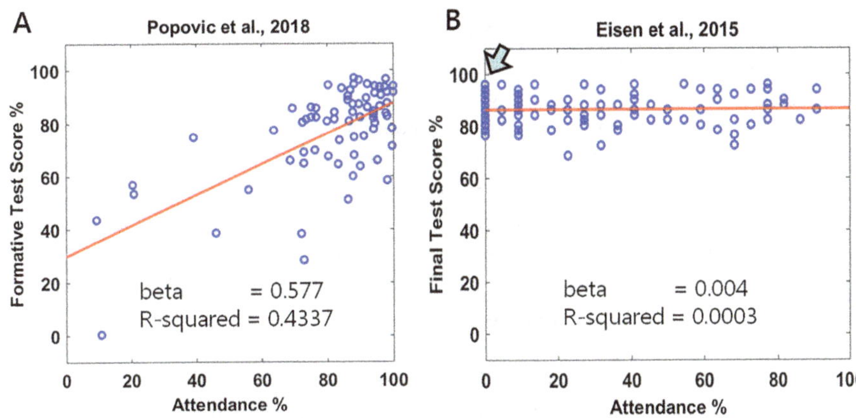

Fig. 12.1 Scatter plots of attendance versus test scores for two different studies. The regressions presented here (solid red lines) were not in the original papers. Each blue circle represents the performance of an individual student. Panel **a** refers to Popovic et al. (2018); the regression is highly significant ($P < 0.001$). Panel **b** refers to Eisen et al. (2015); the regression was not significant. The arrow in the top left of panel **b** highlights a student who scored well in the final exam despite not attending any of the classroom sessions

occurred because the students in this program were graduates, and most will have undergraduate training in the biomedical sciences. Due to their prior study, the postgraduate students may have already had a strong background knowledge and consequently may not have required further lecture attendance in this topic. In support of this, the study showed that the students with zero attendance had higher undergraduate grade point averages (GPA) prior to the course (Kauffman et al., 2018). In a follow-up survey study, Kauffman et al. (2019) showed that lecture PowerPoint slides were very popular with the students, and thus it is possible that with their prior knowledge from undergraduate study, students could glean the material by using PowerPoint slides, rather than having to attend lectures. A possible explanation is that students want to have control over when and how they engage and what they study.

Eisen et al. (2015) found that for students in the University of California at Davis postgraduate medical degree program, there was no link between lecture attendance in a required year 2 introduction to dermatology course and their final exam (see Fig. 12.1, panel b); this was true even after adjusting for sex, age, Medical College Admission Test (MCAT) score, having children at home, and reason for attending class. Interestingly, they surveyed that cohort, and those students who preferred to learn by watching online videos scored significantly higher on the final examination. A study of students in a postgraduate dental degree at Tufts University showed that year 2 students' test scores in six courses were not linked to classroom attendance (Azab et al., 2016).

In contrast with postgraduate degrees in healthcare, most undergraduate students arrive at university at age 18/19. Their previous secondary education will have been much more carefully guided with lower expectations for independence. Their studying may benefit more from the regimentation of group lecture attendance because some of the students may not yet have evolved their own independent study skills. For instance, a range of studies with early years undergraduates (year 1 and year 2 of a first degree, when the mode student age would be 18–20) have shown a positive association between attending lectures in the biosciences and academic outcomes, even when LCR videos are available. Soto and Anand (2009) demonstrated in a year 1 undergraduate cell biology course at San Jose State University that there was a strong link between attendance and exam performance. Likewise, Doggrell (2018) found that in year 2 undergraduate nursing students studying pharmacology at Queensland University of Technology (QUT) in Brisbane, there was a large effect of attendance to exam performance. For year 2 physiology students enrolled in undergraduate BSc and Bachelor of Health Science degrees at University of Adelaide, there was a small ($r = 0.21$) correlation between attendance and exam score (Horton et al., 2012).

An interesting change point may be among year 3 (Y3) undergraduate students, who may represent a transition between early undergraduate cohorts and postgraduate cohorts. The making of highly selected postgraduate cohorts (e.g., predoctoral medical students in the USA) involves both selection and adaptation. By contrast, year 3 undergraduate students will have mostly undergone the adaptation step but not the selection step. Without the selection step, some Y3 students will have failed to develop strong or self-regulated study skills, whereas after postgraduate selection all students will have both strong and self-regulated study skills. Two studies from our research team found opposite results among Y3 students. In a study of 39 undergraduate Y3 students in a pharmacology course as part of a biomedical sciences degree at QUT in Brisbane (Doggrell, 2019), there was a weak positive association between classroom attendance at lectures and performance, and when attendance was categorized into $\geq 50\%$ attendance and $< 50\%$, the marks were significantly higher (with a high effect size) for the high attenders (Cohen's $d > 0.8$, $P = 0.020$); these statistical differences could suggest a two-population model. By contrast, in a study by the same author of 69 undergraduate Y3 students in a diagnostic endocrinology course for a Bachelor of Medical Laboratory Science degree at QUT in Brisbane, there was definitely no positive association between attendance and performance, whether this was sought by correlation or by grouping students as high and low attenders (Doggrell, 2020b). This study's failure to detect a link is particularly notable because the course and its marks included a laboratory component. Thus, there may be differences between cohorts; lecture attendance may be more important for younger and less academically experienced students such as in Y1/Y2 nursing and introductory biology students than for the academic performance of advanced biomedical science/medical laboratory science students who are committed to the topic.

When considering the discrepancies that might cause some cohorts to bypass the link between lecture attendance and exam performance, two key features are motivation and established academic ability (i.e., will versus skill), which may combine as a form of self-regulated learning (McCombs & Marzano, 1990). We propose that cohorts that have less will or skill (e.g., young undergraduates) are most likely to demonstrate a clear link between attendance and performance (see Fig. 12.1, panel a), if only because their unmeasured engagement (i.e., other learning activities besides attendance) will be much less effective for learning. This would be most clear in service teaching, where modules that are "core" for a particular degree are provided by a department other than that which "owns" the degree program (University of Sheffield, 2020); for example, when laboratory scientists are teaching scientific information to students enrolled in a nursing degree, the students may believe it is background material that is not core to their future caring roles (McVicar et al., 2015). In undergraduate nursing students, for example, there is a consistently found link between attendance and exam performance (Brown et al., 1999; Doggrell, 2018; Hammen & Kelland, 1994; Salamonson et al., 2009, but see mixed results in McCarey et al., 2007). This comes with the caveat that even those studies over 20 years old (i.e., pre-lecture capture) suggest that attendance plays only a limited role in performance and that other factors will be as important.

12.2.2 *Differences in the Classroom*

12.2.2.1 Research Pre-2000: Before Video Lecture Capture and Extensive Online Resources

Home study has been revolutionized by new online resources such as video lecture capture. The original research strongly linking attendance to exam performance occurred before video lecture capture or Internet learning resources were commonplace. It seems feasible that discrepancies between older and newer studies could depend on whether appropriate alternate learning resources were available. We would define appropriate resources as those that were tailored to the assessment; tailoring (Lee et al., 2017; Steinel et al., 2019) makes it possible for a student to manage and digest potentially vast information sources, such as a long reading list or the Internet. It stands to reason that if the lectures were in fact the only resources that were tailored to the test, then missing the lectures should be expected to have a negative effect on learning what was on the test. A few examples of historical studies may illustrate this point. Among postgraduate dental students at the University of Tennessee, Newman et al. (1981) found there to be a correlation between attendance and exam performance; this program had mandatory attendance, which resulted in high mean attendance, and that policy also morally links attendance with professionalism (Smith, 2012). Furthermore, among postgraduate medical students at Medical University of South Carolina, where attendance was not mandatory, Sade and Stroud (1982) showed a very clear difference in the overall marks of postgraduate Y1 and Y2 students who rarely attended lectures (<5%) vs. those who nearly always attended lectures (>95%); this extreme division of student categories may imply that the study was observing compliant versus noncompliant students (i.e., a two-population model), rather than an effect of attendance. As such, this study was criticized by Glanz and Fiel (1984) because of the extreme division of student categories and also because the attendance rates were estimated retrospectively by class presidents. In Glanz and Fiel's own study of postgraduate medical students at Temple University, they used sign in sheets during 7 days of classes to measure attendance and found no correlation between specific lecture attendance and the relevant exam question in their summative exams, but they did find a weak positive correlation (i.e., a single-population model) between the overall summative mark and the attendance level. All three of these studies could indicate that missing lectures resulted in the loss of unrecoverable learning opportunities; however, these studies also include elements suggesting that the nonattending students were more generally noncompliant or disengaged.

The combination of several of the above features may lead to unequivocally strong effects linking attendance to exam performance, such as seen in a series of studies by Moore (2003, 2006) at the University of Minnesota's General College. These studies were influenced by (a) historical lack of video resources, (b) undergraduate students in Y1/Y2, (c) students from historically underserved communities, and (d) service teaching of students (nonmajors) who needed to learn principles of biological sciences as a prerequisite for further study. The results showed strikingly strong effects of attendance on both course grade (Moore, 2003) and overall grade point average (Moore, 2006). These large studies (434 and 263 students) concern courses that did not mandate or reward attendance, although it was encouraged. Moreover, attendance was a much better predictor of both course performance and grade point average than standardized aptitude ratings (ACT AAR). Teasing out the role of the lack of video lectures would clarify what this clear result would mean for attendance policy.

12.2.2.2 Availability of Tailored Alternate Learning Resources for Skills Learning

Among contemporary studies, one type of learning session where classroom attendance would provide unique opportunities for learning would be when skills are taught in the class. This would be important in any class where laboratory skills are taught. One such skill is learning to recognize microscopy sections, such as in histology or pathology. Two studies of postgraduate students seem to illustrate this. Shumway et al. (2018) tested 233 students attending a Y2/Y3 oral pathology class in the postgraduate dental program at University of Louisville; oral pathology is strongly dependent on recognition of histological images. It was taught with lectures, a textbook, and handouts but without laboratory sessions; students were allowed to audio-record the lectures, but there were no video recordings, which is notable for a course dependent on image recognition. While the correlation between attendance (taken with student iClickers) and exam performance was small ($r = 0.173$), it was significant ($p = 0.008$). Likewise in a histology class (Y1/Y2) in the postgraduate medical program at University of Michigan, there was a clear benefit to exam scores for students who said they always attended the lectures (Selvig et al., 2015); however, there was no difference between students who never attended and students who said they attended frequently. While lectures were video recorded, laboratory sessions were not. We propose that learning visual recognition skills such as in histology and pathology may have particular needs, such that attendance at classroom sessions may provide uniquely tailored learning materials and guided learning opportunities that cannot be recreated during home study.

12.2.3 Differences in the Learning Environment and Technologies

12.2.3.1 Video Lecture Capture Recordings: Supplement or Substitute?

The automation of digital video recording and storage has permitted the straightforward capture of face-to-face lectures as videos. In many institutions these lecture videos are now routinely made available to students (online) for access or download. Stakeholders now expect universities to provide students with such lecture capture recordings (Beech, 2017) as a route to accessibility and consequent inclusivity. Some of the lecture capture video systems include capturing the face and gestures of the instructor, which provides more expressive and social elements of teachers to the student, which may benefit teaching (Witchel et al., 2020). However, many lecture capture videos are simple screencasts in which the instructor's face is not visible, but their voice is synchronized with the advancing slides from PowerPoint. While students unequivocally love having access to video-recorded lectures (van der Meij & Böckmann, 2020), even simple screencasts, because of the implied freedom from not having to attend some lectures (Kanter, 2012; Zazulia & Goldhoff, 2014), the benefits to students' learning is more difficult to discern. Students insist upon having the videos because they believe having them will aid their studies, while instructors believe that the very existence of the videos will encourage poor study habits among the students (Franklin et al., 2011; Morris et al., 2019). Instructors also have concerns about patient confidentiality, their own intellectual property, and that the videos might be used for performance monitoring and disciplinary procedures (Joseph-Richard et al., 2018). Meanwhile, some students state that they cannot concentrate while sitting in a lecture theatre (Doggrell, 2020b), and others state that the recorded lectures give them psychological support (Traphagan et al., 2010). When testing the objective benefits to student learning, some studies have found a clear benefit (e.g., in psychology courses such as Bos et al., 2016), no discernable benefit (Leadbeater et al., 2013), or a detriment, i.e., a clear association between greater use of recorded lectures and low exam performance (Selvig et al., 2015). Like the discrepant results regarding classroom attendance, these discrepancies on the value of video-recorded lectures also need to be seen in a nuanced way.

The initial expectation when first popularized was for these videos to act as a supplement to the face-to-face lectures, but their ease of use and comprehensive recording encouraged many students to consider them as a substitute for, rather than a supplement to, face-to-face lectures (Bos et al., 2016; Nordmann et al., 2020). Consequently there has been a large literature focusing on the relative benefits of lecture capture recordings versus face-to-face teaching, and the resulting decreases in face-to-face attendance levels (Albon et al., 2020; Banerjee, 2020). Unfortunately, the original idea of supplementation has often been overlooked and the majority of publications lump together as single-population students who use lecture recordings as substitute or supplement. Thus, quantifying the effects of watching LCR videos is also dogged with the problem of two populations. This may well be one of the major contributors to the inconsistencies reported in published research results: the research commingles those students who only use the recordings and do not attend lectures with students who do both: attending face-to-face lectures and using recorded lectures. Compounding this problem has been a general lack of understanding of each student cohort as a heterogeneous entity with variations in student self-regulatory and metacognitive abilities.

For the reasons mentioned above, it is quite difficult to bring together information about supplementary use of lecture recordings. We believe that future research will need to concurrently track student use of online material and also whether or not the same students attend face-to-face lectures; this represents a necessary first step for making sense of exam performance data. There are two ways to deal with the uncertainties of lecture attendance: firstly to examine cohorts where most of the students attend lectures and/or to rely on student feedback (to establish that they used LCRs and attended lectures) and secondly to use lecture attendance recording methods (paper-based lists or electronic monitoring) in addition to survey-based or electronic monitoring of LCR use. Where researchers have used student self-reports, it is sometimes possible to deduce whether they are supplementing or substituting, but this very much depends on the specificity of the questions asked.

12.2.3.2 Evidence from Studies Measuring the Effects of Lecture Capture Videos

A benefit to students who use LCR is seen in the work by Bos et al. (2016) who monitored concurrently online LCR use and student attendance electronically in their psychology classes. Four groups of students were identified: those who did not attend lectures or use LCRs, those who only used LCRs and did not attend lectures (substituters), those who attended lectures and did not use LCRs, and those who used both lectures and LCRs (supplementers). On average the supplementers formed 53% of the student cohort (bearing in mind that a given student might switch between categories). Students in this category performed significantly better on a knowledge-based test than the other groups and thus had an advantage over those who only attended lectures (and did not view LCRs). However, this difference was not found for a subsequent "higher-level" test. The importance of these modest benefits of LCR supplementation is made even less certain because one also needs to know exactly how a given student uses a lecture recording. For example, they may watch the entire recording, watch only small sections, or watch large sections multiple times.

Nordmann et al. (2019) found that less academically capable Y1 (psychology) students gained the most advantage from LCR use when their face-to-face lecture attendance was high. Thus, weaker students appeared to benefit from using the LCRs as a supplement. Owston et al. (2011) found that students who did not access many LCRs and viewed a single small section of each achieved higher grades than those who watched the entire LCR (for each class) several times or viewed the entire recording once and sections several times. The conclusion here was that "better students," with mature learning strategies, make strategic use of the LCRs. On the other hand, weaker students, with less developed learning approaches, use LCRs multiple times in an attempt to understand course material.

Leadbeater et al. (2013) found that the most students downloaded a small number of LCRs and used small sections, while a small minority (~5%) viewed the entire set of lectures, taking up a large proportion of their independent study; this minority had a high proportion of dyslexic and English as a second language students. It was suggested that "short section users" were using LCRs for expanding notes or understanding difficult concepts (supplementary). 40–60% of students reported that they sometimes used LCR to catch up on missed lectures by viewing the entire recording (substitution). In another study (Owston et al., 2011) found that some students did not access many LCRs but viewed a single small section of each, whereas others watched the entire LCR (for each class) several times or viewed the entire recording once and sections several times. Thus, the student population appears to be a heterogeneous group when their behavior toward supplementary LCRs is analyzed. Ebbert and Dutke (2020) investigated the suggestion that varying outcomes described in the LCR literature might be related to discrete differences in how LCRs are used and how this relates to individual students' learning behavior. The large student sample ($n = 1079$) was taken from nine academic disciplines. Five clusters were found dependent on how LCRs were viewed and how many face-to-face lectures were missed. Clusters 3/4 (20/30% of the cohort) used the LCR for about 82% of their lectures and had a high lecture attendance and viewed sections of the LCRs. Clusters 1/2 (52%/62% of the cohort) used LCRs for about 58%/87% of their lectures. They had high lecture attendance and watched entire LCRs. Cluster 5 substituted LCRs as they missed most lectures. It is interesting that clusters 3 and 4 had similar access to LCRs but different perceptions of LCR use for exam revision (84% and 62% agree, respectively). The importance of studies like this is that they provide an insight into cohort heterogeneity and underline the importance of understanding this when interpreting study results regarding student use of LCRs and videos.

Thus, the technical ability to determine exactly how a student uses an individual LCR reveals distinct patterns. The most obvious conclusion is that those students who are the most selective in their LCR use are the most strategic and self-regulated in their learning approach, while those that view entire lectures (sometimes multiple times) are unsure (and possibly anxious) about their understanding resulting from attending the face-to-face lecture.

12.2.3.3 Do Students Have Insight into How They Decide to Study with Technology?

A wealth of other uncertainties still exist in our understanding of the variations in how students use LCRs. For example, references to the word "use" in most cases really means "access," as it is obviously difficult to determine whether a given student is watching the LCR carefully or taking notes. A well-established technique students use to save time is to watch LCR videos at double speed (Emanuel, 2020; Kanter, 2012); this technique would be helpful for supplementers who are searching for a particular ambiguity, but it would be slapdash for substituters who believe that the taught material is mostly "filler." Whether or not these strategic variations will result in learning gains may well depend upon the balance between the academic ability of the student and the knowledge level of the lecture. For example, a student may replay the lecture a second or third time yet still not understand it. Thus, some other method of "bridging the gap" between existing and expected levels of factual or conceptual understanding might be considered.

This leads to the question of how much insight medical students have into what they need to succeed. Among 106 year 1 postgraduate medical students at Texas A&M University, the factors most correlated with objective final marks at the end of the first semester were processes: self-reported time-management ($r = 0.45$) on the Learning and Study Strategies Inventory (LASSI) and self-testing as rated on the LASSI ($r = 0.36$) (West & Sadoski, 2011). These effects on final mark far exceeded the effects of test-taking strategies (0.27), motivation (0.24), concentration (0.23), Medical College Admissions Tests (MCAT, a national admissions test for postgraduate medical study) score (0.14), and undergraduate grade point average (GPA, $r = 0.11$). It is striking that the students' self-rating of time management and self-testing were much more related to exam outcome than objective tests of aptitude

(MCAT) and previous academic success (GPA), given that many medical students who do not want to attend classroom lectures give reasons based on time wasting and efficiency, and many ask for more practice exams and mock questions (Davis et al., 1997). The implication is that the students have insight into activities benefiting their effective performance (Adesope et al., 2017).

12.2.3.4 Third-Party Resources and Practice Test Questions for Studying

To understand why there was no correlation between attendance and exam scores among predoctoral Y2 medical students studying pathophysiology at University of Central Florida (UCF), Kauffman et al. followed up their initial observation (2018) with a study that included detailed surveys (2019), which revealed a great deal that was unique about the current situation for postgraduate medical students in the USA. Kauffman et al. surveyed a subset of their students about the various learning resources supplied; the students were asked, "How important are each of the following course resources to your learning," on a scale of 1 (not important for my learning) to 7 (essential to my learning). Live lectures (mean rating = 3.9) were rated as significantly less useful than lecture PowerPoint slides (6.4), clinical case answers (6.0), self-learning modules (5.8), and recorded lectures (5.4). Live lectures were also rated lower than in person case-based learning (4.2) and team-based learning sessions (4.0), but these differences did not reaching significance. In summary, for this cohort, all live sessions were rated as less useful for their studies than all self-study resources (except for the recommended reading).

Also, among their student cohort at University of Central Florida, the use of third-party study resources (which include exam practice questions) was ubiquitous. The researchers surveyed students and found 15 distinct suppliers of third-party resources in use among their students; 4 of these external sources (Uworld, Pathoma, First Aid, and Sketchy) were in use by over 80% of their students, which means that most students were using at least 2 of them, and some were probably using 3 or more different sources. By using many additional study resources, these students' dependence on any one study source (e.g., live lecturing) is diminished.

This situation in the USA is currently unimaginable for undergraduate medical students in the UK because the cost and structure of medical education is radically different. (1) In the UK there is as yet no standardized national medical qualification exam equivalent to the USA's USMLE Step 1 and Step 2 (this will change in 2024 (General Medical Council, UK, 2020); this means that third-party commercial providers have not been able to aim their products toward a specific examination. (2) British medical schools have tuition fees that are one half the price (<£10,000/ year UK home student compared to UCF of >$25,000 for in state students), so the government loans available to support students making discretionary purchases (i.e., third-party study resources such as Uworld) are concomitantly smaller. Keep in mind that the US medical education is a postgraduate education, so that students who need loans will likely have additional loans to cover their previous undergraduate education as well. These extensive loans in the USA are justified by the fact that similar-level doctors in the USA are paid over threefold more than in the UK (Essaji, 2020). (3) The current passing score for examinations in the UK is much lower. That is, in provincial UK medical schools, a passing score for an exam would be around 50%, whereas in a US medical school, scoring below 75% on a multiple-choice test would imply that a student was at risk. Thus, the current effects of third-party learning resources (e.g., breaking the link between attendance and performance) in postgraduate medical education in the USA have not been seen in undergraduate UK medical students, possibly because the factors driving the use of third-party resources in the USA are not yet found in the UK.

12.3 Revisiting Mechanisms for How Attendance and Performance Are Linked

12.3.1 Mediated Effects: Learning in the Classroom

To help tease apart whether attendance was important for learning, Credé et al. (2010) provided a range of mechanistic models linking attendance to exam performance, each of which provides numerically testable hypotheses. Two of them (see Fig. 12.2) suggest that classroom attendance is a direct cause of improved grade outcomes: the Unique Effects Model (panel a) and the Mediated Effects Model (panel b). We assume that individual difference factors in these two models relate mostly to motivation, academic ability, prior knowledge of the students, and their

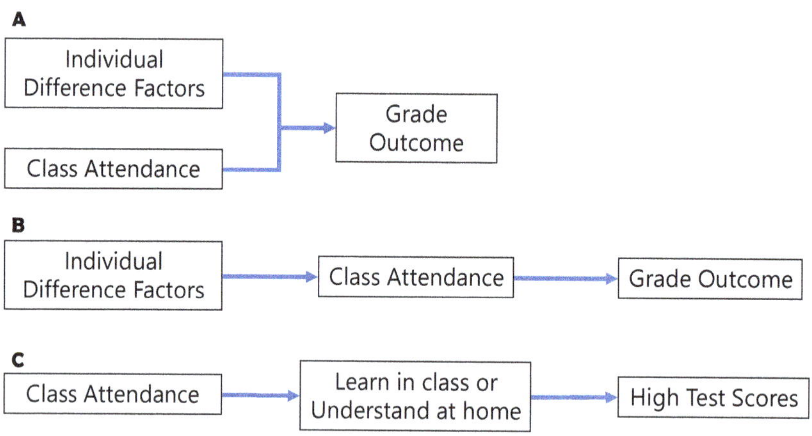

Fig. 12.2 Learning in the classroom. Three proposed mechanistic models to explain the link between classroom attendance and exam performance that incorporate extensive learning in the classroom. The Unique Effects Model (panel **a**) and the Mediated Effects Model (panel **b**) are based on Credé et al. (2010). The Classroom Explanation Model is our mechanistic expansion of these models (panel **c**)

Table 12.3 Valuable aspects of classroom attendance (Witchel et al., 2020)

Adaptive teaching and learning
Rapid teacher-to-student feedback
Rapid student-to-teacher feedback
Recognizing students' confusion and need for clarification
Adapting to students' weaknesses
Skills learning
Communication skills
Manual skills
Social learning
Joining a community of practice
Peer-to-peer learning
Additional social motivation for engagement
"Liveness"
Gestures
Body language
Crowd enthusiasm

ability to self-regulate learning. We propose that the effects of classroom attendance in both of these models could be mediated by causing students (1) to learn knowledge in class or (2) to understand the knowledge while studying at home (panel c). We propose that the Classroom Explanation Model is most relevant for skills learning (including communication skills, laboratory work, and microscopy), problem-solving, and humanities seminars where extensive learning occurs in class.

The Mediated Effects Model suggests that (for some students) the entirety of the learning benefits resulting from motivation and academic ability are mediated through classroom attendance. Credé et al. supported the Mediated Effects Model, based on two pieces of evidence. (1) The sheer weight of evidence and the number of studies showing a correlation between attendance and performance with an average rho of 0.44. They admit that the rho value for different studies has a wide spread (from 0.04 to 0.89) but that the preponderance of evidence appeals to them. (2) Subanalyses of their data set, either by topic (science versus nonscience) or by year, do not seem to change the correlation. In particular, because the year-by-year analysis of data until 2010 still supports the Mediated Effects Model, they conclude:

> it appears that use of online classroom resources and improved textbooks have not decreased the importance of attending class ... [and] does not suggest a dramatic change in attendance–grade correlations at the approximate point of the introduction of Web-based technologies.

We support the idea that many students benefit from both the time spent and the discipline of attending lectures and note-taking (Doggrell, 2020a; Matheson, 2008), despite current resistance to classroom lecturing (Emanuel, 2020; Freeman et al., 2014). We believe that there are a variety of unique educational benefits from the experience of attending a live classroom lecture that watching a video recording might not replicate (see Table 12.3).

These mechanisms suggest that the oft-observed link between attendance and test scores can have a mechanistic basis, which supports Credé et al.'s conclusion that classroom attendance can remain valuable, even in an era of online learning resources. However, the absolute nature of Credé et al.'s conclusions may not hold for postgraduate student cohorts, student cohorts with access to appropriately tailored substitute learning materials, and student cohorts with strong self-regulated learning skills. For these cohorts, the link between information transmitted by classroom lecture attendance and exam performance cannot possibly be absolute, as evidenced by the many research studies showing that some students can perform perfectly well with little or no classroom lecture attendance (see arrow on Fig. 12.1 panel b, (Azab et al., 2016), and Chap. 13 (distance learning) in this volume). For cohorts comprising primarily these exceptional students, the Mediated Effects Model only makes sense when they are learning a skill that requires access to a unique piece of equipment that may only be found in the classroom, clinic, or laboratory.

12.3.2 Support for Studying from the Classroom

The Classroom Explanation Model is only one possible mediator between classroom attendance and exam performance. An alternative is the Emphasis or Support Model (see Fig. 12.3), in which classroom attendance functions to support the student in coping with the available learning resources that may be studied at home (Valverde & Schmidt, 1997). This particular classroom support could be emphasis of the most

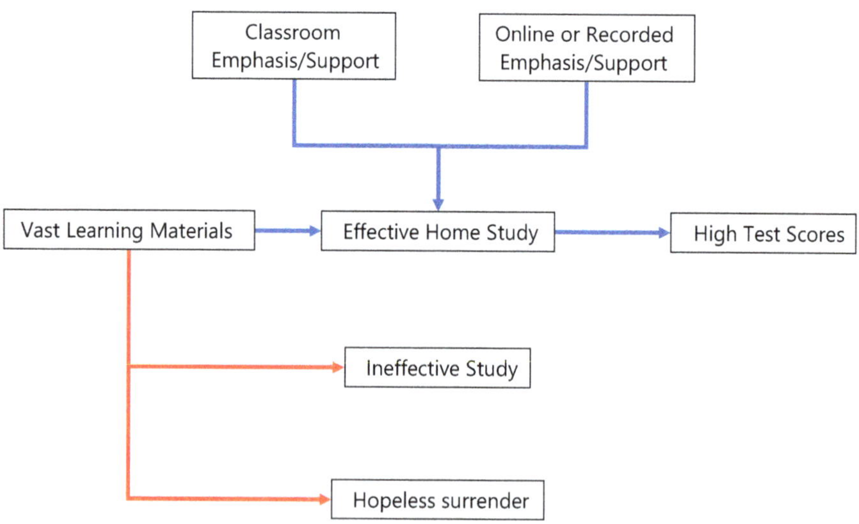

Fig. 12.3 The Emphasis/Support Model linking classroom attendance to exam performance. In this model mostly ancillary learning occurs during the classroom session

relevant knowledge for the test, which would allow students to prioritize their studying. This model linking classroom attendance to exam performance via emphasis would be most appropriate for unrecorded lectures pre-2000 because nowadays such information about prioritization could be provided by the ready availability of PowerPoint slides or even by watching lecture videos at double speed. By contrast, the classroom support could instead be academic confidence building, to provide students with the self-efficacy or self-regulation needed to attack the learning resources; this kind of social or skills support seems more likely to come from small group learning than from large lectures (see Chap. 4 (engagement) in this volume), PowerPoint slides, or videos. Another social benefit from lecture attendance comes from students relying on one another. That is, students often study together and help each other, and a student who misses a single lecture may experience no ill effects because of their relationships with other students; however, if a student misses many lectures, then their relationships with other students may become insufficient to gain help. In their relationships with each other, many students will not tolerate free riders (Hall & Buzwell, 2013).

Credé et al. give less support to the Unique Effects Model (Fig. 12.2, panel b) because the associations between performance and the various unique characteristics that they tested had smaller rho values. In their meta-analysis, the personality construct of conscientiousness (see Cheng & Ickes, 2009) was correlated with performance with a rho = 0.24, time spent studying had a rho = 0.20 linking it to performance, and high school grade point average had a rho = 0.16. While individually these are all small effect sizes, together the performance-enhancing effects of motivation, engagement with studying, and prior academic achievement seem unlikely to be eliminated even among very able students in postgraduate cohorts. So although the effect sizes supporting the academic benefits of student traits independent of attendance (i.e., the Unique Effects Model) are small, mechanistically they are likely to remain even in the discrepant studies. This mechanism needs to be tested more carefully.

12.3.3 Confounding: The Same Motivated Students Attend More and Study More

It is also possible that there is relatively little actual learning that takes place in the classroom and that the apparently causal link between attendance and high test scores is spurious. The model, labelled here as the Confounder Model (Fig. 12.4), is identical to Credé et al.'s (2010) Common Cause Model, except that where we place motivation and engagement, they put "individual difference factors." This model is most appropriate for topics focused on knowledge/theory learning in courses with sufficient accessible home study materials. The Confounder Model is especially relevant to any cohort with two populations incorporating a mixture of disengaged students with highly engaged students.

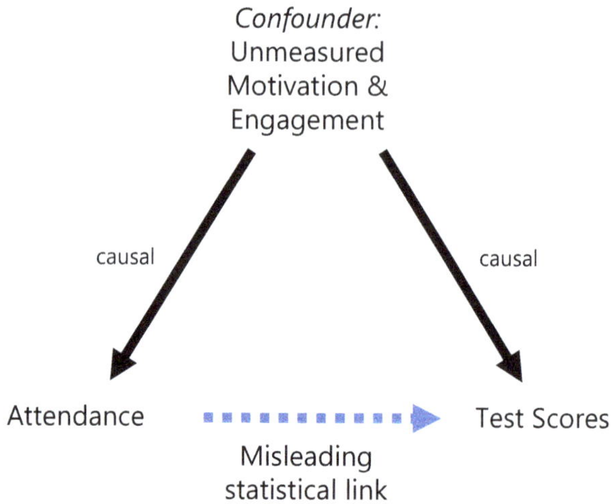

Fig. 12.4 The Confounder Model. This is effectively identical to Credé et al.'s (2010) "Common Cause Model"

Credé et al. rejected the Common Cause Model because they did a meta-analysis of three small studies (none in the life sciences) suggesting that mandatory attendance policies increase exam performance, albeit modestly. Kooker (1976) found a clear benefit of mandatory attendance in an undergraduate Y3/Y4 experimental psychology class that included a final project. Berenson et al. (1992) found a negligible effect of mandatory attendance on class performance on grades in an undergraduate Y1 remedial mathematics (algebra) class for at-risk students (Cohen's $d = 0.15$). Chan and Shum (1997) found no benefit (negative value) for mandatory attendance for final marks in an undergraduate principles of finance class. Another study supporting Credé et al.'s claim is Zhu et al. (2019), which found that mandatory attendance policies directly improved performance in computer programming/mathematics classes associated with skills for computing.

Several other studies on the effects of mandatory attendance in the life sciences are worth noting because they do not support the claim that mandatory attendance improves exam scores. In an undergraduate Y1 Foundations of Health and Disease course at the medical school at University of the West Indies, Cohall and Skeete (2012) observed no improvement in exam scores after they started enforcing attendance. In an undergraduate cell biology class with laboratory sessions at San Jose State University, Soto & Anand (2009) found no difference in class performance between years when mandatory attendance was eliminated. In a postgraduate oral pathology course at University of Louisville dental school, there was no difference found in performance when mandatory attendance was eliminated (Shumway et al., 2018). In all of these courses, a mandatory attendance policy did successfully result in an increase in attendance levels, yet they did not result in an improvement in apparent learning. In an exceptional study, Lamb et al. (2020) eliminated mandatory attendance in the postgraduate medical program at University of Utah and found that attendance at Y1/Y2 case-based learning sessions paradoxically increased, and so did their students' local test scores, passing rates, and scores on the national USMLE

Step 1 test. Of all these studies, only Kooker (1976) strongly supports mandatory attendance policies, and most other studies show that such a policy has little or no beneficial learning effects despite successfully increasing student attendance. We conclude that Credé et al. (2010) were premature in their preliminary rejection of their Common Cause Model, and we suspect that the Confounder Model may be valid in some cases. More research is needed to specifically test whether mandatory attendance policies are beneficial for university students in the biosciences, and these studies should be carefully considered with respect to the cohorts' self-regulation and motivation (i.e., postgraduate, lower level undergraduate, at-risk students). We speculate that some previously observed associations between attendance and exam performance may be coincidentally due to the same highly motivated students both attending lectures and also studying at home more diligently.

12.4 Differences in Statistical or Study Organization

12.4.1 Who Is Being Counted in the Statistics

12.4.1.1 Percentage of Failing Students

If the cohorts from those studies that found a link were mixtures between two populations, one highly engaged and the other disengaged, then one would expect not only a statistical link between attendance and exam performance but also a wide range of exam scores with a reasonable percentage of students having failing exam scores that reflect massive gaps in knowledge. If so, cohorts with a wide spread of scores and many students who fail (all indicating a potential for two populations) would show a clear link between attendance and test scores, whereas cohorts with uniformly high test scores would show no link. Thus, one possible marker for the source of the discrepancies where no link was found is when the cohort has few or no students who have failing test scores.

As an example of a cohort with many failing students, in the classic study showing a link, Moore's undergraduate Y1 cohort (2003) of 434 at-risk students included 22% of students who withdrew or failed and an additional 13% who received a grade of D (i.e., only 65% of Moore's students received a grade or either A, B or C). Although Moore does not list the actual numerical scores the students received on his exams, these failing percentages suggest that the demands of the taught material were at a level that many of his students found to be at or above their abilities. A similar finding occurs in Soto and Anand's (2009) undergraduate cohort, where 29% of their students failed the course (C- or below). Large failure rates are also seen in Horton et al. (2012) (22% of students scored below 50% of the final exam), Doggrell (2018) (mean final exam test score was 57.4% with a standard deviation of 14.0), and Hammen and Kelland (1994) (17% of the Spring 1990 cohort scored below 62% on the exam). So there are many examples where clear links between attendance and performance occur when a substantial percentage of the student cohort score poorly on the exam.

By contrast, in the studies where no clear link between attendance and performance are found, often the number of poorly performing students is small to nonexistent. In (Kauffman et al., 2018) none of the Y2 postgraduate medical students scored below 65% on their exam. Likewise, in Eisen et al. (2015), none of the postgraduate medical students in their cohort scored below 65% on the exam. Similarly in Azab et al. (2016) the mean grades for preclinical science courses in this postgraduate dental cohort was 86.48, while the standard deviation was 5.73, implying that a score below 65% was more than three standard deviations away from the mean. Whereas there are some exceptions to the rule that high scoring cohorts that have few failing students will not have a link between attendance and performance (e.g., Selvig et al., 2015), it occurs often enough that (a) it is worth investigating and (b) authors should clearly consider this when describing their cohort.

12.4.1.2 Removal of Failing Students

When Chan and Shum (1997) investigated the effect of attendance rates on course performance for undergraduate students in a Principles of Finance class at University of Wisconsin-Parkside, they were explicit about the importance of eliminating survival bias from their analysis. Survival bias is also known as selectivity bias or as censored sample bias; the sample is censored when students who withdraw from the course before its completion do not have their final course score included in the analysis. University of Wisconsin-Parkside admitted all students, so this course had a high attrition rate of 10–15%. If these withdrawing students all have a systematic bias (e.g., toward failing the course) rather than being randomly distributed, then if those students are not part of the final analysis, the overall correlation will be biased toward successful students. The advantage of including all students (even those who have withdrawn) in the final analysis is that it will be much more sensitive to a two-population model such as the Confounder Model. The disadvantage of including withdrawn or failing students in the analysis is that if you attempt to produce a correlation by assigning the withdrawn students with a final mark of zero, the withdrawn students will act as outliers that will mask any generalized effects of attendance (or lack thereof) on a single mainstream population, such as in the Classroom Explanation Model.

The importance of deciding whether to include outlier data in linear regression analysis was illustrated in a study of 39 students studying biomedical science (Doggrell, 2019). The only student to fail the course was an outlier in the regression analysis and skewed the results of the analysis. When including the failing student, there was a weak positive correlation between lecture attendance and examination mark and a moderate negative correlation between accessing lecture recordings and examination mark. Without the failing student, there was a moderate positive correlation between lecture attendance and examination mark and no association between accessing lecture recordings and examination mark. Thus, the ability to detect various univariate correlations depends on the selection of the students. Finally, it is possible that insufficient spread of the exam score data (such as when failing or withdrawn students are removed or do not exist) will lead to a failure to statistically detect an underlying correlation with attendance.

12.4.1.3 Weak Associations That Did Not Reach Significance

There are many studies where small numbers of students produced an apparently positive relationship between attendance and performance that did not reach significance, such as in the biomedical course for postgraduate pharmacy studies of Hidayat et al. (2012) or the Y2 biochemistry course of Davis et al. (2012). It is perplexing whether to categorize such results as paltry evidence in favor of a very weak link or evidence against any link. If the relationship between attendance and final performance is complicated by other contributors, such as in the Unique Effects Model, then the statistical link will be masked by those other contributors. For example, in a study by Ikonne et al. (2018) of 100 predoctoral (i.e., postgraduate) osteopathic medicine students at A.T. Still University School of Osteopathic Medicine in Arizona, the correlation coefficient between attendance and aggregate mean course grade was weak (Pearson $r = 0.17$) and did not reach significance ($P = 0.29$). However, after controlling for Medical College Admissions Test score, the trend approached significance ($r = 0.18$; $P = 0.08$) and one of the courses (Neuromusculoskeletal A) did have a significant link ($r = 0.22$; $P = 0.027$). To make sense of these insignificant but positive numerical results in small studies, researchers should report the effect size detected, the proposed value of a meaningful effect size, and the P values, rather than simply relying on the binary cutoff of $P < 0.05$ for insignificance (Curran-Everett, 2020; Wasserstein et al., 2019).

12.4.2 The Role of How Attendance and Performance Are Measured

For year 2 undergraduate medical students learning physiology at University of Donja Gorica, Podgorica in Montenegro, Popovic et al. (2018) found a strong link ($r = 0.56$) between attendance and the final summative grade. However, in these calculations, attendance and participation were explicitly part of the calculation of the final summative grade (20%); this means that it was inevitable that there would be a statistical link between the two metrics. As can be seen in Fig. 12.5, high attendance was necessary but not sufficient for a high final summative grade (note the empty region in the upper left of the scatter plot); this would be expected for a summative grade that incorporates attendance inside it. Popovic also included a plot of formative test scores versus attendance, which also showed a strong correlation,

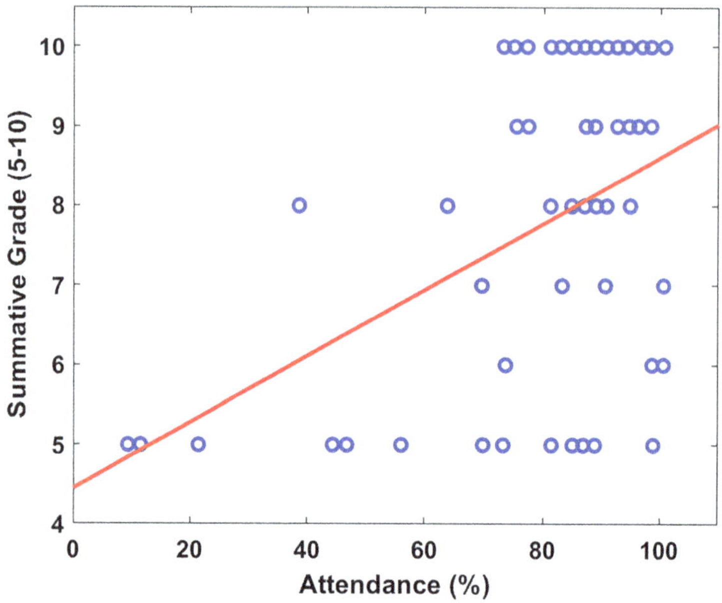

Fig. 12.5 Relationship between final summative grade and attendance for Popovic et al., 2018. The summative grade includes a component for participation, making attendance necessary but not sufficient for attaining a high final mark (the empty area in the upper left of the scatter plot). These final marks relate to the same study as the formative test marks shown in panel **a** of Fig. 12.1

Table 12.4 Power calculations for correlations at different *r* values

r	Sample size
0.15	347
0.20	194
0.25	123
0.45	36

For all of these calculations, alpha = 0.05 (two tailed) and beta = 0.20. Thus, the standard normal deviate for $\alpha = Z_\alpha = 1.9600$, and the standard normal deviate for $\beta = Z_\beta = 0.8416$. $C = 0.5 \times \ln[(1 + r)/(1 - r)]$. Total calculated sample size required $= N = [(Z_\alpha + Z_\beta)/C]^2 + 3$. See Hulley et al. (2013, appendix 6C, page 79) and https://sample-size.net/correlation-sample-size/

but this correlation was mostly driven by outlier points for students performing at the low end (see Fig. 12.1, panel a).

There has been some concern as to whether discrete categorization by attendance of students, followed by statistical tests for group differences, could be misleading compared to correlation, regression, and raw scatter plots. For small studies, searching for group differences improves the sensitivity compared to correlation; Table 12.4 shows some power calculations for correlations at different r values, demonstrating that weak correlations could easily be missed in studies with fewer

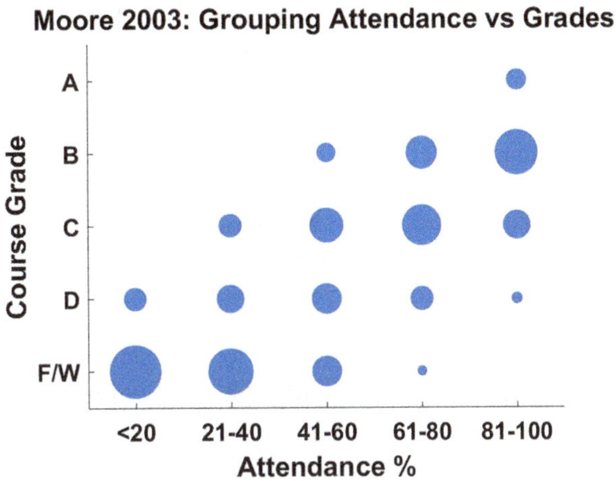

Fig. 12.6 Course grade versus attendance for the undergraduate cohort based on data from Moore (2003). In this bubble plot, the size of each circle represents the number of students in that group

than 200 students. This is not to imply that smaller studies are not worthwhile or unable to detect effects; merely that small positive effects that fail to reach significance when sample sizes are smaller than this are not surprising and should not be considered as evidence against an effect.

Some studies have included both correlations and group differences (Doggrell, 2020b; Doggrell, 2019). If the cutoff points for the categorizations are chosen poorly, it may fail to detect (at a level of significance) genuine effects. For significant effects, calculating group differences provides some useful information; however, for results suggesting that there is no difference, seeing the data in detail is quite important. Scatter plots are beneficial when claiming there is no correlation, and most journals these days suggest that the researchers post their data online for other researchers. It is worth remembering that group differences are ideal for testing two-population models, while correlations are more useful when testing for single-population models. When testing a single-population model by correlating cohorts with high levels of withdrawal, a way to avoid survival bias is to plot categorized groups where each parameter has three to five bins; in Fig. 12.6 we have plotted the data from Moore (2003) to show that for that undergraduate Y1 at-risk cohort, the link between attendance and performance demonstrates a single-population model (e.g., the Classroom Explanation Model).

A final concern regarding attendance measurements are that they are inaccurate. Before the advent of iClickers and other automated systems, the labor and time pressures of accurately recording the attendance of every individual student while simultaneously teaching a large class were problematic. In the past researchers have surveyed students at the end of the course (Selvig et al., 2015), asked class presidents to assess other students' attendance (Sade & Stroud, 1982), sampled a small number of sessions (Doggrell, 2018), asked students to sign in (Cohall & Skeete, 2012), or run a quiz every session and used the quiz to formulate attendance records (Soto & Anand, 2009). This is one problem where education technology plainly will provide the solution. All self-report systems are potentially inaccurate due to memory or social desirability issues, and even sign in sheets can be made inaccurate via the process of forging the signature or having the ID card of another student, so-called buddy punching. Newman-Ford et al. (2008) used a system called UniNanny at the University of Glamorgan to electronically record attendance at the university-wide level. Shumway et al. (2018) have used iClickers to have students sign into live classes. While the possibility of inaccurate attendance might affect the conclusions from some previous research studies, it seems unlikely that this problem is the cause of the main discrepancies in this field.

12.4.3 Limitation: Using Exam Scores as a Proxy for Learning

Most studies that objectively investigate the relationship between attendance and learning analyze examination scores as a surrogate for learning. There have been many arguments in the biological sciences to suggest that exam scores in general, and multiple-choice tests in particular, have limitations in recognizing true understanding and mastery (Momsen et al., 2010; Zheng et al., 2008). Learning in the biosciences is expected to become more project-based and interdisciplinary, which means that the relevant experimental design, quantitative abilities, and communication skills are becoming increasingly difficult to test with standardized pen and paper tests (National Research Council USA, 2004). Science educators accept that no single type of assessment will suffice for all functions for all audiences, which include formative assessment, summative assessment, and program evaluation (National Academies of Sciences, Engineering and Medicine, USA, 2018); understanding the relationship between attendance and learning straddles two apparently different purposes: assessing an individual student's readiness for progression versus the use of individual summative assessments to evaluate teaching programs. The use of summative assessments with specific correct answers is an inevitable element of cost-effective objective research on the benefits of most educational interventions, but we accept that this practice has limitations and is associated with reduced engagement and learning (Gibbs & Simpson, 2005).

12.5 Conclusions and Future Recommendations

12.5.1 Improving Learning Opportunities

Based on these results, there are many possible recommendations to improve learning by making it more active than a traditional lecture. As stated above, the increasing use of technology, especially the availability of lecture recordings, has been blamed for reducing lecture attendance (Edwards & Clinton, 2019; Traphagan et al., 2010). Students assuredly believe that lecture recordings can substitute for attendance (Kanter, 2012), and instructors do not like that the students believe that (Morris et al., 2019). While there seems to be little evidence that using video lectures as a substitute (as opposed to being a supplement) for in person lecture attendance results in an improvement in learning or test scores (Bos et al., 2016; Selvig et al., 2015), there is some evidence that having improved access to practice examination questions leads to better test results and improved learning (Karpicke & Roediger, 2007; Wenger et al., 2009). Rehearsing retrieval is an established mechanism for how regularly using practice questions can lead to improved learning (Szpunar et al., 2013), although replacing lectures with multiple-choice practice questions may do little for the student's ability to generalize ideas or to transfer knowledge between contexts.

More recently, activities such as questioning of students with clickers have been introduced into many lectures to increase the engagement with the students. Studies using clickers in lectures can be marked or unmarked. Clickers in lectures improve lecture attendance in courses which have also shown an association between lecture attendance and academic outcomes. This finding occurs both in courses where the clicker responses are marked (Freeman et al., 2007) or unmarked (Daniel & Tivener, 2016; Gauci et al., 2009). Thus, the use of clickers may promote attendance and possibly strengthen the association between lecture attendance and academic outcomes.

12.5.2 *Improving Lecture Capture*

12.5.2.1 Bookmarks, Links, and Lecture Length

It is worth considering how LCRs can be improved. Although the content of the LCR is fixed, it is nonetheless possible (with some systems) to bookmark certain points in the LCR. For example, the *Tegrity* lecture capture system allows students to bookmark particular slides in the LCR and send linked comments to other students or faculty. This approach could be used to improve or highlight lecture sections (Franklin et al., 2011). On the other hand, the faculty member could (in advance) bookmark sections of the LCR with links to supporting information. Lecture material may well already include "links," but these might only be active when the student is provided with printed lecture notes. At the present time, video recordings of lectures generally will not allow any displayed links to become active; this seems surprising as television viewers have had access to news ticker and teletext technologies such as the BBC's Ceefax since the 1970s. However, as consumers of educational technologies become more demanding, it may become standard to add in active links to the LCR to assist student understanding.

Another possibility for improvement that will also help students to find information more easily would be to shorten the length of LCR videos. The literature suggests that supplementary videos that are shorter in duration (6–20 min) are better for the student because they prevent cognitive overload (Brame, 2016; Crook & Schofield, 2017). Because currently much of online teaching has been adapted from traditional classroom teaching (where travel and attendance can only be justified by longer sessions), many LCRs tend to be the length of the lecture (50–60 min) and may not be very user friendly. Breaking the lecture down in to smaller discrete parts, each with its own integral structure may benefit students.

The most radical approach to LCRs is to reverse the point of view of LCRs as supplementary material by flipping the classroom (see Chap. 6 (Flipped Classroom) in this volume). In a flipped classroom, the LCR is considered the central focus for initial factual learning, and then the additional (possibly live classroom) material would be formulated to supplement or expand upon the video lecture (Emanuel, 2020). While classroom or online group activities (acting out concepts, clicker interactions, etc.) could be one kind of supplement, another would be further videos that incorporate quizzes and other self-directed learning activities that go beyond simply presenting information.

12.5.2.2 Active Learning with Embedded Problem-Solving

Students learn more from videos where relevant quizzing is interpolated within the information presented (Bjork et al., 2013; Szpunar et al., 2013); applying this teaching technique to full-length LCRs could also beneficial. Jing et al. (2016) conducted a study where students viewed a 40 min LCR with interposed quizzes. They found that this strategy improved learning and also improved both short- and long-term assimilation of information. They also found that test group participants reported more lecture-linked thoughts during mind wandering episodes and that these were positively related to final assessment. Thus, problem-based slides could be included at various points in the presentation but not discussed in class. It has been found that insertion of open problem-based content into LCRs improves effectiveness (van der Meij & Böckmann, 2020). In an experimental study, a 29-minute video was divided at salient points by insertion of open-ended questions and compared with a control video. The experimental group spent significantly more time viewing the video than the control group and also replayed the video more often. In addition there was a limited but significant increase in learning (van der Meij & Böckmann, 2020). The important thing here is that no feedback was provided for the embedded questions. The implication is that these questions encourage "extra thinking" where students consider the video content and attempt to link it to their existing knowledge. Thus, it is possible that a similar strategy could be used by inserting problems into LCRs with a view to improving the learning experience of viewers (see short video section related to interposed questions). The benefits of interpolating test questions seems obvious, but not all institutions do this, possibly because some LCR systems do not easily allow for it. Suppliers should add a feature that makes it straightforward for instructors to interpolate such thought questions postrecording.

12.5.3 Future Research

12.5.3.1 Acknowledging the Differences

Students need to engage in their studies in order to learn. Usually that successful engagement is driven by motivation, which for many students is driven by assessment (Wormald et al., 2009). Lecture attendance is only one small element of possible engagement for learning in the biosciences. Perhaps the reason there are dwindling student numbers in some classrooms is because the students do not believe there is a strong connection between attending those lectures and performing well on the assessment; if there were an unequivocal connection between attendance and grades, they would assuredly attend (see Fig. 12.5). Based on the wealth of previous research, we conclude that more and more carefully formulated research needs to be done to clarify the role of classroom attendance on student performance. This is not because the answer is unknown—there is already extensive and thought-provoking evidence in the literature including a large meta-analysis and a systematic review (Credé et al., 2010; Doggrell, 2020a), both of which support the benefits of live lecture attendance on exam performance.

However, the link between attendance and performance is a complicated one, in which there are many moderating factors, and the meaningful interpretation of the data requires a more nuanced description of both the actual policy decision being investigated and the precise educational situation being examined. For example, some studies have demonstrated an inverse relationship between religiously watching LCR videos and performance (Selvig et al., 2015); such a relationship, where increased LCR use (or other measurable engagement) is seemingly correlated with lower exam scores, might relate to an anxiety model (see Fig. 12.7), where students who are anxious about their lack of learning are most likely to watch an

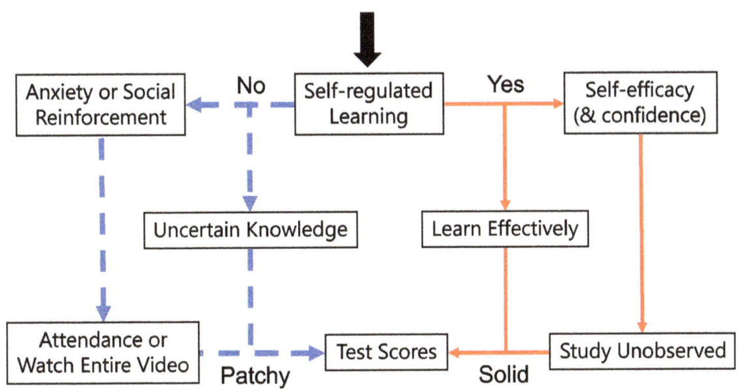

Fig. 12.7 The Anxiety/Social Model that inverts the relationship between measurable engagement and test scores. In this model, higher measurable engagement (attendance or video watching) is apparently linked with slightly lower test scores because it is confounded with lack of self-regulated learning

entire video lecture. Similarly, such a model may result from students who have low self-efficacy or need social reinforcement that comes from attending live lectures (Kauffman et al., 2018). In either case, the effective studying by self-regulated learners may be unmeasurable (e.g., home study), so that apparently measurable engagement is associated with weaker students. Such a counterintuitive model would be most likely for an unusual cohort of able learners studying a topic that has standardized knowledge materials.

The answer for a particular classroom depends on which of the three different issues that relate to this link is being considered: (a) at-risk students, (b) lecture quality, and (c) distribution of credit between final versus ongoing assessments. The recent discrepancies in results suggests that predoctoral (postgraduate) medical/dental students can and should be trusted individually to decide how best to master the factual material presented in lectures (Azab et al., 2016; Eisen et al., 2015; Kauffman et al., 2018; Lamb et al., 2020). However, attendance may still be beneficial where (1) skills are being taught, (2) students are not mature enough or self-regulated enough to manage their own learning, (3) substitute learning resources are not available for home study, or (4) the students are being enculturated into a social community.

12.5.3.2 Mandatory Attendance

While classroom attendance is associated with benefits, mandated classroom attendance at the university level is not supported by most studies (Chan & Shum, 1997; Cohall & Skeete, 2012; Lamb et al., 2020; Shumway et al., 2018; Soto & Anand, 2009). Although mandated attendance does improve attendance, it seems to have very little benefit for learning; it appears that you can lead a horse to water, but you cannot make it drink. The educational value of mandatory attendance policies needs to be more vigorously tested by making it the explicit research question of future studies, as it is currently often wrapped inside other studies.

More research is needed to explain the counterintuitive lack of benefit for mandating classroom attendance at lectures. These experiments are generally difficult to run properly (and ethically), as it is nontrivial to create two equivalent cohorts that either do or do not have required attendance. Even if you have the cohorts and separately educate them, you still have to test them equivalently. This almost certainly would require a national test with a very well characterized question set. This data would then need to be coordinated with survey or interview information in a mixed methods approach, in order to understand how enforced attendance/engagement leads to no learning benefit. The current trend is moving away from universities teaching with lectures, despite their cost efficiencies (Freeman et al., 2014; Laurillard, 2013), so it may be that these experiments will not be properly funded for some time, or even that some policymakers may consider the answers unnecessary.

Given that truancy was at one stage a crime for some students, both the students' details and our society's progress need to be considered carefully. Part of our goal in education is to socialize students, whether into a profession or into a society. We have to balance this with our respect for freedom and thinking for yourself, or more properly as liberty and dissent.

12.5.3.3 At-Risk Students

At-risk students are often identified either by exam performance or by ad hoc reporting of behavioral violations. Attendance monitoring is an effort to use ongoing automated monitoring to predict exam performance, and it simply will not generalize. Most students do need to attend class in order to gain, but technology has created some exceptional situations where some students simply do not need to attend in order to learn basic facts or even concepts. The idea that attendance is consistently a good proxy for exam performance cannot be uniformly generalized.

However, this does not mean that absenteeism at educational events can be ignored. Community engagement and professionalism are genuine educational outcomes, particularly for educations that end with a golden ticket (Eardley et al., 2015). It is a rarely observed fact that "best practice" sought in many professions often changes, sometimes even reversing. Best practice is really mandating community standards and uniformity. The idea that students can learn everything in isolation and then emerge like a chrysalis ready to join a professional cadre seems very unlikely. This harks back to previous arguments about dissent and liberty but is a reminder that even in our age of remote computer meetings, no person is an island.

12.5.3.4 Balancing Ongoing Student Work and Assessments in Final Marks

Should we allow students to receive good marks on the basis of their compliance, even if their skills cannot be demonstrated on a final exam? That is, should a student be allowed to pass primarily on the basis of their coursework, even if they fail the exam? From an assessment perspective, this questions whether we should allow compensation between different elements of a learning process. Another version of that question is should we grant a degree to a student doctor who learned absolutely everything about the thorax but skipped the abdomen. To appropriately respond to such detailed questions, we have to understand what is the meaning and purpose of the particular qualification. A straightforward policy is that a student must pass all the separate elements deemed as essential for a particular degree, and that extra credit on one assessment should not be able to compensate for another assessment that is failed. If we mandate that passing one course will not allow compensation for another failed one, then we can prevent incompetent graduates from practicing.

But this does not solve the wider question of government incentives to universities (Doggrell, 2020c; Ryan, 2004). If more graduates are needed, and the government financially incentivizes universities to achieve completion rates approaching 100% without also mandating national exit exams (Martin, 2015), there is a risk that some institutions will succumb to temptation leading to lowering of standards (Stilwell, 2003). If this is institution-wide, it becomes a kind of corruption. This can only be fixed at the government level. If the government strongly incentivizes degree completion without insisting on quality education, institutional trustees will inevitably respond to financial pressures.

12.5.3.5 Our Research Recommendations

While more research on this topic needs to occur, we propose a range of suggestions for both designing and reporting research on this topic. Most importantly, we believe that research teams should be explicit about the underlying policy they are trying to research. They should also be clear about the model (two-population or single-population) they are looking for. These two models are not mutually exclusive, and there will be situations that are a mixture of both; nevertheless, these models will help to clarify thinking and hypothesis testing. In terms of reporting outcomes, we have a range of suggestions:

1. Report whether the cohort is primarily postgraduate or undergraduate.
2. Report the mode/median age of the students.
3. Report on variables that affect student self-regulation, especially whether students are from an at risk or academically challenged cohort.
4. Explicitly consider the number of students who are plainly failing or have learned less than 60% of the taught material.
5. Report effect size (and the proposed meaningful effect size) in addition to P value.
6. Include some graphical output that shows the variability of the cohort's performance clearly, such as a scatter plot of attendance versus exam score data with a best fit line. For skewed or bimodal distributions, this would be revealing, and it is essential for reports where the association is not statistically significant.
7. Be explicit about testing single-population and two-population models when statistically testing correlations versus group comparisons.
8. Be explicit about the role of skills (as opposed to factual knowledge) in the assessment.
9. Be explicit about the role of laboratories, and if possible remove laboratory performance from reported test scores.
10. Report whether attendance is mandatory and whether students receive credit for it, either explicitly or implicitly (i.e., in daily quizzes that have a summative component).
11. Report the method of capturing data for attendance, and consider explicitly its accuracy.
12. Report the type of test/exam of performance (e.g., constructed versus selected), and consider its relationship to the relevant knowledge.
13. If attendance is mandatory, report pure test scores rather than a final grade that includes attendance or quiz scores that depend on attendance.

14. Report the presence (or absence) of lecture capture.
15. Distinguish between supplementary and substitution use of lecture capture by measuring lecture capture use, attendance, and performance in the same students.

References

Adesope, O. O., Trevisan, D. A., & Sundararajan, N. (2017). Rethinking the use of tests: A meta-analysis of practice testing. *Review of Educational Research, 87*(3), 659–701.

Albon, S. P., Larson, K., & Marchand, J. P. (2020). Lecture capture in pharmacy education at UBC: Updating our understanding. *Currents in Pharmacy Teaching and Learning, 12*(9), 1037–1045.

Alexander, V., & Hicks, R. E. (2016). Does class attendance predict academic performance in first year psychology tutorials. *International Journal of Psychological Studies, 8*(1), 28–32.

An Bord Altrainais. (1994). *Rules for the education and training of student nurses*. An Bord Altrainais.

Astin, A. W. (1993). *What matters in college: Four critical years revisited*. Jossey-Bass.

Azab, E., Saksena, Y., Alghanem, T., Bidle, J. B., Molgaard, K., Albright, S., & Karimbux, N. (2016). Association among dental students' class lecture attendance, use of online resources, and performance. *Journal of Dental Education, 80*(4), 452–458.

Banerjee, S. (2020). To capture the research landscape of lecture capture in university education. *Computers & Education, 160*, 104032.

Beech, D. (2017). *Going for gold: Lessons from the TEF provider submissions*. Higher Education Policy Institute. Available online at: www.hepi.ac.uk. Accessed 12 Jan 2018

Berenson, S. B., Carter, G., & Norwood, K. S. (1992). The at-risk student in college developmental algebra. *School Science and Mathematics, 92*(2), 55–58.

Bjork, R. A., Dunlosky, J., & Kornell, N. (2013). Self-regulated learning: Beliefs, techniques, and illusions. *Annual Review of Psychology, 64*, 417–444.

Bos, N., Groeneveld, C., Van Bruggen, J., & Brand-Gruwel, S. (2016). The use of recorded lectures in education and the impact on lecture attendance and exam performance. *British Journal of Educational Technology, 47*(5), 906–917.

Brame, C. J. (2016). Effective educational videos: Principles and guidelines for maximizing student learning from video content. *CBE—Life Sciences Education, 15*(4), es6.

Brown, B., Graham, C., Money, S., & Rakoczy, M. (1999). Absenteeism and grades in a nursing curriculum. *Michigan Community College Journal: Research & Practice, 5*(2), 81–84.

Brownstein, N. C., Louis, T. A., O'Hagan, A., & Pendergast, J. (2019). The role of expert judgment in statistical inference and evidence-based decision-making. *The American Statistician, 73*(suppl 1), 56–68.

Chan, K. C., & Shum, C. (1997). Class attendance and student performance in principles of finance. *Financial Practice & Education, 7*(2), 58–65.

Charlton, B. G. (2006). Lectures are an effective teaching method because they exploit human evolved "human nature" to improve—Editorial. *Medical Hypotheses, 67*, 1261–1265.

Cheng, W., & Ickes, W. (2009). Conscientiousness and self-motivation as mutually compensatory predictors of university-level GPA. *Personality and Individual Differences, 47*(8), 817–822.

Cohall, D. H., & Skeete, D. (2012). The impact of an attendance policy on the academic performance of first year medical students taking the Fundamentals of Disease and Treatment course. *The Caribbean Teaching Scholar, 2*(2), 115–123.

Colby, J. (2005). Attendance and attainment—A comparative study. *Innovation in Teaching and Learning in Information and Computer Sciences, 4*(2), 1–13.

Credé, M., Roch, S. G., & Kieszczynka, U. M. (2010). Class attendance in college: A meta-analytic review of the relationship of class attendance with grades and student characteristics. *Review of Educational Research, 80*(2), 272–295.

Crook, C., & Schofield, L. (2017). The video lecture. *The Internet and Higher Education, 34*, 56–64.

Cruz, T. E., Riddell, R., Gamaldo, C. E., Strowd, R. E., Oakley, C., & Salas, R. M. E. (2019). The impact of student absences on grade outcomes in a neurology clerkship setting. *MedEdPublish, 8*.

Curran-Everett, D. (2020). Evolution in statistics: P values, statistical significance, kayaks, and walking trees. *Advances in Physiology Education, 44*, 221–224.

Cutler, C. W., Parise, M., Seminario, A. L., Mendez, M. J. C., Piskorowski, W., & Silva, R. (2016). Should attendance be required in lecture classrooms in dental education? Two viewpoints: Viewpoint 1: Attendance in the lecture classroom should be required and viewpoint 2: Attendance should not be required in the lecture classroom. *Journal of Dental Education, 80*(12), 1474–1478.

Daniel, T., & Tivener, K. (2016). Effects of sharing clickers in an active learning environment. *Journal of Educational Technology & Society, 19*(3), 260–268.

Daud, A., Bagria, A., Shah, K., & Puryer, J. (2017). Should undergraduate lectures be compulsory? The views of dental and medical students from a UK university. *Dentistry Journal, 5*(2), 15.

Davis, E. A., Hodgson, Y., & Macaulay, J. O. (2012). Engagement of students with lectures in biochemistry and pharmacology. *Biochemistry and Molecular Biology Education, 40*(5), 300–309.

Davis, M. J., Wythe, J., Rozum, J. S., & Gore, R. W. (1997). Use of World Wide Web server and browser software to support a first-year medical physiology course. *Advances in Physiology Education, 272*(6), S1–S14.

Deane, R. P., & Murphy, D. J. (2013). Student attendance and academic performance in undergraduate obstetrics/gynecology clinical rotations. *Journal of the American Medical Association, 310*(21), 2282–2288.

Doggrell, S. A. (2018). Does attending lectures matter when lecture recordings are available? Results for a preliminary study comparing attending and non-attending nursing students in bioscience and pharmacology. *International Journal of Innovation in Science and Mathematics Education, 26*(7), 1–10.

Doggrell, S. A. (2019). The relationships between lecture attendance or accessing lecture recordings and academic outcomes: Results from a pharmacology course in a biomedical science degree. *International Journal of Innovation in Science and Mathematics Education, 27*(5), 1–12.

Doggrell, S. A. (2020a). A systematic review of the relationship between lecture attendance and academic outcomes for students studying the human biosciences. *International Journal of Innovation in Science and Mathematics Education, 28*(1), 60–76.

Doggrell, S. A. (2020b). No apparent association between lecture attendance or accessing lecture recordings and academic outcomes in a medical laboratory science course. *BMC Medical Education, 20*(1), 1–12.

Doggrell, S. A. (2020c). Descriptive study of how proportioning marks determines the performance of nursing students in a pharmacology course. *BMC Nursing, 19*(1), 1–10.

Eardley, I., Scott, H., & Wilkinson, D. (2015). Why a career in surgery is no longer the golden ticket. *British Medical Journal, 351*, h3685.

Ebbert, D., & Dutke, S. (2020). Patterns in students' usage of lecture recordings: A cluster analysis of self-report data. *Research in Learning Technology, 28*, 2258.

Edwards, M. R., & Clinton, M. E. (2019). A study exploring the impact of lecture capture availability and lecture capture usage on student attendance and attainment. *Higher Education, 77*(3), 403–421.

Eisen, D. B., Schupp, C. W., Isseroff, R. R., Ibrahimi, O. A., Ledo, L., & Armstrong, A. W. (2015). Does class attendance matter? Results from a second-year medical school dermatology cohort study. *International Journal of Dermatology, 54*(7), 807–816.

Emanuel, E. J. (2020). The inevitable reimagining of medical education. *Journal of American Medical Association, 323*(12), 1127–1128.

Essaji, H. (2020). US vs UK doctors salary—Detailed comparison. https://revisingrubies.com/us-vs-uk-doctors-salary/. Accessed 23 Dec 2020.

Feldman, S. (2013, March 14). Should university lectures be compulsory? Student perspective. *The Guardian* newspaper, UK. https://www.theguardian.com/education/mortarboard/2013/mar/14/should-university-lectures-be-compulsory. Accessed 12 Apr 2020.

Franklin, D. S., Gibson, J. W., Samuel, J. C., Teeter, W. A., & Clarkson, C. W. (2011). Use of lecture recordings in medical education. *Medical Science Educator, 21*(1), 21–28.

Freeman, S., Eddy, S. L., McDonough, M., Smith, M. K., Okoroafor, N., Jordt, H., & Wenderoth, M. P. (2014). Active learning increases student performance in science, engineering, and mathematics. *Proceedings of the National Academy of Sciences of the United States of America, 111*(23), 8410–8415.

Freeman, S., O'Connor, E., Parks, J. W., Cunningham, M., Hurley, D., Haak, D., Dirks, C., & Wenderoth, M. P. (2007). Prescribed active learning increases performance in introductory biology. *CBE—Life Sciences Education, 6*(2), 132–139.

Friesen, N. (2011). The lecture as a transmedial pedagogical form: A historical analysis. *Educational Researcher, 40*(3), 95–102.

Gauci, S. A., Dantas, A. M., Williams, D. A., & Kemm, R. E. (2009). Promoting student-centered active learning in lectures with a personal response system. *Advances in Physiology Education, 33*(1), 60–71.

General Medical Council, UK. (2020). *Medical licensing assessment.* https://www.gmc-uk.org/education/medical-licensing-assessment. Accessed 23 Dec 2020.

Gibbs, G., & Simpson, C. (2005). Conditions under which assessment supports students' learning. *Learning and Teaching in Higher Education, 1*, 3–31. ISSN 1742240X.

Glanz, K., & Fiel, S. B. (1984). Effect of attendance at lectures on medical student performance. *Academic Medicine, 59*(6), 516–518.

Hall, D., & Buzwell, S. (2013). The problem of free-riding in group projects: Looking beyond social loafing as reason for non-contribution. *Active Learning in Higher Education, 14*(1), 37–49.

Hammen, C. S., & Kelland, J. L. (1994). Attendance and grades in a human physiology course. *Advances in Physiology Education, 267*(6), S105–S108.

Hanushek, E. A., & Wößmann, L. (2006). educational tracking affect performance and inequality? Differences-in-differences Does evidence across countries. *The Economic Journal, 116*(510), C63–C76.

Harden, N. (2013). The end of the university as we know it. *The American Interest, 8*(3), 54–62.

Hassel, H., & Lourey, J. (2005). The dea (r) th of student responsibility. *College Teaching, 53*(1), 2–13.

Hidayat, L., Vansal, S., Kim, E., Sullivan, M., & Salbu, R. (2012). Pharmacy student absenteeism and academic performance. *American Journal of Pharmaceutical Education, 76*(1), 8.

Horton, D. M., Wiederman, S. D., & Saint, D. A. (2012). Assessment outcome is weakly correlated with lecture attendance: Influence of learning style and use of alternative materials. *Advances in Physiology Education, 36*(2), 108–115.

Hulley, S. B., Cummings, S. R., Browner, W. S., Grady, D., & Newman, T. B. (2013). *Designing clinical research : An epidemiologic approach* (4th ed.). Lippincott Williams & Wilkins.

Ikonne, U., Campbell, A. M., Whelihan, K. E., Bay, R. C., & Lewis, J. H. (2018). Exodus from the classroom: Student perceptions, lecture capture technology, and the inception of on-demand preclinical medical education. *Journal of American Osteopathic Association, 118*(12), 813–823.

Jing, H. G., Szpunar, K. K., & Schacter, D. L. (2016). Interpolated testing influences focused attention and improves integration of information during a video-recorded lecture. *Journal of Experimental Psychology: Applied, 22*, 305–318.

Joseph-Richard, P., Jessop, T., Okafor, G., Almpanis, T., & Price, D. (2018). Big brother or harbinger of best practice: Can lecture capture actually improve teaching? *British Educational Research Journal, 44*(3), 377–392.

Kanter, S. L. (2012). To be there or not to be there: Is attendance really the question? *Academic Medicine, 87*(6), 679.

Karpicke, J. D., & Roediger, H. L. (2007). Expanding retrieval practice promotes short-term retention, but equally spaced retrieval enhances long-term retention. *Journal of Experimental Psychology Learning Memory Cognition, 33*, 704–719.

Kauffman, C. A., Derazin, M., Asmar, A., & Kibble, J. D. (2018). Relationship between classroom attendance and examination performance in a second-year medical pathophysiology class. *Advances in Physiology Education, 42*, 592–598.

Kauffman, C. A., Derazin, M., Asmar, A., & Kibble, J. D. (2019). Patterns of medical student engagement in a second-year pathophysiology course: Relationship to USMLE Step 1 performance. *Advances in Physiology Education, 43*, 512–518.

King, A. (1993). From sage on the stage to guide on the side. *College Teaching, 41*(1), 30–35.

Kooker, E. W. (1976). Changes in grade distributions associated with changes in class attendance policies. *Psychology: A Journal of Human Behavior, 13*(1), 56–57.

Lamb, S., Chow, C., Lindsley, J., Stevenson, A., Roussel, D., Shaffer, K., & Samuelson, W. (2020). Learning from failure: How eliminating required attendance sparked the beginning of a medical school transformation. *Perspectives on Medical Education, 9*(5), 314–317.

Laurillard, D. (2013). *Rethinking university teaching: A conversational framework for the effective use of learning technologies*. Routledge.

Leadbeater, W., Shuttleworth, T., Couperthwaite, J., & Nightingale, K. P. (2013). Evaluating the use and impact of lecture recording in undergraduates: Evidence for distinct approaches by different groups of students. *Computers & Education, 61*, 185–192.

Lee, M. W., Johnson, T. R., & Kibble, J. (2017). Development of statistical models to predict medical student performance on the USMLE Step 1 as a catalyst for deployment of student services. *Medical Science Educator, 27*(4), 663–671.

Lukkarinen, A., Koivukangas, P., & Seppälä, T. (2016). Relationship between class attendance and student performance. *Procedia-Social and Behavioral Sciences, 228*(16), 341–347.

Martin, S. (2015, September 9). Pushback Against Performance Funding: As Australian candidate suggests linking government funds to student retention and job outcomes, university leaders warn of unintended consequences. *Inside Higher Ed (The Australian)*. https://www.insidehighered.com/news/2015/09/09/australian-universities-balk-suggestion-funds-be-tied-retention. Accessed 19 Jan 2021.

Matheson, C. (2008). The educational value and effectiveness of lectures. *The Clinical Teacher, 5*(4), 218–221.

Mattick, K., Crocker, G., & Bligh, J. (2007). Medical student attendance at non-compulsory lectures. *Advances in Health Sciences Education, 12*(2), 201–210.

McCarey, M., Barr, T., & Rattray, J. (2007). Predictors of academic performance in a cohort of pre-registration nursing students. *Nurse Education Today, 27*(4), 357–364.

McCombs, B. L., & Marzano, R. J. (1990). Putting the self in self-regulated learning: The self as agent in integrating will and skill. *Educational Psychologist, 25*(1), 51–69.

McKenzie, K., & Schweitzer, R. (2001). Who succeeds at university? Factors predicting academic performance in first year Australian university students. *Higher Education Research & Development, 20*(1), 21–33.

McVicar, A., Andrew, S., & Kemble, R. (2015). The 'bioscience problem' for nursing students: An integrative review of published evaluations of Year 1 bioscience, and proposed directions for curriculum development. *Nurse Education Today, 35*(3), 500–509.

Momsen, J. L., Long, T. M., Wyse, S. A., & Ebert-May, D. (2010). Just the facts? Introductory undergraduate biology courses focus on low-level cognitive skills. *CBE Life Sciences Education, 9*, 435–440. https://doi.org/10.1187/cbe.10-01-0001

Moore, R. (2003). Attendance and performance. *Journal of College Science Teaching, 32*(6), 367–371.

Moore, R. (2006). The importance of admissions scores and attendance to first-year performance. *Journal of the First-Year Experience & Students in Transition, 18*(1), 105–125.

Morris, N. P., Swinnerton, B., & Coop, T. (2019). Lecture recordings to support learning: A contested space between students and teachers. *Computers & Education, 140*, 103604.

Mustafa, T., Farooq, Z., Asad, Z., Amjad, R., Badar, I., Chaudhry, A. M., Zaman Khan, M. A., & Rafique, F. (2014). Lectures in medical education: What students think? *Journal of Ayub Medical College Abbottabad, 26*(1), 21–25.

National Academies of Sciences, Engineering and Medicine, USA. (2018). *How people learn II: Learners, contexts and cultures*. National Academies Press.

National Research Council, USA. (2004). *Bio2010: Transforming undergraduate education for future research biologists*. National Academies Press.

Newman, S. N., Schuman, N. J., Fields, W. T., & Nunez, L. (1981). Dental students' grades and their relationship to classroom attendance. *Journal of Dental Education, 45*(6), 360–361.

Newman-Ford, L., Fitzgibbon, K., Lloyd, S., & Thomas, S. (2008). A large-scale investigation into the relationship between attendance and attainment: A study using an innovative, electronic attendance monitoring system. *Studies in Higher Education, 33*(6), 699–717.

Nieder, G. L., Parmelee, D. X., Stolfi, A., & Hudes, P. D. (2005). Team-based learning in a medical gross anatomy and embryology course. *Clinical Anatomy, 18*(1), 56–63.

Nordmann, E., Calder, C., Bishop, P., Irwin, A., & Comber, D. (2019). Turn up, tune in, don't drop out: The relationship between lecture attendance, use of lecture recordings, and achievement at different levels of study. *Higher Education, 77*(6), 1065–1084.

Nordmann, E., Küepper-Tetzel, C. E., Robson, L., Phillipson, S., Lipan, G. I., & Mcgeorge, P. (2020). Lecture capture: Practical recommendations for students and instructors. *Scholarship of Teaching and Learning in Psychology*. https://doi.org/10.1037/stl0000190

Owston, R., Lupshenyuk, D., & Wideman, H. (2011). Lecture capture in large undergraduate classes: Student perceptions and academic performance. *The Internet and Higher Education, 14*(4), 262–268.

Popovic, N., Popovic, T., Rovcanin Dragovic, I., & Cmiljanic, O. (2018). A Moodle-based blended learning solution for physiology education in Montenegro: A case study. *Advances in Physiology Education, 42*(1), 111–117.

Ryan, J. E. (2004). The perverse incentives of the no child left behind act. *NYU Law Review, 79*, 932–989.

Sade, R. M., & Stroud, M. R. (1982). Medical student attendance at lectures: Effect on medical school performance. *Academic Medicine, 57*(3), 191–192.

Salamonson, Y., Andrew, S., & Everett, B. (2009). Academic engagement and disengagement as predictors of performance in pathophysiology among nursing students. *Contemporary Nurse, 32*(1–2), 123–132.

Schwartzstein, R. M., & Roberts, D. H. (2017). Saying goodbye to lectures in medical school—Paradigm shift or passing fad. *New England Journal of Medicine, 377*(7), 605–607.

Selvig, D., Holaday, L. W., Purkiss, J., & Hortsch, M. (2015). Correlating students' educational background, study habits, and resource usage with learning success in medical histology. *Anatomical Sciences Education, 8*(1), 1–11.

Shumway, B. S., Bernstein, M. L., Qian, C., Kulkarni, M. Y., & Rai, S. N. (2018). Effect of lecture attendance and prerequisite academic outcomes on dental students' oral pathology performance. *Journal of Dental Education, 82*(3), 306–312.

Smith, L. B. (2012). Medical school and online learning: Does optional attendance create absentee doctors? *Medical Education, 2*(46), 137–138.

Soto, J. G., & Anand, S. (2009). Factors influencing academic performance of students enrolled in a lower division Cell Biology core course. *Journal of the Scholarship of Teaching and Learning, 9*(1), 64–80.

St. Clair, K. L. (1999). A case against compulsory class attendance policies in higher education. *Innovative Higher Education, 23*(3), 171–180.

Steinel, N., Palmer, G. C., Nowicki, E., Lee, E., Nelson, E., Whiteley, M., & Lee, M. W. (2019). Integration of microbiology, pharmacology, immunology, and infectious disease using active teaching and self-directed learning. *Medical Science Educator, 29*(1), 315–324.

Stilwell, F. (2003). Higher education, commercial criteria and economic incentives. *Journal of Higher Education Policy and Management, 25*(1), 51–61.

Subramaniam, B. S., Hande, S., & Komattil, R. (2013). Attendance and achievement in medicine: Investigating the impact of attendance policies on academic performance of medical students. *Annals of Medical and Health Sciences Research, 3*(2), 202–205.

Szpunar, K. K., Khan, N. Y., & Schacter, D. L. (2013). Interpolated memory tests reduce mind wandering and improve learning of online lectures. *Proceedings of the National Academy of Sciences of the United States of America, 110*(16), 6313–6317.

Tower, M., Walker, R., Wilson, K., Watson, B., & Tronoff, G. (2015). Engaging, supporting and retaining academic at-risk students in a Bachelor of Nursing: Setting risk markers, interventions and outcomes. *International Journal of the First Year in Higher Education, 6*(1), 121–134.

Traphagan, T., Kucsera, J. V., & Kishi, K. (2010). Impact of class lecture webcasting on attendance and learning. *Educational Technology Research and Development, 58*(1), 19–37.

University of Sheffield. (2020). *Guidance on the quality assurance of service teaching*. https://www.sheffield.ac.uk/lets/pp/qa/service. Accessed 18 Dec 2020.

Valverde, G. A., & Schmidt, W. H. (1997). Refocusing US math and science education. *Issues in Science and Technology, 14*, 60–66.

van der Meij, H., & Böckmann, L. (2020). Effects of embedded questions in recorded lectures. *Journal of Computing in Higher Education, 33*, 235–254. https://doi.org/10.1007/s12528-020-09263-x

Wasserstein, R. L., & Lazar, N. A. (2016). The ASA statement on p-values: Context, process, and purpose. *The American Statistician, 70*(2), 129–133.

Wasserstein, R. L., Schirm, A. L., & Lazar, N. A. (2019). Moving to a world beyond "p< 0.05". *The American Statistician, 73*(S1), 1–19.

Wenger, S. L., Hobbs, G. R., Williams, H. J., Hays, M. A., & Ducatman, B. S. (2009). Medical student study habits: Practice questions help exam scores. *Journal of the International Association of Medical Sciences Education, 19*(4), 170–172.

West, C., & Sadoski, M. (2011). Do study strategies predict academic performance in medical school? *Medical Education, 45*(7), 696–703.

Westrick, S. C., Helms, K. L., McDonough, S. K., & Breland, M. L. (2009). Factors influencing pharmacy students' attendance decisions in large lectures. *American Journal of Pharmaceutical Education, 73*(5), 83.

Witchel, H. J., Langlands, K., MacQueen, H. A., Kumar, P., Torrens, C., & Hall, S. K. (2020). Online learning versus classroom learning: Questioning who learns what. *Physiology News, 119*, 16–17. https://doi.org/10.36866/pn.119.16. https://www.physoc.org/magazine-articles/online-learning-versus-classroom-learning/

Wormald, B. W., Schoeman, S., Somasunderam, A., & Penn, M. (2009). Assessment drives learning: An unavoidable truth? *Anatomical Sciences Education, 2*(5), 199–204. https://doi.org/10.1002/ase.102

Zazulia, A. R., & Goldhoff, P. (2014). Faculty and medical student attitudes about preclinical classroom attendance. *Teaching and Learning in Medicine, 26*(4), 327–334.

Zheng, A. Y., Lawhorn, J. K., Lumley, T., & Freeman, S. (2008). Application of Bloom's taxonomy debunks the "MCAT myth". *Science, 319*, 414–415. https://doi.org/10.1126/science.1147852

Zhu, L., Huang, E., Defazio, J., & Hook, S. A. (2019). Impact of the stringency of attendance policies on class attendance/participation and course grades. *Journal of the Scholarship of Teaching and Learning, 19*(2), 130–140. https://doi.org/10.14434/josotl.v19i1.23717

Chapter 13
Online Science Education at Scale: Open and Distance Learning, MOOCS, and Other Learning Assets for Theory and Practice

Peter Alston, Terry Gleave, Mark Hirst, and Hilary MacQueen

Abstract In this chapter we examine what is meant by online distance learning. We consider briefly the history and refinement of the early correspondence course model and outline the modern principles of learning design that ensure an enjoyable deep learning experience for students. We also consider the breadth of the distance learning audience and look at some of the barriers to learning that they may experience. The well-intentioned aspirations of truly open education, accessible to all, currently remain unfulfilled, but a good deal of progress has been made in that area. A range of learning offerings exists, from long courses (up to 1/2 of a year of full-time study) through to short learning assets that can be studied in a few hours. We note the situations in which MOOCs (Massive Open Online Courses), SPOCs (Small Private Online Courses), BOCs (Badged Open Courses), and practical assets can supplement or replace traditional face-to-face teaching methods. We use a case study to illustrate the mechanics of producing a successful MOOC. We also describe the issues around teaching practical science online and offer some examples of how this approach can not only be pedagogically successful but can also deliver a learning experience that students enjoy. In particular we point out the need for attention to detail during production, good quality assurance processes to ensure maximum accessibility, and the value of ongoing engagement with, and support of, learners. These features enable the production and delivery of effective, deep, and enjoyable learning.

P. Alston
School of Technology, BPP Education Group, Manchester, UK

T. Gleave
School of Life Sciences, University of Liverpool, Liverpool, UK

M. Hirst · H. MacQueen (✉)
School of Life, Health and Chemical Sciences, The Open University, Milton Keynes, UK
e-mail: hilary.macqueen@open.ac.uk

H. J. Witchel, M. W. Lee (eds.), *Technologies in Biomedical and Life Sciences Education*, Methods in Physiology, https://doi.org/10.1007/978-3-030-95633-2_13

13.1 Introduction

13.1.1 Definitions and Global Context

Distance learning (DL) is a well-established method for teaching students who, for a variety of reasons, are unable to participate in face-to-face teaching at a conventional "brick" educational institution. In the USA distance learning is not always highly thought of, being associated for many people with a transactional process that offers a degree certificate in exchange for money. This view derives from the checkered history of "distance learning" in the USA, where plans to educate war veterans through correspondence courses were corrupted by unscrupulous providers who saw an easy way to make money rather than deliver proper education (Watkinson, 1996; Whitman, 2018). This negative view has persisted to this day, and a recent Gallup poll revealed that only 34% of Americans consider that online courses are good or excellent (Saad et al., 2013). Elsewhere, however (in Canada, Asia, Africa, and Europe), distance learning is seen as a valued way of delivering deep learning to previously disenfranchised students or those who, for a variety of reasons, are otherwise unable to access higher levels of education. In China distance learning is considered the most effective way to address the learning needs of millions of students (the Open University of China has more than four million active students, http://en.ouchn.edu.cn/index.php/about-v2/new-style-university). Even in the USA, which remains reluctant to accept parity between face-to-face and distance learning, organizations including the United States Distance Learning Association (USDLA) https://usdla.org/ endeavor to promote quality distance learning. USDLA lists over 90 quality standards for DL. More widely, the International Council for Open and Distance Education (ICDE) (https://www.icde.org/) exists to bring accessible, quality education to all through open, online, and distance learning. ICDE originated in Canada, a country that, in part due to its geography, has been in the forefront of quality distance education for a number of years. According to their website, ICDE members have more than 15 million students across 6 continents and work in support of UNESCO's recommendations on Open Educational Resources (UNESCO, 2019). This indicates the scale and importance of DL across the world. The growing popularity of distance teaching globally may in part be due to the view that it is a cheap way to deliver education at large scale; however, this is misguided because to deliver quality teaching requires a substantial initial outlay and high ongoing student support costs (Banas & Emory, 1998).

The concept of open learning (OL) is not restricted to a DL approach, but in practice the terms have become closely associated. OL involves teaching students who may have few or no previous educational qualifications, and in the context of this chapter, the teaching is at higher, post-compulsory, level. OL is a major driver of social mobility (Tait, 2013) as it allows people from disadvantaged backgrounds an opportunity to advance their education and hence their careers and life chances. This was the core concept upon which was founded the UK's Open University (OU) in 1969 and subsequently other open universities across the globe. Initially described as the "University of the Air," the OU provided a blend of custom-written teaching material, radio and television broadcasts, and home experiment kits sent through the post. Students were supported by telephone and at occasional face-to-face events. From its inception the OU has had no requirements for formal academic entrance qualifications, thereby allowing many formerly excluded people to be educated at HE level.

In this era of global information technology reach, DL and OL are often conflated as open distance learning (ODL), and much of it is delivered virtually and is referred to as e-learning. The reliance on technology both to deliver and to receive learning is not without its issues, and although a number of platforms have been used to deliver teaching material (Rienties et al., 2020), currently web-based delivery methods appear to offer the best access to the greatest number of people. Most of the OU's teaching is now online, with a seamless blend of text, figures, audio and visual clips, and practical activities, and student support is also delivered largely online. This approach is now considered the gold standard for ODL (Talbot, 2021); however, truly open access to the entire range of teaching materials is difficult to achieve, as we discuss below.

Who are the students who use ODL? They include students who have employment or caring responsibilities and so lack the time or ability to travel in order to attend face-to-face sessions scheduled during the working day, and others whose personal circumstances, such as a disability, may inhibit them from doing so. Some may be prevented from attending face-to-face sessions by virtue of cultural restrictions (e.g., gender-based); others may be imprisoned in secure environments. Some may simply prefer to study at their own pace in isolation and shun the social interactions that go alongside a traditional university experience. But for many students around the world, e-learning is simply the only option as there may be no university in their geographical area, and/or they may lack the financial resources to attend. This last category represents an enormous potential audience whose educational needs are not being met. It should also be borne in mind that even where delivery of educational materials is easy, students also need support to learn, and this introduces a further level of complexity since precise support requirements depend on the student, the learning topic, and the context in which learning is taking place (Banas & Emory, 1998; Pike, 2009; MacQueen & Aiken, 2019).

A large number of open universities cite inclusion and diversity in their mission statements and universities collect large amounts of demographic data about their students, but very few achieve a student demographic that is truly socially inclusive and reflective of their local gender and ethnic context (Tait, 2013; see Inclusivity chapter in this volume). Inclusivity is also impaired by the growing evidence of widespread digital poverty, even in high-income countries. In many areas there is poor or no broadband connectivity, curtailing would-be students' ability to access the intended materials and learning experience (Hannula, 2020; Dietrich et al., 2020). These problems are becoming ever more apparent during the COVID-19 pandemic, where multiple individuals from one household must compete to access limited technology (see COVID chapter in this volume). What availability there is may be confounded by unforeseen factors such as the inability to use video and audio at the same time, physical pressures on the working environment such as sharing a table with other people, noisy or abusive members of the household, and the intrusion of young children or pets (Kazmer & Haythornthwaite, 2001; Pike, 2009). Anecdotal evidence suggests that the majority of learners have to cope with at least one of these barriers, and clearly there remains some way to go before inclusivity can become a reality. However, one approach that education providers can take to help students overcome these barriers is to provide learning in small "bite-sized" chunks that individually can be studied in short periods of time. This approach is the major theme of this chapter.

13.1.2 Teaching Science Online

In spite of the now well-established history of distance-teaching universities, there continues to be debate about the effectiveness of ODL, in particular for the sciences, where practical, hands-on study is fundamental to an understanding of the topics and to the ability to apply learning in other contexts (Potkonjak et al., 2016; Faulconer & Gruss, 2018; Sandrone & Schneider, 2020). The perception that science is difficult, if not impossible, to teach online is prevalent in the sector, and many DL universities do not feature scientific subjects in their curriculum but instead concentrate more on the humanities. One of the OECD goals for students is that they should learn to "think like a scientist," whether or not they are studying science (OECD, 2013). Thinking like a scientist can be unpacked to itemize skills such as critical thinking, evaluation of evidence, problem-solving, and team working skills and does not necessarily include the hands-on aspects of practical science, even though these are important in the development of problem-solving skills (Herodotou et al., 2019). As we consider below, ODL can teach and assess all these skills, evidencing the parity of this approach to other pedagogic strategies. Benchmarking, both by the external examiner system and by professional body accreditation, shows that ODL is a valid approach to teaching right across the range of subject areas (Faulconer & Gruss, 2018; Quality Assurance Agency, 2018). However, the view remains that e-learning is unsuited to teaching traditionally hands-on practical skills. Nonetheless, evidence is emerging to challenge this view (Hirst & MacQueen, 2016; Argles, 2017; Pearson et al., 2019); indeed some studies show that online methods give better results (Herodotou et al., 2018), and some quantitative studies report no significant difference in students' understanding whether they were taught by traditional or online methods (Špernjak & Šorgo, 2018). These findings, and ways of addressing the issues they raise, will be discussed in depth later in this chapter. It is clear that acceptance of online practical science teaching is growing, and with that acceptance comes a burgeoning number of online resources, many of them open-access, for example, resources provided by the Society for Neuroscience (https://neuronline.sfn.org/). Such assets are not without their limitations but can also bring enormous benefit for science students.

13.1.3 Designing Online Learning

Learning analytics deployed for online courses indicate that correct learning design is very important for ensuring a good student learning journey and consequently student retention and success (Olney et al., 2018; Nguyen et al., 2020). Learning design incorporates parameters such as module size, text density, interactivity, and visual attractiveness. These aspects appear particularly relevant for short pieces of learning (Rizvi et al., 2019) where students may be more casual learners and may be harder to engage and retain.

Many shorter modules such as Massive Open Online Courses (MOOCs), Badged Open Courses (BOCs), and Small Private Online Courses (SPOCs) are short pieces of learning focused on specific topics and designed to be studied in isolation. They are sometimes referred to as micro-credentials and are commonly used for continuing professional development (CPD) as they are popular with many employers (Gautier, 2020; Ralston, 2020). To qualify as a micro-credential, there must be some value attached to the learning asset, and this is commonly a certificate or badge that is achieved by passing some form of final assessment. The majority of longer online modules (e.g., 30 or 60 CATS points[1]) are part of an ongoing program of study and, while they may be self-contained, are designed to be studied as part of a longer learning journey such as an undergraduate degree. This has implications for the design of the module and particularly for the assessment components. The purpose of assessment may simply be to test recall, as is found, for example, in some MOOCs, BOCs, and SPOCs (in the case of BOCs, passing the assessment is associated with claiming the badge), but the best kind of assessment leads to a deeper form of learning and needs to be carefully designed from the outset (Hounsell et al., 2007; Whitelock & Cross, 2012). It is difficult to achieve this level of rigorous assessment in short pieces of learning, and the academic value of micro-credentials is therefore not always seen as comparable with longer study materials. However, careful learning design, as described in the Case Study below, can ensure high academic quality and rigor.

[1] CATS (Credit Accumulation and Transfer Scheme) is a UK-based measurement for awarding credit for higher education study that functions analogously to the credit hours/unit system in the USA. In the UK each CATS point represents 10 h of combined study (in class contact plus self-study), so a traditional full-time university study program lasting one semester would entail 60 CATS points, which would be equivalent to 30 points in the European ECTS system and 15 credit hours/units in the US system (all resulting in 600–750 h of total study in and out of class during a 15 week semester). In a typical UK Bachelor's degree that lasts 3 years full time, a student would graduate with 360 CATS points.

13.1.4 Technology and Online Learning

There have been many opportunities and challenges faced by educators in the delivery of education across the disciplines and levels. Paramount among these in recent years relates to the use of technology to support teaching, learning, and assessment activities. In the online learning environment, these activities have rapidly become commonplace as part of an educator's everyday practice, and they offer many advantages over traditional face-to-face instruction (Larreamendy-Joerns & Leinhardt, 2006; Means, 2014). Norton and Wiburg (2003) argued that this move of learning to an online environment is a "defining technology," whereby electronic resources to support learning are now an *expectation*, much in the same way that printed materials were often considered the norm. The sociocultural implications of this shift to online means that educators need to consider how their learners can fully engage in a meaningful learning experience (Alston, 2017).

The move from print delivery to online delivery requires that the pedagogy of teaching, learning, and assessment be reconsidered for the online environment. Material designed for presentation on one platform does not generally transfer to other platforms, and reversioning is time-consuming and resource-intensive. Many technologies have been explored in the past (MacQueen & Thomas 2009), and currently it is widely believed that web-based technologies are preferred because they can be presented on a variety of browser platforms. Most higher education institutions (HEIs) have their own internal virtual learning environments (VLEs) where many learning resources including assessments can be presented, and this approach readily fosters a team spirit among students, improving the learning experience (Pennaforte, 2016; MacQueen & Aiken, 2019). As well as the advantages of VLEs, such as temporal and geographical flexibility and the opportunity to integrate textual, audio, and video materials seamlessly, there are also some disadvantages. VLEs frequently do not permit the conventional face-to-face teaching that is such a strong tradition in education; there is often a lack of immediacy in communication, and, as noted above, setting up and maintaining teaching assets on a VLE is resource-intensive. Furthermore, teaching through a VLE is currently not achievable worldwide, a consideration we turn to now.

Accessibility to online tools is a growing requirement of many legislatures. The core principles that online materials should be perceivable, operable, understandable, and robust are laid out in the Web Content Accessibility Guidelines (WCAG2.1-2, 2018). For designers, this means that navigation through online materials should accommodate students with difficulties with vision, hearing, and mobility, and the onscreen activity and guidance should accommodate differences in levels of understanding. Furthermore, we must be mindful of the geographical differences that exist across the world if ODL is to be truly global. It is noteworthy that even in countries where electricity supply is restricted and thus computers are limited to occasional online access, most people have mobile phones (more than 5 billion people in October 2020, according to datareportal.com). Consequently, there is a school of thought that says that educational material should all be presented in formats accessible by mobile phones. How realistic is this in practice? Mobile phones have small screens and are not easy to use for reading large amounts of text. Much interactivity cannot be reproduced on mobile phones (at present), and learning objects widely used in science teaching, such as interactive diagrams and large tables of data, are completely inaccessible on a phone. These factors contribute to the inequalities in educational opportunity that ODL seeks to address. It is difficult to see how this dilemma can be resolved with current technology, but we now move on to look at some successful examples of ODL using facilities that are widely available.

13.2 The Rise of MOOCs

In 2020, the requirements of, and for, ODL were brought to the fore as the world was presented its greatest challenge/threat in recent years as the coronavirus pandemic impacted on our everyday lives. Since March 2020, the world of education has been precipitated into a seismic shift—or "pivot" as it was commonly referred to—to the delivery of online education to learners across the ages and disciplines. As a consequence of social distancing rules, face-to-face teaching became less commonplace, and there was a greater reliance on the online environment becoming the de facto place for delivering learning. For many educators, this has arguably been a shock to the system, as they stepped away from their traditional teaching methods, and sought to identify ways to adapt them to the new paradigm. Perhaps one of the most notable examples of online education over the past decade has been the growth and variety of choice in a range of freely available, online courses developed by educational institutions around the world—the MOOC.

The origins of the term Massive Open Online Course (MOOC) tend to be recognized in one of two early examples. The first is in a course developed by David Wiley at Utah State University in 2007 (Gulati, 2013), which saw 50 online students from across 8 countries engage in a course alongside 5 students studying face-to-face. However, this particular course is often termed a proto-MOOC, or a precursor to what we see today. The second, which is often the more cited example, is in an open, online course delivered by George Siemens and Stephen Downes at the University of Manitoba in 2008 (Cormier & Siemens, 2010; Masters, 2011; Daniel, 2012; Siemens, 2013; Yuan & Powell, 2013). Initially a paid-for course available to 24 enrolled students, *Connectivism and Connective Knowledge (CCK08)* was also made available online for a wider audience to engage with learning materials and interact in the discussion forums. As a result, the course was able to attract and engage with over 2000 additional students, with Cormier and Siemens (2010, p. 34) noting that they "significantly enhanced the course experience [as they] extended the contributions of the instructors." The evolution of MOOCs is illustrated in Fig. 13.1.

There have been various commentaries regarding the term "MOOC" to better understand *why* it has been used to describe such a course (Masters, 2011; McAuley et al., 2010; Cross et al., 2013; Siemens, 2013) but also noting that it very much aligns with Ivan Illich's (1971) notion that an educational system should enable

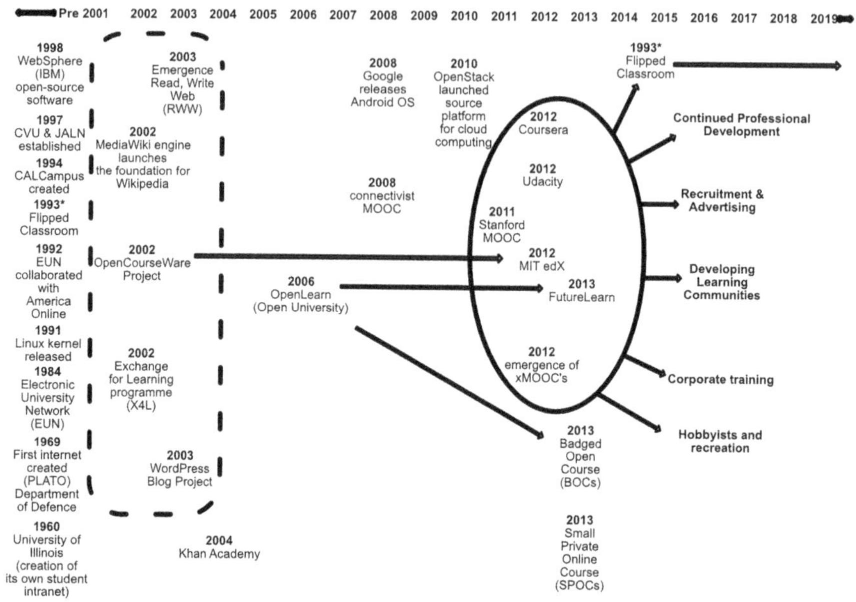

Fig. 13.1 A brief timeline illustrating the origins and developments of the open online education movement, into the more recent growth of MOOC providers

access to resources to those who wish to learn at any time. Indeed, Downes (2012) contends that a key feature of the Manitoba CCK08 course was the use of distributed open resources that support thousands of learners, while also indicating that learners could succeed so long as they were able to connect with each other. It is this latter aspect which has seen two broader pedagogical directions into which MOOCs are often aligned. We describe these below.

xMOOCs—eXtended MOOCs—(Daniel, 2012; Siemens, 2013) are known for their instructivist approach and arguably are the closest to traditional teaching and learning practice, with tutors as "experts" and learners as "consumers." Sebastian Thrun and Peter Norvig's "artificial intelligence" MOOC offered through Udacity is one such example of an xMOOC, attracting upward of 160,000 learners when it first launched in late 2011 (Osvaldo, 2012). xMOOC course structures are typically rigid, with topics explored through short video lectures and interactive activities, and learners typically assessed through automated quizzes. Direct feedback from the tutor is relatively uncommon owing to potentially large class sizes.

cMOOCs are based on Siemens and Downes "connectivist" pedagogy, which views knowledge as a networked state, and learning and the creation of knowledge as the process of collaborating across these networks, adding and removing connections where necessary (Daniel, 2012; Siemens, 2013; Yuan & Powell, 2013). Such connections can be made between learners and learners, learners and materials, and learners and tutors. cMOOCs tend to have largely open activities that learners can interact with, and because of the emphasis on autonomy, learning is inherently more personal since learners, and tutors make use of a range of tools to engage and devise their way through the course.

The one challenge that has constantly faced those involved in designing, developing, and delivering MOOCs is *retention* and how best to engage learners through to course completion. Reich and Ruipérez-Valiente (2019) conducted an analysis of courses delivered by MIT and Harvard from 2012 to 2018, which covered over 5 million learners and almost 13 million course registrations. Considering all MOOC participants, they reported that only 3% of participants engaged on a course actually completed: a figure which stood at almost 6% in 2014/2015. Exploring this further, Reich and Ruipérez-Valiente (2019) also noted that learners engaged with MOOCs with some irregularity, a trend which has been ever-present since 2012. At the height of this retention challenge, they note that while over one million learners engaged in their first MOOC in 2015/2016, only 12% took part in a MOOC in the following academic year. A similar picture is evident in the UK, with a report conducted by MoocLab examining almost 200 MOOCs delivered by 30 providers between 2012 and 2015 suggesting an average completion rate of 12% (McIntyre, 2016). An interesting point to note here was a tentative correlation between a high enrolment rate resulting in a lower completion rate. Goopio and Cheung's (2020) systematic review of empirical studies reporting on the MOOC dropout phenomenon over the period 2008–2018 identified several key issues relating to student attrition in MOOCs, many of which are often cited by others when asked about the retention rates within courses: learner experience, interactivity, the design of the course, technology, language, time, and situation.

One effort to alleviate the retention challenge in online courses has seen an increase in the use of SPOCs (Small Private Online Course). In introducing the term, Fox (2013) argued that he did not see SPOCs as an alternative to the traditional MOOC; rather, a SPOC intends to capitalize on the affordances of a MOOC and can be "used as a supplement to classroom teaching, rather than being a replacement for it" (p. 38). On the face of it, SPOCs are perhaps more closely related to a traditional education setting than MOOCs are, with tutors interacting with learners in small groups and providing feedback at the content level, offering asynchronous and synchronous teaching, and the use of trained moderators to support learners through the course (Uijl et al., 2017). Such characteristics have indeed contributed to the perceived success of SPOCs, with Filius et al. (2018) noting that the smaller group sizes and increased interactions between those involved have contributed to much higher retention rates (>90%) when compared to their MOOC counterparts.

More recently, The Open University (OU) in the UK recognized an opportunity to build on the success of the MOOC, developing an alternative for those who want to upskill in a particular area, while having their learning activities formally recognized. Badged Open Courses (BOCs) are seen by the OU as an evolution of the traditional MOOC, since they enable learners to refresh their knowledge at any time, breaking free of the traditional enrolment, start and finish dates associated with a MOOC, with the badges awarded demonstrating evidence of a commitment to professional development and learning (OpenLearn, 2021). Both SPOCs and

BOCs are popular for CPD because of their tight focus and short study times. The increasing use of blockchain systems to verify badges achieved from learning is leading to truly portable credentials (Badgechain, 2017).

13.3 Why Produce a MOOC?

At their height of exposure, MOOCs were seen by many to be the "new premise" of online learning on a mass scale, offering free courses to the masses in a range of subject areas with an opportunity to receive a certificate on completion, without any particular entry requirements (Coursera, 2021; FutureLearn, 2021; Udacity, 2021; Yuan & Powell, 2013). A key feature in the design of a MOOC is ensuring that the principles of openness in education, the sharing of knowledge, and a supportive learning community are clearly demonstrated (FutureLearn, 2021; Yuan & Powell, 2013). They are often designed to encourage learners to be self-regulated and to take ownership of their learning, rewarding those who engage fully with the course with peer relationships that can develop further once the course is completed.

Much of the wider discussion surrounding MOOCs often centered on the disruptive force they would have on the higher education sector (Daniel, 2012; DiSalvio, 2012; Hyman, 2012; Kirschner, 2012; Yuan & Powell, 2013). MOOCs were seen by many as a disruptive innovation—one which typically combines new technologies that offer the potential to evolve quickly with innovative business models, resulting in a disruption of existing markets and possible supplanting of earlier technologies, though possibly at the cost of lower standards (Christensen, 1997; Adner, 2006; Yuan & Powell, 2013). The number of providers and universities operating in the MOOC space at the time offered learners access to high-quality university-level education within a specific subject. Here, learners had flexibility and the potential to map out their own program of study focusing on aspects that appealed most to them. A key benefit for learners was the position that MOOCs took at the low-cost end of the higher education market, offering low-end disruption and performance at a cheaper price than traditional university study (Christensen & Raynor, 2003; Yuan & Powell, 2013).

A voice of reason at the time acknowledged that the education system is actually highly complex and advised that the use of disruptive innovation theory to describe MOOCs should be applied with caution (Yuan & Powell, 2013). Commercial MOOC providers such as Coursera and Udacity developed new business models and markets, while other universities have made use of MOOC platforms for testing out new forms of online learning (Jarrett, 2012; Bates, 2013). Further critiques of the "disruptive force" that MOOCs present suggests that they are better identified as "disturbing" rather than disruptive, since it would be difficult to replace traditional models of higher education—not least because of the lack of individualized and personalized tutoring in MOOCs (Langen & Bosch, 2014). Others noted that rather than replacing, MOOCs could offer an opportunity to *support* traditional models of delivery, notably through hybrid offerings (Daniel et al., 2015), or in our own experiences (see Case Study), improving outreach (Macleod et al., 2015). Howarth et al., (2016) extended this, suggesting that MOOCs could be best utilized as the starting point in the application process for a student, noting the opportunity for learners to evaluate potential programs of study ahead of committing to a large financial outlay.

There have been many incentives for producing MOOCs such as the extension of brand identity and program exposure. It is now clear that MOOCs, SPOCs, and BOCs are being used to extend, rather than disrupt learning, so much so that an increasing number of higher education institutions (HEIs) worldwide have been reconsidering their online provision and the way in which they deliver traditional degree programs. Shah (2020) notes that one third of learners who registered on a MOOC platform did so in 2020, suggesting that learner engagement with free, open, online courses is still very much on the increase. And just as with continued increase in learner engagement, the number of offerings continues to grow. In 2020 alone, it was reported that over 2800 MOOCs were launched and made freely available to learners, across the multitude of platforms available to access such courses.

13.4 Case Study: Making a Physiology MOOC

Like many disciplines, Physiology often struggles for recognition among 16- to 18-year-old students who are at an age (in the UK) where they are making their choices for undergraduate university study. While many study an integrated advanced (A)-level Biology program, there is often little mention, or recognition, of the wide range of biological sub-disciplines that comprise the subject. Typically, some aspects of Biology such as Biochemistry, Genetics, and Ecology tend to be more clearly defined in the delivery of the curriculum, whereas Physiology tends not to be defined in the same manner. Students are often taught about the functionality of key human organ systems and the importance of homeostasis without comprehending that they fall under the domain of Physiology. Indeed, it has already been reported that students studying human anatomy and Physiology programs often struggle with comprehending the subject, most commonly because of the features inherent within the discipline (Michael, 2007; Sturges & Maurer, 2013; Slominsk et al., 2019). Could this be a case that Biology has become more and more compartmentalized as we have advanced in our knowledge and discoveries? In doing so there is a challenge to highlight to our young learners that there is a need to think more holistically. In doing so we need to integrate this knowledge into a coherent picture (Pinter & Pinter, 1993; Naftalin, 2011).

In 2016, The Physiological Society—the Learned Society for Physiology established in the UK in 1876—published a call for proposals in an attempt to address this issue, looking for interested UK universities who were already working in partnership with the platform provider FutureLearn to work with them on the design and development of an introductory MOOC aimed at 16- to 19-year-old A-level students. For context it is worth noting that a wide variety of subjects can be studied at A-level, with up to five examining boards each setting their own curricula, sometimes across the same subject.

The Society's primary objectives for this introductory MOOC were to enhance and extend learning at A-level (and equivalents), to introduce concepts encountered in the first year of undergraduate study of Physiology and other related degrees, and to support the transition to university study. The Society is a charitable organization with a history of supporting and sustaining the discipline through the advancement of science and education, and its desire to develop a freely available, online course is very much in keeping with its ethos. It is the Society's belief that the development of an introductory MOOC focusing on the subject of Physiology will not only provide learners access to university-led education, particularly those who may otherwise not have access to such opportunities, but could also encourage individuals to engage in further study within the discipline at university level. Having undertaken research to uncover the topics common to all examination boards within the UK, the Society predetermined the content of the MOOC: the heart and circulatory system, the respiratory system, and the nervous system. A key feature of the MOOC was that learners would be exposed to data-handling and analytical activities, helping to increase student confidence in the mathematical aspects of the study of Biology at undergraduate level (National Research Council, 2003).

An internationally renowned and leading research institution, the University of Liverpool, is the UK's leading international provider of wholly online postgraduate education (University of Liverpool, 2021). The university has been involved in teaching and researching in the Life Sciences domain for over 100 years, offering undergraduate and postgraduate programs across the full spectrum of the Biology discipline, with its undergraduate Physiology program having an emphasis on integrating the study of molecular, cellular, systems, and whole body function, distinguishing it from other Biology-related degree programs. Our approach to teaching the program is appreciated by students, with our outstanding commitment to teaching evidenced by National Student Survey (NSS) scores, with a 97% overall satisfaction rate among students in the most recent survey (TheUniGuide, 2021). More broadly, the university was already in a partnership with FutureLearn and had already designed and successfully delivered a series of MOOCs on the FutureLearn platform in the fields of archaeology, engineering, nursing, and psychology.

The proposal for "Physiology: the Science of Life" MOOC mapped against the activities of several UK A-Level Biology specifications and would extend the topics taught to educate and inform learners about the three key areas of Physiology. Upon completion, the MOOC would serve as both an extension activity for A-level students and a primer course for first year undergraduate students. Through the design of the MOOC, we aimed to highlight the importance and relevance of Physiology as a discipline and encouraged learners to consider Physiology or Physiology-related degree programs when applying for university or further study.

13.4.1 The Design Approach

Similar to many MOOCs delivered via the FutureLearn platform, we adopted the "xMOOC" approach to the design of the course, ensuring that the target demographic would have some familiarity with the didactic structure designed into the course, enhanced through the use of key features in the content. Each week started with an interactive, multimedia timeline presenting an historical overview of the system in question, giving learners the opportunity to see a high-level view of the research developments in that area. Interestingly, these historical timeline sections were among those with the highest levels of interaction. Learners were very positive about the historical coverage of the topics and reported enjoying having the chronological development of knowledge in the areas put into context. The timeline was accompanied by an introductory video on the topic by the Lead Educator, along with demonstrations of the equipment used within the discipline to measure specific functions of each physiological system. The use of "talking head" academic and clinical experts was identified as a key differentiator, as we drew upon experts in the field across the UK to initiate discussion activities around some of the chronic diseases associated with each of the systems. We also included the student perspective on studying Physiology at university level.

Aligned to the proposal requirements, a range of interactive activities designed to promote confidence and skill in data-handling and interpretation was also included. Here, we presented learners with data sets produced in equipment demonstration videos and encouraged them to participate in active discussions regarding what the data showed. This strategy enabled students to learn from material in context. The combination of contextualized learning and provision of multiple discussion opportunities specific to the weekly content encouraged active learning while promoting both learner-learner and learner-tutor interactions, resulting in the development of a community of learning (WeiWei et al., 2017; Gamage et al., 2018). While Gameel's (2017) review of four MOOCs with almost 2000 learners found that learner-learner and/or learner-tutor interaction had no impact on learner satisfaction, we felt it crucial as a development team to provide a learning environment that did not leave the intended demographic participants feeling abandoned and alone. Figure 13.2 shows the content of a typical week of study.

13.4.2 Students and Teachers as "Co-creators"

As part of the approach to designing and developing the MOOC, we were keen to involve students in the process, particularly because of the many benefits that such a partnership affords in the co-creation of content (Bovill et al., 2011; Carey, 2013; Brooman et al., 2015). In UK HE, final year undergraduate students are typically required to complete a capstone project as part of their program of study. For many, this involves engagement in an extended piece of research under the close

Week 2: Cardiovascular

Cardiovascular System

This week, we learn about the cardiac cycle, what an ECG is used for, and learn about the effect of exercise on the human cardiovascular system.

2.1 Welcome to Week 2 VIDEO (01:22)

2.2 The human cardiovascular system VIDEO (10:41)

2.3 Cardiovascular physiology through the ages EXERCISE

2.4 Introduction to control of the heart and cardiovascular system VIDEO (09:27)

2.5 An introduction to the ECG VIDEO (08:05)

2.6 Heart and autonomic nervous system VIDEO (09:15)

Cardiovascular System in Action

This is your chance to get involved in the course, and participate in one the clinical tests used to asses cardiovascular function.

2.7 Harvard Step Test VIDEO (06:44)

2.8 How did you score? ARTICLE

Fig. 13.2 An outline of the topics covered in Week 2 of the "Physiology: the Science of Life" MOOC delivered through the FutureLearn platform

supervision of a member of academic staff. For those students at the University of Liverpool interested in pursuing a career in teaching after graduation, we also offer the opportunity to engage in an education-based project, with past examples including the development of resources in collaboration with a school, educational resources for local museums, or the testing of a pedagogical question by questionnaire or evaluation.

The opportunity to engage in this specific project yielded great interest from a number of our final year students, with some engaging in the live demonstrations of specific equipment. However, one student expressed the desire to engage with the MOOC as an educational project, since this would align more with her career aspirations of becoming a teacher. Key to this choice was a passion for the discipline and a welcome opportunity to engage with an initiative that would help to educate people about the discipline, and in the student's words, "correct some of the misconceptions about what the study of physiology actually entails, especially for those who think it is the same as physiotherapy."

The students formed a key part of the development team, as we adopted a staff-student partnership in the creation of content (Bovill et al., 2011, Cook-Sather, 2016). Here, they contributed to the research for the historical timelines and the development of the weekly multiple-choice question (MCQ) assessments designed to test the knowledge and analytical skills of the learners and under the supervision of the academic supervisor developed the storyboards and scripts for the video content used in the final week of the course. This opportunity allowed them to engage in both the writing and production of content for a large online course and has indeed served as excellent preparation for those seeking subsequent teaching careers.

Given the intended audience of the MOOC, we also believed it was important to confirm that the content was pitched at the right level. As such, the University of Liverpool and The Physiological Society engaged with a team of secondary school and college Biology teachers to review the content that was developed. This step proved to be enormously valuable, particularly in relation to the use of specific terminology used in Physiology teaching.

13.4.3 Bringing the MOOC to Life

Within the Faculty of Health and Life Science at the University of Liverpool, there are a number of Technology Enhanced Learning (TEL) academics, each of whom is located in an undergraduate school and offers expertise in pedagogy and technology in relation to teaching, learning, and assessment activities. Being located within a particular school and subject discipline enables the TEL academic to not only get a feel for how the subject is delivered but also oversee a consistent approach in the use of technology to enhance the student experience.

Unlike previous MOOC offerings by the University of Liverpool that made use of a centrally located e-learning team to provide technology and pedagogy expertise, the development of this MOOC primarily drew upon the resources and staff from the faculty, making use of localized knowledge and expertise of the discipline. In this instance, the subject academics involved were able to focus on mapping the course content on to the existing A-level specifications, to ensure that the requirements from The Physiological Society were being met. As part of the planning process, the designated TEL academic was able to provide guidance in content design and delivery involving the use of technology for online learning, while also being cognizant of how it was being used elsewhere in the delivery of university-level

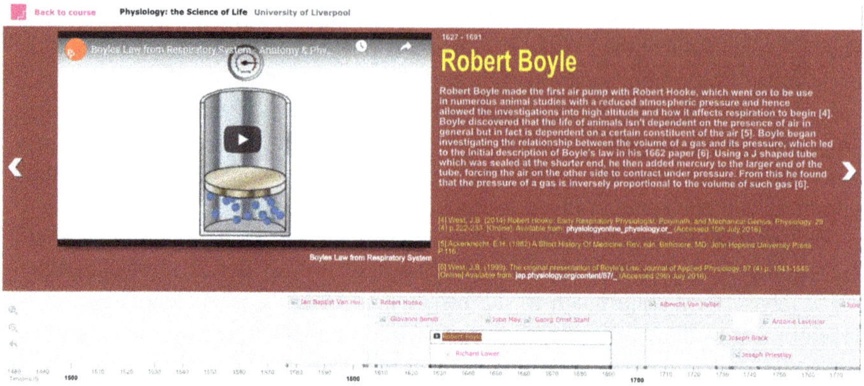

Fig. 13.3 An interactive timeline used as part of Week 1—The Respiratory System

education. This early involvement of the TEL academics ensured a smooth integration of technologies in the project. Had this step been left until later, we would no doubt have missed opportunities for development and increased our workload later in the project.

The benefit of such a collaboration between subject and pedagogy/technology experts meant that the TEL academic was able to act as an interface, or translator, between the academic and technical personnel involved, helping to minimize confusion and misunderstanding regarding technology- or pedagogy-related issues during the course development process. This proved particularly useful when working with our content developer and media specialists, when the subject academics and students were required to produce storyboards and scripts in order to identify suitable locations for the filming of video content.

Included in the delivery of the MOOC was the development of interactive timelines which enabled learners to move through the history of the topic, exploring content links to multimedia for key moments within the discipline. Learners also contributed to the content of the timelines, often adding comments in the live course relating to pioneers they believed should also be included, who were duly added while the course was live. A typical timeline is shown in Fig. 13.3.

To bring the discipline to life, we recorded live demonstrations of clinical assessments related to the material for each of the weeks. Examples include the use of Benedict-Roth Spirometry to measure lung volumes and the mechanics of breathing and the recording of surface potentials to demonstrate nerve stimulation and conduction velocities. To support learner data-handling skills, we incorporated two interactive exercises into the course that reviewed data in different ways. The first was to engage learners in conducting a live Harvard Step Test, which they could undertake safely in their home environment (Uslu & Körükcü, 2020). This afforded learners the opportunity to look at their own physiological responses and then submit them online in real time, to compare their own performance against other learners on the course. The second activity—called "How Fast is Fast?"—used an interactive spreadsheet which enabled learners to conduct an analysis of datasets using simple tools to interpret data and appreciate the speed of a nerve action potential. This facility again provided learners with the opportunity to contextualize the material and obtain feedback instantly (Bell et al., 2011).

Finally, to help set the course content in the context of the discipline as a whole, we engaged with experts in the key themes, interviewing a selection of subject experts and clinicians. Their responses were recorded and packaged for learners to gain an insight into the latest research in the field and perhaps more importantly in relation to the expected learner audience, the journey that each had taken in their career to get to where they were. Such interviews provided an excellent starting point for learner-learner and learner-tutor discussions on the platform.

13.4.4 Quality Assurance (QA) Prior to Release: The Final Checks

Since the recorded content, either film or audio, would require the most time and money to produce and edit, for the QA process, it was crucial that it was reviewed by the collective committees. For this step we produced a separate resource pack for each week of the course that was distributed to those involved in the review process. Each pack contained an outline of the course showing how each week aligned with the overall course, a high-level outline of the week's activities, and the drafts of historical timelines, storyboards, scripts, formative assessments, and any additional content associated with that week's activities. Reviewers were invited to complete an online form to provide their feedback on the content, and comments were then collated and passed back to the MOOC team (subject and TEL academics) to make any necessary changes. This feedback was predominantly based around the use of language which was appropriate to the target audience. Other comments were on the apparent conflict between direct scientific language and the use of a narrative to contextualize the material covered. The narrative approach was preferred by educators of the target audience (Hadzigeorgiou & Schulz, 2019). Following the review and incorporation of feedback, the team set about completing the recording and production of the MOOC material. Prior to its final sign-off and release it was deemed essential for this "learner-ready" material to be vetted one final time. This would be to check for any nonscientific content issues that might have arisen in the production process, such as sound and image quality/clarity and any additional accessibility issues. Upon completion of this process, the final QA element was for our platform (FutureLearn) to run the necessary technological checks to identify any further accessibility issues.

While the QA process did present an increased workload for those involved, we believe it provided the teachers with a sense of ownership and empowerment in the course, since they could directly influence the design of the course so that it met the

Table 13.1 Headline data for all iterations of the *Physiology: the Science of Life* MOOC

Learning parameter	Score
Enrolments	17,535
Activated learners	8485
Activated social enrolments	34%
Completion rate	21.3%
Positive feedback	92.2%

Activated learners, learners who viewed at least two steps; activated social enrolments, percentage of activated learners who posted at least one comment on any step; completion rate, percentage of activated learners who completed >90% of the course, and > 70% in any tests; positive feedback, proportion of learners who responded with a "smiley face" in the weekly sentiment survey

needs of their own learners. Had this QA step been omitted, we could have missed out on valuable insights from these specialists for our target learners.

As a part of the MOOC, we also included areas in the discussion boards for learners to comment on the material and the course and where appropriate the TEL and development team would respond to and address these issues in real time, along with the MOOCs moderation team. This typically involved a team of four or five moderators, and so the commitment per person was in the realms of 1 hour in the morning and again in the afternoon/evening throughout the presentation of the MOOC. This ensured a daily presence online. One particularly rewarding example of user feedback was the themed timelines. Learners were encouraged to suggest historical figures and where their contribution(s) were to the field, and this illustrates the continued commitment to engaging learners as the co-creators of knowledge and content in the course (Bovill et al., 2011; Cook-Sather, 2016). This dynamic approach provided not only a QA opportunity but also a mechanism by which learners and educators could engage with the material in several different ways, facilitating further discussion and interaction with the resource. Learners commented on feeling that they felt supported in their learning journey, and it was refreshing to see live involvement of the lead educator and the course team, contributing to over 90% positive feedback from learners across all iterations of the course (Table 13.1). It is also of note that the completion rate for this MOOC (21.3%) is more than double the average reported retention rates (10%) for MOOCs across the board (see above). This underlines the fact that an engaging MOOC that is well constructed and well supported can deliver a good educational experience that students enjoy.

This case study exemplifies the care and attention to detail that must be brought to bear when constructing an asset for ODL. It is evident that the whole process of production and presentation is resource-intensive, and we would emphasize the need not to cut corners, for fear of producing an asset that is not fit for purpose. We turn now to an example that is arguably even more resource-intensive to produce: assets for teaching practical Biology.

13.5 The Challenges of Teaching Practical Bioscience Skills Online

For the biosciences, engagement with practical or investigational work is an essential component at all levels of education, at all ages, as it is in all other areas of science. The element of scientific enquiry, encapsulated in observation and experimentation, is at the core of scientific practice and, therefore, should be at the core of what we teach. The experience of scientific enquiry for most students consists of practical classes in a laboratory, and for most, this sets their expectations from school through to higher education: science is about being "in the lab"; the sounds, smells, and physical elements, alongside the dropped equipment, spilled reagents, safety glasses, and unwashed laboratory coats. For many, this might also be what stops them continuing with a science education. It often comes as a surprise, therefore, to both students and instructors, that anyone could even consider not teaching science practicals in a lab, instead substituting an online practical. How can anything online replace these more immediate and tactile aspects?

This view of practical science represents a commonly misunderstood aspect of online practical teaching tools: that they should attempt to *replace* being in a lab. They don't; rather, what they do is provide students an exposure to the critical aspects of rigorous scientific enquiry through active learning within a biological scenario. This will be explored in more detail later, but an essential question to start with is why bother; why try to teach practical science online?

13.5.1 Why Try to Teach Practical Science Online?

As we noted above, recent events centered on the COVID-19 pandemic have driven many instructors to abruptly introduce online substitutes for closed laboratories, and this has reignited interest in using online practicals. Beyond this immediate need, there are many ways in which online tools have been deployed. In developing a suite of tools for teaching practical biosciences online, the UK's Open University (OU) utilizes a variety of tools aimed at students studying remotely, asynchronously, and often in large numbers. Here, the use of bespoke tools provides the most affordable and practical way to develop core investigational skills; residential laboratory schools have prohibitive running costs for large groups of geographically dispersed students. The OU's approach has been to blend online practical-based activities with online and print-based teaching material in a blended approach that integrates core biological concepts with practical skills.

In many cases, online activities can be used to prepare students for a hands-on practical class. For example, asking students to work through an online simulation of an instrument can prepare them for performing the equivalent task in the laboratory, reducing student hours in a teaching laboratory, increasing throughput and enabling a more efficient timetable. It can also allow more depth to the lab activities as core

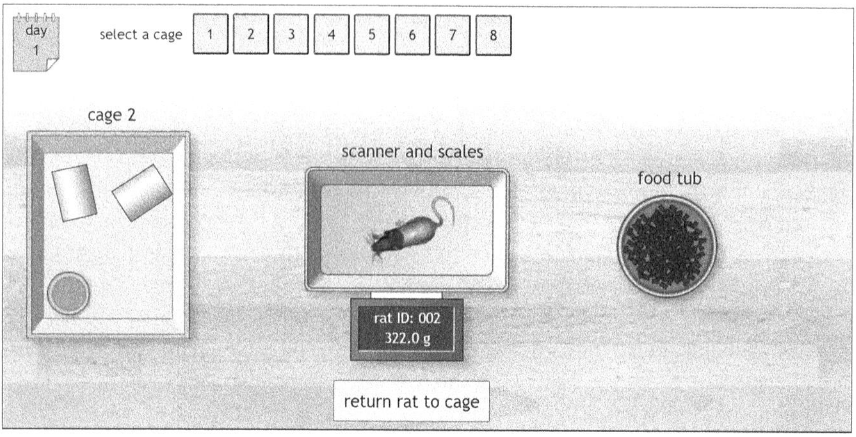

Fig. 13.4 An onscreen screen experiment where students maintain virtual rats ahead of a behavioral experiment. Maintaining records of feeding and animal weights over 7 days develops record-keeping skills

concepts, and operational aspects can be engaged with in advance. Post-laboratory extension studies are another approach. Once the student has spent time in the laboratory, a well-designed online activity can extend their experience with the practical technique and foster structured or free-ranging explorations and extended rounds of data collection, such as is required for project work.

Beyond the positioning of online tools in a curriculum, these tools provide additional advantages. Some can allow student access to rare resources, such as collections of digitized microscope slides, 3D renders of human skulls, or samples with an unusual provenance. Work with animals or their tissues is one area where ethical, safety, and cost issues usually limit their use in teaching, with activities frequently available only to small numbers of undergraduates, often at a project level at institutes with suitable facilities. Online activities based upon samples of real tissues, or presenting complex protocols such as behavioral training, therefore not only maximize the use of rare materials but also broaden the student experience beyond studies that can be provided in their own institute. As an example, in one OU online activity, students are required to engage with a simulated animal husbandry and behavioral training protocol over 7 days; students are responsible for weighing, feeding, and training their virtual rats (Fig. 13.4).

More recently, the ability to directly access high-end, research-grade laboratory instruments has brought a new aspect for online learners. While this approach is more common in other fields such as Astronomy, where access to remote telescopes is almost the norm, access to laboratory instrumentation is still limited to only a few centers. The OU's OpenSTEM laboratories (https://learn5.open.ac.uk/course/view.php?id=2) provides remote access to a suite of live instruments such as the scanning electron microscope (Fig. 13.5) for analysis of samples that are increasingly being used by postgraduate and undergraduate cohorts of students.

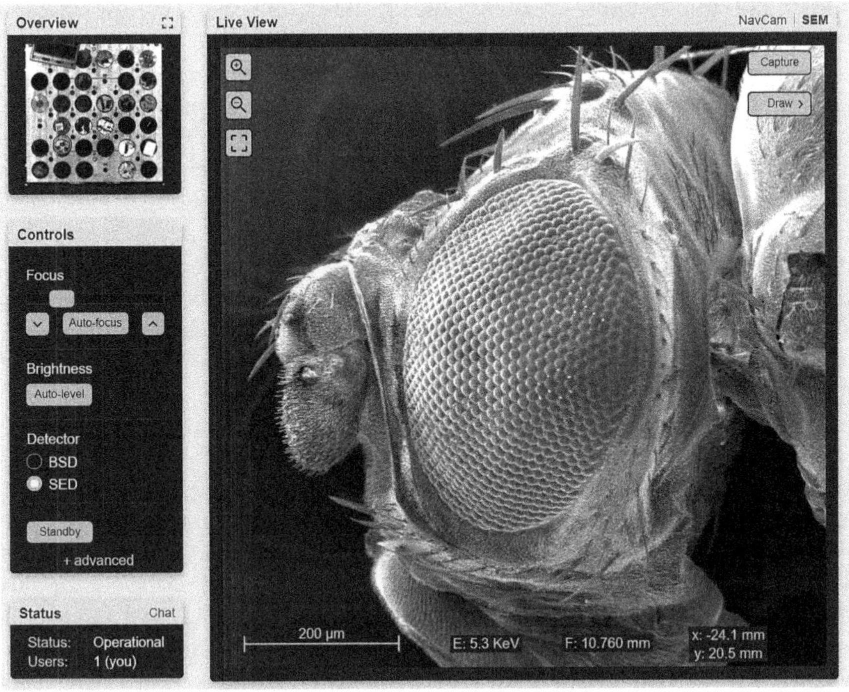

Fig. 13.5 Remote access to live instrumentation enhances a student's experience of working equipment and specimens, such as with an SEM shown here being used to examine the eye phenotype associated with a particular genetic marker in *Drosophila melanogaster*

Whatever the reason to teach online practical bioscience, whether it is by necessity or as a method of enhancing or supplementing existing hands-on classes, it is important to recognize what such tools can and cannot do.

13.5.2 Online Practical Work: What It Is (and What It Isn't)

Online tools to teach practical science cannot replace the hands-on and sensory experiences of being in a laboratory environment. While attempts are being made to emulate or simulate a real-life laboratory experience using 3D technology and haptics, most current examples of online tools aimed at developing scientific enquiry skills involve working onscreen, more frequently on tablet or mobile devices, relying upon cloud-based resources accessed over the Internet. Gamification approaches, which offer more immersive scenarios and the potential to interact using standard gaming interfaces, are becoming more common, particularly at school level (Labster, 2021). Most approaches recognize that the technology acceptance model (TAM;

Davis, 1989) core concepts of ease of use and usability are central to their ability to engage students in their own learning.

The approach used at the OU has been to examine the specific laboratory activity, instrument, or experiment of interest, and to identify which aspects are suitable for the online delivery of an active learning experience of the core biological concepts and investigational skills. For example, microscopy is excellent for extending a student's understanding of biological structures from the organismal through to intracellular level. The simulation of an analytical instrument such as a spectrophotometer, on the other hand, allows students to explore aspects of analytical instrument functionality including accuracy, precision, and the use of standards and controls in measurement and experimentation. While using different practical paradigms, both approaches provide a student with an active, experiential understanding of the key concept of biological variation.

Most research studies use several different experimental approaches to address a question or test a hypothesis. In an online context, this approach can be brought to students in ways that would not usually be possible in a standard laboratory class, due to either limits of available facilities or timetabling. As an example, two online tools are used in an OU module that allows students to examine the physiological changes that occur in brown adipose tissue (BAT) during cold adaptation. In this case, a combination of the spectrophotometric analysis of lipids and proteins is combined with a microscopy study of both white and brown adipose tissues (Fig. 13.6). Such online approaches also allow access to experimental paradigms that go beyond a standard practical laboratory class, allowing longer-term experimentation usually carried out over days or even weeks to be condensed down to hours or even minutes.

When designed well, online tools provide an excellent opportunity to develop students' data recording and observational skills; data generated can feed into statistical analysis and presentation channels just as any other practical work can. This is perhaps one of their strengths, as students can collect data, often in abundance and at their own pace, and analyze it.

In talking about experimental outcomes, however, it is also important to focus attention on the data used within the tools. Where possible, the use of authentic data in online tools provides a more realistic experience for students. This can mean that tools are built around data from previously taught face-to-face practical classes, data collected on similar instruments or even data from actual experimental samples. In addition, where variability can be built in, it often is. For example, numerical data can be randomized to reflect the intrinsic variability seen in an instrument's measurements; its accuracy and precision values can be used to modify data.

As we stated above, a major problem for online teaching is student retention. Thus, in developing and designing any practical study, one key objective is to ensure that students are engaged with the material; online working presents a different set of obstacles and challenges from a hands-on laboratory scenario.

Both approaches aim to make the work engaging, but working online presents the challenge of getting students to think beyond progression through an interactive experiment in a "point and click" manner, a challenge that has yet to be fully

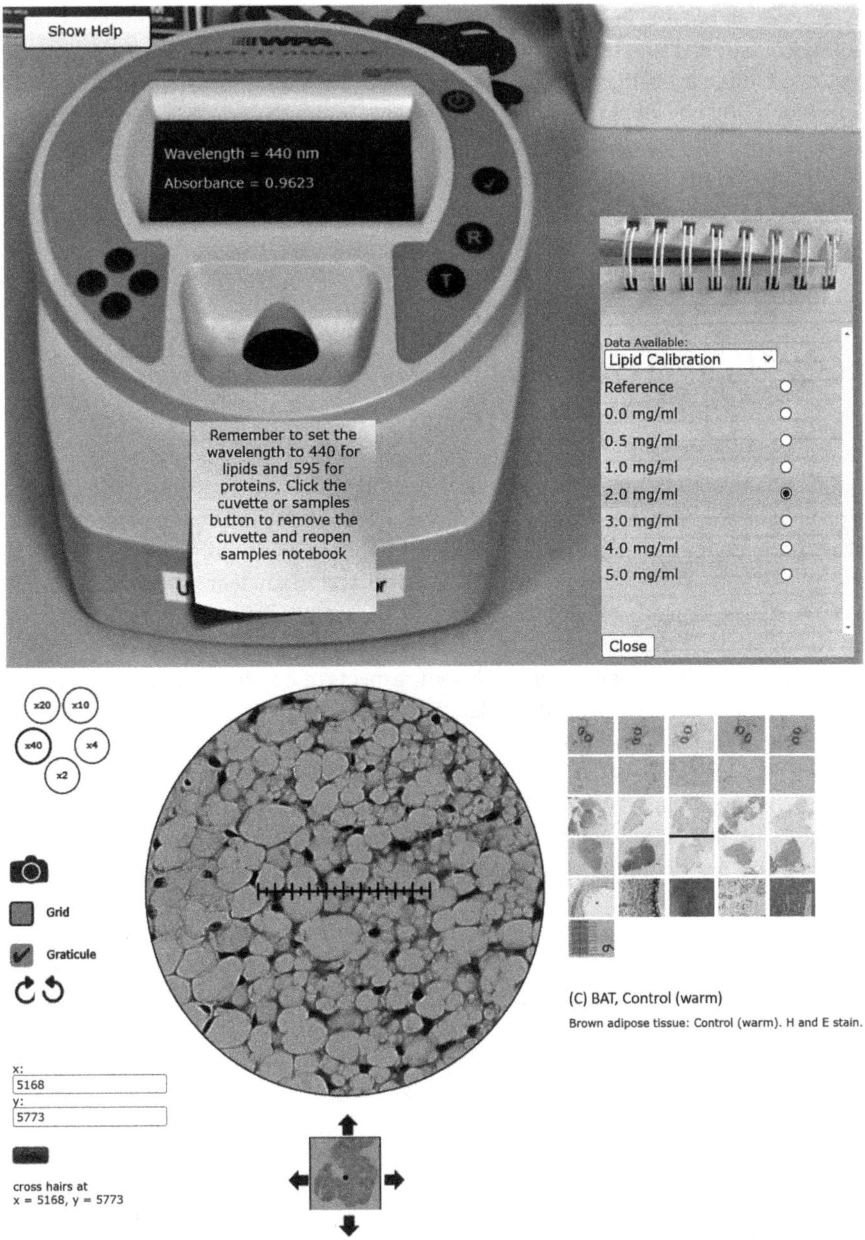

Fig. 13.6 Onscreen tools such as a spectrophotometer (upper panel) and digital microscope (lower panel) can be combined to allow students to integrate data to examine the structural and biochemical changes accompanying BAT adaptation in cold-acclimated rats

overcome, as is ensuring that the tools enhance students' active learning of both biological and investigational principles without obstructing learning with technical barriers. Online tools that increase the student perception of enjoyment or playfulness will likely contribute to student motivation, leading to an improved perception of their ease of use and utility and thereby improve student engagement and satisfaction (Estriegana et al., 2019). There remains a balance that needs to be struck, however, to ensure too much playfulness doesn't get in the way of the intended learning objective (Ejsing-Duun & Karoff, 2014).

13.5.3 Issues to Consider in Developing and Supporting Online Practical Tools

Online practical tools are particularly useful as platforms for students to collect data, especially numerical data. However, their use across large numbers of students, over many presentations of a module, increases the likelihood of occurrences of collusion and plagiarism. The automatic systems used to screen student assessment cannot usually detect copied images or numerical datasets. Clearly, practices such as instructor sign-off of data collected in-lab are not usually available in online tools. Most VLEs used in education can track some aspects of a student's engagement with an activity, but few will record all details of a student's work within an activity. Designing in features that either minimize the opportunity for collusion or recognize when it has happened is an important factor, therefore. This can be achieved by building in significant data variability, by ensuring no two experimenters will ever get the same numerical data, or by ensuring that a tool houses numerous different datasets that are randomly distributed to any one student group, possibly with a record kept.

Onscreen tools used to develop practical skills often provide a greater challenge to meet the WCAG2.1 guidelines as most rely heavily on visual representations of laboratory instruments and data. Ensuring that all students have equitable access to such resources means that the accessibility requirements must be considered from an early stage of design to ensure as similar experiences as possible, whether the tools are accessed using a mouse, touchscreen, or keyboard. Similarly, building in screen reader capability, including visual descriptions of objects, enhances the accessibility of the tools without creating a barrier to understanding by being overly lengthy. Additional consideration must also be given in the distance learning context to providing materials for students within a cohort who cannot access online resource, such as students in secure facilities or with limited access to cloud-based resources. In this case, it is common to provide an offline interactive version that can be run locally, or a paper- or screen-based copy of materials that best facilitates the student being able to complete the activity.

Supporting students as they work within online tools is another major area for consideration. In a laboratory environment, experienced instructors have often learnt to recognize participants who are uncertain about what they are doing and can intervene and support where needed. This on-site presence also allows instructors to highlight key aspects of experimental work, critical stages, or safety issues and to support students with good scientific practice skills such as record keeping and data analysis.

For synchronous remote use, a similar level of support can be provided through shared interfaces where peer support and interaction are available and an instructor presence is easy to establish. The increasing use of screen-sharing platforms for group use (Teams, Zoom, etc.) will certainly serve to increase how online tools are used. Similarly, a student group working together with online tools in a standard face-to-face classroom environment allows the instructor to observe and support students just as they would in a laboratory. Asynchronous support is less straightforward and responsive. The primary means of support in this case is in preempting obstacles and providing inbuilt advice or help screens and the provision of supporting protocols or workbooks. In all cases, building a community around an online activity can help to build social aspects which will contribute to student engagement. Online forums, briefings, and support sessions can also serve to support students, especially if they use short video tutorials for commonly encountered problems.

In feedback from students, the most common reason for not being able to engage with or complete an online activity is technical problems. With an increasingly more web-comfortable population, a lack of confidence working onscreen and poor IT skills are becoming less frequent. However, browser compatibility, software updates, antiviral packages, and broadband speed all still represent serious problems for some students, particularly those who live in remote areas, as we noted above.

13.6 Evaluating the Success of Online Teaching Assets

Success is defined as the accomplishment of an aim or purpose: the good or bad outcome of an outcome or undertaking. This clearly implies that "success" is variable and depends to a large degree on context. In the context of an online teaching resource, a successful outcome might be one or more of the following:

- A measurable gain/output/enhancement of an achieved learning outcome
- The total number of learners that the resource has attracted
- The geographic reach/spread of uptake
- Something more specific, e.g., relevant to the target demographic

As mentioned above, typically only about 10% of learners starting a MOOC will complete the whole resource. Therefore, success may be based on this metric. Perhaps there should be a recognition that learners choose their level of engagement. Learners may get what they need from a particular section of the resource and then disengage, so completion of a section or even a small element could be seen as success. Similarly, success of the resource could be seen in the emergence of

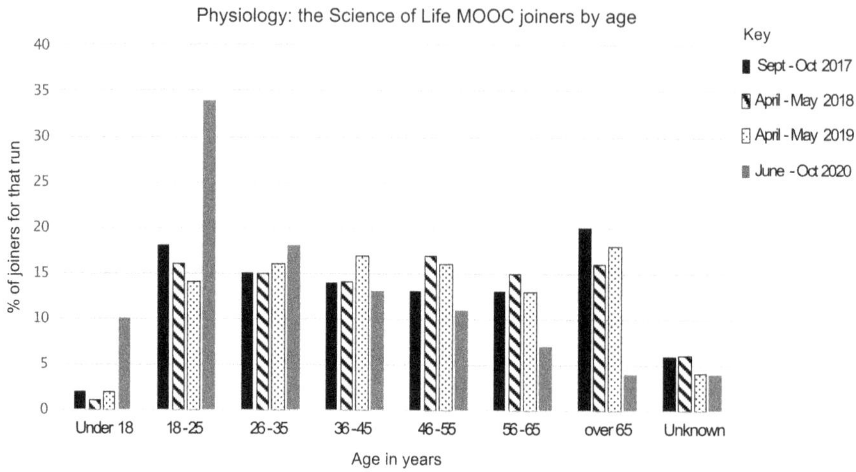

Fig. 13.7 Comparison of the demographics (joiners by age) from four runs of the *Physiology: the Science of Life* MOOC

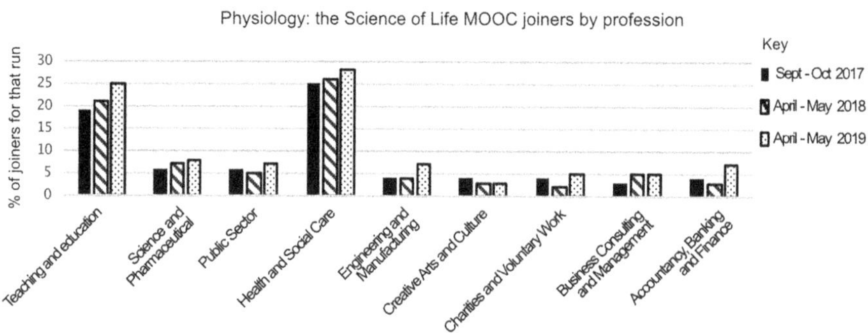

Fig. 13.8 Learners' area of employment from three runs of the *Physiology: the Science of Life* MOOC. This showed whether those who were employed were already in the sector associated with the course

learner-learner or learner-tutor interaction, growing into an active community of practice and learning. Success could also be in the form of a longer-term goal, such as increased recruitment beyond the immediate target area. More surprisingly, success could present itself from "left field," from what may look like an initial failure to meet the intended targets. For example, in the *Physiology: the Science of Life* MOOC, the intended target was A-level students, yet in the early runs, there was an increased number of learners in the upper age brackets (Fig. 13.7).

Further investigation showed that the MOOC was also being used for continuing professional development. This was not the intended target; however, it was certainly a pleasant surprise to see established professionals in education and healthcare using this resource for their own needs (Fig. 13.8). We did see a marked increase in learners from our target demographic in our fourth run (during the COVID-19 pandemic).

There clearly needs to be some flexibility in the evaluation of success for these nontraditional teaching assets. During design of such resources, it is worth considering the intended criteria of success to check that they can be measured effectively and bearing in mind that there will likely be a number of unintended outcomes and unexpected success measures. Success can be discussed in theoretical terms, but it may be more difficult to assess the detailed measures in practice. Multiple-choice questions are a common way to assess MOOCs, BOCs, and SPOCs, and while the best-designed questions can adequately assess recall of content, it is difficult to use them to assess deep learning (Whitelock & Cross, 2012). To do this it is often necessary to move to a more traditional assessment format. For example, a wide range of approaches are used to assess outcomes in learners studying practical teaching tools. Perhaps the most common is the assessment of a scientific report based upon data obtained from a particular study. While this approach also assesses tangential skills such as scientific writing, data analysis, and presentation, it often suffers from a lack of methodological engagement in a technique which students didn't perform in the simulation.

Interestingly, and probably not surprisingly, the views of academics and students about the value of assessment may diverge. While academics may remain wedded to traditional forms of assessment such as report-writing, direct feedback from students themselves indicates that they focus on other aspects of their learning: for example, in the rat training laboratory detailed earlier, having completed the activity students reported positively for three key outcomes (Fig. 13.9). This supports the notion that good student-centered learning is linked to student satisfaction (Wright, 2011) and suggests that as well as modifying teaching styles, teachers must also modify their approach to assessment.

13.7 Conclusions

This chapter has presented an overview of the kinds of teaching resources that are offered online and at scale to populations of students whose reasons for studying are enormously diverse. We have tried to draw out some key themes: the high aspirations underpinning the development of online science teaching at scale, the high level of resource required to produce such learning tools, and the technological barriers that prevent their adoption in a truly open and inclusive way. We have highlighted the situations in which MOOCS, SPOCs, BOCs, and practical assets can supplement or replace traditional face-to-face teaching methods, and we have stressed the need for attention to detail during production, good QA processes to ensure maximum accessibility, and the value of ongoing engagement with and support of learners.

Because they are relatively new, there is as yet scant data around the effectiveness of teaching and learning via MOOCs, BOCs, SPOCs, and practical assets. For

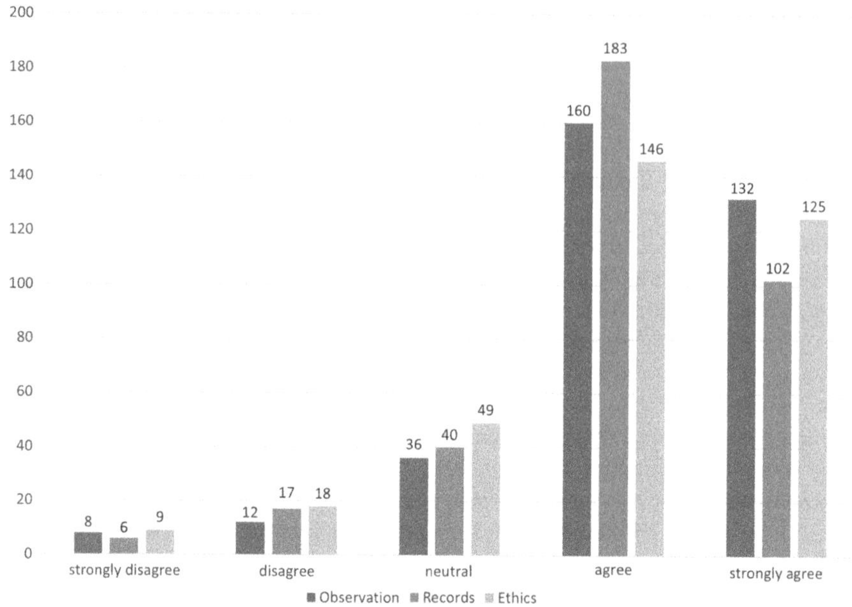

Fig. 13.9 Students report beneficial impact of the use of a virtual rat training laboratory in their observation and record-keeping skills and their understanding of the ethics of working with rats

example, knowing more about the context of students' learning could allow more meaningful evaluation of effectiveness. It would also allow more nuanced support to be built in from the outset. Ultimately it would be useful to find the "sweet spot" between offering adequate support for all students and affordability. We have described some examples of assets that have been successful within the framework and expectations set by the providers, but our experiences may not be generalizable globally, not least because of the high costs involved. This is an active area of research, and we anticipate the emergence of many more case studies, particularly as many providers are rapidly moving toward ODL as a result of the COVID pandemic. It is important to note in this context that interpretation of effectiveness data must be done with care: the methodologies of data collection are not uniform, and the rapidly evolving technology means that longitudinal studies or datasets may not be comparing like with like in terms of the technology available and/or the audience's use of it. The advent of cheaper and more accessible and realistic haptics could transform the usefulness of online practical teaching, and this would be a fruitful area of research.

As online tools become more sophisticated and are integrated with web-based life and at-home tools, they will continue to develop. The potential for a responsive support tool is huge, given the increasing sophistication of web interfaces and artificial intelligence. The future design of online labs capable of recognizing student hesitancy or uncertainty in a mouse or keyboard sequence as effectively as a laboratory instructor does at present is probably a long way in the future, but responsive help tools are likely to be further deployed in the foreseeable future. Such technologies already exist, for example, in a FutureLearn MOOC *Irish 101: an introduction to Irish language and culture*, where correct pronunciation is an important factor, there is an online teacher (from the provider, Dublin City University) able to help in real time, as well as an online help bot (https://www.futurelearn.com/courses/irish-language).

In some areas, the future is already here—where science practical work has broken out of its stronghold of education and into the broader camp of citizenry. Citizen science projects are increasing training and utilizing citizens around the globe to analyze biological data. For example, the Etch-a-Cell project (Etch-a-Cell, 2021) asks users to identify nuclear boundaries in TEM images of cells, enabling a massive dataset to be gathered and collated to understand nucleus structures and boundaries. The Project Discovery has engaged with over 300,000 gamers to annotate the cellular locations of proteins (Sullivan et al., 2018). Similar projects are directed to gather training sets for AI systems to use to eventually recognize biological features by machine learning.

Online learning using onscreen interfaces that allow access to instruments and tools that simulate experiments is becoming more widely used. They excel at delivering targeted teaching of specific aspects of investigative Biology through active learning of the underlying biological concepts while developing good scientific practice. Recognizing their limitations, supporting students appropriately and deploying the tools to maximize their pedagogical effect will serve to enhance student engagement and learning. Hopefully they will also make it fun.

References

Adner, R. (2006). Match your innovation strategy to your innovation ecosystem. *Harvard Business Review., 84*(4), 98–107.

Alston, P. (2017). *Influential factors in the design and implementation of electronic assessment at a research-led university.* PhD thesis, Lancaster University. https://doi.org/10.17635/lancaster/thesis/157.

Argles, T. (2017). Teaching practical science online using GIS: A cautionary tale of coping strategies. *Journal of Geography in Higher Education, 41*(3), 341–352. https://doi.org/10.1080/03098265.2017.1315531

Badgechain. (2017). Available at Innovations in Open Badges & Blockchain | BadgeChain| Innovations in Open Badges & Blockchain | BadgeChain. Accessed 11 March 2021.

Banas, E., & Emory, W. (1998). History and issues of distance learning. *Public Administration Quarterly, 22*(3), 365–383.

Bates, T. (2013). *Harvard's current thinking on MOOCs.* Available from http://tinyurl.com/a2uh86z. Accessed 29 Jan 2021.

Bell, D., Daniels, M., & Lawless, J. (2011). Expeditionary learning: Authentic education in the 21St century. *Ohio Social Studies Review, 47*(1), 11–20.

Bovill, C., Cook-Sather, A., & Felten, P. (2011). Students as co-creators of teaching approaches, course design, and curricula: Implications for academic developers. *International Journal for Academic Development, 16*(2), 133–145. https://doi.org/10.1080/1360144X.2011.568690

Brooman, S., Darwent, S., & Pimor, A. (2015). The student voice in higher education curriculum design: is there value in listening? *Innovations in Education and Teaching International, 52*(6), 663–674. https://doi.org/10.1080/14703297.2014.910128

Carey, P. (2013). Student as co-producer in a marketized higher education system: A case study of students' experience of participation in curriculum design. *Innovations in Education and Teaching International, 50*(3), 250–260. https://doi.org/10.1080/14703297.2013.796714

Christensen, C. M. (1997). *The innovators dilemma: When new technologies cause great firms to fail*. Harvard Business Press.

Christensen, C. M., & Raynor, M. E. (2003). *The innovators solution: Creating and sustaining successful growth*. Harvard Business Press.

Cook-Sather, A. (2016). Undergraduate students as partners in new faculty orientation and academic development. *International Journal for Academic Development, 21*, 151–162. https://doi.org/10.1080/1360144X.2016.1156543

Cormier, D., & Siemens, G. (2010). Through the open door: Open courses as research, learning, and engagement. *EDUCAUSE Review, 45*(4) 30–39. Available from https://er.educause.edu/articles/2010/8/through-the-open-door-open-courses-as-research-learning-and-engagement. Accessed 29 Jan 2021.

Coursera. (2021). Courses | Coursera. Available from http://www.coursera.org/courses. Accessed 29 Jan 2021.

Cross, J., Siemens, G., Downes, S., De Coutere, B., Griffith, T., Finnern, M. et al. (2013) Business and MOOCs. *Google Hangout*. Available from http://youtu.be/DGaUfWkJdi4. Accessed 29 Jan 2021.

Daniel, J. (2012) Making sense of MOOCs: Musings in a maze of myth, paradox and possibility. *Journal of Interactive Media in Education*, 2012. Available from http://www-jime.open.ac.uk/article/2012-18/. Accessed 29 Jan 2021.

Daniel, S., Cano, E., & Cervera, M. (2015). The future of MOOCs: Adaptive learning or business model? *Universities and Knowledge Society Journal, 12*(2015), 64–73. https://doi.org/10.7238/rusc.v12i1.2475

DataReportal. https://datareportal.com/global-digital-overview. Accessed 21 Jan 2021.

Davis, F. (1989). Perceived usefulness, perceived ease of use, and user acceptance of information technology. *MIS Quarterly, 13*(3), 319–340. Available from: https://doi.org/10.2307/249008. Accessed 10 March 2021.

Dietrich, N., Kentheswaran, K., Ahmadi, A., Teychené, J., Bessière, Y., Alfenore, S., Laborie, S., Bastoul, D., Loubière, K., Guigui, C., Sperandio, M., Barna, L., Paul, E., Cabassud, C., Liné, A., & Hébrard, G. (2020). Attempts, successes, and failures of distance learning in the time of COVID-19. *Journal of Chemical Education, 97*(9), 2448–2457.

DiSalvio, P. (2012) Pardon the disruption... innovation changes how we think about higher education. *The New England Journal of Higher Education*. Available from https://nebhe.org/journal/disruptive-innovation-changing-how-we-think-about-higher-education/. Accessed 29 Jan 2021.

Downes, S. (2012). *The rise of MOOCs [blog post]. Stephen's Web*. Retrieved from http://www.downes.ca/post/57911. Accessed 12 Jan 2021.

Ejsing-Duun, S., & Karoff, H. (2014). Gamification of a higher education course: What's the fun in that? *Proceedings of the European Conference on Games-based Learning, 1*, 92–98.

Estriegana, R., Medina-Merodio, J., & Barchino, R. (2019). Student acceptance of virtual laboratory and practical work: An extension of the technology acceptance model. *Computers & Education, 135*, 1–14. https://doi.org/10.1016/j.compedu.2012.12.014

Etch-a-cell. (2021). *Etch a Cell — Zooniverse*. Accessed 29 Jan 2021.

Faulconer, E., & Gruss, A. (2018). A review to weigh the pros and cons of online, remote, and distance science laboratory experiences. *The International Review of Research in Open and Distance Learning, 19*(2). https://doi.org/10.19173/irrodl.v19i2.3386

Filius, R., de Kleijn, R., Uijl, S., Prins, F., van Rijen, H., & Grobbee, D. (2018). Challenges concerning deep learning in SPOCs. *International Journal of Technology Enhanced Learning, 10*(1–2), 111–127.

Fox, A. (2013). From MOOCs to SPOCs. *Communications of the ACM, 56*(12), 38–40.

FutureLearn. (2021). *FutureLearn explained.* Available from https://www.futurelearn.com/using-futurelearn. Accessed 29 Jan 2021.

Gamage, D., Perera, I., & Fernando, S. (2018) Increasing interactivity and collaborative ness in MOOCs using facilitated groups: A pedagogical solution to meet 21st century goals. *IEEE Global Engineering Education Conference (EDUCON).* 17–20 April, 2018. Available at IEEE Xplore. Accessed 11 March 2021.

Gameel, B. (2017). Learner satisfaction with massive open online courses. *American Journal of Distance Education, 31*(2), 98–111. https://doi.org/10.1080/08923647.2017.1300462

Gautier, T. (2020). The value of micro-credentials: The employer's perspective. *The journal of competency-based. Education, 5*(2), e01209. https://doi.org/10.1002/cbe2.1209

Goopio, J., & Cheung, C. (2020). The MOOC dropout phenomenon and retention strategies. *Journal of Teaching in Travel & Tourism.* https://doi.org/10.1080/15313220.2020.1809050

Gulati, A. (2013). An overview of massive open online courses (MOOCs): Some reflections. *International Journal of Digital Library Services, 3*(4), 37–46.

Hadzigeorgiou, Y., & Schulz, R. (2019). Engaging students in science: The potential role of "narrative thinking" and "romantic understanding". *Frontiers in Education, 4*, 38. https://www.frontiersin.org/article/10.3389/feduc.2019.00038

Hannula, L. (2020, April 22). Weak internet cripples productivity for 1/3 of work-from-home employees. *Whistle Out.* Available from https://www.whistleout.com/Internet/Guides/weak-internet-a-problem. Accessed 11 March 2021.

Herodotou, C., Muirhead, D., Aristeidou, M., Hole, M., Kelley, S., Scanlon, E., & Duffy, M. (2018). Blended and online learning: A comparative study of virtual microscopy in higher education. *Interactive Learning Environments.* https://doi.org/10.1080/10494820.2018.1552874

Herodotou, C., Sharples, M., Gaved, M., Kukulska-Hulme, A., Rienties, B., Scanlon, E., & Whitelock, D. (2019). Innovative pedagogies of the future: An evidence-based selection. *Frontiers in Education, 4*, 113. https://doi.org/10.3389/feduc.2019.00113

Hirst, M., & MacQueen, H. (2016). Practical biology at a distance: How far can we go with online distance learning? In D. Kennepohl (Ed.), *Teaching science online: Practical guidance for effective instruction and lab work. Online learning and distance education.* Stylus Publishing.

Hounsell, D., Falchikov, N., Hounsell, J., Klampfleitner, M., Huxham, M., Thomson, K., & Blair, S. (2007, November). Innovative assessment across the disciplines: An analytical review of the literature. *HEA.* https://doi.org/10.3389/feduc.2020.00128

Howarth, J., D'Alessandro, S., Johnson, L., & White, L. (2016). Learner motivation for MOOC registration and the role of MOOCs as a university 'taster'. *International Journal of Lifelong Education, 35*(1), 74–85. https://doi.org/10.1080/02601370.2015.1122667

Hyman, P. (2012). In the year of disruptive education. *Communications of the ACM, 55*, 20–23. https://doi.org/10.1145/2380656.2380664

Illich, I. (1971). *Deschooling society.* Harper & Row.

Jarrett, J. (2012). *What are MOOCs and why are education leaders interested in them?* Available from http://www.huffingtonpost.com/impatient-optimists/what-are-moocs-and-why-ar_b_2123399.html. Accessed 29 Jan 2021.

Kazmer, M., & Haythornthwaite, C. (2001). Juggling multiple social worlds: Distance students online and offline. *American Behavioral Scientist, 45*(3), 510–529.

Kirschner, A. (2012). A Pioneer in online education tries a MOOC. *The chronicle of higher education.* Available from http://chronicle.com/article/A-Pioneer-in-Online-Education/134662/. Accessed 29 Jan 2021.

Labster. (2021). Virtual Lab Simulation Catalog | Labster|Virtual Lab Simulation Catalog | Labster. Accessed 29 Jan 2021.

Langen, F., & Bosch, H. (2014). Massive open online courses: Disruptive innovations or disturbing inventions? *Open Learning: The Journal of Open, Distance and e-Learning, 28*(3), 216–226. https://doi.org/10.1080/02680513.2013.870882

Larreamendy-Joerns, J., & Leinhardt, G. (2006). Going the distance with online education. *Review of Educational Research, 76*(4), 567–605. https://doi.org/10.3102/00346543076004567

Macleod, H., Haywood, J., Woodgate, A., & Alkhatnai, M. (2015). Emerging patterns in MOOCs: Learners, course designs and directions. *Association for Educational Communications and Technology, 59*, 56–63.

MacQueen, H., & Aiken, F. (2019). Supporting distance-taught students in the workplace. *Higher Education, Skills and Work-Based Learning, 10*(1), 49–60. https://doi.org/10.1108/HESWBL-04-2019-0048

MacQueen, H., & Thomas, J. (2009). Teaching biology at a distance: Pleasures, pitfalls and possibilities. *American Journal of Distance Education, 23*(3), 139–150. https://doi.org/10.1080/08923640903080505

Masters, K. (2011). A brief guide to understanding MOOCs. *The Internet Journal of Medical Education, 1*(2). Available from http://ispub.com/IJME/1/2/10995. Accessed 29 Jan 2021.

McAuley, A., Stewart, B., Siemens, G., & Cormier, D. (2010). *The MOOC model for digital practice.* Available from http://davecormier.com/edblog/wp-content/uploads/MOOC_Final.pdf. Accessed 29 Jan 2021.

McIntyre, C. (2016). *UK MOOC report.* Retrieved from https://www.mooclab.club/Reports/UK%20MOOC%20Report%202016.pdf. Accessed 10 Dec 2020.

Means, B. (2014). Technology and education change: Focus on student learning. *Journal of Research on Technology in Education, 42*(3), 285–307.

Michael, J. (2007). What makes physiology hard for students to learn? Results of a faculty survey. *Advances in Physiology Education, 31*, 34–40. https://doi.org/10.1152/advan.00057.2006

Naftalin, R. (2011) Opinion: The decline of physiology. *The scientist.* Available from http://www.the-scientist.com/?articles.view/articleNo/29658/title/Opin-ion-The-decline-of-physiology/ 26 Feb. 2017. Accessed 11 March 2021.

National Research Council, USA. (2003). *BIO2010: Transforming undergraduate education for future research biologists*. National Academies Press.

Nguyen, Q., Rienties, B., & Richardson, J. (2020). Learning analytics to uncover inequality in behavioural engagement and academic attainment in a distance learning setting. *Assessment and Evaluation in Higher Education, 45*(4), 594–606. https://doi.org/10.1080/02602938.2019.1679088

Norton, P., & Wiburg, K. (2003). *Teaching with technology: Designing opportunities to learn.* Harcourt Brace.

OECD. (2013). *PISA 2015 draft science framework.* Available from: https://www.oecd.org/pisa/pisaproducts/Draft%20PISA%202015%20Science%20Framework%20.pdf. Accessed 11 March 2021.

Olney, T., Rienties, B., & Toetenel, L. (2018). Gathering, visualizing and interpreting learning design analytics to inform classroom practice and curriculum design: A student-centred approach from the Open University. In J. M. Lodge, J. C. Horvath, & L. Corrin (Eds.), *Learning analytics in the classroom: Translating learning analytics research for teachers* (pp. 71–92). Routledge. https://www.taylorfrancis.com/books/9781351113038/chapters/10.4324/9781351113038-6

OpenLearn. (2021). *Badged courses.* Retrieved from: https://www.open.edu/openlearn/get-started/badges-come-openlearn. Accessed 10 Jan 2021.

OpenSTEM Labs. The OpenSTEM Labs | Faculty of Science, Technology, Engineering & Mathematics

Osvaldo, R. (2012) MOOCs and the AI-Stanford like courses: Two successful and distinct course formats for massive open online courses. *European Journal of Open, Distance and E-Learning.* Available from: https://eric.ed.gov/?id=EJ982976. Accessed 11 March 2021.

Pearson, V., Lister, K., McPherson, E., Gallen, A.-M., Davies, G., Colwell, C., Bradshaw, K., Braithwaite, N., & Collins, T. (2019). Embedding and sustaining inclusive practice to support disabled students in online and blended learning. *Journal of Interactive Media in Education, 1*, 4. https://doi.org/10.5334/jime.500

Pennaforte, A. P. (2016). Organizational supports and individuals' commitments through work-integrated learning. *Higher Education, Skills and Work-Based Learning, 6*(1), 89–99. https://doi.org/10.1108/HESWBL-07-2015-0038

Pike, A. (2009). Developing online communities to support distance learning in secure environments. In:*7th International Conference on Education and Information Systems, Technologies and Applications: EISTA 2009,* 10–13 Jul 2009, Orlando, Florida, USA. Available from A-Pike-Developing_online_communities_to_support_DL_in_secure_environments.pdf (open.ac.uk). Accessed 11 March 2021.

Pinter, G., & Pinter, V. (1993). Is physiology a dying discipline? *Physiology (Bethesda), 8*, 94–95. https://doi.org/10.1152/physiologyonline.1993.8.2.94

Potkonjak, V., Gardner, M., Callaghan, V., Mattila, P., Guetl, C., Petrović, V. M., & Jovanović, K. (2016). Virtual laboratories for education in science, technology, and engineering: A review. *Computers & Education, 95*, 309–327.

Quality Assurance Agency for Higher Education. (2018). *UK quality code for higher education.* www.qaa.ac.uk/quality-code. Accessed 18 Feb 2021.

Ralston, S. J. (2020). Higher Education's microcredentialing craze: A postdigital-Deweyan critique. *Postdigital Science and Education, 1*, 19.

Reich, J., & Ruipérez-Valiente, J. (2019). The MOOC pivot. *Science, 363*, 130–131. https://doi.org/10.1126/science.aav7958

Rienties, B., Køhler, S., & Herodotou, C. (2020). Defining the boundaries between artificial intelligence in education, computer-supported collaborative learning, educational data mining, and learning analytics: A need for coherence. *Frontiers in Education (Early access), 5*, 128.

Rizvi, S., Rienties, B., Rogaten, J., & Kizilcec, R. (2019). Investigating variation in learning processes in a FutureLearn MOOC. *Journal of Computing in Higher Education, 32*, 162–181. https://doi.org/10.1007/s12528-019-09231-0

Saad, L., Busteed, B., & Ogisi, M. (2013). In US, online education rated best for value and options. Available at: In U.S., Online Education Rated Best for Value and Options (gallup.com) Accessed 11 March 2021.

Sandrone, S., & Schneider, L. (2020). Active and distance learning in neuroscience education. *Neuron, 106*(6), 895–898.

Shah, D. (2020). *The second year of the MOOC: a review of MOOC stats and trends in 2020.* Available from: https://www.classcentral.com/report/the-second-year-of-the-mooc/. Accessed 29 Jan 2021.

Siemens, G. (2013). Massive open online courses: Innovation in education? In McGreal, R., Kinuthia, W., & Marshall, S. (eds) *Open educational resources: Innovation, research and practice.* Commonwealth of Learning and Athabasca University. Available from https://www.oerknowledgecloud.org/archive/pub_PS_OER-IRP_CH1.pdf. Accessed 29 Jan 2021.

Slominsk, T., Grindberg, S., & Momsen, J. (2019). Physiology is hard: a replication study of students' perceived learning difficulties. *Advances in Physiology Education, 43*(2), 121–127.

Špernjak, A., & Šorgo, A. (2018). Differences in acquired knowledge and attitudes achieved with traditional, computer-supported and virtual laboratory biology laboratory exercises. *Journal of Biological Education, 52*(2), 206–220. https://doi.org/10.1080/00219266.2017.1298532

Sturges, D., & Maurer, T. (2013). Allied health students' perceptions of class difficulty: The case of undergraduate human anatomy and physiology. *Internet J Allied Health Sci Pract, 11*, 1–10.

Sullivan, D., Winsnes, C., Akesson, L., Hjelmare, M., Wiking, M., Schutten, R., Campbell, L., Leifsson, H., Rhodes, S., Nordgren, A., Smith, K., Revaz, B., Finnbogason, B., Szantner, A., & Lundberg, E. (2018). Deep learning is combined with massive-scale citizen science to improve large-scale image classification. *Nature Biotechnology, 36*, 820–828. https://doi.org/10.1038/nbt.4225

Tait, A. (2013). Distance and e-learning, social justice and development: The relevance of capability approaches to the mission of open universities. *The International Review of Research in Open and Distance Learning, 14*(4), 1–18. https://files.eric.ed.gov/fulltext/EJ1017505.pdf

Talbot, D. (2021). Available at Five reasons why online and blended learning could be a game-changer for education and training (fenews.co.uk). Accessed 11 March 2021.

TheUniGuide. (2021). *Study Human Physiology at University of Liverpool.* The University of Liverpool. Available from: https://www.theuniguide.co.uk/university-of-liverpool-l41/courses/human-physiology-bsc-hons-2021-fc932a65bc06. Accessed 02 March 2021.

Udacity. (2021). *Program catalog.* Available from http://www.udacity.com/courses/all. Accessed 29 Jan 2021.

Uijl, S., Filius, R., & Ten Cate, O. (2017). Student interaction in small private online courses. *Medical Science Educator, 27*(2), 237–242.

UNESCO Recommendations on Open Educational Resources. (2019). http://portal.unesco.org/en/ev.php-URL_ID=49556&URL_DO=DO_TOPIC&URL_SECTION=201.html. Accessed 29 Jan 2021.

University of Liverpool. (2021). *About us.* The University of Liverpool. Available from: https://www.liverpool.ac.uk/about/. Accessed 2 March 2021.

Uslu, S., & Körükcü, M. (2020). The outcomes of constructivist learning environments from the perspectives of secondary school students. *International Education Studies., 13*(8), 16–28. Accessed 8 March 2021.

Watkinson, J. (1996). Education for success: The international correspondence schools of Scranton, Pennsylvania. *The Pennsylvania Magazine of History and Biography, 120*(4), 362–363.

WCAG2.1 Principles. (2018). Web Content Accessibility Guidelines (WCAG) 2.1 (w3.org) Accessed 29 Jan 2021.

WeiWei G., Ayub, E., Wong, S., & Lim, C. (2017). The importance of teacher's presence and engagement in MOOC learning environment: A case study. In *IEEE Conference Nov, 2017; IEEE Language: English, Database: IEEE Xplore Digital Library.*

Whitelock, D., & Cross, S. (2012). Authentic assessment: What does it mean and how is it instantiated by a group of distance learning academics? *International Journal of e-Assessment, 2*(1), Article 9. http://journals.sfu.ca/ijea/index.php/journal/article/view/31

Whitman, D. (2018). *The cautionary tale of correspondence schools.* newamerica.org. Available at www.newamerica.org/education-policy/reports/cautionary-tale-correspondence-schools/. Accessed 11 March 2021

Wright, G. (2011). Student-centered learning in higher education. *International Journal of Teaching and Learning in Higher Education, 23*(1), 92–97.

Yuan, L. & Powell, S. (2013). MOOCs and open education: Implications for higher education. JISC CETIS. Available from https://publications.cetis.org.uk/wp-content/uploads/2013/03/MOOCs-and-Open-Education.pdf. Accessed 29 Jan 2021.

Chapter 14
Social Online Learning: Leveraging Social Media and Web-Based Co-creation to Drive Learning

Emma Rengasamy and Duncan Cole

Abstract Social media use in higher education is becoming increasingly common. The tools available to the educator are sophisticated and numerous, and it can be challenging to know which ones to use and how to use them to best effect. We begin by asking what we want to achieve in educating our students and how social media may enable this. We argue that Christensen's "jobs-to-be-done" theory can help clarify faculty requirements and the needs of students. We then explore connectivism an example of a pedagogy that takes explicit account of our digitally connected world.

We move on to consider the practical issues confronting the use of social media for educational purposes, including the issue of boundaries, which are of importance in setting expectations and standards of behavior. We explore factors that can affect engagement with social online learning and behaviors such as lurking and suggest ways in which this can be understood and mitigated. We discuss the darker side of social media, such as trolling, and how it might be dealt with. Social media has the potential to set faculty-student relations on a more equal footing, and we look at how co-creation can be used to leverage this and drive learning.

Our final section examines practical challenges to implementing a social media platform as part of a module or course. We make recommendations on how to approach this and indicate how further research may help us gain a deeper understanding and develop solutions for some of the problems we highlight.

E. Rengasamy
Department of Public Health and Primary Care, University of Cambridge, Cambridge, UK

D. Cole (✉)
Centre for Medical Education, Cardiff University School of Medicine, Cardiff, Wales, UK
e-mail: coleds1@cardiff.ac.uk

H. J. Witchel, M. W. Lee (eds.), *Technologies in Biomedical and Life Sciences Education*, Methods in Physiology, https://doi.org/10.1007/978-3-030-95633-2_14

14.1 Introduction

Social media has become ubiquitous in modern life and is increasingly used in educational settings, including the biomedical and life sciences. The advent of Web 2.0 technologies, which allow for active online participation, has transformed the use of the Internet from being largely a way to access a huge repository of information to one where small- and large-scale social interactions can occur synchronously and remain archived indefinitely for the speakers, the wider community, and researchers (Johnson, 2006).

Faculty and students in higher education approach social media technologies from different perspectives and with different needs and concerns. Our aim in this chapter is to review the field with two perspectives in mind: describing some of the educational phenomena that have arisen and speculate on some of the educational opportunities that the technology could enable. We have framed our approach using Christensen's jobs-to-be-done theory, which translates the user (or customers) "needs" into specific tasks (or jobs) that the person needs to get done (Christensen et al., 2016).

We propose that for faculty the jobs-to-be-done are to:

- Understand the technologies available and what they can do in order to see how they may be used to enhance the effectiveness of a module or course.
- Design and implement a module or course that successfully develops student learning networks or communities of practice, either wholly or in part in an online space.
- Successfully co-create with students a module or course consisting of online learning materials and experience.
- Support students who may find online learning and social interaction intimidating or difficult to engage with.

And for students the jobs-to-be-done are to:

- Successfully develop a digital "identity" and negotiate online spaces to ensure psychological safety within an online group.
- Successfully learn in an online environment while enjoying the experience.
- Access the online spaces with minimal barriers or technology constraints.

We will address these objectives through a review of the literature, aiming to provide evidence-based recommendations for successful strategies and through case studies of our own experience as both faculty and students.

We start by considering learning theories that can be used to guide the design of social online learning in order to help frame the discussion and highlight areas of practice that may be beneficial and where problems may lie. We move on to more specific instructional design aspects as we develop the theoretical aspects into practical activities. As part of this, we discuss scaffolding of online learning tasks, collaborative learning activities, and student-led (peer-peer or near-peer) activities. This is closely associated with issues of engaging students and faculty in online communities, an area where many interventions struggle, and we consider how to develop an effective learning group, either fully online or as part of a blended learning approach.

We discuss how to encourage students to develop personal learning networks, as a complementary approach to specifically designing interventions that build a community of practice. This may include a network of both students and faculty across institutions, and this leads us on to a discussion of co-creation, a process that involves faculty and student collaboration in the production of online learning resources.

The final challenge is how to implement successful pilot interventions in the wider course, and we draw lessons from the literature and our own experience to provide some recommendations.

To close, we provide a discussion of where we see the challenges remaining, and outline where further research could address some of the outstanding issues in the field.

14.2 Learning Theory and Instructional Design

With the introduction of Web 2.0 systems, the "learning ecology" (Williams et al., 2011) has significantly widened, such that web-based learning has expanded to include not just extensive resources and single-user e-learning courses but also the facility to discuss and interact with others either synchronously or asynchronously via text (e.g., discussion boards) or via audio or video conference platforms such as Zoom. A decade ago, interactivity in online learning courses was quite variable and remained low in at least a third of courses reviewed by Cook et al. (Cook et al., 2010). There was, however, significant breadth in course design in terms of the approaches used – for example, tutorials, virtual patients, and text-based discussions. Interactivity is important for achieving learning outcomes effectively in web-based learning (Cook et al., 2010) and is therefore something that should be consciously designed in, along with practice exercises, feedback, and repetition. Interactivity, of course, may or may not involve discussion, but it is a cornerstone of social learning, and peer discussion in particular seems to be important in driving behavior and patient-oriented outcomes (Cook et al., 2008).

As with any learning, whether it is in the biomedical life sciences or in other disciplines, good instructional design can contribute to a successful outcome, particularly in more formal settings, although such considerations may not be as important in the case of less formal learning such as apprenticeship-style learning (Sadler et al., 2010). Learning theory is important in directing the design of a module or course, although there are few learning theories that explicitly take account of web-based learning and social media in particular. Conceptualization of learning using social media by medical educators has been linked to two dominant theories – constructivism and connectivism (Flynn et al., 2015). The former is well-known in biomedical and life science education, where social constructivism is the basis of approaches such as problem-based learning (PBL) (Bate et al., 2014).

Connectivism (Siemens, 2005) may be less well-known in these fields but is highly relevant to web-based social learning. It focuses on the interactions of learners with a community and learning materials; materials and learning that are found to be useful are shared with others, and those less useful are rejected (Siemens, 2005). Connectivism recognizes that web-based learning is characterized by a huge array of available information and opinions and that learning occurs both within the remit and direction of traditional institutions and outside of this in wider web-based communities. Connectivism conceives such institutional and noninstitutional communities and the wide array of information sources available on the web as occurring within a "learning network", where each learning community or organization is a node in the network and their interactions the network edges. Learning is seen to emerge from the interactions of the nodes, rather than being seen as something that just happens internally in the learner. Of course, each node also consists of a network of individuals, for example, in a learning community, and the network of which the learning community is part is also part of a larger network. And each individual's brain is a network of neurons – each network then being part of another and consisting of other networks. This is therefore conceptualized as a "nested" network structure in which learning (in its different guises) emerges from the interactions within the networks at each level.

Interestingly, connectivism leads to the assertion that knowledge also resides outside of each learner, in the learning community itself, and also beyond this in wider networks. This concept of knowledge existing "outside" of individual learners is known as distributed knowledge (Downes, 2009). This can be helpful when considering social learning as it draws our attention to the essential nature of this type of learning, i.e., its dependence on other learners, on the wider community, and on linking and sharing sources of information. It also emphasizes how knowledge conceptualized in this way self-organizes and emerges from the multiple interactions of the participants, including the teacher. In one sense this means that there is a fundamentally unpredictable element to social and networked learning, but it could be argued that this is true of any type of learning, and the difference here may simply be one of degree. We should also note that although connectivism was developed with web-based networks in mind, and with a specific recognition of how learning occurs in a digital context, many elements of the theory are not dependent on technology and are just as relevant without reference to the web at all.

One issue that such a conception of learning may face is where the boundaries of the networks lie. If we take a systems-thinking perspective on learning (Cabrera et al., 2015), consideration of the boundaries of the learning activities is important: we are familiar with this in the form of learning outcomes, learning community ground rules, and the purpose of the activities learners are being asked to undertake. Without clear boundaries, learners may become lost and unable to focus on what they are trying to achieve, and potentially destructive behaviors may arise and be inadequately dealt with. This is a familiar issue with large social media companies, where boundaries may not be clear or are easily breached and which may not be consistently upheld and enforced. We deal with some of the issues this raises in more detail below.

14.2.1 Implications of Connectivism for the Design of Online Social Learning

When adopting connectivist principles, there are several important implications that arise for learning design. The first is the acknowledgement that networks are important for learning, both within a learning community itself, but also with the wider community. This suggests we should design learning activities that enable students to connect with and share learning and resources online with one another and also with the wider scholarly community (Bell, 2009). Social media can assist in this process. We also need to ensure that learning communities are diverse and inclusive, which helps promote the widest range of views and perspectives and thus broadens learning (Siemens & Conole, 2011). As educators, connectivism also suggests we should be tolerant of the wide-range learning that can take place both within a defined learning community and with the community's wider interaction with scholars in the field—the role in this context is one of facilitator and coach, promoting critical thinking, discussion, and creativity. However, we must acknowledge that such an open approach to learning can attract those with negative behaviors, and thus the ideals of such a philosophy must be tempered by setting and enforcing standards of behavior. This may be challenging in practice, as it is often only the platform provider that can enforce sanctions on individuals who contravene their behavior policy, particularly if the group that is affected is a public one that anyone can join.

Whether we deliver learning in-person or online will impact delivery and dictate constraints. This can be important to recognize, as Fawns et al. (2019) point out: learning does not just happen "online" but wherever the learners are situated and is driven by their interactions with peers, teachers, and learning materials. This important reminder that social online learning still happens in physical spaces prompts us to consider not only how we design the online space itself but also how we take account of learner context. Thus, we need to consider where students will be when engaging in online learning activities; who else is there (e.g., other learners, family members, or no one); local technology constraints such as bandwidth, differing time zones; and ultimately how this might impact on their engagement and progress with their learning.

Thus, there are several considerations and activities we may wish to adopt when following a connectivist approach to learning design. We must also acknowledge that there are a number of challenges both staff and students face when using social media in learning, including variable learner participation (which may occur for a variety of reasons), technical issues, and demands on time (Cheston et al., 2013). There may also be a tension between the inherent value placed on participation when using Web 2.0 technologies for social online learning, and the assessment of learners as individuals (Bennett et al., 2012), which may cause difficulties for both faculty and students. The alignment of curricula that include such learning designs with assessment practices therefore needs careful consideration.

14.3 Engaging with Online Social Learning: Opportunities and Challenges

14.3.1 Case Study: A student's Experience of Online Content Curation

The medical course at Cardiff University, especially in the preclinical years, consisted of a case-based learning approach. As part of this, each group of students over the 2-week case would be required to respond to a clinical scenario through generating learning objectives, discussion, and self-guided learning. An important part of this was individual research to feed back to the group during the case.

The location of appropriate resources was a key element to the success of achieving those objectives and therefore of the group. Often, a difficulty was the vast and overwhelming amount of information available. It would be tempting to use sources that everyone else was using simply because it appeared "easier" to access. This would mean that we were less adventurous with what sources and ultimately what information we obtained.

The integration of a new tool "Scoop.it" was introduced to students (see Cole et al. (2017) *for further details). The aim was simple: a curation platform was provided to enable students to not only locate resources for each "case" but for the group to view. Curation involves the searching for information relevant to a specified topic or clinical scenario, selection of appropriate and high-quality resources, and the sharing of these with a wider group with a commentary (see* Fig. 14.1 *for an overview of the curation concept). The wider group or audience can then engage with the resource and enter into a discussion should they so wish. The benefits of introducing this were significant: students were firstly encouraged to engage with online content; they were encouraged to look at types of content they may not originally consider and in the process of reviewing and curating resources were assisted in the development of critical analysis skills.*

The interactive element was interesting, allowing students to comment on others' findings. However, a component that set it apart was the ability for faculty to be involved in students' curation. Faculty could also comment on the usefulness of a source or simply give encouragement. This was a useful feature allowing students to start to rationalize the vast number of information sources.

Students initially engaged well – 28 of the 30 groups used the curation tool at least once – but the majority remained hesitant, with only a third of the year posting to the curation platform. One reason for this may have been the unfamiliarity with

Fig. 14.1 The process of online content curation. Note the iterative nature of the process. The more actively engaged in the social aspects of the process the curator is, the more valuable the collection is likely to be, as this will help direct the topic content through discussion with the audience. These activities can be undertaken by both faculty and students, either as an individual or as a group. For more details see Cole and Cunningham (2014) and Cole and Jones (2019a)

the new curation tool. Many students opted instead to use more familiar tools such as "Facebook" and "WhatsApp" to engage with their groups. Students would often create private "Facebook" groups to discuss the group case-based learning scenario or stick to familiar and well-used "wiki" services provided by the University. When I think back to why we did this, it was simple: these tools were familiar, easy to use, and easy to access, and most students had an account. Interestingly, this has shifted in the last few years to more practical collaborative working approaches such as Google Sheets or the use of instant messaging groups such as WhatsApp groups.

When creating an online learning platform, students need to feel that it is accessible; otherwise they feel reluctant to spend time to learn how to use a challenging system. Some difficulties may have rested with the fact that "Scoop.it" was new and unknown; perhaps the benefits of the tool weren't fully appreciated by students; or perhaps students felt it was too time-consuming to learn and use the tool effectively.

14.3.2 Setting up an Online Social Learning Community: Encouraging Students and Faculty to Take Part

One of the first steps of creating any social online learning community is encouraging participation. This has many facets and requires consideration of a wide range of issues. These include ensuring both students and faculty are comfortable with the technology, are able to access the online platform easily, know what is expected, feel psychologically safe with the environment and activities, and find the experience enjoyable (Salmon, 2013; Bottanelli et al., 2020).

In the initial stages, when planning activities and developing the online learning community, it is important to ensure transparency (Bottanelli et al., 2020) particularly if this is the first time participants have been involved in this type of group. From both a faculty and student perspective, delving into the unknown is not very appealing unless we know further details. Therefore, at the outset, it is useful to explain the key benefits, expectations, time commitments, and training that may be involved and how it may integrate into a course or wider learning of the participants. The benefits should be clearly stated and disseminated as widely as possible to ensure maximum engagement. At this stage, it is helpful to think about potential difficulties so that students and staff can best be supported.

Selecting the learning platform to be used can be challenging. We have found that considering the precise "job-to-be-done" when using a technology, as well as fully evaluating its features against these requirements, is a constructive way to approach this. There are guides available to help with this process: for example, with online content curation, as described in the case study, we have developed Best Practice guidance (Cole & Jones, 2019a) and Technology Reviews (Cole & Jones, 2019b) to help faculty new to this area to plan their approach and select an appropriate online platform (see Table 14.1 for a summary). Ultimately, we are aiming to ensure the "right fit" in terms of cost, accessibility, and interaction. Done well, this can be an important step in encouraging both faculty and students engage, and piloting a platform with both these groups can be a helpful way of gaining early feedback and identifying potential problems.

The roles of the lead faculty member and staff "champions" should be considered carefully. Engaging with an online community can involve a shift in ways of working for both staff and students, and therefore the instructor's role becomes more important in making it an inclusive and welcoming environment. There is a growing research base aimed at establishing the best ways to ensure the environment is nurturing and "caring" so that students can achieve their goals (Soper & Ukot, 2016). From our experience, from the student perspective, this can often be

Table 14.1 Useful social learning website and platforms mentioned in the chapter. This table lists several websites that either can be used for social online learning or contain examples of the output of this type of learning

Website / platform	Type	Link	Description
Diigo	Content curation	https://www.diigo.com	Website that allows you to collect social bookmarks of webpages, organise them into "outliners", post virtual "sticky notes" on webpages and share these via social networks
Facebook	Social media	https://www.facebook.com	Well-known platform for posting and sharing among "friends" and in groups
MS Teams	Social media / document sharing / video conferencing	https://teams.Microsoft.com/edu start	Versatile applications that can operate like Facebook, and allow videoconferencing, document sharing and links with other Office 365 applications
Pearls4Peers	Blog	https://pearls4peers.com	Blog sharing co-created medical education material from a unit in the United States
Scoop.it	Content curation	https://www.scoop.it	Content curation site that enabling sharing of social bookmarks with commentary in an attractive magazine-style layout. Topics can be shared in multiple ways
Twitter	Social media	https://twitter.com	Well-known social media platform using "microblogs" of 240 characters. Has been used educationally for journal clubs, case-based learning, and critical thinking
Wakelet	Content curation	https://wakelet.com	Content curation platform that collects social bookmarks and allows curation into topics with commentaries, and display in multiple different ways.
Wix	Blog hosting	https://www.wix.com	Blog-hosting service, useful for beginners as it helps organise blog layout and has useful guides and prompts.
Zoom	Video conferencing	https://zoom.us	Easy-to-use video conferencing platform that allows sharing of screens, breakout rooms, chat and polling while meetings are ongoing.

overlooked as the practicalities take precedence. Confidence with online interactions may take time to build, and small actions such as checking in on students, perhaps through a student facilitator or getting regular feedback to ascertain how they are finding it, are likely to result in improved engagement. In a study by Xu et al. (2020), the authors showed the importance of empathizing with students: understanding that some may be bored or feel disconnected or lonely online can shift the facilitator's approach from sharing of knowledge and content to discussions that address these reactions, which can then better meet the needs of that particular individual or group. Xu et al. (2020) used "student leaders" to achieve this, which may have helped to avoid student inhibitions due to hierarchy and had successful results.

Practice Points

- Take time in selecting the most appropriate online platform – this may not necessarily be new or innovative but could simply be one that is familiar and accessible but also effective for your purpose.
- Ensure the key benefits are communicated to both students and faculty.
- Think about interaction – what type of interaction is desired? How much? Who will facilitate?
- Planning – consider using the "jobs-to-be-done" approach to evaluate how the platform meets the needs of its users.
- Consider how the online community appears to students, e.g., is it inclusive? Is it caring and empathetic? This may be particularly important for cohorts that do not know each other well.
- Consider the use of a lead instructor and faculty "champions" to engage others.
- Consider recruiting a student leader to help run and moderate the online community.
- Make it enjoyable!

14.3.3 Digital Literacy and Information Overload

Digital literacy is defined as a set of "skills and strategies" in a digital environment (Eshet-alkalai, 2004). Rather than a single concept, it works on a number of levels and acts much like a developmental process (JISC, 2017). From a student perspective, there is often an expectation that they will have a generally high level of digital literacy as they have grown up using online spaces. However, this can be a misconception (Bennett et al., 2012), as the selected platform(s) may be unfamiliar and using them can require students to work in new ways online. A number of students may indeed be expert "digital natives" who use technology and social media frequently and are at ease with the engaging socially in an online environment. Leveraging this expertise can help the group as a whole (Ng, 2012), for example, through having "student champions" who act as ambassadors in disseminating knowledge and benefits of such platforms and acting as role models (Terry et al., 2019).

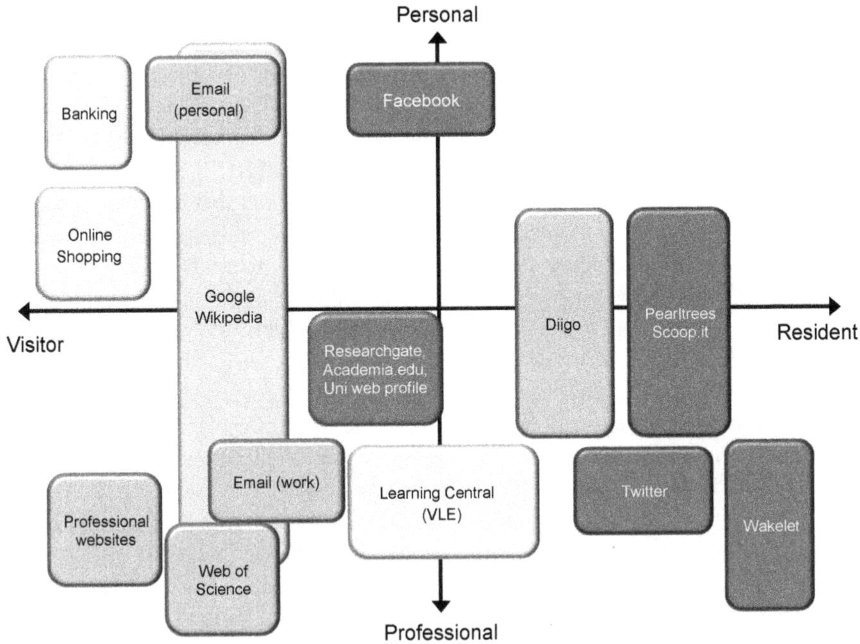

Fig. 14.2 Visitor and Resident mapping of the online activities of one of the authors (DC). See text for details on the concept. The blocks represent a personal classification of the type of use of the website named – they do not represent importance or time spent. Mapping is not intended to be a fully accurate representation of how all people use the web but a starting point for reflection and discussion

In helping students understand and reflect on how they use the Internet, David White's "Visitors and Residents" paradigm can be useful and has certain advantages over the digital native and immigrant dichotomy (White & Le Cornu, 2011). In this context, a "Visitor" behavior is one where a minimal digital trace is left, for example, when browsing the Internet for information, or reading emails. On the "Resident" end of the spectrum, a substantial "digital footprint" may be left in a public-facing forum or website, such as a blog, or public group on a social media platform. It can be useful to map personal use of the Internet along this spectrum and to include another axis on a personal to professional continuum (see Fig. 14.2 for an example from one of the authors). There is no right or wrong way to engage on the online, but a mapping exercise such as this can be a helpful way to encourage students and faculty to reflect on and discuss how they personally interact in an online environment and why. We have used this approach in our practice to open a discussion of facilitators and barriers to using the Internet in a particular way, including for online social learning.

As recognized in the case study above, one challenge to the effective use of the Internet for learning is the wealth of information readily available. From a students' perspective, this can be a positive in that it is easy to find information on a topic, but conversely it can be overwhelming and induce anxiety. Several "information pathologies" have been described in the literature, including information overload, reference overload, and information anxiety (see Bawden & Robinson, 2009, for a review). With the vast number of online resources now available, these are very real problems, and selecting those that are relevant or useful requires a further dimension to understanding and navigating what Polizzi (2020) calls the "broader digital environment." Faculty should therefore keep in mind that even when digital literacy appears to be high in their students, they may nevertheless benefit from extra training or support in finding and accessing the best materials to support their learning and in engaging with peers and faculty online.

Tackling information overload was one of the drivers to introducing online content curation using Scoop.it in our institution (Cole et al., 2017; Case study above; Fig. 14.1) which facilitated the sharing and critical review of online resources between both students and staff, thus potentially reducing the effort required to find good-quality resources relevant to learning. However, engagement was not uniform across the student or faculty cohorts, which may in part have been due to varying levels of digital literacy but was also influenced by a variety of other factors including familiarity of the platform, uncertainty of the value of using such approaches, and dislike or suspicion of social media (for many reasons), which can all lead to an unwillingness to spend the time and effort required to use the platforms effectively and to use them to interact with others. Curation may however be one way to engage both the "power users" who wish to interact online and post materials to the curation website and also the "lurkers" who do not wish to interact much but will still benefit from viewing the material posted as it may save them time and effort trawling the web themselves.

14.3.4 Professionalism

We noted above the issue of boundary setting and the problem of destructive and antisocial behavior on social media. This is related to the concept of professionalism, which focuses on the behaviors appropriate to a given professional group and defines standards that should be adhered to. This is well-known in fields such as medicine, where organizations such as the General Medical Council in the UK produce and uphold standards for doctors and medical students and have produced guidance on the use of social media (General Medical Council, 2013). When entering such regulated professions, students may struggle with adhering to professional boundaries due to the mixing of professional and personal online identities (Daigle, 2020), which can result in inappropriate behavior.

Online professionalism has been studied particularly in students on healthcare courses where there is such professional regulation. In dentistry, students used social media extensively and were aware of potentially inappropriate behavior online (Kenny & Johnson, 2016) and were conscious of privacy and security settings and the implications of inappropriate behavior for their own Fitness to Practice. Despite this, students did admit to being involved in, and almost all had been exposed to, inappropriate behavior online. This example demonstrates a high level of insight among students but varying levels of adherence to standards expected of a professional. Similar findings were observed in nursing students (Duke et al., 2017), where students had high awareness of e-professionalism and of potential privacy issues.

Aiming to encourage a high standard of online professionalism, some authors have suggested that faculty role models may be a good way to demonstrate correct and appropriate use of social media (Ross & Myers, 2017). In practice this could mean faculty acting as professionalism champions as well as demonstrating high standards of engagement and professional behavior online. However, given that Duke et al. (2017) found that faculty were more cautious in using social media for educational purposes than their students and were less likely to use privacy features, this may be an approach that may not be easily implemented in practice.

Student professionalism when online may not only impact their course but also their ongoing career. Daigle (2020) suggests that teaching online professionalism should be explicitly tied to the values of the profession. These skills are best taught and developed early on in a course and in a way that makes them accessible and relatable to the role of a professional (Sinclair et al., 2015). Ness et al. (2014) found that as pharmacy students progressed through their course, they were increasingly aware of the image they needed to portray as newly graduated professionals, as they felt it may affect employability. Emphasizing these aspects to students early and consistently may therefore be an effective strategy for encouraging professional behavior.

14.3.5 The Impact of Trolling on Student and Faculty Engagement

Social media, despite our best efforts, still has a dark side. We only need to read or listen to the daily news to hear of the misuse of social media. This includes "trolling" behavior where individuals may seek to provoke or upset others online. It is hard to pinpoint the exact reasons why a particular individual trolls, but studies suggest it may be for entertainment (trolls often think the behavior is funny), due to boredom, or for revenge (Griffiths, 2014). Peer pressure may also have a role, for example, if a young person's friends also troll (Griffiths, 2014).

There is little literature on the direct impact of trolling on students and faculty, but it can clearly be destructive and lead to those being abused and bullied in this manner to withdraw from online activities. Studies examining how those subject to this behavior perceive trolls found that they actively conceived trolls as conflict- or attention-seeking, vicious, or uneducated (Maltby et al., 2016). Interestingly, the first of these, viewing the troll as conflict- or attention-seeking, was associated with less emotional distress when exposed to such behavior and thus suggests that framing the troll and their behavior in this way may help develop resilience to it.

Practice Points: Suggestions for Ground Rules in the Use of Social Media

1. Create a comfortable environment – keep it friendly:
   (a) Think before you post and be aware of what others are posting.
   (b) Be respectful.
   (c) If you are unsure if the post is suitable, check with someone before posting.
   (d) If you notice trolling behavior, do not engage and report it immediately (BMA, 2021).
2. Take control:
   (a) You are the "master of your presence" (Isis et al., 2019).
   (b) Consider the image you wish to portray.
3. Maintain boundaries (General Medical Council, 2013):
   (a) Think about maintaining professional boundaries online.
4. Consider privacy settings (Health and Care Professions Council, 2017).
   (a) Remember your post can be shared widely so think about the content carefully before you post.
5. Keep it confidential.
   (a) Avoid posting any confidential information.

14.3.6 Active and Passive Engagement Online

When people engage online, including with learning communities, many different types of behaviors may be seen, ranging from more active behaviors such as those typical of "the power user" to more passive attitudes such as those exhibited by "lurkers" (Cranefield et al. 2015). Research into student online behavior often looks at how engaged students are in a quantitative manner, e.g., number of logins, number of posts, or number of interactions. This will miss more passive behaviors and may carry the assumption that there is nonengagement. The term "lurker" refers to a user who rarely participates actively but may read and explore the content. "Lurking" often has negative connotations, but in fact research has shown that there are many benefits (Cranefield et al., 2015). "Lurkers" make up to 90% of an online community according to Nielsen (2006) with only approximately 1% of the online community actually contributing content, and thus this type of online behavior is the norm rather than the exception. We must, therefore, explicitly acknowledge this when designing social online learning activities.

So why do students "lurk"? There may be several reasons, including lack of confidence and wishing to avoid humiliation, a desire to remain anonymous, lack of technological expertise, personal preference, being "shy," not wishing to be attacked, or simply not seeing the importance of contributing. According to Mazuro and Rao (2011), "lurking" may be a necessary first step to becoming a more active member of the online community. In addition, while "lurking" may be passive, students may prefer to learn in this manner, through observing others. Nevertheless, lurking behaviors are complex and may be a mixture of active and passive learning; Koutropoulos et al. (2019) interestingly describe it as "engaging silently."

Factors driving these varied behaviors include personality traits, self-efficacy, or students' needs. Gender can also play a part in online learning, particularly with regard motivation: research has shown that females, and in particular students who were older with family responsibilities, gained more from working in an online setting (McSporran & Young, 2001; Sullivan, 2001). There is also evidence that culture may play a role in how students behave online. For example, learners from some societies, e.g., Confucian cultures, prefer more face-to-face contact and clear instructions, while those from other societies, e.g., Western cultures, may prefer a more self-directed approach to online learning (Kang & Chang, 2016). Students from different cultures may therefore perceive the online student-teacher relationship differently, which can in turn affect their online behavior. This was especially the case with how students saw the teacher, some viewing the online faculty lead in a more authoritative manner and others seeing them more in a facilitative role. Conversely, students from different cultures may also see different barriers to online learning whether this is motivation, control, or engagement (Lim, 2004).

14.3.7 The Influence of Personality on the Use of Social Media

There is a growing body of research looking at whether different personality types have any bearing on how we use social media. This often examines the "big five" personality traits of extraversion, agreeableness, openness, conscientiousness, and neuroticism (Roccas et al., 2002). These traits may directly relate to values or behaviors of the individual and thus may have an impact on how we behave online. Traits, values, and behaviors are so closely linked that using an individual's "digital footprint" (i.e., the traces left online such as posts, comments, tweets, etc.), we can deduce their personality profile (Azucar et al., 2018). Gil De Zuniga et al. (2017) aimed to categorize traits indicating positive or negative online behaviors: openness and neuroticism were linked to more negative behavior online, while traits such as conscientiousness were more associated with positive behavior. The authors suggest that this may be due to conscientiousness and extraversion being linked to more confident manners of engaging with others, engaging more readily and attending to details, but also willingness to empathize with others. They argue that this drew such individuals to use social media not just to keep up to date but also connect with others. In contrast, they found that those with higher neuroticism tended to use social media less, but Vaid and Harari (2020) found that neuroticism was linked to more open use of social media. Neurotic traits have also been linked to more passive online behavior, less frequent posting, smaller online networks, and a more cautious approach to portrayal on social media (Bowden-Green et al., 2020).

14.3.8 Strategies to Engage Students in Social Online Learning

We have discussed a wide range of factors that can influence whether and how students may engage with social online learning. Understanding how students may perceive online communities is useful as we can predict how they may interact, and appreciating some of the reasons they may not wish to contribute is key to this. We have found that using Christensen's jobs-to-be-done theory (Christensen et al., 2016) can help to design appropriate learning environments by focusing on what both students and faculty need to achieve (i.e., what their jobs-to-be-done are). This includes not just the high-level jobs we outlined at the start of the chapter but also supply chain jobs (what other jobs need to be done in order to achieve the main job) and emotional and social jobs associated with these more functional jobs. We feel it is emotional and social jobs that are particularly important, given the issues raised in the preceding sections. We should also recognize that students may have different lists of jobs-to-be-done or may hold some jobs as more important than others, as a result of their personality or past experience. We therefore recommend that students from a variety of backgrounds are surveyed to gain a broad perspective.

A further way engagement can be encouraged is through "student facilitators" who can act as a bridge between faculty and students. For instance, student facilitators can help with the cohesiveness of the group and assist with guiding and scaffolding (Xie et al., 2014), as well as with deadline reminders, clarifications of instructions, or providing tips and prompts. Such a prompt from a peer may help those who are less confident or unsure to feel "safe" (Hew & Cheung, 2008). This has been taken further in the concept of "shadow modules" (Scott et al. 2014) where student-led groups run in parallel to the taught module and students discuss content, curate and share resources, and create their own. Faculty interact with groups mainly via the student leads, reviewing resources, and suggesting questions or topics for discussion. This has been shown to be highly beneficial to students who engage (Rutherford et al., 2017), but there is still an issue with variable participation. Appreciating that "lurkers" can still gain from the experience may be reassuring from a faculty perspective: we do not have to become disheartened if this behavior predominates.

Consideration should also be given to oversight and support of student moderators, as they may encounter problems or spend too much time focusing on online activities. In the shadow module approach, student leads are linked to a member of faculty who can not only provide academic guidance but can supervise and support the student lead as well (Rutherford et al., 2017).

There may be concerns from both faculty and students who are e-moderating with regard to the time this may require of them. This can vary from 5 to 45 min per asynchronous online interaction, which occur every few days, for a 16 student group – see diary entries from an e-moderator in Salmon (2011) for more details – or possibly longer depending on the course. Students value highly a timely turnaround in responding to their posts and submissions in such online environments, which should ideally be 48 hours or less (Thomas et al., 2004). Careful instructional design can help with reducing the amount of time required for such activities, as can ensuring an optimal moderator to student ratio: 1 moderator to 15–20 students appears to be about right (Salmon, 2011).

Practice Points

- Ensure students and staff are aware of the ground rules, expectations, and aims of the new initiative, and consider boundaries from the outset.
- Consider the type of monitoring you may wish to use in the online community and how frequent this should be, e.g., will there be a faculty moderator or student moderator?
- Consider how you will promote online safety, e.g., ensure privacy settings meet the expectations of the users so that they feel comfortable and safe, and ensure the group is inclusive and welcoming.
- Engage with hesitant learners – Try to understand how they feel about the experience. How does their background impact their engagement approach? How might they wish to engage?
- Consider the online behaviors of your students: Are they predominantly "lurkers" or "power users"? Might they need more scaffolding?
- Consider using curation as a way of engaging lurkers and directing them to high-quality resources.

14.4 Co-creation Between Students and Faculty

14.4.1 What Is Co-creation?

Co-creation, at its core, involves working with others to produce a product. It is commonly used in many fields, from business to healthcare. When an architect designs a building for any client, it would be rather unusual for the client to not be involved in the design process. In fact, it is the norm in this field for the client and architect to work together to produce the final design. In this example there is a "consumer" (client) and "producer" (architect) who are working together to create a bespoke product for the client. The architect cannot do this as well or as efficiently without the involvement of the client, as they do not have a complete understanding of the client's needs. Likewise, the client cannot produce the product without the skills of the architect. Thus, both parties are needed if a design is to be produced to the highest standards and that meets the client's needs. This is the central point of co-creation – that all parties bring something of value to the process and together produce a product that is of a higher quality than any individual party could do alone.

Co-creation is also routinely used within the healthcare setting, although it is not commonly known by this name. In this environment, patients and care providers make decisions together regarding treatment pathways, particularly in complex cases (Légaré et al., 2010). Benefits of this approach include a fresh perspective, partnership between healthcare provider and patient, and shared decision-making (Stiggelbout et al., 2012; Batalden et al., 2016). Thus, in this context, co-creation yields more meaningful engagement for both parties but also allows benefits such as transparency and trust to develop (Prahalad & Ramaswamy, 2002).

Co-creation in an educational context has grown in popularity due to its collaborative approach to learning (Dollinger et al., 2018). It offers a unique approach for teaching faculty to collaborate with students to produce content such as practice questions, blogs, or e-learning resources and also to develop curricula. Co-creation also ensures that students play an active role in teaching and learning which facilitates the shift from a passive to an active learner role (Bovill, 2020). Co-creation is not simply about the output but runs deeper as a pedagogical approach, offering a number of benefits to those that participate in such projects. Similar to patient benefits in shared decision-making in a healthcare context, learner benefits include improved engagement, shared decision-making, and learner empowerment. Könings et al. (2020) have developed a framework that emphasizes such benefits and also points to the benefits for others: for faculty in improving their teaching, for curriculum stakeholders in improving course design, as well as wider benefits for those not directly involved in the co-creation process. Involving learners in course design in this way may not only improve their attendance (see attendance Chaps. 11 and 12 in this volume) but may also improve their performance (Brooman et al., 2015).

14.4.2 Initiating and Maintaining Co-creation Projects

Co-creation projects may involve a very different way of working for both faculty and learners. To be successful, teachers and learners must be prepared to participate on a more equal footing than they may be used to, and they need to make best use of this shared power in creating the output. At the start of any co-creative project, it is important to create a forum to begin a dialogue between students and faculty so that such issues can be addressed (Vallentin-Holbech et al., 2020).

Prerequisites for a co-creation project should include a foundation of trust, knowledge, and respect between all parties and also reciprocity (Cook-Sather et al., 2014). Participants should be aware of, and set appropriate expectations for, the co-creative process itself and also for the output. This requires that there is a mutual understanding of the autonomy and influence both students and faculty will have (Martens et al., 2020).

After this foundation comes the planning stage, many have found that training, and in particular early training, of students can help increase their confidence in taking part. If we take the example of question-writing as a co-creative approach, students clearly benefit from training in question-writing (Jobs et al., 2013). This can be achieved through well-known pedagogical approaches such as "scaffolding" which aim to break down the task to make it more accessible to students and help them structure their learning and activities (Jumaat & Tasir, 2014). This training can also help students feel accountable for their contributions and work and ultimately encourage them to take part in future co-creational activities.

Maintaining engagement requires ongoing work and commitment by the co-creation team focusing on the team dynamics and ensuring students are well-supported. Co-creation can be readily supported through the use of modern web applications allowing document sharing, blogging, and communication platforms. These can be leveraged not only for discussions between team members but also to facilitate the co-creation process itself.

14.4.3 Outputs of Co-creation

The outputs of co-creation depend on the specific interests and requirements of faculty and students. Common outputs include blog creation, question-writing, and production of learning content such as modules, general curriculum design, and wikis. We discuss the first two of these outputs as examples below.

14.4.3.1 Question-Writing

Question-writing as a student-led output of co-creation can be a useful way to engage learners. Approaching question-writing as a co-creative project has benefits for faculty as well: it provides a valuable insight into students' progress, learning strategies, and understanding of particular topics or topics they may be struggling with (Sircar & Tandon, 1999). While question-writing is popular, sustaining engagement can be difficult. Several common barriers include time (as question-writing by its nature is time-consuming), unfamiliarity with the process, and motivation.

There are number of question-writing tools available on the web (see Table 14.2 for examples). One tool that aims to make question-writing more interactive is *PeerWise* (Denny et al., 2008a), which enables students to create questions for self-assessment and sharing among their peers. There is a built-in gamification element where students rate the quality of the questions their peers have written. Students submit their own questions, and their peers and faculty can then review and comment on them. Depending on this feedback, students receive a rating, promoting active engagement and a drive to improve.

PeerWise has been adopted by many institutions internationally for many different courses including Medicine and Biomedical Science (Bottomley & Denny, 2011). The competitive element has been shown to work in several case studies: Worm and Buch (2014) found that quiz modules or tools that instigated competition increased efficiency, usage, and self-awareness among students. This assessment of

Table 14.2 Question and flashcard writing and sharing websites: examples. This table lists several websites that use sharing functionality and other social media and gamification techniques focused on the creation and use of quizzes and flashcards

Website / platform	Type	Link	Description	Student tracking	Sustainability
Anki	Flashcard sharing	https://apps.ankiweb.net/	Flashcard creation and sharing with a rating system that feeds into learning repetition frequency. Sharing decks can be public, or limited to a specific group.	No	Free, downloadable application to desktop/laptop and also smartphones. Decks of cards are saved as a file. Syncing and sharing can be done via a web application.
Chegg Prep	Flashcard sharing	https://www.chegg.com/flashcards	Flashcard creation and sharing. This used to be StudyBlue. Can be used in flashcard or quiz mode.	No	Free. No clear option to download flashcard decks or print them from within the website. Can use an app for testing.
Mentimeter	Interactive presentations and quizzes	https://www.mentimeter.com/	Online platform for creating interactive elements for presentation, feedback, and quizzes that are particularly effective when run live.	Yes – and allows download of answers to an Excel spreadsheet	Free and paid options. Allows upload of existing Powerpoint slides and sharing of questions between team members. Downloadable quiz results, but not full slide sets.
Peerwise	Question sharing	https://peerwise.cs.aukland.ac.nz	Question writing and sharing platform for students, with gamification elements and feedback mechanisms built-in.	Yes, and has a system of rewards for engagement and performance.	Free software, able to bulk upload students to a course, but not questions.
Quizlet	Quiz and flashcards	https://quizlet.com/	Instructor generated learning materials and quizzes that students can engage with in multiple ways, either as a conventional quiz, flashcards, or live team learning. Uses gamification in some applications.	Yes	Free and paid versions. Allows upload of existing materials and questions, and also export and printing of sets.

questions can facilitate deeper learning benefiting students' critical analysis skills as well as engagement with the subject (Denny et al., 2008b). Doyle et al. (2020) point to improved learning and examination performance as a further benefit, and Lubicz-Nawrocka (2018) found that shared learning benefitting both students and faculty but also developed students' cognitive, interpersonal, and adaptive skills.

14.4.3.2 Blog Creation

Blogs are not a new concept, but they have been used widely in co-creation projects across different fields of study. One example is *Pearls4Peers* an open-access platform on which faculty and students based at a hospital department in the USA create educational blog posts (Manian & Hsu, 2019). In this example, the blog was used by learners to address clinical questions they had encountered during their placement ward rounds. The project involved students and faculty at different levels in blog post production, a clear practical example of partnership. This approach was successful, and students enjoyed the challenge of producing content and as a result not only increased in examination confidence but also developed skills in collaborative working and writing.

14.4.4 Benefits of Co-creation

We have discussed some of the benefits of a co-creational approach above; it is clear that co-creation improves motivation, engagement, and knowledge of students on a course program. There have been mixed results with regard impact on examination performance, and this most likely reflects the varying methods and outputs studied. Many investigations have been centered on question-writing; Foos (2009) found that it was an important study technique and students performed better in exams. Walsh et al. (2015) found that question-writing approaches through use of the *PeerWise* platform proved valuable in examination preparation and that this may be affected by the cohort targeted for such interventions. They found that the more junior the cohort, the better the response. Jobs et al. (2013) found that lower-performing students performed better in exams as a result of co-creative question-writing, but this appeared to be due to memorizing the questions themselves rather than because participation promoted deeper learning strategies.

The nature of co-creation means that students can benefit from a deeper learning experience. One example may be the use of decision-making in designing their own questions. Further to this, it can also promote appreciation of the assessment process and thus improve assessment literacy. Overall, co-creation can provide a useful learning strategy and benefit both students and faculty alike (Lubicz-Nawrocka, 2018).

14.4.5 Student and Faculty Perceptions of Co-creation

So, what do both parties think of such approaches to learning? We mentioned earlier that a dialogue is needed between both faculty and students. But what does this entail, and does it always work? From a teaching perspective, some faculty believe that students must be motivated for co-creation to work; others felt that students lacked confidence and the requisite knowledge to participate (Martens et al., 2020). Thus it makes sense that some faculty members maintain that while co-creative approaches are a partnership, faculty should have the final say (Martens et al., 2020). And while they may be able to co-create small portions of the course content, for students to be able to get involved in co-creation projects with a greater impact such as curriculum design or assessment, they need to have training.

The student perception is more mixed. While most students have positive experiences of co-creation, they cite the need for initial training, support, and an incentive for doing so. Walsh et al. (2015) found that the use of the *PeerWise* platform discussed earlier was for many students enjoyable which contributed to how they perceived learning and engaged. They also found that while students found this approach engaging, motivational, and enjoyable, some students had concerns about the quality of questions produced and valued faculty involvement.

14.4.6 Barriers and Limitations Faced by Co-creation Projects

So what is stopping us from using such approaches more widely? There are a number of barriers, which include power differentials, time commitment, prior student knowledge, and communication issues.

An important finding from attempts to implement effective shared decision-making in healthcare settings is the effect that the power differential between doctors and patients can have on inhibiting patients from making their voice fully heard (Joseph-Williams et al., 2014). Particular care therefore needs to be taken in establishing trusting and equal relationships, so that patients are assured that they will not "displease" the doctor or healthcare professional and suffer poorer-quality care as a result. A similar dynamic is at play in the learner-teacher relationship and needs to be effectively addressed if co-creation efforts are to be successful (Könings et al., 2020). They advocate similar interventions as are used in healthcare settings to place relationships on a more equal footing, and they recognize that this can take time as both learners and teachers may be stepping outside of their traditional roles, which may be uncomfortable. An interesting effect of using digital technologies as part of this process is that it appears to be easier to break down such barriers using this medium (Blau & Shamir-Inbal, 2018), and thus developing online co-creation groups may be more effective than those that use a solely face-to-face approach.

Faculty need support and training within the organization to use any tools required and students need training for the task at hand. Training should be step-wise and aim to build confidence and competence so that tools can be used effectively: training in question-writing and blog writing, or the use of wikis can help engagement (Grainger et al., 2018). There are mixed thoughts on who should have the overall control in a co-creative project, but some argue that having faculty oversee this works best (Martens et al., 2020).

Practice Points

- Ensure that staff and students feel supported through regular and open communication channels, e.g., regular updates, troubleshooting meetings, and check-ins.
- Support students with training in the focus of the co-creation activity, e.g., question-writing master-classes.
- Take a step-by-step approach giving students gradually more responsibility, e.g., use a scaffolding approach such as starting with question prompts and discussion and then moving on to writing full blog posts.
- Make clear the benefits to students and faculty alike, e.g., for students this may be question-writing skills or critical analysis, and consider use of certificates of participation or incentives.

14.5 Embedding Social Online Learning into the Curriculum

Embedding social online learning in curricula can be challenging. Figure 14.3 illustrates some of the key points to consider. To help us start, it is useful to think of the purpose of using a specific social online learning tool – the "job-to-be-done" – what output we hope to gain from it and where it fits within the course or curriculum (Sandars, 2006; Cole & Jones, 2019a). For example, the learning platform may itself serve as a resource to students or act as an avenue to provide signposting to other resources. Clarity regarding the job-to-be-done early on helps us understand what format may be best: a blog may be the best option if each student is required to produce a piece of work; or perhaps a wiki would be more suitable if a continually updated class resource is being built.

14.5.1 Planning

Bovill (2019) discusses the importance of the planning stages and suggests we need to consider the focus of co-creational activity, numbers of students involved, context, year of study, scale of activity, and whether it will involve all the cohort or selected students. Considering how much interaction would work best and what type of interaction from the faculty is needed is important. This could be an observational

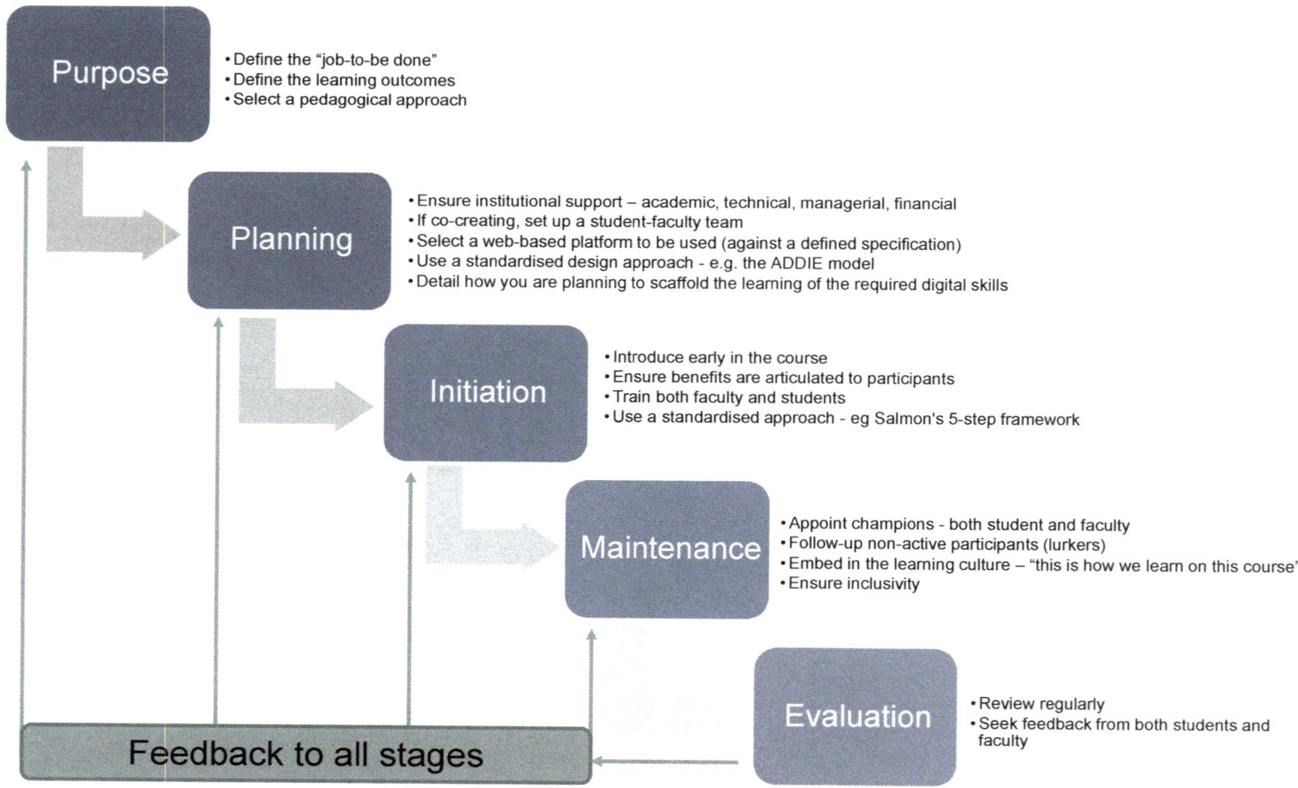

Fig. 14.3 Framework for implementing social online learning into a course. This figure summarizes the several of the conclusions in the chapter and arranges them into a practical framework that can be used to help direct the process of introducing such learning into an existing course. Note the importance of evaluation and feedback – getting such a new and complex learning experience embedded is challenging and is likely to need considerable effort to review and modify the activities

role or an information seeking role through curating tools. It may instead be more of a content-focused interaction needing a clear output such as via a blog or via production of assessment materials. Thinking about these aspects can help fine-tune exactly how we can embed it into the course so that it becomes practical, useful, and also a positive interaction for both students and faculty.

Early integration helps introduce both staff and students slowly to new tools and platforms and adopts a "gentler" approach. This can be done in many ways such as through training or preparation activities. This is where scaffolding becomes important. Scaffolding is crucial as it not only helps to provide a framework and instructions for students to follow but also helps faculty to outline the expectations of students (Jumaat & Tasir, 2014). It also helps make the tool and content more accessible to students by breaking down a large project into smaller, more manageable, tasks.

14.5.2 Initiation

Salmon (2013) suggests learners should be guided through five distinct developmental stages in order to achieve success. Each stage has specific technical support and e-moderating requirements which help support and guide learners in each stage, although it is recognized that learners may have different needs and progress at different rates and indeed may cycle through the process on many occasions, for example, when a new platform or social online learning group is set up. The stages are (from Salmon, 2013):

1. **Access and motivation** – here learners must have a reason to go online and engage, and there must be as few technical barriers to this as possible.
2. **Online socialization** – in this stage learners start to connect to others and to faculty and start to develop their digital identities.
3. **Information exchange** – students share knowledge and support one another's' learning.
4. **Knowledge construction** – at this stage, learner discussions become more collaborative and team-orientated, and knowledge begins to be co-constructed in addition to being shared.
5. **Development** - learners are now comfortable working online and are able to make full use of their skills, including reflecting on the process itself and generating their own goals and ideas.

Practically, this can include a range of different activities (Jumaat & Tasir, 2014) such as question prompts, introductory posts, initial activities to engage, and starting a discussion. It can also extend to more concrete support such as guides and technological or web support for using the platform. Such scaffolding allows the students to gradually take on the shift in responsibility and they can move on to take more of a student-led role.

A similar approach may be needed for faculty as well, as many may not have the skills or confidence in this type of learning either. Indeed, there may well be resistance from some faculty, an issue that must be actively tackled if such approaches to learning are to be successfully embedded. This may take time, and a flexible approach is recommended to ensure that all staff feel ready to incorporate this into their own methods of teaching. Training faculty also helps to ensure that platforms are used to their full potential and staff do not miss any key features (Sandars, 2006).

14.5.3 Maintenance

Sustaining a new initiative is often the most challenging aspect of a project, and social online learning is no different. It may in fact be even more challenging due to the issues outlined above and the need to invest substantial amounts of time in the online community. As the cohort changes, tutors may change, and the learning focus may change too; there may be a different emphasis in evolving curricula, and if there is insufficient faculty or student engagement, it may fail (Cole et al., 2017). In some cases, directors of learning may be unsure of the benefits and of how such technologies might be used (Wittich et al., 2017) which may generate subtle resistance. Planning for a sustainable change is therefore crucial, and this involves comprehensive initial training and ongoing communication. We mentioned earlier that this should be like a "dialogue" and this becomes even more important in maintaining its effectiveness.

Another important facet is support across different levels. This may refer to wider institutional support from the organization so that staff feel acknowledged in terms of the time and additional work facilitating an online community may involve. However, this may also be social support which entails the initial training but also regular ongoing discussions with staff.

This topic becomes relevant during uncertain times such as our current pandemic – indeed there has been an explosion of online learning as a result of COVID-19 lockdowns, and advice has been rapidly distributed as to how to manage this (Bottanelli et al., 2020). Long term, the online learning community could form part of a wider program or perhaps extending to building a series on online learning communities.

Practice Points

- Plan carefully from the outset – be clear on the "job-to-be-done" and the place of social online learning in the curriculum.
- Gain support from the institution – ensure sufficient support is available for technical assistance and staff and student training.
- Use a framework such as that developed by (Salmon, 2013) to help scaffold learning and online socialization.

(continued)

- Be aware that many staff and students may feel threatened and unskilled in this way of working – ensure they are well supported, for example, through encouraging early and staged engagement activities and recruiting technology champions.
- Sustaining a new intervention requires time and effort – ensure frequent communication and review, and seek regular feedback from both staff and students, and be prepared for the time commitment required.

14.5.4 Case Study: Faculty Perspective on Introducing Virtual Case-Based Learning (vCBL) During the COVID-19 Pandemic

During the COVID-19 pandemic, the undergraduate medical degree program in my institution, where I am deputy director, had to make significant changes to the way we delivered case-based learning (CBL), an approach similar to problem-based learning (PBL) that runs throughout our curriculum, but with different formats and emphases according to the year in which it is delivered. During the spring of 2020, all of our students had to be pulled from their clinical placements, and all of our course delivery was moved online. The most radical change was in the delivery of year 3 CBL, which we will focus on here.

*Year 3 in our 5-year course is the first year with extended clinical placements. In Phase 1 (years 1 and 2), students are predominantly university-based, with 1 day a week or fortnight in a community or hospital setting learning clinical and communication skills. In year 3 they have three 8-week clinical placements, with the rest of their time based in the university. During each of the 8 weeks on placement, they have a "case of the week" that is used to structure their learning. They are provided with a case vignette and a Cardiff Framework grid that prompts them to ask relevant questions, and they are expected to make notes using this prior to attending the first case group meeting of the week. At the first meeting the students discuss the case, facilitated by a tutor, making use of both the seven-step PBL method they use in Phase 1 CBL and the Cardiff Framework grid. They note learning points and learning outcomes at the end of the session, and two students are tasked with presenting patient cases for the next meeting. The cases are based on the same presentation (*e.g., *acute abdominal pain) that is the focus of the case of the week. The group meet for a second time at the end of the week, and the patient cases are discussed using the same frameworks as the virtual case.*

During spring 2020, the CBL described above could not be delivered in-person and had to be delivered online. In addition, our students did not yet have key worker status and therefore could not attend clinical placements, although a significant proportion did volunteer for various roles in the NHS. CBL sessions were usually staffed by senior and junior clinicians working for our NHS partner organizations, and we therefore had a reduced faculty as these individuals could no longer facilitate these sessions due to the COVID-19 workload. We therefore had to adapt our approach to one that was much more student-led.

We set up virtual CBL (vCBL) with two video conferencing meetings per week per group of five students and asked them to use MS Teams to host both video conferencing, chat, and file-sharing, as this platform was supported by our institution. The sessions were student-led, with no staff facilitator, and they were asked to appoint a Chair and Scribe for each of their meetings. This follows a similar arrangement as their Phase 1 CBL, where at the end of year, two student groups are effectively autonomous and need little faculty input. In addition, we asked them to complete two tasks for each case and submit them after their second meeting: the first was to identify a web-based resource they had found helpful for the case and to upload a link and commentary on this to a MS Form. The second was for each student to produce a Single Best Answer MCQ and upload these to PeerWise and to select one from the group as a whole to submit via *an MS Form. Faculty then reviewed the website commentaries and curated these into a searchable spreadsheet that was made available to the whole year and reviewed and edited the MCQs and produced two quizzes that were delivered in the middle and end of the module. Students were also supported by a variety of web-based resources and we ran online "virtual clinics" where clinicians discussed a variety of cases from their practice.*

The change was successful and students engaged well with the approach as evidenced by the outputs they produced and shared. They valued the autonomy and the creation of learning materials for their peers, and the co-creation with faculty helped to drive engagement and learning. Of course, it could not substitute for clinical learning on placement, but our evaluation data of delivering CBL in this way encouraged us to integrate it into the course as a permanent feature, although it has required alteration as it is now part of a blended learning approach rather than wholly online. Students remain in their groups all year and, and as they are not all based in the same hospital or primary care practice, they are able to share their learning, clinical experience, and insights into local healthcare needs that vary from area to area.

14.6 Looking to the Future: What Do We Need to Successfully Adopt Online Social Learning in the Biomedical and Life Sciences?

There is no doubt that the COVID-19 pandemic has accelerated change in the area of social online learning. This is true across the higher education sector, including biomedical and life sciences. Virtual classrooms have become the norm and a requirement for fast broadband access a necessity for learning. The use of Zoom and Teams for video conferencing has exploded. There is a specific challenge for the biomedical and life sciences, however, with regard to the practical aspects of the subjects they encompass, such as learning practical skills at the lab bench, cannot be entirely replaced by online learning. Leaving this issue aside, what is the role of social online learning for these subjects, and how do we decide how to best use this?

We have covered a wide range of applications in this chapter and provided some examples of how we have used these principles in our practice and our experiences of implementation. The evidence base, while extensive in certain areas, is very limited in others, particularly with regard to effectiveness and implementation in life science courses. The medical education literature provides more evidence, including some useful guides and detailed reviews, but there remain significant gaps. We are still lacking in clear evidence that using new learning theories such as connectivism in our design leads to improved learning in this context and that using online collaborative methods unequivocally leads to deeper learning of content.

An important point here is that online social learning has benefits other than improved content learning – it promotes collaborative working skills, reflection on professional practice, and developing ways to interact with others where there is a power differential. At present, it is not clear what other benefits there may be, or how this impacts on success in the future working lives of learners. There is a social and psychological challenge in working and learning in this way; while frameworks exist to help navigate this and design appropriately scaffolded learning activities, it is not clear how less confident learners and those with lower digital literacies could best be guided and mentored or how we effectively manage situations where students have been exposed to threatening or abusive behavior online. That the considerations remain a prominent issue is highlighted by the fact that "lurking" remains the default participation mode for many learners.

Therefore, we feel that while social online learning may provide great benefits for both learners and faculty, there are significant challenges to introducing and embedding these approaches in biomedical and life science courses, such that undertaking a project in this area should be done carefully and with a clear goal in mind. We feel it is important that there is sufficient institutional support, particularly from a technical and training perspective and that there is sufficient time available to successfully complete student and staff socialization into the online environment. The COVID-19 pandemic has forced both learners and faculty to adopt online technologies as a necessary consequence of national lockdowns, which has driven implementation but not necessarily made it easier. There remains a need to develop digital literacies and pedagogical knowledge to higher level in faculty in order to make the most of these developments. There is a real opportunity over the coming years to develop both online and blended learning approaches that make best use of online social learning, but this will not happen unless those designing and implementing curricula understand the online tools available to them and how to best leverage their strengths to promote learning.

Acknowledgments We would like to thank Amber Moorcroft for useful discussions regarding the content of the chapter.

References

Azucar, D., Marengo, D., & Settanni, M. (2018). Predicting the big 5 personality traits from digital footprints on social media: A meta-analysis. *Personality and Individual Differences, 124*-(December 2017), 150–159. https://doi.org/10.1016/j.paid.2017.12.018

Batalden, M., et al. (2016). Coproduction of healthcare service. *BMJ Quality and Safety, 25*(7), 509–517. https://doi.org/10.1136/bmjqs-2015-004315

Bate, E., et al. (2014). Problem-based learning (PBL): Getting the most out of your students - their roles and responsibilities: AMEE guide no. 84. *Medical Teacher, 36*(1), 1–12. https://doi.org/10.3109/0142159x.2014.848269

Bawden, D., & Robinson, L. (2009). The dark side of information: Overload, anxiety and other paradoxes and pathologies. *Journal of Information Science, 35*(2), 180–191. https://doi.org/10.1177/0165551508095781

Bell, F. (2009). Connectivism: A network theory for teaching and learning in a connected world. *Educational Developments, 10*(3). Available at: http://usir.salford.ac.uk/2569/. Accessed 15 Feb 2021.

Bennett, S., et al. (2012). Implementing web 2.0 technologies in higher education: A collective case study. *Computers and Education, 59*(2), 524–534. https://doi.org/10.1016/j.compedu.2011.12.022

Blau, I., & Shamir-Inbal, T. (2018). Digital technologies for promoting "student voice" and co-creating learning experience in an academic course. *Instructional Science, 46*(2), 315–336. https://doi.org/10.1007/s11251-017-9436-y

BMA. (2021). *Ethics of social media use for doctors*. Available at: https://www.bma.org.uk/advice-and-support/ethics/personal-ethics/ethics-of-social-media-use. Accessed 25 March 2021.

Bottanelli, F., et al. (2020). Science during lockdown – From virtual seminars to sustainable online communities. *Journal of Cell Science. Company of Biologists Ltd.* https://doi.org/10.1242/JCS.249607

Bottomley, S., & Denny, P. (2011). A participatory learning approach to biochemistry using student authored and evaluated multiple-choice questions. *Biochemistry and Molecular Biology Education, 39*(5), 352–361. https://doi.org/10.1002/bmb.20526

Bovill, C. (2019). A co-creation of learning and teaching typology: What kind of co-creation are you planning or doing? *International Journal for Students as Partners, 3*(2), 91–98. https://doi.org/10.15173/ijsap.v3i2.3953

Bovill, C. (2020). Co-creation in learning and teaching: The case for a whole-class approach in higher education. *Higher Education, 79*(6), 1023–1037. https://doi.org/10.1007/s10734-019-00453-w

Bowden-Green, T., Hinds, J., & Joinson, A. (2021, June). Understanding neuroticism and social media: A systematic review'. *Personality and Individual Differences, 168*, 110344. https://doi.org/10.1016/j.paid.2020.110344

Brooman, S., Darwent, S., & Pimor, A. (2015). The student voice in higher education curriculum design: Is there value in listening? *Innovations in Education and Teaching International, 52*(6), 663–674. https://doi.org/10.1080/14703297.2014.910128

Cabrera, D., Cabrera, L., & Powers, E. (2015). A unifying theory of systems thinking with psychosocial applications. *Systems Research and Behavioral Science, 32*(5), 534–545. https://doi.org/10.1002/sres.2351

Cheston, C. C., Flickinger, T. E., & Chisolm, M. S. (2013). Social media use in medical education. *Academic Medicine, 88*(6), 893–901. https://doi.org/10.1097/ACM.0b013e31828ffc23

Christensen, C. M., et al. (2016). *Competing against luck: The story of innovation and customer choice*. Harper Business.

Cole, D., & Cunningham, A. M. (2014). How to . . . Curate online resources. *Education for Primary Care, 25*(3), 176–178.

Cole, D., et al. (2017). Using social media to support small group learning. *BMC Medical Education, 17*(1), 201. https://doi.org/10.1186/s12909-017-1060-7

Cole, D. S., & Jones, R. M. (2019a). *Best practice in online content curation in higher education.* Available at: https://www.cardiff.ac.uk/learning-hub/view/best-practice-in-online-content-curation-in-higher-education. Accessed 15 Feb 2021.

Cole, D. S., & Jones, R. M. (2019b). *Online content curation tools: Technology reviews.* Available at: https://www.cardiff.ac.uk/learning-hub/view/online-content-curation-tools-technology-reviews. Accessed 25 Jan 2021.

Cook, D. A., Erwin, P. J., & Triola, M. M. (2010, August 13). Computerized virtual patients in health professions education: A systematic review and meta-analysis. *Academic Medicine, 85*(10), 1589–1602. https://doi.org/10.1097/ACM.0b013e3181edfe13

Cook, D. A., et al. (2008). Internet-based learning in the health professions - a meta-analysis. *Jama-Journal of the American Medical Association, 300*(10), 1181–1196. https://doi.org/10.1001/jama.300.10.1181

Cook-Sather, A., Bovill, C., & Felten, P. (2014). *Engaging students as partners in learning and teaching: A guide for faculty*. Jossey-Bass.

Cranefield, J., Yoong, P., & Huff, S. L. (2015). Rethinking lurking: Invisible leading and following in a knowledge transfer ecosystem. *Journal of the Association for Information Systems, 16*(4), 3.

Daigle, A. (2020). Social media and professional boundaries in undergraduate nursing students. *Journal of Professional Nursing, 36*(2), 20–23. https://doi.org/10.1016/j.profnurs.2019.08.007

Denny, P., Hamer, J., & Purchase, H. (2008a). PeerWise: Students sharing their multiple choice questions. In *Proceeding of the Fourth International Workshop on Computing Education Research (ICER '08); Sydney, Australia, 2008 Sept 6–7* (pp. 51–58). Association for Computing Machinery.

Denny, P., Luxton-Reilly, A., & Hamer, J. (2008b). The PeerWise system of student contributed assessment questions. *Conferences in Research and Practice in Information Technology Series, 78*, 69–74.

Dollinger, M., Lodge, J., & Coates, H. (2018). Co-creation in higher education: Towards a conceptual model. *Journal of Marketing for Higher Education, 28*(2), 210–231. https://doi.org/10.1080/08841241.2018.1466756

Downes, S. (2009). Learning networks and connective knowledge. in Yang, H. H. and Yuen, S. (eds) *Collective intelligence and E-learning 2.0: Implications of web-based communities and networking*, pp. 1–26. Hershey, PA: IGI Global: https://doi.org/10.4018/978-1-60566-729-4.ch001.

Doyle, E., Buckley, P., & McCarthy, B. (2020). The impact of content co-creation on academic achievement. *Assessment and Evaluation in Higher Education.* https://doi.org/10.1080/02602938.2020.1782832

Duke, V. J. A., et al. (2017). Social media in nurse education: Utilization and E-professionalism. *Nurse Education Today, 57*(June), 8–13. https://doi.org/10.1016/j.nedt.2017.06.009

Eshet-alkalai, Y. (2004). Digital literacy: A conceptual framework for survival skills in the digital era. *Journal of Educational Multimedia and Hypermedia; Norfolk, 13*(1), 93–106.

Fawns, T., Aitken, G., & Jones, D. (2019). Online learning as embodied, socially meaningful experience. *Postdigital Science and Education, 1*(2), 293–297. https://doi.org/10.1007/s42438-019-00048-9

Flynn, L., Jalali, A., & Moreau, K. A. (2015). Learning theory and its application to the use of social media in medical education. *Postgraduate Medical Journal, 91*(1080), 556–560. https://doi.org/10.1136/postgradmedj-2015-133358

Foos, P. W. (2009). Effects of student-written questions on student test performance. *Teaching of Psychology, 16*(2), 77–78. https://doi.org/10.1207/s15328023top1602_10

General Medical Council. (2013, April). *Doctors' use of social media*. General Medical Council, pp. 1–4. Available at: http://www.bmj.com/content/325/7372/1090.1.

Gil De Zuniga, H., et al. (2017). Personality traits and social media use in 20 countries: How personality relates to frequency of social media use, social media news use, and social media use for social interaction. *Cyberpsychology, Behavior and Social Networking, 20*(9), 540–552. https://doi.org/10.1089/cyber.2017.0295

Grainger, R., et al. (2018). Medical students create multiple-choice questions for learning in pathology education: A pilot study. *BMC Medical Education, 18*(1), 1–8. https://doi.org/10.1186/s12909-018-1312-1

Griffiths, M. D. (2014). Adolescent trolling in online environments: A brief overview. *Education and Health, 32*(3), 85–87.

Health and Care Professions Council. (2017). *Guidance on social media.* pp. 1–9. Available at: http://www.hpc-uk.org/assets/documents/10005190Guidanceonsocialmedia-draftforconsultation.pdf.

Hew, K. F., & Cheung, W. S. (2008). Attracting student participation in asynchronous online discussions: A case study of peer facilitation. *Computers and Education, 51*(3), 1111–1124. https://doi.org/10.1016/j.compedu.2007.11.002

Isis, D., et al. (2019). *Social Media for the Physiologist.* American Physiological Society. Available at: https://www.physiology.org/careers/mmCareersMentorProfessional-NetworkingIdentifying-Networking-Opportunities/Social-Media-for-the-Physiologist?SSO=Y. Accessed 25 March 2021.

JISC. (2017). *Developing digital literacies | Jisc, Jisc.* Available at: https://www.jisc.ac.uk/guides/developing-digital-literacies. Accessed 25 Jan 2021.

Jobs, A., et al. (2013). *Question-writing as a learning tool for students-outcomes from curricular exams.* Available at: http://www.biomedcentral.com/1472-6920/13/89.

Johnson, G. M. (2006). Synchronous and asynchronous text-based CMC in educational contexts: A review of recent research. *TechTrends, 50*(4), 46–53. https://doi.org/10.1007/s11528-006-0046-9

Joseph-Williams, N., Edwards, A., & Elwyn, G. (2014). Power imbalance prevents shared decision making. *BMJ (Online), 348*(May), 1–3. https://doi.org/10.1136/bmj.g3178

Jumaat, N. F., & Tasir, Z. (2014). Instructional scaffolding in online learning environment: A meta-analysis. In *Proceedings - 2014 International Conference on Teaching and Learning in Computing and Engineering, LATICE 2014* (pp. 74–77). IEEE Computer Society. https://doi.org/10.1109/LaTiCE.2014.22.

Kang, H., & Chang, B. (2016) Examining culture's impact on the learning behaviors of international students from Confucius culture studying in Western online learning context. *Journal of International Students, 6*(3), 779–797. Available at: http://jistudents.org/. Accessed 25 January 2021.

Kenny, P., & Johnson, I. G. (2016). Social media use, attitudes, behaviours and perceptions of online professionalism amongst dental students. *British Dental Journal, 221*(10), 651–655. https://doi.org/10.1038/sj.bdj.2016.864

Könings, K. D., et al. (2020). Learner involvement in the co-creation of teaching and learning: AMEE guide no. 138. *Medical Teacher*, pp. 1–13. https://doi.org/10.1080/0142159X.2020.1838464.

Koutropoulos, A. et al. (2019, June). Rethinking lurking. *Learn Magazine.*

Légaré, F., et al. (2010). Interventions for improving the adoption of shared decision making by healthcare professionals. In *Cochrane database of systematic reviews.* John Wiley & Sons. https://doi.org/10.1002/14651858.cd006732.pub2

Lim, D. H. (2004). Cross cultural differences in online learning motivation. *Educational Media International, 41*(2), 163–175. https://doi.org/10.1080/09523980410001685784

Lubicz-nawrocka, T. (2018). From partnership to self-authorship: The benefits of co-creation of the curriculum. *International Journal for Students as Partners, 2*(1), 47–63.

Maltby, J., et al. (2016). Implicit theories of online trolling: Evidence that attention-seeking conceptions are associated with increased psychological resilience. *British Journal of Psychology (London England: 1953), 107*(3), 448–466. https://doi.org/10.1111/bjop.12154

Manian, F. A., & Hsu, F. (2019). Writing to learn on the wards: Scholarly blog posts by medical students and housestaff at a teaching hospital. *Medical Education Online, 24*(1), 1565044. https://doi.org/10.1080/10872981.2018.1565044

Martens, S. E., et al. (2020). Mind the gap: Teachers' conceptions of student-staff partnership and its potential to enhance educational quality. *Medical Teacher, 42*(5), 529–535. https://doi.org/10.1080/0142159X.2019.1708874

Mazuro, C., & Rao, N. (2011). Online discussion forums in higher education: Is "lurking" working? *International Journal for Cross-Disciplinary Subjects in Education, 2*(2), 364–371. https://doi.org/10.20533/ijcdse.2042.6364.2011.0051

McSporran, M., & Young, S. (2001). Does gender matter in online learning? *ALT-J, 9*(2), 3–15. https://doi.org/10.1080/0968776010090202

Ness, G. L., Sheehan, A. H., & Snyder, M. E. (2014). Graduating student pharmacists' perspectives on e-professionalism and social media: Qualitative findings. *Journal of the American Pharmacists Association, 54*(2), 138–143. https://doi.org/10.1331/JAPhA.2014.13188

Ng, W. (2012). Can we teach digital natives digital literacy? *Computers and Education, 59*(3), 1065–1078. https://doi.org/10.1016/j.compedu.2012.04.016

Nielsen, J. (2006). *Participation inequality: The 90-9-1 rule for social features*. Available at: https://www.nngroup.com/articles/participation-inequality/. Accessed 18 January 2021.

Polizzi, G. (2020). Digital literacy and the national curriculum for England: Learning from how the experts engage with and evaluate online content, *152*, 103859. https://doi.org/10.1016/j.compedu.2020.103859

Prahalad, C. K., & Ramaswamy, V. (2002). The co-creation connection. *Strategy and Business, 27*, 50–61. Available at: https://www.strategy-business.com/article/18458 Accessed 15 Feb 2021.

Roccas, S., et al. (2002). The big five personality factors and personal values. *Personality and Social Psychology Bulletin, 28*(6), 789–801. https://doi.org/10.1177/0146167202289008

Ross, J. G., & Myers, S. M. (2017). The current use of social Media in Undergraduate Nursing Education a Review of the literature. *Comput Inform Nurs, 35*(7), 338–344. https://doi.org/10.1097/CIN.0000000000000342

Rutherford, S. M., Limorenko, G., & Amici-Dargan, S. L. (2017). 'Shadow module leaders'-student experiences as peer-teachers and facilitators of peer-assisted learning. Presented at: Ireland International Conference on Education, Dublin, Ireland, 24–26 Apr 2017. https://orca.cardiff.ac.uk/101345/1/Rutherford%20Short%20Abstract%20IICE%202017%20Revised.pdf. Accessed 28 Feb 2022.

Sadler, T. D., et al. (2010). Learning science through research apprenticeships: A critical review of the literature. *Journal of Research in Science Teaching, 47*(3), 235–256. https://doi.org/10.1002/tea.20326

Salmon, G. (2011). *E-moderating: The key to teaching and learning online* (3rd ed.). Routledge.

Salmon, G. (2013). *E-tivities: The key to active online learning* (2nd ed.). Routledge.

Sandars, J. (2006). Twelve tips for using blogs and wikis in medical education. *Medical Teacher, 28*(8), 680–682. https://doi.org/10.1080/01421590601106353

Scott, J. L., Moxham, B. J., & Rutherford, S. M. (2014). Building an open academic environment - a new approach to empowering students in their learning of anatomy through "shadow modules". *Journal of Anatomy, 224*(3), 286–295. https://doi.org/10.1111/joa.12112

Siemens, G. (2005). Connectivism: A learning theory for the digital age. *International Journal of Instructional Technology and Distance Learning, 2*(1), 3–10. Available at: http://www.itdl.org/journal/jan_05/article01.htm

Siemens, G., & Conole, G. (2011). Special issue - connectivism: Design and delivery of social networked learning. *International Review of Research in Open and Distance Learning, 12*(3), I–IV.

Sinclair, W., McLoughlin, M., & Warne, T. (2015). To twitter to woo: Harnessing the power of social media (SoMe) in nurse education to enhance the student's experience. *Nurse Education in Practice, 15*(6), 507–511. https://doi.org/10.1016/j.nepr.2015.06.002

Sircar, S. S., & Tandon, O. P. (1999). Involving students in question writing: a unique feedback with fringe benefits. *Adv Physiol Educ, 22*(1), 84–91.

Soper, T., & Ukot, E. (2016). Social presence and cultural competence in the Online Learning Environment (OLE): A review of literature. *American Journal of Health Sciences (AJHS), 7*(1), 9–14. https://doi.org/10.19030/ajhs.v7i1.9692

Stiggelbout, A. M., et al. (2012). Shared decision making: Really putting patients at the centre of healthcare. *British Medical Journal, 344*, e256.

Sullivan, P. (2001). Gender differences and the online classroom: Male and female college students evaluate their experiences. *Community College Journal of Research and Practice, 25*(10), 805–818. https://doi.org/10.1080/106689201753235930

Terry, J., et al. (2019). Improving the digital literacy competence of nursing and midwifery students: A qualitative study of the experiences of NICE student champions. *Nurse Education in Practice, 34*, 192–198. https://doi.org/10.1016/j.nepr.2018.11.016

Thomas, B., et al. (2004). Student perceptions of effective e-moderation: A qualitative investigation of e-college Wales. *Paper presented at the Networked Learning Conference,* Lancaster University, UK.

Vaid, S. S., & Harari, G. M. (2020). Who uses what and how often? Personality predictors of multiplatform social media use among Young adults. *Journal of Research in Personality, 91*, 104005. https://doi.org/10.1016/j.jrp.2020.104005

Vallentin-Holbech, L., et al. (2020). Co-creating a virtual alcohol prevention simulation with young people. *International Journal of Environmental Research and Public Health, 17*(3), 1–12. https://doi.org/10.3390/ijerph17031097

Walsh, J. L., Denny, P., & Smith, P. E. (2015). Encouraging maximal learning with minimal effort using PeerWise. *Medical Education, 49*(5), 521–522. https://doi.org/10.1111/medu.12720

White, D. S., & Le Cornu, A. (2011). Vistiors and residents: A new typology for online engagement. *First Monday, 16*(9), 7. https://doi.org/10.5210/fm.v16i9.3171

Williams, R., Karousou, R., & Mackness, J. (2011). Emergent learning and learning ecologies in web 2.0. *International Review of Research in Open and Distance Learning, 12*(3), 39–59. https://doi.org/10.19173/irrodl.v12i3.883

Wittich, C. M., et al. (2017). E-learning in graduate medical education: Survey of residency program directors. *BMC Medical Education, 17*, 114. https://doi.org/10.1186/s12909-017-0953-9

Worm, B. S., & Buch, S. V. (2014). Does competition work as a motivating factor in e-learning? A randomized controlled trial. *PLoS One, 9*(1), e85434. https://doi.org/10.1371/journal.pone.0085434

Xie, K., Yu, C., & Bradshaw, A. C. (2014). Impacts of role assignment and participation in asynchronous discussions in college-level online classes. *Internet and Higher Education, 20*, 10–19. https://doi.org/10.1016/j.iheduc.2013.09.003

Xu, B., Chen, N. S., & Chen, G. (2020). Effects of teacher role on student engagement in WeChat-based online discussion learning. *Computers and Education, 157*, 103956. https://doi.org/10.1016/j.compedu.2020.103956

Chapter 15
The Role of Educational Technology on Mitigating the Impact of the COVID-19 Pandemic on Teaching and Learning

Sandra Arango-Caro, Lisa L. Walsh, Emma R. Wester, and Kristine Callis-Duehl

Abstract The COVID-19 pandemic forced higher education institutions around the world to find ways of maintaining education while adhering to public health guidelines. As classes went online, instructors struggled to balance new technology demands with the stress of teaching during an emergency, hereafter referred as emergency remote teaching (ERT). This chapter presents teaching tools and techniques, assessment strategies, emergent technologies, and recommendations that apply to the COVID-19 ERT. Videoconferencing tools promote students' sense of community when used to foster real-time conversations between students. However, when they are used as a lecturing platform, these tools can increase the digital divide. Using a combination of synchronous and asynchronous tools can provide the best of both worlds, allowing students to effectively communicate with each other with the freedom of accessing course material when time and social demands necessitate alternative learning means. While some instructors struggled with academic integrity and technical issues during exams, others reflected on the pedagogical purpose of student assessments and developed new assignments to better measure learning outcomes. The accelerated application of emergent technologies and practices has minimized the educational disruption through adaptive learning and extended reality. Although these tools are mostly experimental, they will play a key role in the transformative change of postsecondary education. Effective training in online teaching is essential to maintain student engagement and vital assistance and should be disseminated internationally. Digital inequalities need to be addressed by identifying and remedying "digital desserts." With the growing rate of education disruptions due to climate change, disease, and war, institutions must become proactive in their planning for future emergencies.

S. Arango-Caro · L. L. Walsh
Donald Danforth Plant Science Center, St. Louis, MO, USA

E. R. Wester · K. Callis-Duehl (✉)
Donald Danforth Plant Science Center, St. Louis, MO, USA

Saint Louis University, St. Louis, MO, USA
e-mail: KCallis-Duehl@danforthcenter.org

H. J. Witchel, M. W. Lee (eds.), *Technologies in Biomedical and Life Sciences Education*, Methods in Physiology, https://doi.org/10.1007/978-3-030-95633-2_15

15.1 Emergency Remote Teaching (ERT)

In March 2020, the coronavirus disease 2019 (COVID-19) was declared a pandemic, and higher education institutions around the world scrambled to maintain education while adhering to public health guidelines (Crawford et al., 2020). Zhang et al. (2004) define e-learning as "technology-based learning in which learning materials are delivered electronically to remote learners via a computer network." Within e-learning, there are two categories, asynchronous and synchronous modalities. Asynchronous learning also often referred to as "online teaching" is commonly implemented through a combination of email, pre-recorded videos, and discussion boards. Synchronous or "virtual learning" is implemented through a combination of videoconferencing and chat platforms (Zoom, Webex, Skype, etc.; Hrastinski, 2008). What faculty and students experienced during the spring 2020 term due to COVID-19, however, was not precisely e-learning. For example, in some cases, instructors had only a few days to transfer their class to an online platform, which does not reflect well-researched or properly planned e-learning pedagogy (see Chap. 13 in this volume). This situation is referred to as rapid transition to remote teaching (Hill, 2020; Fig. 15.1) or emergency remote teaching (ERT; Hodges et al., 2020). In this chapter, we will discuss techniques used in ERT, which draws from e-learning but involves additional considerations due to its emergency learning situation.

The ERT transition during the current COVID-19 pandemic occurred in multiple phases and, as of December 2020, continues to develop (Flaherty, 2020; Hill, 2020; Fig. 15.1). The initial transition from in-person to online formats is considered phase

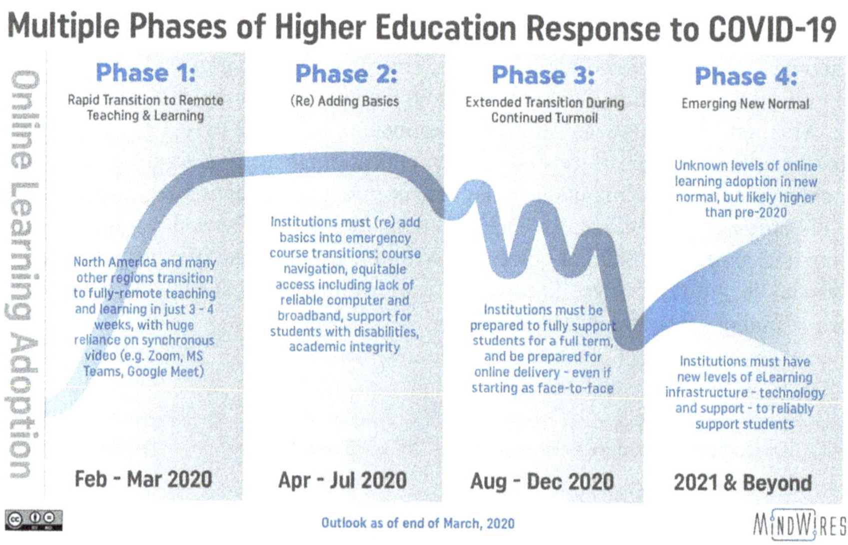

Fig. 15.1 A general trend of how emergency remote teaching due to COVID-19 changed over spring and fall semesters of 2020 (Hill, 2020)

1 of COVID-19 ERT. During this phase, many faculty members had neither experience with online teaching nor the time necessary to develop effective practices as they converted their classes to online formats (Hodges et al., 2020). In addition, faculty faced challenges that came with teaching students who lacked proper equipment or Internet access, students who had to get jobs because of the financial fallout, and students who were caring for or losing loved ones to the pandemic (Walsh et al., 2021a). During phase 2 (April–July 2020) of COVID-19 ERT, instructors started to address these disparities in access and technology and devise approaches to address pedagogical shifts toward e-learning practices. In phase 3, which spans the fall 2020 semester, institutions lent their support, both to students and to instructors, to ensure that the transition survived and thrived. As COVID-19 ERT continues from 2020, instructors could implement a phase 4 to improve how online classes are administered and help shape the "new normal" (Fig. 15.1). We will address what we have learned about the implementation of technology tools for higher education science teaching during COVID-19 ERT from phase 1 to 3 and discuss ways technology use will influence higher education in phase 4.

15.1.1 History of Emergency Remote Teaching (ERT)

To best anticipate the future, we must learn from the past. We must not only learn from what occurred during COVID-19 ERT but take lessons in ERT events that occurred prior to the current pandemic. Previous pandemics and natural disasters that caused disruptions to in-person learning can offer many lessons. Case studies below are like the COVID-19 response because they were rapid responses to emergencies in the digital age. However, these examples were geographically and temporally limited compared to the response to the COVID-19 pandemic.

In 2003, Hong Kong closed schools for weeks to curb the spread of the severe acute respiratory syndrome virus (SARS), moving education online. The rapid speed of school closures was similar during the COVID-19 pandemic, leaving teachers feeling unprepared, shocked, and frustrated with the lack of guidelines from the government and schools' administrations (Fox, 2004). In a survey of 60 primary and secondary teachers, most felt that they did not have adequate training to leverage the Internet for mandatory distance education. The most technology-savvy teachers wanted to learn how to leverage technology to improve student learning, while other teachers requested training on the technology itself (Fox, 2004). Student needs quickly became the focus for instructors, while McNaught (2004) argued that innovative action by faculty helped quell faculty unease. Science and biomedical instructors highlighted the fact that the SARS crisis revealed deficiencies in their departments' pedagogical approaches and created new challenges to tackle in an approach they felt that would enrich teaching and learning in the future (McNaught, 2004).

In response to the 2006 war between Israel and Hezbollah, Empire State College suspended a traditional 10-day visit by members of their faculty to Lebanon for their

Fig. 15.2 "Rhodes Must Fall" student protest in South Africa, 2015. Photograph by Desmond Bowles, distributed under a CC BY-SA 2.0 license

Lebanon Residence Program. The college replaced scheduled in-person lectures by sending DVDs, since Internet access was scarce at the time, to Lebanon. This ERT response did not negatively impact faculty or students because the majority of the semester was already taught remotely through online instruction (SchWeber, 2008). This example demonstrates the resiliency afforded to instructors and teachers in emergencies if they are already familiar with distance education.

In April 2009, flooding caused Valley City State University (VCSU) in North Dakota to shutter its doors and offer the last 3 weeks of classes online. VCSU had no preexisting emergency plan but required that all students own a laptop as part of their standard criteria for enrollment. Additionally, all courses were required to have a learning management system (LMS) site. Despite the established infrastructure, faculty found it difficult to reconfigure their classes to an online format, while students struggled with Internet access in conditions following the flooding. The flooding event highlighted the importance of all faculty members being familiar with the university's LMS (Young, 2009) before it becomes necessary. Now, VCSU has an emergency management team that plans for potential floods each year. When VCSU announced their emergency transition to online classes in March 2020, the VCSU president felt confident the school was prepared: "Valley City State University has a long history of online teaching; in fact, during the Sheyenne River flood of 2009, students finished the semester online after the campus closed following Easter break" (Urness, 2020).

In response to student protests (e.g., the decolonizing "Rhodes Must Fall" protest and socio-financial "Fees Must Fall" protest; Fig. 15.2) from 2015 to 2017, South African universities experienced temporary campus closures, typically near the end of each academic year. During educational disruptions, faculty and students faced a myriad of financial, temporal, and emotional problems, as well as infrastructural constraints (Czerniewicz et al., 2019). Researchers attempted to interview faculty but had low participation because of the turmoil. All 16 faculty who were interviewed indicated they lacked motivation because of the politicized nature of the educational disruption—it was a controversial decision by the institution to continue education as students protested the cost and colonial underpinnings of their university. In this circumstance, leveraging technology to continue education undermined student protests for financial and cultural improvement. The context of this ERT may be partially responsible for the instructors' negative views of blended learning, with most viewing blended learning as pedagogically inferior to the standard level of face-to-face instruction. Faculty also believed students did not like blended learning, pointing to solitary and virtual aspects as sources of discontent, resulting in reduced student output and effort (Czerniewicz et al., 2019).

Previous ERT experiences provide useful information on how to address educational challenges faced during the COVID-19 pandemic while highlighting areas for improvement moving forward. Higher education researchers face difficulties collecting data because of tumultuous events that coincide with ERT (e.g., McNaught, 2004; Czerniewicz et al., 2019). Instructors are often shocked and overwhelmed at the onset of ERT (Fox, 2004; McNaught, 2004; Czerniewicz et al., 2019). Anecdotally, instructors with a stronger technology background were less concerned with logistics and more focused on higher learning and online pedagogy (Fox, 2004). While disruptions to education due to geopolitical conflict, disease, and natural disaster events could be lessened through computer-based learning, there are limitations to its utility when the disruption is due to the institution or political movements led by students (Czerniewicz et al., 2019). A common theme across education disruption events is the desire to learn from an ERT experience to continue improving teaching into the future (Fox, 2004; McNaught, 2004).

15.2 Teaching Strategies Employed During COVID-19 ERT

15.2.1 Synchronous Learning

As students moved into ERT situations, many instructors found themselves learning how to use new technological tools. There are a wide variety of synchronous learning tools available on the market; below we will describe and discuss the most popular tools that have been used during the COVID-19 pandemic.

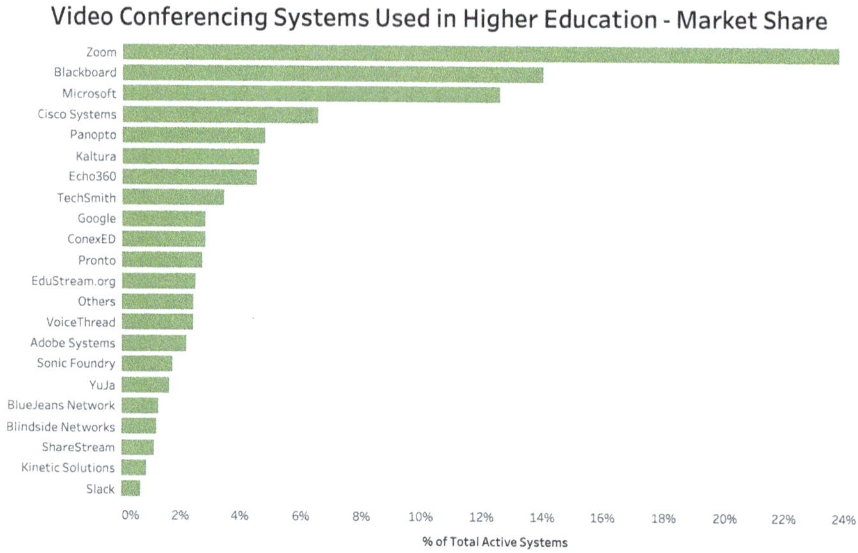

Fig. 15.3 The percentage of market share of each of 22 popular synchronous videoconferencing software platforms for June 2020. The most popular four constitute over 50% of the synchronous tools used in higher education (Menard, 2020)

15.2.1.1 Videoconferencing Tools

While there are a wide range of platforms used for synchronous learning (Fig. 15.3), we will only be describing, comparing, and contrasting the four most popular platforms: Zoom, Blackboard Learn, Microsoft Teams, and Cisco's Webex Teams (Menard, 2020).

- *Zoom*

The most widely adopted videoconferencing software during COVID-19 ERT has been Zoom, with over 200 million daily users (Fig. 15.3; Menard, 2020). Zoom is used for audio/videoconference meetings, teaching classes, maintaining office hours, creating student presentations, etc. (Bellan, 2020). Advantages include features like screen sharing, screen recording, team chats, and searchable history. Zoom also allows instructors to place students into "breakout rooms," allowing group work or small group discussions. Small groups in breakout rooms can work on case studies or create projects with Google Docs and report back to the large group (Saidy & Sura, 2020). Participation in small-group collaborative learning correlates with deeper learning, increased teamwork, and a greater sense of community among students (Brindley et al., 2009). As Zoom rose in popularity, it was plagued with several security and privacy issues, including widely publicized hacking or "Zoombombing" incidents (Redden, 2020). However, there are steps instructors can take to make Zoom safer. Password-protecting meetings, especially with a unique one-time password, are the best option to prevent unwanted attendees from entering the meeting (Zoom, 2020). After the meeting starts, locking the meeting prevents unwanted users from participating, and switching off participant screen sharing can reduce disruptions from outsiders (Zoom, 2020). "Zoom fatigue" is recognized as feeling overwhelmed or mentally drained by the duration or frequency of videoconferencing sessions. Using music during videoconferencing can help build classroom engagement and reduce "Zoom fatigue" (Halpin & Crowther, 2021; for more details on music, see Chap. 7 in this volume).

- *Blackboard*

The Blackboard Learn learning management system (herein referred to as Blackboard) offers a platform with both synchronous and asynchronous options (see Asynchronous section). The synchronous mode offers live interaction through Web conferencing with tutors and peers. The videoconferencing uses a Web browser interface, eliminating the need to download and install software. Like Zoom, the Blackboard platform offers a chat function and breakout rooms for students. These breakout rooms offer the same functionality and thus the same benefits and drawbacks as those used in Zoom. In pre-COVID-19 studies, Blackboard's synchronous functions were found to be an acceptable alternative to in-person learning, with the largest being limitations on writing mathematical equations (Tonsmann, 2014); note that this is also an issue in Zoom. Students have also reported positive experiences using the Blackboard platform during COVID-19 ERT, with over 87% of students indicating they felt that synchronous tools in Blackboard enhanced their learning experience (Chen et al., 2020).

- *Microsoft Teams*

Microsoft Teams (herein referred to as Teams) is an increasingly common education communication platform, with an 894% market growth in educational use since the start of the COVID-19 pandemic (Microsoft, 2020). Each member of a faculty has his or her own channel for carrying out classes, with the ability to access other individuals' channels for collaborative projects. One of the main advantages of Teams is the leveraging of the Microsoft ecosystem. For example, OneNote, a multipart, note-taking tool, is built into Teams for a variety of lessons and activities. It has built-in meeting features to hold classes and share documents while integrating PowerPoint presentations, interactive websites, videos, e-books, etc. Teams, built on Office 365's cloud-based enterprise, delivers advanced security through organization-wide and team-wide two-factor authentication, single sign-on, and encryption of data both in transit and in storage (Holzapfel, 2020). Teams has also recently added breakout rooms as an additional feature (Microsoft website).

- *Cisco Webex Teams*

Cisco Webex Teams (hereby referred to as Webex) is a videoconferencing and Web conferencing software developed by Cisco's subsidiary Webex. It allows users to create secure virtual workplaces (Cisco website). Using Webex, students and faculty can initiate and join meetings instantly without needing to schedule them in advance. It provides native support for external video, and up to 200 video users can join a team meeting via computers, Cisco video devices, or industry standard SIP endpoints. Webex's ability to integrate with third-party applications like OneDrive and Google Cloud facilitates an uninterrupted workflow, giving it a slight edge over other platforms. Teleconferencing is one of the application's most prominent features, allowing students without stable Internet access to call into class through their phone instead of relying on video devices or the Internet. Webex offers end-to-end encryptions of all data and messages before sending them to the cloud, which is helpful in protecting the identity of both sender and receiver (Webex website).

No matter which videoconferencing software an institution or instructor employs, using the full range of features, from breakout rooms to note-taking, can help instructors meet e-learning goals of facilitating interactions and collaborative learning. During the current ERT, these tools are essential to provide students with the ability to interact in real time.

- *A Comparison of Platforms*

Clearly each of the above-described videoconferencing platforms has its advantages and disadvantages. While IT departments applaud the security built into Microsoft Teams and Webex (Sevilla, 2020), Zoom tends to be the preference for ease of use for cross-institutional collaboration due to variation in institutional access to Microsoft products or cost considerations (Unify Square, 2020). Specific institutions may standardize to one specific platform, but often universities opt to support multiple different platforms so that faculty and students can choose which one suits their needs. In fact, instead of deciding between Teams and Zoom, the answer might be both, since they integrate well together, as long as the institution has funds to provide both (Unify Square, 2020).

- *Other Tools*

In addition to traditional synchronous lectures utilizing videoconferencing, another notably successful synchronous teaching method used during the pandemic is a form of gamification. Gamification is the use of game design elements outside of a game environment (Robson et al., 2015). Faculty use avatars to teach classes in virtual worlds like Second Life (an online three-dimensional community) using virtual classrooms, with communication tools such as "Chat" and "Speech" functions (Saidy & Sura, 2020). This technology makes learning fun through exciting problem-solving activities that offer awards (Saidy & Sura, 2020). Heim and Holt (2021) also used game design during an undergraduate ecology course to promote content learning, access, and collaboration, while engaging and assessing students. Hosting a virtual class-based poster session also is an alternative to face-to-face presentations and assessments. Holt et al. (2020) used the Mozilla Hubs virtual reality platform to have a synchronous poster presentation from undergraduate students during their ecology course. Several student benefits were identified from this event such as content learning, peer and local audience interactions, and authentic feeling of a professional session.

15.2.1.2 Communication Tools

In addition to the standard email form of communication, faculty turned to social media; text messaging systems, such as WhatsApp, Snapchat, and GroupMe; as well as basic text messaging, to reach students during the COVID-19 pandemic. Saidy and Sura (2020) recommended using multiple communication methods to send each message. This multimodality messaging ensures that the digital divide does not prevent students from receiving class information. Communicating regularly with students during COVID-19 ERT is also important to ensure students feel the instructor is accessible, as well as giving students a sense of connection to the class. In a survey of students across 23 institutions in the spring semester of 2020, some students noted that the effective use of communication tools, such as email or text messages, during ERT was essential to keeping them on track (Wester et al., 2021). An example of a typical positive student response included the following:

> I also really liked emailing my professors with questions because they got back to me very timely and with really good answers. It just streamlined my learning and helped me get the answers that I needed very easily.
> — Undergraduate A

However, the majority of students found electronic communication tools were more difficult to navigate when seeking answers to their questions (Wester et al., 2021):

> One of the problems I found is I have less communication with my professor, because its [sic] tough to explain a problem via email.
> — Undergraduate B

The pervasiveness of social media presents an opportunity for students to collaborate on projects inside and outside of the traditional classroom. The integration of social media within the learning environment offers students self-agency in learning and career planning (O'Brien, 2020), as well as strengthening a sense of community in virtual classes (Hsu & Ching, 2011). Classroom-specific Twitter threads can be used to provide course updates and facilitate academic conversations in a manner familiar to students (Burden, 2014). Likewise, Facebook groups can be used to share ideas and offer support, ask questions, and participate in discussions, all of which have been shown to strengthen virtual student learning communities (Schoper & Hill, 2017). Integration of social media through inclusive technologies such as the smartphone can prevent low retention rates typical in e-learning environments during the COVID-19-induced closures.

15.2.1.3 Using Synchronous Tools Effectively

While instructors initially employed synchronous tools to mimic the classroom-learning setting, it is important to understand that the use of synchronous tools does not aim to replicate the classroom setting; rather, it fosters real-time interactions within the class (Huang & Hsiao, 2012; Akarasriworn & Ku, 2013;

Yamagata-Lynch, 2014). Therefore, treating synchronous tools as direct substitutes for in-person class runs counter to the evidence regarding effective e-learning, but they can be important in facilitating real-time conversations and interactions within a classroom.

When used effectively, synchronous tools can have a wide range of benefits for students in online learning, including reduced virtual isolation for online learners (Hrastinski, 2008; Akarasriworn & Ku, 2013), feelings of connection among users (Huang & Hsiao, 2012), as well as increased virtual presence (Yamagata-Lynch, 2014). For example, videoconferencing software allows professors to invite guest presenters from across the globe who can provide unique experiences for students.

As Jaggars (2014) pointed out, online learning's lack of face-to-face contact with instructors can lead students to feel that they are "teaching themselves" (p. 31) the course content, but synchronous delivery can ameliorate this perception. One primary benefit, for instance, is immediate feedback, where students can ask questions and receive instant answers (Skylar, 2009; Martin et al., 2012). For example, effective use of synchronous tools during COVID-19's ERT had a greater positive impact on student learning and improved their perception of content difficulty when compared to asynchronous education alone in a calculus-based physics class (Guo, 2020). A further benefit is spontaneity of discussion like that experienced in a campus-based classroom (Yamagata-Lynch, 2014). Therefore, synchronous tools are highly effective in fostering connectivity through conversation within a class, allowing students to work together to solve problems, discuss topics, or engage in scientific argumentation (Huang & Hsiao, 2012; Akarasriworn & Ku, 2013; Yamagata-Lynch, 2014). These, in turn, lead to improved student attitudes toward class and increased sense of belonging (Fatani, 2020).

Despite the possibility of increasing students' sense of belonging through synchronous learning, these tools can also engender disparities within the class, especially under ERT such as the COVID-19. Synchronous delivery could be considered antithetical to online education's main benefit of "anytime, anywhere learning" (Skylar, 2009, p. 71). Time zone differences (Tunceren et al., 2015) and conflicting schedules can prevent students from attending synchronous sessions (Huang & Hsiao, 2012). Additionally, multiple studies have reported technical difficulties, such as audio problems forcing students to resort to the chat function, hindering videoconferencing (Cornelius, 2014; Correia, 2020). In a survey of students across 23 institutions during the COVID-19 transition to ERT, students described problems using the Internet as the biggest technological barrier to access online or virtual course materials, with hardware issues being a close second. Responses such as "I experienced many technology issues including slow Internet connections and not enough computer RAM to take exams with the monitoring software required by professors" (Wester et al., 2021) typified student feedback. Poor audio and video quality can cause delays and interfere with communication (Akarasriworn & Ku, 2013; Correia, 2020) and the delivery of learning content (Huang & Hsiao, 2012). Weak Internet connections further inhibited successful online experiences (Lavolette et al., 2010; Rudd & Rudd, 2014), and students who access synchronous software from a workplace computer may encounter issues with network firewalls (Michael,

2012). Although Internet and home broadband use has steadily increased over time (Pew Research Center 2020), one-third of adult respondents to a study on technology use during COVID-19 had difficulty accessing work or class due to Internet connectivity issues (Hannula, 2020). Two-thirds said video calls froze or disconnected. In another COVID-19-related study, students reported a range of digital divide disparities related to synchronous learning, from "technical difficulties and Internet issues not working" to "I now have to look at getting my own computer and I cannot get the Internet where I live" (Wester et al., 2021). Furthermore, issues with accessibility to synchronous learning can create a dichotomy within a class of the "haves" (e.g., have reliable Internet access) and the "have nots" (e.g., living in a time zone that makes attending class difficult). Synchronous instruction can also lead to inconsistent accessibility to learning materials for students with special learning needs or disabilities (Lederman, 2020). One way to increase accessibility is to record live class meetings or lectures and post them online along with transcripts and other materials. In fact, students who were exposed to both asynchronous, pre-recorded lectures and recorded synchronous lectures disproportionately preferred the synchronous recorded option (Dingwall, 2020), due to a decreased sense of belonging associated with asynchronous learning.

Another challenge with synchronous video classes during the COVID-19 pandemic was the limited number of students who turn their cameras on. Castelli and Sarvary (2021) surveyed students in their introductory biology course and found that the main reason students keep their video cameras off was their appearance. Additional common reasons included having other people in the background and poor Internet connections, both disproportionately impacting underrepresented minoritized students (Castelli & Sarvary, 2021). Using their survey results, Castelli and Sarvary (2021) developed a list of inclusive strategies to encourage camera use, including explaining why camera use is beneficial to student education and establishing camera use as a social norm though they suggest not making cameras mandatory.

15.2.2 Asynchronous Learning

To adjust to COVID-19, schools used a variety of online learning platforms for asynchronous classes, the most common being Blackboard and Canvas, which together accounted for over 60% of the LMS usage in higher education prior to the pandemic (Feldstein, 2018). Asynchronous learning in digital formats is superior to asynchronous courses where instructors sent homework to students to allow students to work independently (Ramo et al., 2021). Unlike independent non-virtual asynchronous work, learning management systems allow instructors to periodically check on student progress, and students can work at their own rate. Teachers can create an entire classroom online, providing significant flexibility.

15.2.2.1 Canvas

Canvas is a massive open online course platform used by instructors to connect to students, manage grades and assignments, and contain course information. It allows students to submit assignments electronically, and instructors can even set up reminders for upcoming due dates. It also facilitates communication between instructors and students through emails, discussion boards, and chats. It is used by thousands of universities throughout the USA, where it is the fastest-growing learning management system despite having a startup cost for institutions (Baldwin & Ching, 2019). To help facilitate proper use of the system, Canvas recently published a user checklist to guide the design of quality online courses, which can lower the learning curve of creating an online course (Baldwin & Ching, 2019).

15.2.2.2 *Blackboard*

Like Canvas, Blackboard is a collaborative virtual learning environment and learning management software. It is used by more than 70% of universities and colleges in the USA (Alokluk, 2018). As previously mentioned, it can be used for both synchronous and asynchronous classes. Blackboard has capabilities to deliver personalized, learner-focused activities, but it must be used effectively to have positive contributions to asynchronous learning. This can be challenging when only 33% of teachers have previous experience teaching online (Walsh et al., 2021a). Thus, instructors need to spend time learning about all the many features and tools Blackboard can bring to an online learning environment.

15.2.2.3 Others

Instructors can also use platforms with currently smaller markets or a narrower set of tools, such as Kahoot!, which allows instructors to make fun games through a series of multiple-choice questions, to make learning interesting (United Nations Educational, Scientific, and Cultural Organization [UNESCO], 2020). While they are not directly applicable to classroom instruction, there are also self-directed learning content platforms such as Code it!, which teaches children basic coding concepts; Duolingo, which can support language learning; and Discovery Education, which provides free education resources. These learning platforms can be used to supplement or enhance the more traditional learning management systems.

15.2.2.4 Using Asynchronous Tools Effectively

Using asynchronous tools can provide more flexibility than synchronous online classes. It allows for interaction regardless of time barriers and location and can easily be changed to the limited range of previous studies. For example, one study found that students preferred lecture delivery through live lectures online that could be recorded for later viewing asynchronously (McPhee & Lyon, 2020). An alternative to the hybrid synchronous lecture with asynchronous recording is to provide lectures delivered through a series of short videos posted online, which allows more efficient and flexible time management by students and instructors (Daniel, 2020).

Despite students' doubts over possibilities for peer connection in remote learning, McPhee and Lyon (2020) found that, during the transition, most students felt supported by their instructors and that students valued clear and consistent communications and tools during their online course. Maintaining contact with students is especially important during ERT, and even starting with small but frequent emails helps students feel connected and cared for (Daniel, 2020). During these initial communications, focusing on explicit expectations is vital to student engagement since students are engaging in more self-directed and monitored learning (Ramo et al., 2021). Instructors can choose to use platforms that use blogs and discussion boards, or they can implement structured teams to facilitate social connections during the online asynchronous courses (Flener-Lovitt et al., 2020). Forums, chats, and other communication apps also supply online discussion spaces for students to continue their education even when an in-person environment is not possible (Goh & Sandars, 2020).

A disadvantage of having completely online asynchronous classes is that it can also create a digital divide in some communities that have a lack of Internet or electricity and technology (see Address Digital Inequalities section). When students lack stable access to the Internet, even using asynchronous learning tools becomes a barrier to education and requires creative solutions (UN, 2020). Some institutions and cities have figured out ways to provide students with Internet access (e.g. Wi-Fi hotspots). Some institutions also provided device loans for students who did not have appropriate technology to access online learning (UN, 2020). Keeping this in mind, instructors who have students without stable Internet or few technological resources can choose an asynchronous learning management platform that is built to be used on basic mobile phones such as Cell-Ed (which allows offline options) or KaiOS which can give basic, cheap mobile phone smartphone capabilities. Systems that have strong offline functionality include Kolibri, Rumie, and Ustad Mobile (UNESCO, 2020). Though these platforms are less common, they can help bridge this gap. It is important to remember that different students have diverse needs, and whatever platform or management tool chosen should be flexible to accommodate these needs (Daniel, 2020). Additionally, instructors should make a bigger effort to foster a social environment, so students feel connected and have a sense of belonging.

15.2.3 Special Considerations for Teaching Labs or Hands-On Activities

During the COVID-19 pandemic, hands-on learning has been limited, especially for experiential instruction in the biological, environmental, chemical, and medical fields, which lean heavily on laboratory practice and field or clinical experiences (Barton, 2020; Goh & Sandars, 2020; Holme, 2020; Remtulla, 2020). A variety of alternative means were used to challenge this limitation. Faculty and students were creative in adjusting to facilitate meeting learning objectives and engaging in scientific inquiry from a distance (Herzog & Mawn, 2020; Pennisi, 2020). Videos and simulations allowed students to conduct their own laboratory activities (Herzog & Mawn, 2020; Pennisi, 2020). Some instructors carried out science demonstrations online and described experiments and results in detail. Such results were used as experimental data to write reports and generate discussions, using tools such as Webex, Zoom, and Google Meet to promote critical thinking (Barton, 2020; Herzog & Mawn, 2020; Pennisi, 2020). Course-based undergraduate research experiences (i.e., students doing science research as part of a lab course) continued during remote learning by leveraging synchronous technology to facilitate collaboration within the class (Majka et al., 2021). Instructors conducted field activities that were livestreamed to instruct students remotely (e.g., techniques of coastal ecology; Barton, 2020; Pennisi, 2020). Lab kits with detailed instructions were sent to students to conduct lab activities from home as independent research projects (Arango et al., unpublished data; Herzog & Mawn, 2020). During the COVID-19 pandemic, Project EDDIE (Environmental Data-Driven Inquiry and Exploration; projecteddie.org), which creates curricular modules around open-access large datasets to teach quantitative literacy and scientific concepts, adapted modules for online delivery involving both synchronous and asynchronous components (Meixner et al., 2021). By doing so, these educators have shifted the focus of experiential instruction away from experimental design and execution toward data analysis and interpretation, an aspect of experimental design that tends to be a lower priority for instruction.

Instructors also produced creative online labs to allow students to engage in product creation (e.g., video games), data collection, and/or data analysis (Rholl, 2021; Heim & Holt, 2021). In the medical field, multiple options were offered to students. Software programs with interactive screens were available to study anatomy, some using 3D models that could be dissected virtually, allowing portability using mobile and laptop applications (e.g., Anatomage; Remtulla, 2020). Novel methods, such as the use of wearable devices like Snapchat glasses, made livestreaming surgeries more practical (Remtulla, 2020). Such technology may well become standard in the case of future canceled clinical placements (Remtulla, 2020). Virtual ward rounds for consultants and students were facilitated by staff members physically present at hospitals who maneuvered a computer with a camera—"webcams on wheels" (Remtulla, 2020; Youle, 2020). Telemedicine for remote objective structured clinical examination was implemented to allow students to interact with patient actors via videoconferencing (e.g., "TeleOSCE") (Remtulla, 2020). Medical educators coached students remotely using real-time mobile video tools and apps (Goh & Sandars, 2020).

These instructional strategies face several challenges related to technology and pedagogy as group and collaborative work is limited (Barton, 2020; Herzog & Mawn, 2020). However, as these strategies evolve and emergent technologies develop, students will have better ways to experience remote hands-on instructional experiences. Adaptive learning (e.g., artificial intelligence, learning analytics) and extended reality (e.g., augmented and virtual reality) are at the forefront of educational technologies, improving rapidly to help find solutions to challenges faculty and students face. See section on "Forward Facing."

15.2.4 Virtual Exams and Assessments

As in-person courses were rapidly transferred online due to COVID-19, instructors had to face the challenge of online student assessments. In some cases, academic institution closures happened just days before scheduled midterm or final exams. Under these circumstances, some institutions canceled exams or made them optional, both to give students a chance to adjust to their new class modality and to provide instructors with more time to reshape their courses or obtain licenses for proctoring software (Dietrich et al., 2020; Holton, 2020). While canceling midterm exams was appreciated by more than 60% of students, some felt it prevented them from identifying gaps in their knowledge and put even more pressure on them to succeed in the final exam (Dietrich et al., 2020).

The education literature strongly supports the need to reflect on essential student learning outcomes and to evaluate the purpose of student assessments during emergencies to adjust assignments or develop new assessments (Daniel, 2020; Dietrich et al., 2020). For example, by focusing on sensemaking rather than rote memorization in their online courses, organic chemistry instructors kept 2020 student outcomes consistent with previous years (Stowe et al., 2020). The following section highlights a variety of aspects that directly affected student assessments during the COVID-19 ERT.

15.2.4.1 Implementing the Exam

In traditional science exams, students are often asked open-response questions which professors and teaching assistants mark up to provide feedback. Because of the logistics of grading exams in large courses, online exams were often written as multiple-choice questions to allow automated grading (Dicks et al., 2020). If the exam was hosted on the LMS, this also allowed for randomizing the question order or selecting from a pool of questions (Dicks et al., 2020). In smaller courses or courses with enough graders, open-ended questions were included on exams, which students could download as a PDF, complete, and upload as an annotated PDF or photograph of written answers to their LMS (Dietrich et al., 2020; Stowe et al., 2020). To reduce technical issues with uploading and downloading, students were allotted more time (30–60 min) to complete their exams (Dietrich et al., 2020; Stowe et al., 2020). Grading teams leveraged videoconferencing to ensure consistency while grading and entered feedback directly into the LMS (e.g., SpeedGrader option in Canvas; Stowe et al., 2020). Viva exams, which test clinical competence in medical school students, were difficult to administer but could be implemented with online video calls to orally test the student's decision-making, critical thinking skills, and clinical reasoning (Khan & Jawaid, 2020).

Students responded positively to these adjustments and recommended that instructors provide students with a practice exam to test the exam technology and identify and address any issues (Dicks et al., 2020; Holton, 2020; Petillion & McNeil, 2020). In one course where students had previously been encouraged to elaborate on their short answers, the exam format changed to multiple choice; instructors eased student concerns by uploading responses to their frequently asked questions surrounding the final exam answers (Dicks et al., 2020). Technical issues occurred at low rates for online exams, but instructors should closely monitor their email during exams to promptly help students who encounter problems (Holton, 2020; Stowe et al., 2020).

Because of concerns about the impact that a single set-time exam would have on working students and students in different time zones (Petillion & McNeil, 2020), some instructors chose to offer long time windows in which students could begin a set-time exam (Dicks et al., 2020). After allowing a 24- and 36-h window for exams, instructors recommended either a 12-h window maximum or a handful of discrete-time windows to reduce instructor workload and stress (Dicks et al., 2020).

15.2.4.2 Evidence of Cheating

Based on identical answers or improved assessment scores, faculty felt that cheating increased after transitioning online (Dietrich et al., 2020; Rupnow et al., 2020). Some faculty believed students were cheating because of a lack of discipline and maturity, while others thought it was pressure to continue performing well during the pandemic (Rupnow et al., 2020). According to another source, "closed-resource, proctored assessments did not work because it was nearly impossible to monitor academic misconduct" (Dingwall, 2020). When asked why cheating during the COVID-19 ERT might occur more frequently than in traditional classes, reasons STEM students provided included the lack of proctor, lack of other students in the room, easy access to notes, and ability to use the Internet and homework solution sites (e.g., Chegg and Course Hero; Walsh et al., 2021b).

15.2.4.3 Proctoring Software

Many of the same methods to cheat in person are used by students online, and some of the options available to proctor online exams mirror in-person proctoring. This includes third-party hires to watch students via webcam and software installed onto student computers to record and track student behavior during exams. Software can also be used to block Web browsing and track incidences and durations of dropped Internet connections (Dietrich et al., 2020; Holton, 2020). Some proctoring software facilitate online exam continuity when a student experiences an Internet disruption (e.g., Smart Exam Monitor; Jamieson, 2020).

Algorithmic test proctoring software programs (e.g., ExamSoft, Respondus, ProctorU, etc.) were designed to record students and automatically detect students cheating online. Although many universities use this software, little evaluation has been focused on the ethics and potential gatekeeping of such software. Automated proctoring software flags possible academic misconduct, but it falls on the instructor to use the recorded video to verify that cheating occurred and to address the student directly (Swauger, 2020).

Algorithms used for proctoring software have been programmed to look for suspicious behavior, but these settings are discriminatory due to the potential for misinterpretation. Behaviors including loud noises or moving in and out of the Web camera's frame are considered suspicious, but this would be common for parents, guardians, and those caring for sick relatives, as well as for students who must take medication or cannot sit for extended periods. Algorithms that track eye movement discriminate against blind and autistic students, while the hyperactivity of ADHD students is also commonly flagged as "suspicious." Facial recognition, which is used to confirm that the correct student is taking the exam, is also deeply flawed and racially biased when attempting to correctly identify black and brown students (Swauger, 2020). Students recommended the avoidance of test-proctoring software, as they felt it negatively impacted their test anxiety and academic performance (Petillion & McNeil, 2020). For additional scrutinization of automated proctoring software, we recommend Swauger (2020).

15.2.4.4 Discouraging Cheating: Software Alternatives

Proctoring software is too expensive for many institutions (Nguyen et al., 2020), and in addition to other concerns, it cannot be used by students who do not have computers or reliable Internet connections (Dietrich et al., 2020). Additionally, students have provided examples of work-arounds to cheat the proctoring software. For example, "Even with proctoring systems such as lockdown browsers, I have heard of methods students use to cheat such as having notes on their lap or not using a completely blank sheet of scratch paper," and "I know some people who smear their webcam to decrease the quality of proctored exams so their eyes can move around to cheat more easily" (Walsh et al., 2021b).

One alternative to traditional exams that reduces issues of academic dishonesty is the virtual group oral exam. In one example, a biochemistry course, collaborative learning, and small teams of students were established early in the course. Exams were held over Zoom, exam questions were screen shared to students, and students responded positively to the team building and direct feedback. The instructor felt that "the alignment of team learning activities with group oral exams helped students engage in deeper learning and build a supportive community in a time of crisis" (Goodman, 2020).

Another alternative is to write higher-order thinking questions. Such questions are written using specifics and narratives, so students can use the Internet as a resource but must still think critically and apply the information to reach the correct answer. This makes the assessment a learning tool and a more realistic replication of how scientists operate (Nguyen et al., 2020). If they are written with enough detail, exam questions can be made multiple choice to automate grading in large classes. For specific examples of questions, see Nguyen et al. (2020). The author points out:

> Instead of trying to 'eliminate' cheating, it is perhaps better to be more accommodating and structure exams so that students using 'unauthorized' resources will not have a large advantage over those who do not.
>
> — Nguyen et al. (2020)

After students provided information on how other students were planning on cheating closed-book exams, one professor made the exam open book and reported that students were grateful for this change (Dingwall, 2020). Because some students felt that unproctored, closed-book tests led to academic dishonesty, they expressed preference for final projects or open-book exams as superior learning assessments (Petillion & McNeil, 2020). This is consistent with previous research, which demonstrates that open-book exams reduce the likelihood of cheating and improves student learning and critical thinking skills (Williams & Wong, 2009).

15.2.4.5 Alternatives to Exams

Students surveyed at multiple institutions expressed their concern regarding the lack of student input in course decisions during and after the transition online (Jamieson, 2020; McPhee & Lyon, 2020). Students reported that their learning and well-being improved when instructors were flexible with assignment deadlines and allowed students to have a choice in type and weight of assessments (McPhee & Lyon, 2020). This sort of student input can be collected using a variety of poll options. Some LMS programs offer asynchronous polling directly (e.g., a quiz form in Canvas can be modified into graded or ungraded survey). Polling can also be implemented directly from videoconferencing software (e.g., Zoom) or through clickable links that update on the instructor's presentation slide (e.g., Menti). In one case, an instructor used extra credit to incentivize students to provide weekly feedback via their LMS, so the professor could make constructive adjustments after the transition online.

Anecdotally, the instructor found this created a positive academic environment and improved communication from students (Hwang, 2020).

Most students struggled with motivation after the transition online, leading to a noticeable decrease in student engagement (McPhee & Lyon, 2020; Perets et al., 2020; Wester et al., 2021). Students faced nonacademic challenges including anxiety, stress, feelings of uncertainty, and caring for family members, as well as the academic concern for finding a quiet place to study (McPhee & Lyon, 2020; Ramachandran & Rodriguez, 2020). To promote peer-to-peer interactions that can lessen feelings of isolation during a pandemic, instructors can design learning assessments around class discussion and can assign learning communities in their LMS (e.g., groups function in Canvas) and hold either synchronous discussions with video software such as Zoom or hold asynchronous discussions via discussion boards on LMS or similar software (e.g., Slack, Google Docs). Posing questions and responding to their peers can be made mandatory for students, providing a broad template for learning assessments (McPhee & Lyon, 2020).

Virtual group projects with real-world applications and stakeholders can replace exams to further improve student motivation (Dietrich et al., 2020; LaMack et al., 2020; Petillion & McNeil, 2020). For example, capstone engineering courses organized a public virtual showcase that allowed students' friends and family to attend, along with industry members invited by the professor (Jamieson, 2020; LaMack et al., 2020). The breakout room function of videoconference software was used for each group (Jamieson, 2020). The faculty felt this virtual showcase resulted in turnout and quality comparable to previous in-person events (LaMack et al., 2020). Faculty who implemented group projects recommend no more than three students per group, and that faculty should encourage videoconferencing to facilitate collaboration. They also noted that faculty should be aware that group projects tend to be more time-consuming for students than individual assignments and should adjust deadlines accordingly (Dietrich et al., 2020).

Faculty in a variety of disciplines observed positive student outcomes during COVID-19 when they used a greater percentage of formative assessments over summative assessments. Homework and low-stakes quizzes helped ameliorate the pressure students felt about exams during the COVID-19 ERT (Dietrich et al., 2020), and some students expressed a desire for more low-stakes assessments with timely and frequent feedback to practice applying their knowledge (McPhee & Lyon, 2020; Perets et al., 2020; Petillion & McNeil, 2020). In courses such as chemistry, where student practice leads to proficiency, promoting student practice with frequent quizzes in a remote setting achieved a degree of success (Crucho et al., 2020). Rather than a standalone quiz format, some faculty posed quiz questions during their presentation slides during lecture or embedded videos (Crucho et al., 2020; Gray, 2020). In one example, even though the quiz was not graded, roughly 70% of students participated, and students who took quizzes had higher final exam scores than those who did not. Additionally, quiz averages were positively correlated with exam scores (Fig. 15.4), mirroring formative and summative assessment relationships observed outside ERT events (Krasne et al., 2006; Kibble, 2007). In an optional survey, 95% of students indicated that they appreciated quizzes (Crucho

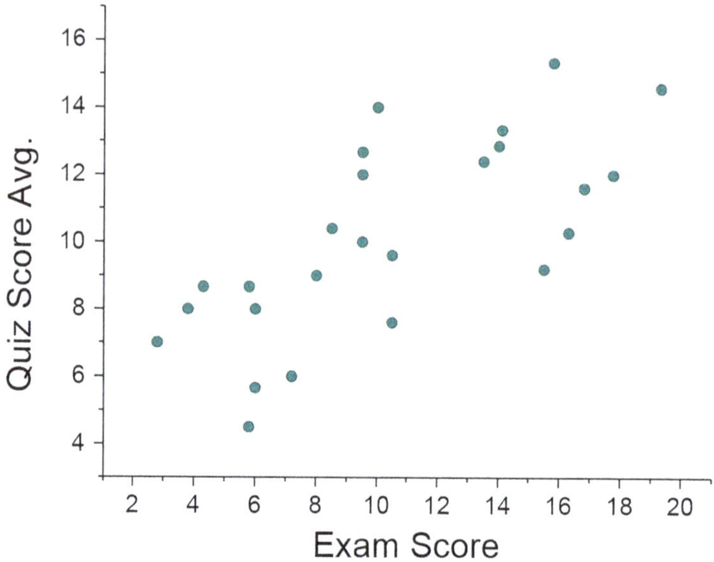

Fig. 15.4 Scatter plot of student scores for zero-stakes quizzes vs. a high-stakes exam in a remote organic chemistry course. Scatter plot illustrates the positive correlation between the two assessment types (Crucho et al., 2020)

et al., 2020). In another case, the professor used embedded lecture quizzes for students to verify their asynchronous attendance by uploading pictures of their answers to the class LMS (see Chaps. 11 and 12 in this volume). Student submissions were reviewed every day, providing feedback without being graded. This allowed for more opportunities to address misconceptions, and student grades improved relative to the face-to-face part of the semester (Gray, 2020). In emergencies, zero-stakes quizzes may be less stressful than frequent, graded quizzes, although, in one case, students disliked frequent low-stakes quizzes out of concern that they would miss deadlines because they were juggling so much remotely (Ramachandran & Rodriguez, 2020).

15.3 Future Trends in Educational Technology

Application and innovation in technology increased rapidly during the COVID-19 pandemic, minimizing disruption to teaching and learning in life sciences and medical fields (Goh & Sandars, 2020; Remtulla, 2020). Familiar online technologies and practices are now more frequently used or implemented in underdeveloped and developing countries, such as India and Pakistan, that are shifting from face-to-face traditional pedagogical approaches (e.g., LMS, online videos, blogs, webinars, podcast, livestreaming, telemedicine, etc.; Adnan & Anwar, 2020; Dhawan, 2020;

Wadhwa et al., 2020). Emergent technologies and practices are playing a key role minimizing the educational disruption. The accelerated application of many new types of emergent technologies with substantial new developments is an essential component of the transformative change that postsecondary education is currently experiencing (Brown et al., 2020; Goh & Sandars, 2020). Being in the third year of the COVID-19 pandemic, returning to previous educational approaches is infeasible, but challenges remain in finding alternatives to the hands-on learning critical to education in life sciences and medicine (Goh & Sandars, 2020; Remtulla, 2020).

Many emergent technologies have the potential to transform higher education's optimization of pedagogy through adaptive learning and extended reality (see Chap. 17 in this volume). These approaches have become a priority for research and implementation during the COVID-19 pandemic (Brown et al., 2020; Goh & Sandars, 2020). However, these approaches face a major challenge due to their affordability, which is discussed in the next section on The Future of Education During Emergencies. Adaptive learning (AL) is a type of personalized learning in which adaptive technology allows students to access a wide range of online instructional resources that foster success and provide educators with learning data (Brown et al., 2020; Goh & Sandars, 2020; see Chap. 5 in this volume). Using AL allows the instructor to move from delivering content via traditional face-to-face teaching to a form of coaching during active learning exercises (Brown et al., 2020). To implement AL, it is critical that instructors have the appropriate resources and that students attend appropriate courses and possess an appropriate level of learning (Brown et al., 2020). Three types of emergent technologies support AL in life sciences and medical education: artificial intelligence (AI), learning analytics, and next-generation digital learning environments (Chan & Zary, 2019; Wartman & Combs, 2019; Brown et al., 2020; Goh & Sandars, 2020). The other approach, extended reality (XR), offers a blend of physical and virtual environments or even provides fully immersive virtual experiences. Augmented reality and virtual reality, some of the best-known XR technologies, are increasingly emerging as educational tools in life sciences and medical fields due to their potential to replace standard instruction during the pandemic while improving student performance and engagement (Brown et al., 2020; Goh & Sandars, 2020; Nesenbergs et al., 2021). These technologies have great applications for skills-based competencies and virtual hands-on practices and provide access to learning for students with disabilities (Brown et al., 2020; Goh & Sandars, 2020). Mixed reality, a hybrid system that involves physical and virtual elements, also has a high potential to play an important role in education (Tang et al., 2020). Other more recent technologies include haptic simulations that replicate physical sensations of touch (Brown et al., 2020; Goh & Sandars, 2020). For more details on simulations, see Chap. 16 in this volume. In the following paragraphs, it is explained in more detail some of these emergent technologies within both adaptive learning and extended reality approaches of pedagogical optimization.

15.3.1 Artificial Intelligence (AI) and Machine Learning (ML)

In a broad sense, AI allows machines to execute tasks intelligently, while machine learning (ML) allows machines to learn and adapt through repetitive processes (Brown et al., 2020). Thus, AI applications can create learning content and assessments that interact with students through text and voice in an adaptive fashion, such as through LMSs and student information systems (Brown et al., 2020; Goh & Sandars, 2020). These educational services are starting to generate emergent courseware products with adaptive learning pathways unique to each student's needs, including providing student feedback through "virtual teaching assistant" applications. On the instructor side, courseware products assist with designing the curriculum, test generation, and validity, as well as flagging students at academic risk and detecting plagiarism using ML (Brown et al., 2020). In medical education, AI helps students learn anatomy, and surgical simulation apps may be used to help surgeons maintain their skills (e.g., About Touch Surgery; Remtulla, 2020; see Chap. 16 in this volume). Although AI and ML are reinventing the process of teaching and learning with autonomous classrooms, intelligent tutoring systems, learning analytics, and automated decision-making, it is necessary that we evaluate the capacity of these technologies to replicate social, emotional, and cognitive qualities of human teachers and students (Selwyn, 2019).

15.3.2 Learning Analytics

Adaptive learning uses applications that measure, collect, analyze, and report data about the progress and success of students, along with registrar records, financial systems, and institutional research units. These data are essential to help educators track the progress and trajectory of individuals and groups of students, especially during educational disruptions (Brown et al., 2020; Goh & Sandars, 2020). Rapid, automated analysis of these data can reduce the time it takes to identify areas of concern during an emergency, provide needed support, and foster resilience in students (Brown et al., 2020). However, learning analytics present some challenges about data quality and gaps, privacy, and ethical considerations. Biased data could affect students by causing them to be labeled as students "at risk," or data on their academic performance could be biased based on gender, race or ethnicity, or income (Wilson et al., 2017; Brown et al., 2020). Thus, educators require new levels of data literacy and interpretation to personalize instruction and achieve more equitable outcomes (Wilson et al., 2017; see Chap. 2 in this volume). The Comprehensive Assessment of Team Member Effectiveness (CATME) Smarter Teamwork is an example of a system where instructions can be personalized (thus, reducing bias toward students) and where students' performance is tracked beyond the individual level by working in teams (Ohland et al., 2012).

15.3.3 Next-Generation Digital Learning Environment (NGDLE)

The next-generation digital learning environment (NGDLE) is an all-in-one digital learning platform with everything in one place including apps, learning services, and adaptive learning management tools (Lynch, 2018; Brown et al., 2020). The NGDLE has five components: (1) interoperability and integration; (2) personalization; (3) analytics, advising, and learning assessment; (4) collaboration; and (5) accessibility and universal design (Lynch, 2018). Those components allow for submission, testing, data management, timetabling, learning analytics, multimedia, learning support, collaboration and communication tools, and tools for sharing materials (Lynch, 2018). NGDLEs are transforming institutions' learning ecosystems by providing a more flexible learning experience to more students (e.g., IMS Global LTI and Blackboard Collaborate Ultra; Brown et al., 2020). Thus, NGDLE is a very promising approach to address and adapt to ERTs to cover several educational challenges at once.

15.3.4 Augmented Reality (AR)

Augmented reality is one of the extended reality tools, which overlays physical objects and places with virtual content (Brown et al., 2020). In medical education, this technology has been used to understand the 3D structure inherent in anatomy with 3D holograms (e.g., postoperative anatomy of organ transplant patients; Kroese & Leiden University Medical Centre [LUMC] 2019). AR has also been used in ecological education for field trip exploration of forest ecosystems (Hewitt, 2017). Some of the AR apps are designed for mobile devices (e.g., phones and tablets), such as apps that display 3D molecular models using free software and mobile AR escape lab activity apps (Eriksen et al., 2020; Estudante & Dietrich, 2020; Sung et al., 2020). In chemistry lab AR, the Ariel application has been designed to connect students to scientific instruments (An et al., 2020). All these examples have helped students to reduce anxiety, improve intellectual accessibility, understand complex structures and biological phenomena, and increase motivation and engagement (Kroese & Leiden, 2019; An et al., 2020; Eriksen et al., 2020; Estudante & Dietrich, 2020; Salar et al., 2020). For more details on augmented reality, see Chap. 17 in this volume.

15.3.5 Virtual Reality (VR)

Virtual reality, another extended reality tool, offers immersive experiences that involve the manipulation and interactions with virtual objects within an entirely virtual environment, trying to simulate or replicate real-life experiences that can be

Fig. 15.5 Virtual reality experience using headset and controllers. Photograph by Mariellesterner, distributed under a CC BY-SA 4.0 license

delivered through headsets or mobile devices (Fig. 15.5; Brown et al., 2020). VR environments in the medical field reconstruct aspects of the clinical environment and simulate clinical learning (e.g., 360° video for the emergency care ABCDE approach; Schipper & LUMC, 2019; Zweifach & Triola, 2019; Goh & Sandars, 2020; Remtulla, 2020). VR use in medical training improved learning and accuracy (Samadbeik et al., 2018). Web-based virtual platforms also offer opportunities for undergraduate students to have virtual field trips to remote forests and rural areas, replacing field-based learning for ecology and environmental chemistry courses (Hewitt, 2017; Fung et al., 2019). Undergraduate students' use of VR activities on interactive molecular dynamics and infrared spectroscopy demonstrated motivational improvement and learning gains (Bennie et al., 2019; Ferrell et al., 2019; Dunnagan et al., 2020). Visualizations, immersion, and accessibility via smartphones offered by AR and VR technology suggest they could be leveraged by a broad audience in future ERT events to improve science and field laboratory classes, as well as help medical clinics with training. A review of Radianti et al. (2020), however, considers the application of VR in higher education at a low-maturity level due to several gaps found. Those gaps include rarely considering learning theories in the development of VR apps to support learning outcomes and evaluating educational VR applications focusing on the usability of the VR apps instead of the learning outcomes. VR experiences also have mostly been part of experimental and development research rather than of actual teaching. Radianti et al. (2020) recognized that the development of VR technologies and their application in higher education teaching are very promising due to the high interest in researching a variety of domains to apply VR tools for educational purposes. For more details on virtual reality, see Chap. 17 in this volume.

15.4 Forward Facing: The Future of Education During Emergencies

Despite historical cases of pandemics and educational disruptions (see Emergency learning and teaching section), higher education institutions across the globe were unprepared for the rapid transition online triggered by the COVID-19 pandemic (Commonwealth of Learning [COL], 2020; Crawford et al., 2020; Daniel, 2020). In 2009, concerns over the H1N1 virus led many US institutions to evaluate their digital infrastructure, curate plans for potential distance education, and train faculty to be prepared to convert their courses to an online modality (Ekmekci & Bergstrand, 2010). With globalization increasing the likelihood of future pandemics (Bedford et al., 2019) and climate change causing more frequent extreme weather events (Banholzer et al., 2014), it is no longer enough for institutions to troubleshoot disruptions as they appear, as in the 2009 case of H1N1. Instead, institutions must equip staff and faculty with skills and technology to rapidly pivot teaching methods and must have an established network that promotes resilience in both employees and students (Brown et al., 2020). Here, we outline recommendations for institutions going forward, largely organized based on Commonwealth of Learning (COL, 2020) guidelines.

15.4.1 Develop and Invest in Infrastructure and Support

Direct investment into e-learning infrastructure can combat the technological divide and unequal access among students (Njenga & Fourie, 2010). An analysis of blended learning at one university showed that student satisfaction was best predicted by the availability of university resources (Calderon et al., 2012). Another study showed that the top faculty-identified needs for successful e-learning are multimedia development support and real-time help desks (Chow & Croxton, 2017). Universities are encouraged to make expenditures related to Internet access necessary for hybrid approaches (Mac Keogh, 2001; Guri-Rosenblit, 2006). Universities need to hire or contract support staff for IT to ensure real-time tech support, which is essential to successful instruction (Chow & Croxton, 2017). By investing in technology infrastructure and support, institutions of higher education will be prepared for the rapid transition necessary during an emergency that impacts educational access.

Development and access to emergent technologies, particularly for teaching and learning, are typically costly and require numerous experts (Brown et al., 2020; Goh & Sandars, 2020; Remtulla, 2020). Adaptive learning technology is still at an early stage of development, and XR is promising if there is high fidelity, but this results in a higher financial burden (Brown et al., 2020). However, XR equipment costs will decrease over time, while its capabilities will increase, making it more accessible to residential and remote learners as network capacities improve (Zweifach & Triola, 2019; Brown et al., 2020). Reducing costs and increasing the number of users of emergent technologies will require the support of commercial providers willing to collaborate with academic institutions (Dhawan, 2020; Organisation for Economic Co-operation and Development [OECD], 2020).

15.4.2 Be Proactive and Organized

Institutions must establish networks of support, provide routine but brief updates for faculty on technology available, and outline a division of labor across departments for future closures (Daniel, 2020). Institutions should have a plan that expedites clear and consistent communication with students, along with reassurance that they will have education continuity (Daniel, 2020). Colleges should ensure that departments have an established emergency plan. This plan would include staff drop-in hours for students to email or call for help troubleshooting issues (COL, 2020). These staff must be familiar with all campus resources relevant to ERT, including remote health and wellness support, as well as technology and discipline-specific equipment loan services (Anstey et al., 2020; Barton, 2020; Daniel, 2020). Each department should have an established network to share and disseminate ideas and best practices for teaching their subject (COL, 2020). Professional development should be holistic, promote a sense of community, and foster four skill groups: pedagogical, technical, professional, and survival. While the first three skill groups are common in STEM professional development, "survival" is rare. Topics that fall under "survival" include physical and mental health, resilience, time management, conflict resolution, and multicultural competence (Dragisich, 2020). An increasingly important goal in higher education is the promotion of diversity and equality both inside and outside the classroom. The COVID-19 pandemic revealed the important connection between well-being and learning. Steps to reduce stress and promote mindfulness and self-care should be actively shared by administrators to faculty and by faculty to their students (Currie, 2020). In a professional development course offered to graduate students, the instructor received positive feedback after they assigned self-care homework and organized weekly discussions centered on affirmations and mindfulness (Dragisich, 2020; Table 15.1). Crucial to this implementation is effective professional development for faculty and staff around issues of mental health and self-care (Brown et al., 2020; see section below on effective training).

Common themes emphasized by both faculty and students that were instrumental in successful ERT courses during the COVID-19 pandemic included flexibility, frequent and clear communication, and clear guidance and structure for students (Anstey et al., 2020; McPhee & Lyon, 2020). Unsurprisingly, these are tenets of successful traditional online teaching (Moore, 1993). In some cases, flexibility must be allotted at the institution level, such as extending the final exam period to 3 weeks to allow for appropriate evaluation measures to be implemented (Dicks et al., 2020).

Table 15.1 Weekly professional development schedule offered to graduate teaching assistants during COVID-19 to foster survival skills (Dragisich, 2020)

Check-in	Objective	Topics covered
#1	Creating opportunities for connection by looking for small ways to work on self-care for resilience	Benefits of mindfulness meditation The Practice Sitting with an intention Self-compassion Positive self-talk
#2	Creating opportunities for connection by looking for small ways to work on self-care for resilience	Teaching and learning in our current environment Teaching challenges (e.g., inability to read the room) Learning challenges (e.g., access to quiet spaces, anxiety) Coping with trauma Support solutions
#3	Creating opportunities for connection by looking for small ways to work on self-care for resilience	Building/Cultivating resilience What is resilience? Types of stress Combating stress via self-regulation, rewriting, and support systems Compassion fatigue

15.4.3 Evidence of Effective Training

Professional development that provided instructors with both technological skills and pedagogical best practices proved to be important for maintaining quality education during the COVID-19 pandemic. A faculty survey collected shortly after schools were closed found that student-centered teaching practices became significantly less frequent after the transition online. This drop in student-centered practice was less significant in faculty with high technological pedagogical content knowledge, leading researchers to conclude that faculty needed support in both technical pedagogy and online pedagogy to best support students (Walter et al., 2021). By August 2020, faculty who never received training in online teaching were more likely to struggle with maintaining or fostering a sense of community in their ERT course, and they were the only faculty surveyed to struggle with student engagement (Fig. 15.6; Walsh et al., 2021a). In place of training, one faculty member noted that a "tremendous amount of 'help' broadcast out was piles of links. [sic] Had to dig through them, wasting time," demonstrating that institutions need to be succinct and clear in their assistance to faculty during emergencies. However, faculty often recounted making stronger personal connections with their students. In some cases, faculty served as social workers, helping create or find resources for their students to stay afloat (Walsh et al., 2021a). Higher education must promote crisis-specific training to increase faculty awareness and comfort level

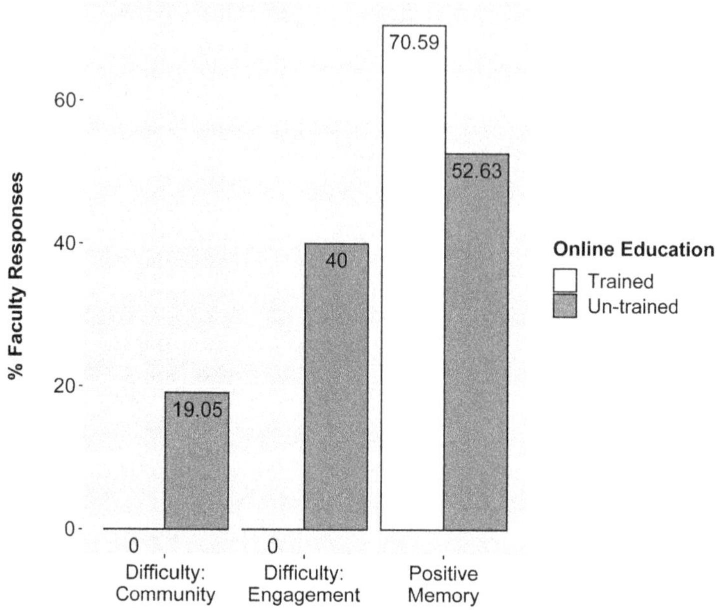

Fig. 15.6 Polarizing responses for faculty with and without training in online education due to COVID-19: difficulty promoting a sense of community, difficulty with student engagement, and their memorable moment during COVID-19 ERT was positive (Walsh et al., 2021a)

in helping students in need, especially during ERT events. Just as importantly, they must aid faculty experiencing trauma and secondhand trauma:

> When my student told me that 4 family members had died in the last two weeks, could she have extensions, she needed to plan the funerals. It made the crisis really personal. I still haven't completely shaken the feeling and continue to wonder if she's okay, how her family is moving on. —Walsh et al. (2021a)

Giving instructors professional training is crucial and has been shown to improve teachers' acceptance of technology for online and virtual instruction and achieve the desired outcome and impact of student learning (Zhai et al., 2018; Zhai et al., 2019; Goh & Sandars, 2020; Toquero, 2020). For this, it is essential that educators have the time and skills to effectively deploy the technology in the online academic environment (Brown et al., 2020; Khan & Somdatta, 2020). Therefore, teacher training capabilities for multiple modes of e-learning should be expanded (Njenga & Fourie, 2010). Training is essential to the effective delivery of electronic physics instruction (Afzal et al., 2015) and has been demonstrated to lead to instructors' enthusiastic acceptance of mobile technology for teaching (Zhai et al., 2018; Zhai et al., 2019; Walsh et al., 2021a). The use of emergent technologies also needs to align with instructors' practices (e.g., learning objectives and content, instructional design, and context; Brown et al., 2020; Goh & Sandars, 2020). This requires providing digital learning opportunities to educators on how to teach online, how to effectively use collaborative platforms, and how to use technology to share their resources, provide feedback to students, and receive peer feedback (Zweifach & Triola, 2019; Goh & Sandars, 2020; OECD, 2020). Educators thus become testers of different digital learning solutions, creators and curators of learning content, and evaluators of which technologies foster deeper student learning (Goh & Sandars, 2020; OECD, 2020). Effective implemented solutions resulting from these tests and evaluations by educators should be documented to share and analyze internationally (Goh & Sandars, 2020; OECD, 2020; Toquero, 2020).

15.4.4 Using the Best of Both Worlds: Synchronous and Asynchronous Instruction

A careful balance needs to be struck between synchronous and asynchronous instruction. Initial feedback showed that Tufts University students did not like lectures on Zoom, while they did like small-group instruction and the use of breakout rooms on the platform (Flaherty, 2020). In medical education, online lecture tools were found to accelerate knowledge acquisition, manage stress better, and improve learning outcomes when compared to in-person lectures (Remtulla, 2020). With asynchronous instruction, students need high executive-functioning skills to set and maintain the level of time management necessary when they are used to a much more regular schedule (Flaherty, 2020). Students reported higher satisfaction (McPhee & Lyon, 2020) and higher student gains (Ramo et al., 2021) from a combination of synchronous and asynchronous instruction. Additionally, undergraduate learners may be more accustomed to predictable routines and instructional methods associated with in-class learning. While synchronous delivery with tools such as Zoom more closely approximates that environment, flexibility is essential to maintain high class accessibility. A hybrid model that incorporates both synchronous and asynchronous activities can serve to keep learners on track while providing some flexibility in the engagement with other content or assessments (Flaherty, 2020).

15.4.5 Rethink Online Assessments

Most students responded well to instructors who changed exams to take-home exams and weighed the final exam less heavily than originally planned (McPhee & Lyon, 2020). If assessments change, the instructor should provide students with a clear rubric or completed example to ensure clarity of each major assignment (McPhee & Lyon, 2020). Faculty should use assignments as an opportunity to promote student interactions (LaMack et al., 2020; McPhee & Lyon, 2020) and should be flexible and empathetic while communicating effectively with clear and consistent assignment plans (McPhee & Lyon, 2020; Vielma & Brey, 2020). When possible, instructors should highlight how assessments help students develop real-world skills (Vielma & Brey, 2020). For more details on assessment and technology, see Chap. 8 in this volume.

15.4.6 Address Digital Inequalities

A disadvantage of completely online classes is that this situation can create a digital divide in communities that lack or have limited access to high-speed Internet, bandwidth, and devices such as smartphones and laptops. This is especially true during the pandemic, which has exacerbated issues of inequality and access to online classrooms (Barnes, 2020; Dietrich et al., 2020; Goh & Sandars, 2020; see Chaps. 11 and 12 in this volume). Additional digital inequalities brought by COVID-19 include the autonomy of technology use (location of access and freedom to use it), social support networks (support from experimented users), and user experience (time to get familiar with the technology and retaining benefits; Beaunoyer et al., 2020; Goh & Sandars, 2020; Remtulla, 2020; Selwyn & Jandrić, 2020). This digital divide was especially stark when contrasting high-income countries with low- and middle-income ones (Brown et al., 2020; COL, 2020; Dhawan, 2020; Goh & Sandars, 2020; Remtulla, 2020).

Key mitigation strategies include identifying "digital deserts" and technology appropriate in local contexts (Beaunoyer et al., 2020; Goh & Sandars, 2020). In areas with limited Internet connectivity, governments and school districts had to get creative on how they provided instructional materials (United Nations [UN], 2020). Some institutions used more traditional distance learning models (sending books and work home with students), while others tried to figure out ways to provide students with Internet access, such as paying for their home Internet or sending buses into low-income areas to serve as a Wi-Fi hotspot for students during the day. Many institutions in developing areas also sent out tablets for students to use during remote learning (UN, 2020). Instructors can choose a management platform that is built to be used on basic and cheap mobile phones (e.g., Cell-Ed and KaiOS) or systems with strong offline functionality (e.g., Kolibri, Rumie, and Ustad Mobile). Many of these platforms also are available in multiple languages (UNESCO, 2020). Virtual reality developers are also exploring ways to offer access to education to students who cannot physically attend class (Dunnagan et al., 2020), which should give students even more options. Increasing access to open educational resources (OER) for both educators and students across the world is essential (Ferrell et al., 2019; Brown et al., 2020; Dietrich et al., 2020). This can be implemented if governments and international collaborations share and translate existing OER (Goh & Sandars, 2020; OECD, 2020). For more details on inclusivity and technology, see Chap. 2 in this volume.

15.4.7 *Digital Integration and Analytics*

With students learning online, the students' metadata is a powerful tool to understand student behavior and provide just-in-time learning. Currently, even the most sophisticated learning management software cannot provide student metadata in an easily accessible and digestible form to facilitate instructors' understanding of where students struggle (e.g., Blackboard website, Canvas website; Juárez Santiago et al., 2020). This lack of data analytics is even more apparent in videoconferencing software platforms, which provide little in the way of feedback on student behavior (e.g., Zoom, Microsoft, and Webex; Ahuja, 2020). With the power of face recognition, eye tracking, speech pattern analysis, and vocabulary recognition, it is technically feasible for an instructor to monitor "breakout" rooms to detect productive versus unproductive conversations (Ahuja, 2020). Further improving access to this information would provide instructors with the tools necessary to engage with students that may need additional instruction, facilitation, or motivation, thus providing just-in-time learning for students. While LMS can generate reports on student usage, such as the number of "hits" on asynchronous assignments (Glass, 2020), the data generated in reports is often overwhelming or looks at a class' overall performance instead of identifying students who may need additional assistance. Viewing a class roster with analytics collected across all platforms used in the course would allow an instructor to more easily identify which students need additional help and which educational intervention applies. As emergent technologies of adaptive learning evolve, AI, ML, learning analytics, and next-generation digital learning environments (NGDLEs) will offer institutions a learning ecosystem in which the management of student and faculty metadata could be a major source to improve learning.

Pedagogical research, data collection, and evidence-based practices focused on e-learning should be expanded. Student feedback can be motivated by effective communication mechanisms integrated into a student's online learning space (Hatziapostolou & Paraskakis, 2010; Jara & Mellar, 2010) and has been shown to be of immense value in improving blended course quality (Calderon et al., 2012). Mechanisms to directly combat high dropout rates for e-learners must be developed, including communication with lower-achieving students (Njenga & Fourie, 2010). Universities should organize a system to analyze feedback data, identify problem points, delegate responsibility for addressing them, and report back to students on resulting actions (Watson, 2003). Institutions should also integrate easy-access course feedback into virtual learning management systems to "close the feedback loop" (O'Brien, 2020) taking into consideration student privacy issues.

15.4.8 Privacy and Ethical Considerations

As part of adaptive learning, learning analytics face several student data privacy and ethical challenges. For example, as referenced above, when providing students with feedback, doing so in an online setting comes with a different set of privacy concerns than those in a classroom. Potential misuse of student data stored on cloud-based systems is under scrutiny, especially when intervention decisions are based on performance metrics for individualized learning (Brown et al., 2020). Governments and institutions need to have a proactive approach to protect student and employee data, carefully choosing partners for data exchange and finding a balance between emergent educational technologies and privacy and ethical considerations (Brown et al., 2020).

15.5 Conclusion

Technology use during emergency remote teaching, combined with the evidence-based pedagogy of e-learning and considering factors that affect students' digital divide (Correia, 2020), minimizes the disruption to life sciences and medical education even when face-to-face classes are not feasible and builds a scaffold of support for instructors and students (Brown et al., 2020). During COVID-19 ERT, we saw evidence of growing educational disparity in terms of the "digital divide," where students who did not have access to technological resources were unable to access education to the same degree as those with devices and reliable connections (Correia, 2020). We also saw benefits of technology use as an "anytime, anywhere" form of education that provided increased access to education for some students, such as part-time students (Goh & Sandars, 2020). Benefits will only continue to improve at an accelerated pace with technological advancements leading to further novel technologies (Bozkurt & Sharma, 2020; Remtulla, 2020).

Acknowledgment We would like to thank our reviewers, Durin Booth and Rebekka Darner, for their prompt and informative feedback. We would also like to thank the *Journal of Microbiology and Biology Education* editorial staff, including Miriam Segura-Totten, for connecting us with groundbreaking research during the COVID-19 pandemic.

References

Adnan, M., & Anwar, K. (2020). Online learning amid the COVID-19 pandemic: Students' perspectives. *Journal of Pedagogical Sociology and Psychology, 2*(1), 45–51.

Afzal, M. T., Safdar, A., & Ambreen, M. (2015). Teachers perceptions and needs towards the use of e-learning in teaching of physics at secondary level. *American Journal of Educational Research, 3*(8), 1045–1051.

Ahuja, A. (2020, July 31). *Development of a virtual classroom teaching effectiveness observation rubric* [conference presentation abstract]. 10th Annual SABER Meeting, Virtual.

Akarasriworn, C., & Ku, H.-Y. (2013). Graduate students' knowledge construction and attitudes toward online synchronous videoconferencing collaborative learning environments. *Quarterly Review of Distance Education, 14*(1), 35–48.

Alokluk, J. A. (2018). *The effectiveness of blackboard system, uses and limitations in information management*. Scientific Research Publishing.

An, J., Poly, L.-P., & Holme, T. A. (2020). Usability testing and the development of an augmented reality application for laboratory learning. *Journal of Chemical Education, 97*(1), 97–105. https://doi.org/10.1021/acs.jchemed.9b00453

Anstey, M. R., Blauch, D. N., Carroll, F. A., Gorensek-Benitez, A. H., Hauser, C. D., Key, H. M., Myers, J. K., Stevens, E. P., Striplin, D. R., Holck, H. W., Montero-Lopez, L., & Snyder, N. L. (2020). #DavidsonTrue: Transitioning to remote teaching while maintaining our values as a liberal arts college during the COVID-19 pandemic. *Journal of Chemical Education, 97*(9), 2800–2805.

Banholzer, S., Kossin, J., & Donner, S. (2014). The impact of climate change on natural disasters. In Z. Zommers & A. Singh (Eds.), *Reducing disaster: Early warning systems for climate change* (pp. 21–49). Springer.

Barnes, S. J. (2020). Information management research and practice in the post-COVID-19 world. *International Journal of Information Management, 55*, 102175.

Barton, D. C. (2020). Impacts of the COVID-19 pandemic on field instruction and remote teaching alternatives: Results from a survey of instructors. *Ecology and Evolution, 10*, 12499–12507.

Beaunoyer, E., Dupéré, S., & Guitton, M. J. (2020). COVID-19 and digital inequalities: Reciprocal impacts and mitigation strategies. *Computers in Human Behavior, 111*, 106424.

Bedford, J., Farrar, J., Ihekweazu, C., Kang, G., Koopmans, M., & Nkengasong, J. (2019). A new twenty-first century science for effective epidemic response. *Nature, 575*, 130–136.

Bellan, R. (2020, March 24). *What you need to know about using Zoom*. Forbes. https://www.forbes.com/sites/rebeccabellan/2020/03/24/what-you-need-to-know-about-using-zoom/?sh=2033a0923284

Bennie, S. J., Ranaghan, K. E., Deeks, H., Goldsmith, H. E., O'Connor, M. B., Mulholland, A. J., & Glowacki, D. R. (2019). Teaching enzyme catalysis using interactive molecular dynamics in virtual reality. *Journal of Chemical Education, 96*(11), 2488–2496.

Bozkurt, A., & Sharma, R. C. (2020). Emergency remote teaching in a time of global crisis due to coronavirus pandemic. *Asian Journal of Distance Education, 15*(1), i–vi.

Brindley, J. E., Blaschke, L. M., & Walti, C. (2009). Creating effective collaborative learning groups in an online environment. *International Review of Research in Open and Distance Learning, 10*(3), 1–18.

Brown, M., McCormack, M., Reeves, J., Brooks, D. C., Grajek, S., Alexander, B., Bali, M., Bulger, S., Dark, S., Engelbert, N., Gannon, K., Gauthier, A., Gibson, D., Gibson, R., Lundin, B., Veletsianos, G., & Weber, N. (2020). *2020 EDUCAUSE horizon report, teaching and learning edition*. EDUCAUSE.

Burden, T. (2014, January 14). K-12 teachers uncertain about how to connect with students and parents via social media, reveals university of phoenix survey. *Business Wire*. https://www.businesswire.com/news/home/20140114005604/en/K-12-Teachers-Uncertain-Connect-Students-Parents-Social

Calderon, O., Ginsberg, A. P., & Ciabocchi, L. (2012). Multidimensional assessment of pilot blended learning programs: Maximizing program effectiveness based on student and faculty feedback. *Journal of Asynchronous Learning Networks, 16*(3), 23–37.

Castelli, F. R., & Sarvary, M. A. (2021). Why students do not turn on their video cameras during online classes and an equitable and inclusive plan to encourage them to do so. *Ecology and Evolution, 00*, 1–12.

Chan, K. S., & Zary, N. (2019). Applications and challenges of implementing artificial intelligence in medical education: Integrative review. *JMIR Medical Education, 5*(1), e13930.

Chen, C.-C., Dpbinson, T., & Kent, S. (2020). Students' perspectives on the impact of blackboard collaborate on Open University Australia (OUA) online learning. *Journal of Educators Online, 17*(1), 41–54.

Chow, A. S., & Croxton, R. A. (2017). Designing a responsive e-learning infrastructure: Systemic change in higher education. *American Journal of Distance Education, 31*(1), 20–42.

Commonwealth of Learning [COL]. (2020). *Guidelines on distance education during COVID-19*. Burnaby: COL. http://oasis.col.org/handle/11599/3576.

Cornelius, S. (2014). Facilitating in a demanding environment: Experiences of teaching in virtual classrooms using web conferencing. *British Journal of Educational Technology, 45*(2), 260–271.

Correia, A.-P. (2020). Healing the digital divide during the COVID-19 pandemic. *Quarterly Review of Distance Education, 21*(1), 13–21.

Crawford, J., Butler-Henderson, K., Rudolph, J., Malkawi, B., Glowatz, M., Burton, R., Magni, P. A., & Lam, S. (2020). COVID-19: 20 countries' higher education intra-period digital pedagogy responses. *Journal of Applied Learning & Teaching, 3*(1), 9–28.

Crucho, C. I. C., Avó, J., Diniz, A. M., & Gomes, M. J. S. (2020). Challenges in teaching organic chemistry remotely. *Journal of Chemical Education, 97*(9), 3211–3216.

Currie, H. N. (2020). Mindful Well-being and learning. *Journal of Chemical Education, 97*(9), 2393–2396.

Czerniewicz, L., Trotter, H., & Haupt, G. (2019). Online teaching in response to student protests and campus shutdowns: Academics' perspectives. *International Journal of Educational Technology in Higher Education, 16*, 43.

Daniel, J. (2020). Education and the COVID-19 pandemic. *Prospects, 49*, 91–96.

Dhawan, S. (2020). Online learning: A panacea in the time of COVID-19 crisis. *Journal of Educational Technology Systems, 49*(1), 5–22.

Dicks, A. P., Morra, B., & Quinlan, K. B. (2020). Lessons learned from the COVID-19 crisis: Adjusting assessment approaches within introductory organic courses. *Journal of Chemical Education, 97*(9), 3406–3412.

Dietrich, N., Kentheswaran, K., Ahmadi, A., Teychené, J., Bessière, Y., Alfenore, S., Laborie, S., Bastoul, D., Loubière, K., Guigui, C., Sperandio, M., Barna, L., Paul, E., Cabassud, C., Liné, A., & Hébrard, G. (2020). Attempts, successes, and failures of distance learning in the time of COVID-19. *Journal of Chemical Education, 97*(9), 2448–2457.

Dingwall, S. (2020). Lessons learned from active engagement in a large-enrollment introductory biochemistry course during a remote quarter. *Journal of Chemical Education, 97*(9), 2749–2753.

Dragisich, V. (2020). Wellness and community modules in a graduate teaching assistant training course in the time of pandemic. *Journal of Chemical Education, 97*(9), 3341–3345.

Dunnagan, C. L., Dannenberg, D. A., Cuales, M. P., Earnest, A. D., Gurnsey, R. M., & Gallardo-Williams, M. T. (2020). Production and evaluation of a realistic immersive virtual reality organic chemistry laboratory experience: Infrared spectroscopy. *Journal of Chemical Education, 97*(1), 258–262.

Ekmekci, O., & Bergstrand, J. (2010). Agility in higher education: Planning for business continuity in the face of an H1N1 pandemic. *SAM Advanced Management Journal, 75*(4), 20–30.

Eriksen, E., Nielsen, B. E., & Pittelkow, M. (2020). Visualizing 3D molecular structures using an augmented reality app. *Journal of Chemical Education, 97*(5), 1487–1490.

Estudante, A., & Dietrich, N. (2020). Using augmented reality to stimulate students and diffuse escape game activities to larger audiences. *Journal of Chemical Education, 97*(5), 1368–1374.

Fatani, T. H. (2020). Student satisfaction with videoconferencing teaching quality during the COVID-19 pandemic. *BMC Medical Education, 20*(1), 396.

Feldstein, M. (2018). *Canvas surpasses blackboard learn in US market share*. https://eliterate.us/canvas-surpasses-blackboard-learn-in-us-market-share/.

Ferrell, J. B., Campbell, J. P., McCarthy, D. R., McKay, K. T., Hensinger, M., Srinivasan, R., Zhao, X., Wurthmann, A., Li, J., & Schneebeli, S. T. (2019). Chemical exploration with virtual reality in organic teaching laboratories. *Journal of Chemical Education, 96*(9), 1961–1966.

Flaherty, C. (2020, April 29). *Zoom boom*. Inside Higher Ed. https://www.insidehighered.com/news/2020/04/29/synchronous-instruction-hot-right-now-it-sustainable

Flener-Lovitt, C., Bailey, K., & Han, R. (2020). Using structured teams to develop social presence in asynchronous chemistry courses. *Journal of Chemical Education, 97*(9), 2519–2525.

Fox, R. (2004, December 5–8). *SARS epidemic: Teachers' experiences using ICTs* [paper presentation]. In Beyond the comfort zone: Proceedings of the 21st ASCILITE Conference, (pp. 319–327).

Fung, F. M., Choo, W. Y., Ardisara, A., Zimmermann, C. D., Watts, S., Koscielniak, T., Blanc, E., Coumoul, X., & Dumke, R. (2019). Applying a virtual reality platform in environmental chemistry education to conduct a field trip to an overseas site. *Journal of Chemical Education, 96*(2), 382–386.

Glass, G. (2020, October 8). 6 ways in which an LMS is central to an effective e-learning strategy. *NEO BLOG*. https://blog.neolms.com/6-ways-in-which-an-lms-is-central-to-an-effective-e-learning-strategy/

Goh, P.-S., & Sandars, J. (2020). A vision of the use of technology in medical education after the COVID-19 pandemic. *MedEdPublish, 9*(10), 49.

Goodman, A. L. (2020). Can group oral exams and team assignments help create a supportive student community in a biochemistry course for nonmajors? *Journal of Chemical Education, 97*(9), 3441–3345.

Gray, S. D. (2020). Embedded video questions as a low-stakes assignment during the remote learning transition. *Journal of Chemical Education, 97*(9), 3172–3175.

Guo, S. (2020). Synchronous versus asynchronous online teaching of physics during the COVID-19 pandemic. *Physics Education, 55*(6), 065007.

Guri-Rosenblit, S. (2006). Eight paradoxes in the implementation process of e-learning in higher education. *Distances et saviors, 2*(4), 155–179.

Halpin, P., & Crowther, G. (2021). Tunes in the Zoom room: Remote learning via videoconference discussions of physiology songs. *Journal of Microbiology & Biology Education*, in press.

Hannula, L. (2020, April 22). *Weak internet cripples productivity for 1/3 of work-from-home employees*. Whistle Out. https://www.whistleout.com/Internet/Guides/weak-internet-a-problem.

Hatziapostolou, T., & Paraskakis, I. (2010). Enhancing the impact of formative feedback on student learning through an online feedback system. *Electronic Journal of e-Learning, 8*(2), 111–122.

Heim, A.B. & Holt, E. A. (2021). From bored games to board games: Student-driven game Design in the Virtual Classroom. *Journal of Microbiology & Biology Education*, in press.

Herzog, J. A., & Mawn, M. (2020, May 27). Teaching lab-based courses online & remote: From 'are you kidding me?' to 'this is effective!' ASM.org, 2020. *American Society for Biology*. https://asm.org/Articles/2020/May/Teaching-Lab-Based-Courses-Online-Remote-From-Are

Hewitt, N. (2017, July 25). *Immersive tools for field-based geoscience education in large undergraduate courses: Pacific Spirit Forest augmented reality field trip and alpine ecosystem virtual reality tours*. The University of British Columbia. https://tlef.ubc.ca/view/tlef-proposals/entry/364/?pagenum=2

Hill, P. (2020, March 31). *Revised outlook for higher ed's online response to COVID-19*. PhilOnEdTech. https://philonedtech.com/revised-outlook-for-higher-eds-online-response-to-covid-19/

Hodges, C., Moore, S., Lockee, B., Trust, T., & Bond, A. (2020, March 27). *The difference between emergency remote teaching and online learning*. EDUCAUSE Review.

Holme, T. A. (2020). Introduction to the journal of chemical education special issue on insights gained while teaching chemistry in the time of COVID-19. *Journal of Chemical Education, 97*(9), 2375–2377.

Holt, E. A., Heim, A. B., Tessens, E., & Walker, R. (2020). Thanks for inviting me to the party: Virtual poster sessions as a way to connect in a time of disconnection. *Ecology and Evolution, 10*(22), 12423–12430.

Holton, A. J. (2020). Implementation of an emergency multisection online general chemistry curriculum in response to COVID-19 pandemic. *Journal of Chemical Education, 97*(9), 2878–2883.

Holzapfel, B. (2020, June 15). *What educators have learned from remote learning prepares them for the new school year*. Microsoft Blog. https://educationblog.microsoftcom/en-us/2020/06/three-months-later-what-educators-have-learned-from-remote-learning-prepares-them-for-the-new-school-year/.

Hrastinski, S. (2008). A study of asynchronous and synchronous e-learning methods discovered that each supports different purposes. *Educause Quarterly, 31*(4), 51–55.

Hsu, Y.-C., & Ching, Y.-H. (2011). Microblogging for strengthening a virtual learning community in an online course. *Knowledge Management & E-Learning: An International Journal, 3*(4), 585–598.

Huang, X., & Hsiao, E. L. (2012). Synchronous and asynchronous communication in an online environment: Faculty experiences and perceptions. *Quarterly Review of Distance Education, 13*(1), 15–30.

Hwang, C. S. (2020). Using continuous student feedback to course-correct during COVID-19 for a nonmajors chemistry course. *Journal of Chemical Education, 97*(9), 3400–3405.

Jaggars, S. S. (2014). Choosing between online and face-to-face courses: Community college student voices. *American Journal of Distance Education, 28*(1), 27–38.

Jamieson, M. V. (2020). Keeping a learning community and academic integrity intact after a mid-term shift to online learning in chemical engineering design during the COVID-19 pandemic. *Journal of Chemical Education, 97*(9), 2768–2772.

Jara, M., & Mellar, H. (2010). Quality enhancement for e-learning courses: The role of student feedback. *Computers & Education, 54*(3), 709–714.

Juárez Santiago, B., Olivares Ramírez, J. M., Rodríguez-Reséndiz, J., Dector, A., García García, R., González-Durán, J. E. E., & Ferriol Sánchez, F. (2020). Learning management system-based evaluation to determine academic efficiency performance. *Sustainability, 12*(10), 4256.

Khan, A., & Somdatta, P. (2020). Bloom of online teaching in the era of Covid 19 pandemic: Boon or bane for medical education. *National Journal of Medical and Allied Sciences, 9*(2), 1–5.

Khan, R. A., & Jawaid, M. (2020). Technology enhanced assessment (TEA) in COVID 19 pandemic. *Pakistan Journal of Medical Sciences, 36*(COVID19-S4), COVID19-S108-S110.

Kibble, J. (2007). Use of unsupervised online quizzes as formative assessment in a medical physiology course: Effects of incentives on student participation and performance. *Advances in Physiology Education, 31*(3), 253–260.

Krasne, S., Wimmers, P. F., Relan, A., & Drake, T. A. (2006). Differential effects of two types of formative assessment in predicting performance of first-year medical students. *Advances in Health Sciences Education, 11*, 155–171.

Kroese, J., & Leiden University Medical Centre [LUMC]. (2019, October 30). Seeing clearly: How augmented reality can help medical students understand complex transplants. *Centre for Innovation*. https://www.centre4innovation.org/stories/augmented-reality-app-leiden-medical-students-transplants/

LaMack, J. A., dos Santos, I., & Imas, O. (2020). Maintaining a rewarding biomedical engineering capstone experience for students in remote learning situations. *Biomedical Engineering Education*. https://doi.org/10.1007/s43683-020-00019-8

Lavolette, E., Gose, E., Huang, E., & Venable, M. (2010). Comparing synchronous virtual classrooms: Student, instructor, and course designer perspectives. *TechTrends, 54*(5), 54–61.

Lederman, D. (2020, March 25). *The shift to remote learning: The human element*. Inside Higher Ed. https://www.insidehighered.com/digital-learning/article/2020/03/25/how-shift-remote-learning-might-affect-students-instructors-and

Lynch, M. (2018, April 18). What is the next generation of digital learning environments? *The Tech Edvocate*. https://www.thetechedvocate.org/what-is-the-next-generation-of-digital-learning-environments/#:~:text=EDUCAUSE%20explains%2C%20%E2%80%9CThe%20next%20generation,focused%20on%20learning%20rather%20than

Mac Keogh, K. (2001). National strategies for the promotion of on-line learning in higher education. *European Journal of Education, 36*(2), 223–236.

Majka, E. A., Guenther, M. F., & Raimondi, S. L. (2021). Science bootcamp goes virtual: A compressed, interdisciplinary online CURE promotes psychosocial gains in STEM transfer students. *Journal of Microbiology and Biology Education, 22*(1), 22.1.24.

Martin, F., Parker, M. A., & Deale, D. F. (2012). Examining interactivity in synchronous virtual classrooms. *The International Review of Research in Open and Distributed Learning, 13*(3), 228–261.

McNaught, C. (2004). Using narrative to understand the convergence of distance and campus-based learning during the time of SARS in Hong Kong. *Educational Media International, 41*(3), 183–193.

McPhee, S., & Lyon, K. A. (2020). Student voices on remote education in the COVID-19 era: Recommendations for fall based on student self-reported data. *UBC Remote Pedagogy Report, 2020*, 1–16.

Meixner, T., Ciancarelli, B., Farrell, E. P., Silva, G. D., Josek, T., Meister, P., Soule, D. C., & Darner, R. (2021). *Teaching data literacy online using Project EDDIE's climate change module*. AGU Fall Meeting.

Menard, J. (2020, June 21). *Video conferencing systems used in higher education*. LISTedTECH. https://www.listedtech.com/blog/video-conferencing-systems-used-in-higher-education.

Michael, K. (2012). Virtual classroom: Reflections of online learning. *Campus-Wide Information Systems, 29*(3), 156–165.

Microsoft. (2020, April 29). *Earnings Release FY20 Q3*. [Press release]. https://www.microsoft.com/en-us/Investor/earnings/FY-2020-Q3/press-release-webcast

Moore, M. G. (1993). Theory of transactional distance. In D. Keegan (Ed.), *Theoretical principles of distance education* (pp. 22–38). Routledge.

Nesenbergs, K., Abolins, V., Ormanis, J., & Mednis, A. (2021). Use of augmented and virtual reality in remote higher education: A systematic umbrella review. *Education Sciences, 11*(1), 8.

Nguyen, J. G., Keuseman, K. J., & Humston, J. J. (2020). Minimize online cheating for online assessments during COVID-19 pandemic. *Journal of Chemical Education, 97*(9), 3429–3435.

Njenga, J. K., & Fourie, L. C. H. (2010). The myths about e-learning in higher education. *British Journal of Educational Technology, 41*(2), 199–212.

O'Brien, D. J. (2020). Feynman, Lewin, and Einstein download Zoom: A guide for incorporating e-teaching of physics in a post-COVID world. arXiv, arXiv:2008.07441.

Ohland, M. W., Loughry, M. L., Woehr, D. J., Bullard, L. G., Felder, R. M., Finelli, C. J., Layton, R. A., Pomeranz, H. R., & Schmucker, D. G. (2012). The comprehensive assessment of team member effectiveness: Development of a behaviorally anchored rating scale for self- and peer evaluation. *Academy of Management Learning & Education, 11*(4), 609–630.

Organization for Economic Co-operation and Development [OECD]. (2020, March 23). *Education responses to COVID-19: Embracing digital learning and online collaboration*. OECD. https://www.oecd.org/coronavirus/policy-responses/education-responses-to-covid-19-embracing-digital-learning-and-online-collaboration-d75eb0e8/

Pennisi, E. (2020, July 15). During the pandemic, students do field and lab work without leaving home. *Science*. https://www.sciencemag.org/news/2020/07/during-pandemic-students-do-field-and-lab-work-without-leaving-home

Perets, E. A., Chabeda, D., Gong, A. Z., Huang, X., Fung, T. S., Ng, K. Y., Bathgate, M., & Yan, E. C. Y. (2020). Impact of the emergency transition to remote teaching on student engagement in a non-STEM undergraduate chemistry course in the time of COVID-19. *Journal of Chemical Education, 97*(9), 2439–2447.

Petillion, R. J., & McNeil, W. S. (2020). Student experiences of emergency remote teaching: Impacts of instructor practice on student learning, engagement, and Well-being. *Journal of Chemical Education, 97*(9), 2486–2493.

Pew Research Center. (2020). *Internet/broadband fact sheet* [fact sheet]. Pew Research Center Internet & Technology. https://www.pewresearch.org/internet/fact-sheet/internet-broadband/.

Radianti, J., Majchrzak, T. A., Fromm, J., & Wohlgenannt, I. (2020). A systematic review of immersive virtual reality applications for higher education: Design elements, lessons learned, and research agenda. *Computers & Education, 147*, 103778.

Ramachandran, R., & Rodriguez, M. C. (2020). Student perspectives on remote learning in a large organic chemistry lecture course. *Journal of Chemical Education, 97*(9), 2565–2572.

Ramo, N. L., Lin, M., Hald, E. S., & Huang-Saad, A. (2021). Synchronous vs. asynchronous vs. blended remote delivery of introduction to biomechanics course. *Biomedical Engineering Education, 1*, 61–66.

Redden, E. (2020, March 26). 'Zoombombing' attacks disrupt classes. *Inside Higher Ed.* https://www.insidehighered.com/news/2020/03/26/zoombombers-disrupt-online-classes-racist-pornographic-content

Remtulla, R. (2020). The present and future applications of technology in adapting medical education amidst the COVID-19 pandemic. *JMIR Medical Education, 6*(2), e20190.

Rholl, D. A. (2021). Complex Systems in a Complex Time: Two flexible activities using online models to study infections disease transmission and code complex systems simulations. *Journal of Microbiology Education, 22*(1), 22.1.26. https://doi.org/10.1128/jmbe.v22i1.2397

Robson, K., Plangger, K., Kietzmann, J., McCarthy, I., & Pitt, L. (2015). Is it all a game? Understanding the principles of gamification. *Business Horizons, 58*(4), 411–420.

Rudd, J., & Rudd, J. (2014). The value of video in online instruction. *Journal of Instructional Pedagogies, 13*(1), 1–7.

Rupnow, R. L., LaDue, N. D., James, N. M., & Bergan-Roller, H. E. (2020). A perturbed system: How tenured faculty responded to the COVID-19 shift to remote instruction. *Journal of Chemical Education, 97*(9), 2397–2407.

Saidy, C., & Sura, T. (2020). When everything changes over night: What we learned from teaching the writing practicum in the era of Covid-19. *Teaching/Writing: The Journal of Writing Teacher Education, 9*(1), 21.

Salar, R., Arici, F., Caliklar, S., & Yilmaz, R. M. (2020). A model for augmented reality immersion experiences of university students studying in science education. *Journal of Science Education and Technology, 29*(2), 257–271.

Samadbeik, M., Yaaghobi, D., Bastani, P., Abhari, S., Rezaee, R. A., & Garavand, A. (2018). The applications of virtual reality technology in medical groups teaching. *Journal of Advances in Medical Education & Professionalism, 6*(3), 123–129.

Schipper, D., & LUMC. (2019, December 4). Time to act - emergency care: Practicing the ABCDE approach in virtual reality. *Centre for Innovation.* https://www.centre4innovation.org/stories/time-to-act-emergency-care-practising-the-abcde-approach-in-virtual-reality/

Schoper, S. E., & Hill, A. R. (2017). Using Facebook to promote a virtual learning community: A case study. *Journal of Learning Spaces, 6*(1), 34–39.

SchWeber, C. (2008). Determined to learn: Accessing education despite life-threatening disasters. *Journal of Asynchronous Learning Networks, 12*, 1–7.

Selwyn, N. (2019). *Should robots replace teachers?* Polity Press.

Selwyn, N., & Jandrić, P. (2020). Postdigital living in the age of Covid-19: Unsettling what we see as possible. *Postdigital Science and Education, 2*(3), 989–1005.

Sevilla, G. (2020). *Zoom vs. Microsoft teams vs. Google meet: A videoconferencing face off.* PCMag UK. https://uk.pcmag.com/how-to-work-from-home/125661/zoom-vs-microsoft-teams-vs-google-meet-a-videoconferencing-face-off

Skylar, A. A. (2009). A comparison of asynchronous online text-based lectures and synchronous interactive web conferencing lectures. *Issues in Teacher Education, 18*(2), 69–83.

Stowe, R. L., Esselman, B. J., Ralph, V. R., Ellison, A. J., Martell, J. D., DeGlopper, K. S., & Schwarz, C. E. (2020). Impact of maintaining assessment emphasis on three-dimensional learning as organic chemistry moved online. *Journal of Chemical Education, 97*(9), 2408–2420.

Sung, R.-J., Wilson, A. T., Lo, S. M., Crowl, L. M., Nardi, J., St. Clair, K., & Liu, J. M. (2020). BiochemAR: An augmented reality educational tool for teaching macromolecular structure and function. *Journal of Chemical Education, 97*(1), 147–153.

Swauger, S. (2020). Our bodies encoded: Algorithmic test proctoring in higher education. In J. Stommel, C. Friend, & S. M. Morris (Eds.), *Critical digital pedagogy (chapter 6)*. Hybrid Pedagogy Inc.

Tang, Y. M., Au, K. M., Lau, H. C. W., Ho, G. T. S., & Wu, C. H. (2020). Evaluating the effectiveness of learning design with mixed reality (MR) in higher education. *Virtual Reality, 24*, 797–807.

Tonsmann, G. (2014). A study of the effectiveness of blackboard collaborate for conducting synchronous courses at multiple locations. *InSight: A Journal of Scholarly Teaching, 9*, 54–63.

Toquero, C. M. (2020). Challenges and opportunities for higher education amid the COVID-19 pandemic: The Philippine context. *Pedagogical Research, 5*(4), em0063.

Tunceren, L., Kaur, A., Mullins, C., & Slimp, M. (2015). Technology review: Personalizing the online enterprise of college learning through synchronous activity. *The Community College Enterprise, 21*(2), 103–106.

Unify Square. (2020). *Microsoft teams vs Zoom: How to determine the right fit*. Unify Square. Available at https://www.unifysquare.com/blog/microsoft-teams-vs-zoom/. Last Accessed 12 Feb 2021.

United Nations [UN]. (2020). *Policy brief: Education during COVID-19 and beyond*. United Nations. https://unsdg.un.org/resources/policy-brief-education-during-covid-19-and-beyond

United Nations Educational, Scientific, and Cultural Organization [UNESCO]. (2020). *Distance learning solutions [fact sheet]*. UNESCO. https://en.unesco.org/covid19/educationresponse/solutions

Urness, S. (2020, March 13). *VCSU response to COVID-19 virus*. News Dakota. https://www.newsdakota.com/2020/03/13/vcsu-response-to-covid-19-virus/

Vielma, K., & Brey, E. M. (2020). Using evaluative data to assess virtual learning experiences for students during COVID-19. *Biomedical Engineering Education, 1*, 139–144.

Wadhwa, N., Khatak, S., & Poonam. (2020). Online versus offline mode of education – is India ready to meet the challenges of online education in lockdown? *Journal of the Social Sciences, 48*(3), 405–413.

Walsh, L. L., Arango-Caro, S., Wester, E. R., & Callis-Duehl, K. (2021a). Training faculty as an institutional response to COVID-19 emergency remote teaching supported by data. *CBE–Life Sciences Education, 20*, Article 3. www.lifescied.org/doi/10.1187/cbe.20-12-0277

Walsh, L. L., Lichti, D., Wester, E. R., Zambrano, C., Sodhi, J. S., Moon, S., & Callis-Duehl, K. (2021b). Why and how students cheat more frequently online: Perspectives during COVID-19 pandemic. *International Journal for Educational Integrity, 17*(1), Article 23. https://doi.org/10.1007/s40979-021-00089-3

Walter, E., Bailey, M., Fernandez, P., & Gill, A. (2021). Suddenly online: Exploring postsecondary teaching, attitudes, technology, and faculty mental wellbeing in Spring 2020. *National Association for Research in Science Teaching Meeting, Virtual*. [Conference presentation abstract]

Wartman, S. A., & Combs, C. D. (2019). Reimagining medical education in the age of AI. *AMA Journal of Ethics, 21*(2), 146–152.

Watson, S. (2003). Closing the feedback loop: Ensuring effective action from student feedback. *Tertiary Education and Management, 9*(2), 145–157.

Wester, E. R., Walsh, L. L., Arango-Caro, S., & Callis-Duehl, K. (2021). Student engagement declines in STEM undergraduates during COVID-19 driven remote learning. *Journal of Microbiology & Biology Education, 22*(1), 2385.

Wilson, A., Watson, C., Thompson, T. L., Drew, V., & Doyle, S. (2017). Learning analytics: Challenges and limitations. *Teaching in Higher Education, 22*(8), 991–1007. https://doi.org/10.1080/13562517.2017.1332026

Williams, J. B., & Wong, A. (2009). The efficacy of final examinations: A comparative study of closed-book, invigilated exams and open-book, open-web exams. *British Journal of Educational Technology, 40*(2), 227–236.

Yamagata-Lynch, L. C. (2014). Blending online asynchronous and synchronous learning. *The International Review of Research in Open and Distance Learning, 15*(2), 189–212.

Youle, R. (2020, June 15). The technology which is allowing hospital consultants to do ward rounds from home. *WalesOnline*. https://www.walesonline.co.uk/news/health/hospitals-swansea-bay-coronavirus-news-18421879

Young, J. R. (2009, April 17). In case of emergency, break tradition—Teach online. *The Chronicle of Higher Education*. https://www.chronicle.com/article/in-case-of-emergency-break-tradition-teach-online/

Zhai, X., Li, M., & Chen, S. (2019). Examining the uses of student-led, teacher-led, and collaborative functions of mobile technology and their impacts on physics achievement and interest. *Journal of Science Education and Technology, 28*, 310–320.

Zhai, X., Zhang, M., & Zhang, X. (2018). Understanding the relationship between levels of mobile technology use in high school physics classrooms and the learning outcome. *British Journal of Educational Technology, 50*(2), 750–766.

Zhang, D., Zhao, J. L., Zhou, L., & Nunamaker, J. F. (2004). Can e-learning replace classroom learning? *Communications of the ACM, 47*(5), 75–79.

Zoom. (2020). Privacy & Security for Zoom video communications [fact sheet]. *Zoomorphology*. https://zoom.us/docs/en-us/privacy-and-security.html

Zweifach, S. M., & Triola, M. M. (2019). Extended reality in medical education: Driving adoption through provider-centered design. *Digital Biomarkers, 3*(1), 14–21.

Part IV
The Future and Research

Chapter 16
The Unpredictable Future of High-Fidelity Patient Simulation in Biomedical Science Education: The Price Must Be Right

David M. Harris and Daniel Franceschini

Abstract Educational simulations provide a platform for producing learning experiences by simulating a model of reality or an environment. They can vary in realism or fidelity and come with different price tags. The first third of this chapter provides a summary table and then compares and contrasts five different categories of simulations in terms of cost, capabilities, and benefits. Comparatively, high-fidelity patient simulators (HFPS) are expensive in terms of monetary costs and faculty resources. The second third of this chapter provides a background with a history and critique on the state of HFPS research with biomedical students. Pioneering days suggested that students favored this tool, and short-term learning goals were achieved. Studies showed that it was feasible to use this technology with biomedical students. The next time phase of HFPS research aimed to expand this research though arguably at a much slower pace than original studies. The final third of this chapter proposes studies and research directions that can leverage benefits of HFPS while justifying resources required and high costs.

16.1 Overview of Categories of Simulations

Simulation, in general, involves an artificial representation of a real-world process to broaden knowledge and enhance experiential learning (Flanagan et al., 2004). Simulation can involve the use of simulators, which are the medium for the experience to occur. In biomedical education, there is a wide range of simulators that include software programs, task trainers, simulated patients (SPs), virtual patients, and manikins or patient simulators (Alinier, 2007). It is critical to recognize that simulators are not only physical materials but can also include activities that occur in minds of learners as they role-play or imagine a situation.

There are three broad types of simulation activities in biomedical science. The first is *lab creations* that allow users to perform actions that are as close as possible to

D. M. Harris (✉) · D. Franceschini
College of Medicine, University of Central Florida, Orlando, FL, USA
e-mail: David.Harris2@ucf.edu; DanFran@ucf.edu

H. J. Witchel, M. W. Lee (eds.), *Technologies in Biomedical and Life Sciences Education*, Methods in Physiology, https://doi.org/10.1007/978-3-030-95633-2_16

real life. One example is high-fidelity patient simulators (HFPS) or manikins that are used in physician, military, and paramedic training. In these, learners must observe and make decisions as if they were performing on a real person. Other examples of educational simulations involve biochemistry or cell biology where a student can sit at a virtual lab bench and interact in 3D and the student may move the mouse of the computer to move equipment around the lab as if they were there. An example is a virtual laboratory where students were shown to streak bacteria onto an agar place, move them to an incubator, and then culture bacteria colonies (Makransky et al., 2016). Lab recreations are mostly used when the emphasis is on technique such as students using simulations as a pre-lab to come prepared for the wet lab. Another example includes a simulator in which learners use a realistic appearing arm to practice intravenous catheter insertion before moving onto real patients. Interestingly, learning techniques via simulation may not translate into increasing student confidence (Reinhardt et al., 2012).

The second group of educational simulations includes *conceptual experiences* that allow learners to view or do things that are not possible. These include mathematical modeling that is not possible in real life or visualizing evolution that takes place across thousands of years. Additional simulations in the group may include depicting how molecules interact, which is not visible. An example may include having students manipulate a DNA sequence with a chosen primer and adjusting the temperature (SimBio). This places the emphasis on the difficult conceptual aspect of DNA sequencing with something that is not truly visible to a learner. The third group is *model construction*, which has students use modeling to construct their own simulations. The idea is that when students construct the mathematics or rules underlying a simulation, they can better understand assumptions and results that emerge.

These three different groups of simulations have different strengths and weaknesses which we will aim to elucidate in this chapter. Conceptual experiences are best used when there is a solid knowledge base established. Model construction is often confusing for lower-level students, and many students have difficulty relating rules or mathematics they put into models to real life.

Another feature of simulators is how real or authentic the simulation looks and feels for the user. The term fidelity has been used in simulation and is defined as "the degree to which a particular simulator or manikin represents a live patient or scenario" (Costluc, 2021b): the more similar to a live patient or event, the higher the fidelity. Typically, there are three types of fidelity in simulation-based education (SBE): low, mid, and high. Table 16.1 illustrates different types of simulators ranging from low to high fidelity and pros and cons associated with each one. Table 16.1 also includes some of the ideal type of learning associated with each simulator as well as costs.

Table 16.1 A comparison of pros and cons with different types of simulators. Lastly, examples or each are provided

Simulator	Pro(s)	Con(s)	Example(s)
Software/computer-based programs	1. Low cost 2. Good for knowledge acquisition 3. Individual pace of learning 4. Widely accessible 5. Good for entry-level or lower-level students	1. Low fidelity 2. Difficult for experiential learning	1. Cardiac responses to drugs (Langendorff setup) 2. Ion channel in nerves
Virtual patients	1. Individual pace of learning 2. Low cost 3. Lower-level students	1. Limited experiential learning 2. Lack of physical interactivity	1. Avatars 2. Clinical cases and interactive patient scenarios
Task trainer	1. Good for procedural tasks 2. Low cost	1. Low fidelity 2. Lack of emotional responses	1. Heart sound simulator 2. Intravenous catheter insertion
Standardized patients	1. Good for communication and interpersonal skills 2. Provide emotional and realistic responses	1. Hard to mimic pathology 2. Mid fidelity	1. Actors
High-fidelity patient simulators	1. High fidelity 2. Higher-level skills such as decision-making and critical thinking 3. Emotional and sensory components of learning	1. High cost 2. Difficult access 3. Lack of human interaction 4. High fidelity	1. Manikins

16.1.1 *Software and Computer-Based Programs*

Software and computer-based programs tend to be lower in cost than other simulators in biomedical education making it an attractive platform for knowledge acquisition and low-level learning. These include virtual labs such as those offered by SimBio which is a mixture of tutorial style as well as workbook style lab recreations. A typical physiology simulator computer program may be a Langendorff setup where a series of drugs can be added to an isolated heart. The simulator can illustrate changes in heart rate and contractility when different drugs are added or when preload is altered. Another simulator lets a learner control the amplitude and the number of action potentials to visualize changes in skeletal muscle force (Physiology; Tchaptchet et al., 2013; Braun & Tchaptchet, 2020). These scenarios provide the learner the ability to control the situation, work at their own pace, and repeat, as necessary. We will consider software and computer-based simulations in greater detail later in this chapter.

16.1.2 Virtual Patients

Virtual patients are a specific type of computer program that simulates real-life clinical scenarios that includes history taking and physical examinations, which can help train students to make clinical decisions. What separates these from abovementioned software programs is that these scenarios can unfold as students ask questions or obtain information (Bateman et al., 2012). Similar to the aforementioned software simulations, these are relatively low cost and allow for paced learning. Virtual patients are low fidelity, and while many schools use virtual patients for case-based learning, there is still a lack of human interactivity.

16.1.3 Task Trainers

Task trainers are designed and extremely useful for learning a specific procedure or for meeting a skill competency. Harvey, a cardiac simulator, is one of the most well-known, and studies show that students who train on these simulators can detect simulated heart sounds as well as those on real patients better than untrained students (Butter et al., 2010). Other task trainers that are commonly used include intravenous (IV) catheter placement, lumbar puncture, and airway management. The fundamental learning theory behind the success of task trainers is that when the learner deliberately practices and repeats the procedure, it inculcates safe procedure as well as building confidence. Because task trainers are geared toward a specific procedure and are often limited to a body section, there is no emotional response with these tools, and they are lower in fidelity than other simulators.

16.1.4 Standardized Patients

Standardized patients (SPs) are healthy individuals or actors who portray a patient with a clinical problem or situation to assess or teach clinical skills for healthcare profession learners. The main advantage of SPs is the authentic humanness, so that communication or interpersonal skills can be learned. Furthermore, there can be emotions portrayed in a scenario to introduce learners to them in a safe environment. The disadvantage of SPs is that pathologies cannot be assessed. For example, a healthy SP will not be able to portray a fever, abnormal heart sound, or wheezing in the lung.

16.1.5 High-Fidelity Patient Simulators

High-fidelity patient simulators (HFPS) represent the most technologically advanced and feature-rich simulators currently available in the healthcare simulation industry. These HFPS manikins have features and functionality that add additional realism to any given scenario and enable a deeper level of learner immersion than previously mentioned categories. These sophisticated simulators can replicate real clinical encounters by enabling the operator either to program a wide variety of physiological parameters into the software in advance of the scenario or to run the manikin "on the fly." For programmed scenarios, the manikin can be "taught" to recognize certain medications, treatments, or physical actions and respond to those interventions automatically. For example, a manikin programmed with an anatomical airway obstruction (the tongue blocking the airway) can automatically recognize the head-tilt-chin-lift maneuver being done by the learner and then respond accordingly with a return of spontaneous respirations. Another example would be programming that same manikin to properly respond to medication administration. Here, when a syringe with a pre-programmed radio frequency identification (RFID) barcode is placed within the medication administration area on the manikin's arm (the inside of the elbow where an IV would typically be placed), the manikin's technology can "read" that label, recognize that the label is programmed to represent a certain medication, and then respond accordingly by making necessary physiological adjustments. With this type of technologically advanced functionality built-in, HFS can be very costly and require advanced training to both operate and maintain.

16.1.6 Cost to Benefit Ratio for Simulators

A 2 × 2 model depicted in Table 16.2 can be utilized to help to think about where simulation falls in relation to cost and benefit. If something can generate high benefit at minimal cost, it is optimal. However, it is important to note that to get some high benefits, costs may have to be elevated as well. These increased costs are not necessarily a disadvantage if benefits are great enough. Let us look at some examples.

One of the areas where simulators have had the greatest impact is the aviation industry. There are all types of trainers focused on stick and rudder control (Ortiz, 1994; Stark, 1989), instrument training, and crew resource management (Weiner

Table 16.2 Combination of benefit and cost

	High cost	Low cost
High benefit	Procedural safety: High-fidelity patient simulation	Clinical problem-solving: Virtual patient
Low benefit	Inspiration and public engagement One-off simulator sessions	Conceptual familiarity: Software simulations

et al., 1993). The benefit of simulators is apparent to consumers as the likelihood of dying in a commercial airplane crash is 1 in 19.8 million (planecrashinfo.com). From a return on investment perspective, the benefit is high, which justifies the high cost; Federal Aviation Administration (FAA)-qualified simulators cost up to $10 million (Parsons, 2019), and the global simulator market revenue is expected to be 7.7 billion dollars by 2025 (Parsons, 2019). Prices of simulator training are high enough to incentivize airline manufacturers to try to develop new planes for which simulator training is not necessary. In an effort to avoid such stimulator training costs for airlines (purchasers of new aircraft), Boeing's 737 Max was designed to appear similar to their older models, and they promised airlines and regulators that pilots would not need simulation training (Krietoff & David, 2020). Unfortunate consequences of these cost savings were evident when planes from Indonesia and Ethiopia crashed due to pilots not knowing how to override an automation system failure; the fleet of these jets was grounded for over 18 months, resulting in about $20 billion in direct costs to Boeing (Isidore, 2020). Obviously, finding the balance of identifying the right outcome or benefit and supporting or justifying it with the right cost, which can sometimes be high, is paramount when viewed as a return on investment.

Educators can take a similar cost-benefit approach to educational interventions, especially with simulators. The return or benefits for simulators can be defined as educational outcomes. For example, from Table 16.1, for computer-based simulators, the educational outcome could be low-level learning for the user. The investment or cost would be the price of software and student time on task. If the learning was focused on fundamental knowledge that a student must know before continuing, it can be considered high benefit. An example is that a student needs to understand anatomical relationships of the heart and lungs before explaining pressure differences of the left and right ventricle. If the learner were guided to a free or low-cost simulator online, this would obviously fall in the high-benefit, low-cost square. Even if low-level learning is obtained at a low cost, it may be considered advantageous.

It is well known that high-fidelity patient simulators come with a significant upfront cost that ranges from $10,000 to $250,000 depending on the brand and features (Costluc, 2021a). At the University of Central Florida, we pay an additional $6500 per year or so on each HFPS for maintenance agreements. Therefore, to balance this high cost, there needs to be a very high benefit to justify this intervention.

Evidence-based medical education guides have helped establish best practices for the use of simulation in medical education. Issenberg et al. concluded in a 2005 Best Evidence in Medical Education (BEME) systemic review that HFPS *successfully* enhanced medical education if provided in right conditions (Issenberg et al., 2005). Benefits of HFPS were variable in papers reviewed; the two most beneficial practices of HFPS education were the educational feedback and repetitive practice. Integration of HFPS into the curriculum should enhance learning, as opposed to simply adding on another technology. Such curricular innovations require strategically finding which skills can benefit (Motola et al., 2013).

Another HFPS feature that seems beneficial to learning is the ability to have an intense, realistic scenario, but within a controlled environment. This allows learners to make, analyze, and correct errors without endangering patients (Issenberg et al., 2005).

Another potential benefit of using HFPS in education is that it provides one of the remaining avenues where students may be able to "see" physiology. Many physiology labs where animals or fellow students were used to collect and analyze data have been removed due to time and ethical reasons. HFPS provide an environment where students can visualize physiology in a patient and then make hypotheses based on assessing the patient. Assessment of the patient can integrate many skills such as heart and lung auscultation and other physical examination skills. Other skills such as team training may also provide a significant benefit that cannot be assessed in other places in the curriculum (McGaghie et al., 2010; Salas et al., 2005).

Admittedly, there are high costs associated with HFPS, and there appear to be educational gains and benefits that can be achieved. The BEME reviews associated with simulation-based education question the rigor and the quality of extant data on HFPS in medical education (Issenberg et al., 2005). The trend in biomedical education appears to use HFPS earlier and earlier in the curriculum. While I (DH) have made previous claims about benefits of HFPS in pre-clerkship training in medical education, even our team has insufficient evidence of its long-term career benefits (Harris et al., 2014; Nguyen et al., 2017). This chapter will aim to provide background on the utilization of HFPS in biomedical education, where the research stands and where it needs to go to justify the high cost.

16.2 Research on High-Fidelity Patient Simulators

16.2.1 Pioneering Work

The first computerized manikin, Sim One, was developed in the 1960s, but the high cost prohibited widespread adoption. Harvey was developed in 1968 and was a much more affordable simulator, which helped to train students to decipher heart and lung sounds. This remains a popular manikin for the distinct goal of training students to listen to these sounds. The first high-fidelity patient simulators that were highly computerized surfaced in the mid-1990s from companies such as Laerdal and METI. This led to early adopters who raced to "buy" into the stock market for the pre-clerkship medical student.

Euliano, at the University of Florida, was the first person to our knowledge to introduce the METI HFPS for first-year medical students. Euliano suspected that the HFPS could be introduced to replace physiology experiments that were previously done with animals to cover topics such as pulmonary function tests and blood gas analysis. Euliano designed workshops where students were expected to check the airway, breathing, and circulation (the ABCs) followed by a physical examination in the initial studies. Students developed differential diagnoses and performed an intervention to monitor changes in vital signs (Euliano, 2000). The intervention's focus was on respiratory physiology to cover typical topics such as hypoventilation, pneumothorax, and dead space. Over 90 % of the students rated these exercises as very good or excellent. In 2001, Euliano extended these studies to include cardiovascular physiology focusing on the Valsalva maneuver and Starling's law (Euliano, 2001). Besides being well liked, she also found that students had a significant increase in confidence and in most quiz question performance after working with the HFPS. It is natural that the initial studies with HFPS focused on student perceptions as these are relatively easy to obtain. Euliano provided the first data to suggest that these tools could be useful for learning, although no control group was used. However, Euliano was indeed one of the pioneers regarding HFPS and its utility to teach physiology.

Another pioneer for advancing studies of HFPS into the pre-clerkship years is James Gordon. Harvard Medical School began to introduce HFPS to pre-clerkship students by simulating an asthmatic patient. Second-year medical students were led by the anesthesia team to direct the treatment and pharmacology to treat this patient. The HFPS allowed clinical cases, often used in problem-based learning (PBL) to "come to life" (Gordon et al., 2004). As predicted, students were very enthusiastic about this technology and the use of the HFPS. In 2006, Gordon provided the first evidence that there may be a performance improvement in students that participated in HFPS sessions. Importantly, the simulation session in this study was an add-on. While both cohorts of students received discussion on the topic of myocardial infarction, the intervention group used the HFPS to "bring the case to life" (Gordon et al., 2006). This improvement was maintained 1 year out from the study. In a 2010 article in *Academic Medicine*, Gordon noted the paucity of studies reporting the use of HFPS to teach basic medical sciences to early medical student learners and to advance the Flexner ideal of integration of clinical and basic science (Gordon et al., 2010).

Another early pioneer in studying the use of HFPS in the pre-clerkship medical student was Michael T. Fitch. His early work aimed to determine whether HFPS could be utilized in a large group setting teaching autonomic physiology and neuropharmacology. Ninety-eight percent of students rated these sessions as very good or outstanding (Fitch, 2007). This feasibility study was followed up 2 years later with data suggesting that reinforcement with HFPS improved performance on a posttest on two of four questions. On the summative exam, they performed over 80 % on the factual recall questions (Heitz et al., 2009).

16.2.2 The Next Wave of High-Fidelity Patient Simulator Research (2010s)

The pressure of institutions to utilize HFPS as much as possible to justify the high costs is real. My first educational research study (DH) was to implement HFPS into a medical physiology course with first-year medical students in Drexel's PBL curriculum. The HFPS session would utilize the manikin to "bring to life" one of the longtime paper case patients who had congestive heart failure (Harris et al., 2012). The focus of the study was a challenging topic for medical students, cardiovascular function curves. At a strategic portion in their paper case when treatment was necessary, students in their PBL groups were brought to the simulation center as they attempted to treat our patient, Mr. Patel. To come up with a treatment plan, students were asked to assess Mr. Patel and draw cardiovascular function curves in his decompensated state and use knowledge of cardiac mechanics to develop treatment strategies. A pretest/posttest model was used on their knowledge of cardiovascular function curves, and to no surprise, students performed significantly better on most questions following the simulation activity. Student perception data also indicated that students valued this experience.

At this point in time, most of the aforementioned studies focused on the feasibility of implementing HFPS into curricula, knowledge-level learning outcomes, and student perceptions of the activities. These are good basic first steps, but in essence, it took 12 years of research to find evidence for easily measurable benefits. The next set of studies focused on how HFPS could be implemented with other pedagogies such as physical examination and Web-based simulations. Swany used a crossover study design where students performed physical examinations on either each other or a SimMan. The two groups crossed over as a control. There was an increased learning benefit for students that performed the physical examination at a mid-test compared to those students who did exams on themselves (Swamy et al., 2013). After both groups did exams on SimMan, there was no difference in posttest scores suggesting that the simulator experience was responsible for the improvement. One limitation of that study was that it only included a total of 24 students though. In 2011, Cendan used a similar crossover study design but used Web-based simulations and HFPS in two groups (Cendan & Johnson, 2011). Data suggested that ideally, simulation activities would be preceded by the Web-based instruction to help form a knowledge base. This compared two cohorts of 20 students each. These studies were steps in the right direction in trying to discover how best to supplement standard, traditional pedagogies with HFPS. In 2020, Larsen introduced a crossover design comparing PBL and HFPS sessions with topics of hypertensive emergency and a chronic obstructive pulmonary disease exacerbation. They found that students who did the simulation performed better on factual knowledge than PBL counterparts (Larsen et al., 2020). Crossover designs help to strengthen studies as there was at least some sort of internal control.

Studies of the mid-2010s remain focused on student perception and knowledge acquisition largely. Some were focused on how to implement simulation into existing curricula. As stated before, one of the motivating factors for the use of HFPS with pre-clerkship medical students is to provide a clinically relevant scenario. Ginzburg designed a study where first- and second-year medical students participated in a HFPS scenario, followed by a debrief containing both a basic scientist and a clinician (Ginzburg et al., 2017). The study's findings measured subjective perceptions, and students felt it was important in providing relevance to basic science and helped to develop clinical reasoning. Another study by Alluri set out to compare lecture versus simulation in immediate learning and delayed learning. It had a relatively small sample size (20 students), where students participated in a combination of two simulation sessions and two lectures on a series of cardiovascular topics. Students took a series of a pretest, a posttest, and a delayed posttest 5 weeks later. Both groups of students improved on the posttest following both the simulation and the lecture intervention. However, only groups with simulation had an improved performance 5 weeks later on the delayed posttest (Alluri et al., 2016). Although studies were designed appropriately and provided some data, there was not much advancement from previous studies. Students perceived HFPS as useful for integration and basic knowledge acquisition could be shown to be improved after a delay, which is similar to Gordon's original study.

Next, let us zoom to 2020 in the HFPS research world and pre-clerkship medical student training. A series of papers were published that illustrated the role of HFPS in basic sciences. Petrizzo was the first to our knowledge that used the HFPS setting to illustrate immunology concepts such as hypersensitivity. The data from her study showed that students valued HFPS in their studies, at least on that subject (Petrizzo et al., 2020). Myers et al. hypothesized that adding a supplemental simulation session would improve first-year medical students' performance on pulmonary physiology written tests. This group showed a significant increase in students that were in the group that had the HFPS supplementation (Meyers et al., 2020). A survey indicated that students found it very valuable, and over 90 % of students agreed that it helped them understand pulmonary physiology.

16.2.3 High-Fidelity Patient Simulation for Undergraduate Biomedical Students

For our purposes here, undergraduate biomedical students are described as students not yet in health profession training such as medical, nursing, or other health professional education. They do not have a formal clinical component of their education. The research in the use of HFPS in this group has been extremely limited and thus very inconclusive to its effectiveness.

Dr. Richard Helyer at the University of Bristol in the UK has been a proponent for the use of HFPS to teach basic sciences. Since at least 2006, Bristol has used simulation as a core of their physiology curriculum coupled with traditional lectures, laboratories, and tutorials. Scenarios are selected to show key principles underlying homeostatic mechanisms. Senior students can also explore deeper integrated physiology scenarios, such as altitude or diving to test their knowledge. Helyer states in his 2016 paper that they feel HFPS is still underutilized and offer some possibilities of why, such as lack of accurate physiological models and insecurity of faculty of using the technology (Helyer & Dickens, 2016).

There have been some nice descriptions of some of the activities that have been designed for undergraduate physiology. Cesari et al. described the use of a HFPS to help students understand the physiology underlying CO poisoning (Cesari et al.,

2006). The strength is that this type of activity fosters critical thoughts as students created experiments independently and also troubleshot problems. This work was further described in a 2010 *Medical Education* article about implementing HFPS into premedical education (Hyatt & Hurst, 2010). They used the manikin as a laboratory tool to ask open-ended questions. They reported positive student reviews of this laboratory experience.

Earlier papers on HFPS in undergraduate students were very descriptive in nature and lacked any learning outcome data. In 2014, my colleagues and I provided the first learning outcome data for undergraduates (Harris et al., 2014). In this study, students participated and observed in a series of two HFPS scenarios (asthma and hemorrhage). To determine whether students learned from the session, a pretest/posttest design was chosen, and there was a significant increase after HFPS activities suggesting a learning effect. It is important to note that the pretest and posttest were identical, and results could be due to a "testing" effect. Also, there was no control group to determine if the intervention was the cause of the increase. These limitations highlight some of the challenges that exist in educational research. Ideally, different but equivalent tests would be used instead of identical tests. This would help elucidate learning from the activity as opposed to the testing effect. Also, it is difficult to have an ideal control group in classroom settings because accrediting bodies often mandate an equivalent experience. In 2016, Chen et al. described how they used HFPS as authentic clinical cases to help students understand pathophysiology (Chen et al., 2016). The main part of their paper focused on student perceptions of their learning, which were very favorable toward HFPS. There was an improvement noted in marks of students from cohorts that participated in HFPS compared to previous students. Although this provides some data supporting enhanced learning, there was also lack of a control or intervention group as the data was compared to historic data. Therefore, it is possible that this improvement in performance could be due to introducing active learning pedagogies that underlie the improvement and right shift in marks. This has been supported by Freeman et al. in 2014 (Freeman et al., 2014).

In 2018, Fragapane et al. performed the first known study to compare three groups of biomedical undergraduate students (Fragapane et al., 2018). The groups were a simulation group, a read-only group, and a control group that had no intervention. Tasks were related to asthma and idiopathic pulmonary fibrosis, topics that should have been new to most of the students. All groups were given a series of three quizzes: one before the simulation, one after the simulation, and one a week after. The control group had quizzes at similar time points with no reading or simulation. All groups had similar baseline scores, but only the simulation and read-only groups improved as expected on the second quiz. One week later, all three groups performed equally suggesting that there was no difference in learning between the simulation and the read-only group. Additionally, the data suggests that there is no long-term retention with either group as both were equal to the control group. Limitations were that it was a one-time event and at a level higher than their baseline knowledge, which could have an effect. However, the data does suggest that educators need to be careful of how to utilize simulations or gauge the cognitive level of simulations.

It has been about 20 years since the original Euliano paper on the use of HFPS in pre-clerkship medical students, and, honestly, there has not been significant progress in the use of HFPS in biomedical students. One conclusion that is most consistent across studies is that students perceive HFPS as valuable or perceive learning in surveys. Interestingly, research on retrieval practice and its effects on learning by Roediger and Karpicke in 2006 showed that learners were poor assessors of their own learning (Roediger & Karpicke, 2006). Learners who repeatedly studied performed worse than those who utilized retrieval practice, yet repeated study participants predicted that they would perform better. It is important to recognize that this phenomenon is not limited to only our learners. A systemic review by Davis et al. indicated that physicians are limited in their ability to self-assess as well. In fact, the worst physicians in regard to self-assessment were the ones who were overconfident and had the least skills, which represents a dangerous situation for patient safety (Davis et al., 2006). Based on these observations of learning, it is critical that future research must have outcomes other than student perception due to its inherent flawed nature.

The easiest data to obtain in studies is at the knowledge level of Bloom's taxonomy, and this usually consists of pretests and posttests. Consistently, it has been observed that student knowledge increases after exposure to HFPS. Gordon et al. have shown long-term retention in medical students over the course of the year, but there are no other studies (Gordon et al., 2006). Most studies also lack control groups or comparison with other pedagogies or lower-cost simulators. Other potential pedagogies that could be compared with HFPS include software simulations that are generally low cost. Patient videos or avatars could be used to get some humanistic element. Learning effects may be similar between some of these lower-cost pedagogies and HFPS. The price tag of HFPS in terms of financial cost and faculty and staff resources is very large. Is this small improvement or similar increase in knowledge worth the investment? Probably not, so what do we need to make it better?

16.3 The Future: Where Does High-Fidelity Patient Simulator Research Need to Go?

In order to make HFPS a better value, the cost can be decreased, or the benefit can be increased. The remainder of this chapter looks at some possibilities for the future of HFPS.

16.3.1 Lower the Cost of HFPS

Technology typically decreases in price over time. For example, one of the first cell phones that became commercially available was the Motorola DynaTAC in 1993 at a price of about $4000 (not adjusted for inflation). As phones evolved from simply being mobile to minicomputers or smart phones, prices continually dropped over time; nowadays, an analogous mobile phone would be 1/5 the price, do 5 times as much, have 5 times the battery life, and be 1/5 the weight. Unfortunately, over the years, HFPS have *defied* this trend.

As stated previously in the chapter, the price of the HFPS ranges up to $250,000 with maintenance agreements at about $6500 each per year. This makes it a challenge to many of the international biomedical faculty to consider the use of HFPS. The high cost of the manikins has fueled the motivation for manikins to be used more to justify their cost. The cost of the actual manikin will probably not go down until enough competitors come to the market to drive down prices.

Another large "cost" is the faculty and staff resources that are necessary for many of the HFPS sessions. The hourly price of some faculty and staff can easily turn a 30-min face-to-face session with students into a very pricy event. Table 16.3 shows

Table 16.3 Schematic of 4-h rotation for 120 students

		HFPS1	HFPS2	HFPS3	HFPS4
8:00-8:10	**pre-brief**	Faculty A	Faculty B	Faculty C	Faculty D
8:10-8:25	**activity**	6 students	6 students	6 students	6 students
8:25-9:05	**debrief**				
8:40-8:50	**pre-brief**	Faculty E	Faculty F	Faculty G	Faculty H
8:50-9:05	**activity**	6 students	6 students	6 students	6 students
9:05-9:45	**debrief**				
9:20-9:30	**pre-brief**	Faculty A	Faculty B	Faculty C	Faculty D
9:30-9:45	**activity**	6 students	6 students	6 students	6 students
9:45-10:25	**debrief**				
10:10-10:20	**pre-brief**	Faculty E	Faculty F	Faculty G	Faculty H
10:20-10:35	**activity**	6 students	6 students	6 students	6 students
10:35-11:15	**debrief**				
10:50-11:00	**pre-brief**	Faculty A	Faculty B	Faculty C	Faculty D
11:00-11:15	**activity**	6 students	6 students	6 students	6 students
11:15-11:55	**debrief**				

faculty resources necessary at the University of Central Florida College of Medicine to rotate our 120 medical students through a HFPS activity in a 4-h block. Students work in groups of six for these activities for a total of 20 groups. There are four HFPS (HFPS1–4) that are placed in patient exam rooms. Each HFPS is operated by a technician and is utilized five times in shown time allotments to get all students through. Because of time overlap in the 4-h time frame, we need to have at least eight facilitators, which are mostly faculty and some fourth-year medical students when available. Standardized patients, who are trained actors, are utilized to address some learning objectives such as asking for a medical interpreter. This requires a standardized patient trainer to oversee this aspect of the activity. From a space perspective, we occupy the Clinical Skills Center on this day as well as a small lecture room to hold the introduction to the case and the debriefing sections. Because it is important to train and standardize the facilitator experience, the same six faculty members have consistently contributed through the years and consist of two physiologists (associate and full professor), one immunologist (assistant professor), one simulation nurse educator, one pediatric nephrologist, and one family physician. All faculty members will be utilized for about 4-h total time. A rough estimate of this single learning event requires eight faculty members, four specialist technicians, four trained actors, and the supervisor for 4 h, so each of our 120 students receives the equivalent over 30 min of one-to-one staff contact time, which does not include preparation time, accommodation, or purchase and maintenance of manikins. To include one 30-min simulation experience to the curriculum, the cost adds up to the thousands of dollars easily, and we do not pay for the space or utilization of the equipment. We currently do this 5 times in the first year.

To reduce these costs, we would like to propose some research questions. Can students be trained to enter the HFPS room and work independently, without the need for a facilitator or faculty member? In this example, perhaps during the first one or two simulation sessions, a faculty member would be needed to orient students and help them to progress, but future HFPS could have the faculty member removed from the room as students begin to function independently. A faculty member could always be viewing from the outside; they would only intervene in extreme circumstances. Alternatively, students could be allowed to make mistakes and "fail" the simulation. Many researchers in the simulation field have opined that students should always be set up for success in the HFPS setting. However, what do you remember best as a life learner? It is your mistakes. For many times, the most challenging part of the sessions is training faculty adequately so that they will let students make a mistake and give a wrong treatment. We program our manikins to react to mistakes at UCF although we do not allow them to "die." During simulations, it is stressed that students hypothesize what they believe a treatment will result in. Often, this slows them down enough to think more thoroughly about their decision and anticipate physiological effects. Sometimes, they continue with their faulty reasoning and give an incorrect treatment, so the manikin responds in a way that reveals their error. This opens another opportunity for learning. The HFPS environment needs to be nonthreatening for students to think, consider their choices, and, frankly, make wrong decisions. So, to lower costs of HFPS activities, faculty facilitators could be removed from rooms over time, and data could be collected to test whether there was harm or detrimental effects to learners. This could be beneficial to learners as it provides a safe environment for opportunities to fail which may improve realistic lifelong learning.

16.3.2 Use HFPS to Develop Soft Skills

Every year on the first day of medical school at UCF, there is a white coat ceremony for incoming students that includes a short talk from the dean. She asks students to be the physician they would envision to be the caretaker of one of their loved ones. She then asks them to call out traits of this doctor, which she terms the "good doctor." Students typically can fill a small white board with the traits. Characteristics that include good communicator, empathetic, resilient, honest, dedicated, ethical, inquisitive, hardworking, team player, enthusiastic, creative, open-minded, reliable, knowledgeable, good listener, and responsible are just a few. While this is an example from medical students, this expectation would be consistent with any career path a biomedical student pursues. Importantly, only one of these traits is related directly to knowledge. Yet, most biomedical classroom teaching results in relaying knowledge. While the limited data states that HFPS can be used for knowledge acquisition and at the very least is equally efficient, the value of HFPS can be increased by addressing these other skills.

To our knowledge, my co-workers and I (DH) were one of the first to study how to use HFPS to develop teamwork and communication skills in first-year medical students. In 2015, we designed an experiment where first-year medical students were introduced to the TeamSTEPPS (Team Strategies and Tools to Enhance Performance and Patient Safety) principles (TeamSTEPPS). TeamSTEPPS was a curriculum developed by the Agency for Healthcare Research and Quality in response to bombshell report, *To Err Is Human*, that highlighted issues of medical error in the USA (Institute of Medicine, USA, 2000).

Each HFPS in a four-part series focused on one principle of TeamSTEPPS: leadership and team structure, mutual support, situational monitoring, and communication (Reinemann, 2015, Med Sci Educ). Each HFPS activity includes physiology learning goals, such as understanding controlled variables/effectors and pulmonary function testing, as well as "triggers" inserted to address a TeamSTEPPS principle. For example, the first HFPS was a heat exhaustion patient, but students had to recognize the janitor as part of the "team," so they had to choose to interact with this person to obtain pertinent history. Another example is at the end of the COPD HFPS, a foreign language-speaking family member entered the room as students had to monitor the situation and order a medical interpreter. A rubric was developed using TeamSTEPPS principles as observable activities, and there was a significant improvement in most groups between their first and last HFPS experience in teamwork and communication skills. This study also showed that you could use either a faculty facilitator or a student observer to run the debrief portion of the HFPS activity and get similar improvements. In fact, the observer role for students allowed them to see how was their group compared to others as far as role assignments and structures and gave them the ability to transfer or not transfer approaches from other teams. This study is important for two reasons. The first reason is that it showed that teamwork and communication skills could be improved and assessed in a HFPS for teams of medical students. Secondly, it supports the idea that some learning can be facilitated equally as well by peers, which could reduce costs of faculty needed for HFPS. Overall, this is an example of value added in learning and a potential reduction in costs.

At most medical education conferences, countless teachers state that critical thinking is lacking in students. Educators need to do more to elicit and test it, and students need to think more conceptually (Michael et al., 2017). This is a very consistent message from almost every educator across disciplines. However, upon return to the home campus, many instructors tend to fall back into the content delivery mode worrying that if students do not hear it from me, they will not hear it (Levin, 2012). If educators are spending significant time on content delivery, they are losing opportunities to develop soft skills such as critical thinking. In general, fostering critical thinking can be a challenge due to a small number of rubrics that are currently available for assessing or observing it. Interestingly, HFPS may be a way to help develop critical thinking skills.

In 2016, my co-workers and I designed a feasibility study to see if group critical skills could be observed and quantified in HFPS sessions (Nguyen et al., 2017). A rubric was designed and named the Critical Thinking Skills Rating Instrument and was based on the IDEAS framework for problem-solving (Facione, 2015). The I stood for "identify problems and set priorities," the D stood for "determine relevant information and deep understanding," the E stood for "enumerate options and anticipate consequences," the A stood for "assess and make a preliminary decision," and the S stood for "scrutinize and self-correct as needed." Faculty observers rated on a 5-point scale (0–4) with descriptors for each value. Over a series of HFPS sessions, students significantly improved their scores suggesting an improvement in group critical thinking skills. Although this study was limited in the sense that it was only focusing on group critical thinking skills, it provides an instrument that could be used for more studies, even with other pedagogies to assess critical thinking skills (Fig. 16.1).

How do you know if a student can critically think? Visually it is tough with current assessments that many of us use, especially for basic science knowledge. However, when teachers speak with a student, most professors gain a quick sense of whether a student "gets it" or not and at what level by the way they ask or answer a question. This conversation usually will uncover knowledge gaps, poor application, and/or misconceptions. For example, an important topic in cardiovascular physiology is the concept of mean arterial pressure. So often, an instructor may ask a student about it, and many will just state that mean arterial pressure is 1/3 systolic blood pressure + 2/3 diastolic blood pressure. A struggling student may be so caught up in memorizing the equation they do not understand why it is weighted toward the diastolic number. They also may not recognize that the Ohm's law version of this

Instructions for Faculty Member: Please circle the appropriate numeric for the phrase that best describes the actions you observe of the learner groups' performance from the recorded simulation, and include additional notes as appropriate.

Critical Thinking Skills Rating Instrument				
IDENTIFY the Problem and Set Priorities				
Example Prompting questions: *What is going on? What is the chief problem? What is the most important issue?*				
Not able to identify chief problem(s) of the patient	Able to identify chief problem(s), but was unable to prioritize key issues.	Able to identify chief problem(s), and was able to prioritize key issues with substantial (≥ 2 times) prompting	Able to identify chief problem(s), and was able to prioritize key issues with minimal (<2 times) prompting	Able to identify chief problem(s), and was able to prioritize key issues without prompting
0	1	2	3	4
DETERMINE Relevant Information and Deepen Understanding				
Example Prompting questions: *How do the results relate to your patient? Do we all agree? Why do you think that?*				
Not able to verbally express correct physiological concepts underlying patient's condition	Able to verbally express some physiological concepts correctly and some incorrectly underlying a patient's condition	Able to verbally express physiological concepts underlying patient's condition correctly with substantial prompting (≥ 2 times)	Able to verbally express physiological concepts underlying patient's condition correctly with minimal prompting (< 2 times)	Able to verbally express physiological concepts underlying patient's condition correctly without prompting
0	1	2	3	4
Not able to assimilate information from sources	Able to gather information from sources but not able to find relationships	Able to assimilate information from sources and find relationships with substantial prompting (≥ 2 times)	Able to assimilate information from sources and find relationships with minimal prompting (<2 times)	Able to assimilate information from sources and find relationships without prompting
0	1	2	3	4
ENUMERATE Options and Anticipate Consequence				
Example Prompting Questions: *Why are you choosing this option? What are the options? What will occur as a result of your choice?*				
Not able to discuss options for treatment	Able to discuss options for treatment, but was not able to anticipate consequence of treatments	Able to discuss options for treatment, and was able to anticipate consequences of treatments with substantial prompting (≥ 2 times)	Able to discuss options for treatment, and was able to anticipate consequences of treatments with minimal prompting (< 2 times)	Able to discuss options for treatment and was able to anticipate consequences of treatments without prompting
0	1	2	3	4
ASSESS the Situation and Make a Preliminary Decision				
Example Prompting Questions: *How did you come up with that conclusion?*				
Not able to come up with diagnosis	Able to diagnose patient but was unable relate physical exam and lab findings with diagnosis	Able to diagnosis patient and can relate physical exam and lab findings with diagnosis with substantial (≥ 2 times) prompting	Able to diagnose patient and can relate physical exam and lab findings with diagnosis with minimal promptings (< 2 times)	Able to diagnose patient and can relate physical exam lab findings with diagnosis without prompting
0	1	2	3	4
SCRUTINIZE the Process and Self-Correct as Needed				
Example Prompting Questions: *Did the patient improve or decline? Why?*				
Not able to determine whether the patient was improving or declining	Was able to determine whether the patient was improving or declining but did not provide possible reasons	Was able to determine whether the patient was improving or declining and provided possible reasons with substantial prompting (≥ 2 times)	Was able to determine whether the patient was improving or declining and provided possible reasons with minimal prompting (< 2 times)	Was able to determine whether the patient was improving or declining and provided possible reasons without prompting
0	1	2	3	4

Adapted from Ideas A 5 Step Critical Thinking General Problem Solving Process. Facione, P. (2015). Critical Thinking: What It Is and Why It Counts. http://www.insightassessment.com/

Comments:

Fig. 16.1 Critical Thinking Skills Rating Instrument (from Nguyen et al., 2017, Adv Physiol Educ 41:604)

variable is more critical to understanding the role of mean arterial pressure in the cardiovascular system. If a teacher meets with a student, this can usually be uncovered and rectified. However, finding the time to do this for usually over 100 medical students is not feasible. Other pedagogies that allow an instructor to visualize a student's thought process are needed. HFPS may be one platform for this assessment to occur.

Our group has unpublished data on an experiment that attempted to decipher an individual student's thought process and confidence in a HFPS setting. Students participated in a HFPS in which the patient presents in diabetic ketoacidosis. This was the last in a series of four HFPS and took place in the second to last week of the module. They were in groups of six students, but facilitators slowed down the HFPS session at strategic time points to allow students to enter individual answers to prompts administered through a mini iPad. The first prompt was about resources available to them in the room, which most students could identify. The second prompt was for individual students to identify physiological problems that need to be addressed as well as prioritization and justification. Only about one third of the students were able to correctly identify, prioritize, and justify the order of addressing hypotension firstly and hyperglycemia secondly. Despite this low score, the confidence of individual students was high, around 81 %. Students could interact and discuss the patient, and the majority of groups was able to come up with the correct priority list. Next, there was another pause where students were first asked individually which intravenous fluid they would administer and secondly what other medications might they administer. While 74 % of students chose normal saline to support blood pressure, only 25 % were able to correctly identify insulin as the other drug to give. Most students chose additional drugs, the most commonly one being bicarbonate to treat this patient. The importance of this study is that it allowed us to "visualize" a student's thought process or approach to a patient. Secondly, it allowed us to address other learning issues such as the overconfidence and the overmedication of patients. Although this was a pilot study to see if we could assess soft skills issues in individuals, more studies like this are needed to advance the HFPS field.

With the USMLE Step 1 licensing examination converting to pass/fail, it should motivate faculty to assess other domains besides knowledge (Murphy, 2020). HFPS can be a good platform to assess other domains which could justify HFPS's high costs. One of the most prudent challenges is the lack of rubrics for standardization and consistency. As written previously in this chapter, new rubrics were created and modified from other resources to initially assess soft skills. A manuscript by Boulet (2008) nicely points out pros and cons of different types of assessment tools that could be used in HFPS. More research is necessary to ensure reliability and validity of rubrics of soft skills, and HFPS may provide the suitable venue to do so.

16.3.3 Capitalize on the Engagement Aspect of HFPS to Facilitate Learning

"Engagement is a precondition for learning. No learning happens until students agree to become engaged with the material" (Jason, 2017). Finding places to engage our learners is paramount if educators desire deeper learning. The likely theoretical framework supporting the use of HFPS for engagement is based on Kearsley and

Schneiderman's engagement theory (Kearsley & Shneiderman, 1998). The basis of this theory is that learning activities occur in a group context and are project based and an authentic focus. Although technology is not always necessary, Kearsley and Schneiderman believe that it can facilitate engagement. Finding ways to measure engagement and understand why certain pedagogies such as HFPS seem more engaging should be a topic of future research.

Years ago, there was a workshop at the International Society of Simulation in Healthcare meeting that described a unique way that HFPS was used in their curriculum. The approach had students as teams create a HFPS scenario based on a limited scope, such as types of shock. They would work with programmers within a short time frame (~30 min) to come up with their scenario and bring it to life. Another team would then attempt that scenario to try to figure out the diagnosis and treatment of that patient. The workshop attendees, which consisted of many physicians, did this same exercise. My group did a scenario that included a leaky dialysis line, so the team attempting the HFPS had to determine if the patient was in shock as well as the error causing these signs and symptoms. The other group designed a scenario where my group had to diagnose malignant hyperthermia. Importantly, participants found this extremely fun and engaging. The learning activity described above certainly fits into those criteria of engagement theory. It allows students to work at higher levels of learning in the create domain of Bloom's taxonomy. By letting students create scenarios, it may also present areas of misconceptions and provide insight into how novices view these scenarios. Studies could seek to understand why this type of activity is engaging, as this is probably the main reason why students consistently state that they "like" HFPS across virtually every published studied.

16.3.4 Compare, Supplement, or Replace HFPS with Virtual Reality

The world of virtual reality is here and described more extensively in Chap. 17 (in this volume) within this textbook. The market is extremely volatile in that companies are coming and going. There are standalone glasses as well as instruments that tether to gaming consoles. They range in price from $400 to over a thousand dollars (Greenwald, 2021). With this technology, a user could be transferred to any environment in the comfort of their couch. If the COVID-19 pandemic has taught us anything, it is that many of the things that we always did in person do not *have* to be done in person. If students had virtual reality capability for learning in medical education, much of it could be done with flexibility of time. Additionally, learning could be "paced" in that students could reach certain developed milestones or competencies at their own speed. Virtual reality sessions could be created to be challenging depending on a student's level of learning. Virtual reality sessions could be adapted to create opportunities for students to compare diseases or clinical presentations. Students could receive feedback or replay the scenario almost instantly. The major question that needs to be considered is what is the role of HFPS if this rising technology exists?

Another potential for changing the trajectory of simulation's future is to reimagine how simulation is delivered to learners. Two significant challenges for simulation have been interrelated challenges of scale and time. Traditionally, simulations target smaller group sizes and are scheduled around faculty, instructor, and/or facilitator calendars. This results in simulations being delivered during the standard workday span between 7 am and 5 pm, the busiest time for both clinicians and students in most settings. This results in a compressed simulation session, a shortened debrief, and a significant number of faculty required to process large groups of learners. Further complicating this delicate yet complex model are the per-faculty cost and a wide variety of student-to-faculty ratios. Leveraging technological advances such as AR, VR, and XR could enable many simulations to be redesigned as remote or virtual simulations to be experienced either synchronously, asynchronously, or any combination thereof. Thinking outside the box may help increase simulation's recognized benefits, yet this approach is not widely accepted as a viable option because many educators and institutions feel the weight of tradition and continue providing what they already know.

16.3.5 Investigate At-Home Simulation for Students

The COVID-19 worldwide pandemic of 2020 may have forced simulation professionals to think outside the box. As a result of an abrupt, almost universal shift from in-person (face-to-face) learning to remote (virtual) learning, simulation centers around the world were forced to adapt or face imminent closure in some cases. This abrupt retooling of the entire simulation industry may have precipitated at-home or on-demand simulation-based learning. The future of simulation may not necessarily be rooted in rapidly advancing technology alone; however, this perfectly viable method of simulation distribution may certainly better position simulation *for* the future. Since not all simulations are manikin based, HFPS may be facing an uncertain future. However, as technology evolves at lightning speed, HFPS may indeed find its way into homes of learners. Much like the high-tech gaming industry brought simulated sports into living rooms around the globe, a healthcare-focused version of this industry can certainly become a tangible possibility, especially in pandemic times.

16.3.6 Establish a Set of Standards for HFPS Research and Use

Although we simply cannot appropriately or adequately train tomorrow's healthcare professionals via 100 % remote learning, the ability to reimagine HFPS is here. One such approach to reinvention would be to establish national (or international) standards to quantify simulation's undervalued benefits. For example, the simulation center of the future may include grading algorithms that set numerical standards for the performance of psychomotor skills, resulting in a definitive performance score (DPS) for any given skill set that can be evaluated by simulation. Take an anesthesia residency for example; here, all residents entering the program must complete a simulated scenario where critical thinking and intubation skills are the main objectives. This simulation assignment can be designed for any time, on-demand access based on any given number of available slots (including and outside of traditional business hours). A resident selects 9 pm to visit the sim lab; upon arrival, the resident swipes their badge to activate the simulation and begin the recording. Upon activation, the resident proceeds to review all pertinent information about the anesthesia case, develops an action plan, and implements that plan by demonstrating the appropriate knowledge, skills, and abilities (KSA) relevant to the case. An intubation is eventually required as part of the scenario, and the resident attempts to intubate the manikin and does so on the second attempt. The scenario ends and the resident swipes out, ending the simulation and stopping the recording. The entire scenario is graded based on the programmed algorithms, and the recording of that session along with the automated score is sent to the residency program director who reviews the video at their convenience and validates the score or adjusts it accordingly. This same approach can be tailored to practically any licensed provider, caregiver, or student, and their performance can be compared to the standard as set forth by the authorizing entity. Although this uses the example of a resident, there is no doubt this standardization could be established in earlier medical school learners. Establishing such standards could also serve as a longitudinal reference to track personal growth and development in the given skill set and to set minimum achievement standards for that content.

Part of these standards would also include developing standards to provide feedback, which is critical to any HFPS experience, especially in basic science learning. Feedback must be learning stage appropriate. What this means is that if we are conducting a HFPS with a first-year medical student, we probably should not expect deep clinical decision-making capabilities. Feedback could address why a group chose a certain treatment strategy to get at underlying thought processes. The focus should be on the group process instead of becoming a lecture on the topic of the HFPS session. It should also be very pointed and concise. We have witnessed faculty trying to get too many learning objectives out of one HFPS session. Some objectives can be as simple as follows: ask for an interpreter. Research questions that can be addressed are how to deliver optimal feedback. Do you need to replay a video of the team's efforts? Should we follow a debriefing session with another HFPS immediately to reinforce our feedback? Do faculty members have to lead these debriefs, or can they simply design how to do it?

Technology is advancing much more rapidly than our ability to determine how best to use it. One thing that computers cannot do well is probing and finding underlying confusions or misconceptions. They can identify a very binary right or wrong answer but have more trouble identifying if it is a knowledge gap or a misapplication of a concept. Faculty are still needed to go through informal and formal assessments to make decisions and direct their resources and time. We have proposed some alternative ways that HFPS could be utilized and studied. The overall aim of these recommendations and proposals is to help create a more advantageous cost to benefit analysis.

Acknowledgments We would like to acknowledge Drs. Eli Meir and Christopher Gallagher for their thoughtful insight and input into this chapter.

References

Alinier, G. (2007). A typology of educationally focused medical simulation tools. *Medical Teacher, 29*(8), e243–e250. https://doi.org/10.1080/01421590701551185

Alluri, R. K., Tsing, P., Lee, E., & Napolitano, J. (2016). A randomized controlled trial of high-fidelity simulation versus lecture-based education in preclinical medical students. *Medical Teacher, 38*(4), 404–409. https://doi.org/10.3109/0142159X.2015.1031734

Bateman, J., Allen, M. E., Kidd, J., Parsons, N., & Davies, D. (2012). Virtual patients design and its effect on clinical reasoning and student experience: a protocol for a randomised factorial multi-Centre study. *BMC Medical Education, 12*, 62. https://doi.org/10.1186/1472-6920-12-62

Boulet, J. R. (2008). Summative assessment in medicine: The promise of simulation for high-stakes evaluation. *Academic Emergency Medicine, 15*(11), 1017–1024. https://doi.org/10.1111/j.1553-2712.2008.00228.x

Braun, H. A., & Tchaptchet, A. (2020). Virtual physiology. Retrieved 31 May, 2021, from http://virtual-physiology.com/.

Butter, J., McGaghie, W. C., Cohen, E. R., Kaye, M., & Wayne, D. B. (2010). Simulation-based mastery learning improves cardiac auscultation skills in medical students. *Journal of General Internal Medicine, 25*(8), 780–785. https://doi.org/10.1007/s11606-010-1309-x

Cendan, J. C., & Johnson, T. R. (2011). Enhancing learning through optimal sequencing of web-based and manikin simulators to teach shock physiology in the medical curriculum. *Advances in Physiology Education, 35*(4), 402–407. https://doi.org/10.1152/advan.00061.2011

Cesari, W. A., Caruso, D. M., Zyka, E. L., Schroff, S. T., Evans, C. H., Jr., & Hyatt, J. P. (2006). Study of physiological responses to acute carbon monoxide exposure with a human patient simulator. *Advances in Physiology Education, 30*(4), 242–247. https://doi.org/10.1152/advan.00063.2006

Chen, H., Kelly, M., Hayes, C., van Reyk, D., & Herok, G. (2016). The use of simulation as a novel experiential learning module in undergraduate science pathophysiology education. *Advances in Physiology Education, 40*(3), 335–341. https://doi.org/10.1152/advan.00188.2015

Costluc, N. (2021a). How much does a human patient simulator cost? Retrieved from https://www.healthysimulation.com/30911/human-patient-simulator-cost/

Costluc, N. (2021b). What is fidelity in simulation? Retrieved from https://www.healthysimulation.com/30181/what-is-fidelity-in-simulation/

Davis, D. A., Mazmanian, P. E., Fordis, M., Van Harrison, R., Thorpe, K. E., & Perrier, L. (2006). Accuracy of physician self-assessment compared with observed measures of competence: a systematic review. *JAMA, 296*(9), 1094–1102. https://doi.org/10.1001/jama.296.9.1094

Euliano, T. Y. (2000). Teaching respiratory physiology: Clinical correlation with a human patient simulator. *Journal of Clinical Monitoring and Computing, 16*(5–6), 465–470. https://doi.org/10.1023/a:1011401011887

Euliano, T. Y. (2001). Small group teaching: Clinical correlation with a human patient simulator. *Advances in Physiology Education, 25*(1–4), 36–43. https://doi.org/10.1152/advances.2001.25.1.36

Facione, P. A. (2015). *Critical thinking: What it is and why it counts*. Measured Reasons.

Fitch, M. T. (2007). Using high-fidelity emergency simulation with large groups of preclinical medical students in a basic science course. *Medical Teacher, 29*(2–3), 261–263. https://doi.org/10.1080/01421590701297334

Flanagan, B., Nestel, D., & Joseph, M. (2004). Making patient safety the focus: Crisis resource management in the undergraduate curriculum. *Medical Education, 38*(1), 56–66. https://doi.org/10.1111/j.1365-2923.2004.01701.x

Fragapane, L., Li, W., Ben Khallouq, B., Cheng, Z. J., & Harris, D. M. (2018). Comparison of knowledge retention between high-fidelity patient simulation and read-only participants in undergraduate biomedical science education. *Advances in Physiology Education, 42*(4), 599–604. https://doi.org/10.1152/advan.00091.2018

Freeman, S., Eddy, S. L., McDonough, M., Smith, M. K., Okoroafor, N., Jordt, H., & Wenderoth, M. P. (2014). Active learning increases student performance in science, engineering, and mathematics. *Proceedings of the National Academy of Sciences of the United States of America, 111*(23), 8410–8415. https://doi.org/10.1073/pnas.1319030111

Ginzburg, S. B., Brenner, J., Cassara, M., Kwiatkowski, T., & Willey, J. M. (2017). Contextualizing the relevance of basic sciences: Small-group simulation with debrief for first- and second-year medical students in an integrated curriculum. *Advances in Medical Education and Practice, 8*, 79–84. https://doi.org/10.2147/AMEP.S124851

Gordon, J. A., Alexander, E. K., Lockley, S. W., Flynn-Evans, E., Venkatan, S. K., Landrigan, C. P., ... Safety, G. (2010). Does simulator-based clinical performance correlate with actual hospital behavior? The effect of extended work hours on patient care provided by medical interns. *Academic Medicine, 85*(10), 1583–1588. https://doi.org/10.1097/ACM.0b013e3181f073f0

Gordon, J. A., Brown, D. F., & Armstrong, E. G. (2006). Can a simulated critical care encounter accelerate basic science learning among preclinical medical students? A pilot study. *Simulation in Healthcare, 1 Spec no*, 13–17. https://doi.org/10.1097/01266021-200600010-00005

Gordon, J. A., Oriol, N. E., & Cooper, J. B. (2004). Bringing good teaching cases "to life": a simulator-based medical education service. *Academic Medicine, 79*(1), 23–27. https://doi.org/10.1097/00001888-200401000-00007

Greenwald, W. (2021). The best VR Headsets for 2021.

Harris, D. M., Bellew, C., Cheng, Z. J., Cendan, J. C., & Kibble, J. D. (2014). High-fidelity patient simulators to expose undergraduate students to the clinical relevance of physiology concepts. *Advances in Physiology Education, 38*(4), 372–375. https://doi.org/10.1152/advan.00063.2014

Harris, D. M., Ryan, K., & Rabuck, C. (2012). Using a high-fidelity patient simulator with first-year medical students to facilitate learning of cardiovascular function curves. *Advances in Physiology Education, 36*(3), 213–219. https://doi.org/10.1152/advan.00058.2012

Heitz, C., Brown, A., Johnson, J. E., & Fitch, M. T. (2009). Large group high-fidelity simulation enhances medical student learning. *Medical Teacher, 31*(5), e206–e210. https://doi.org/10.1080/01421590802637990

Helyer, R., & Dickens, P. (2016). Progress in the utilization of high-fidelity simulation in basic science education. *Advances in Physiology Education, 40*(2), 143–144. https://doi.org/10.1152/advan.00020.2016

Hyatt, J. P., & Hurst, S. A. (2010). Novel undergraduate physiology laboratory using a human patient simulator. *Medical Education, 44*(5), 523. https://doi.org/10.1111/j.1365-2923.2010.03651.x

Institute of Medicine (US) Committee on Quality of Health Care in America. (2000). *To Err is Human: Building a safer health system*. L.T. Kohn, J. M. Corrigan, M. S. Donaldson (eds.) Retrieved from Washington DC: National Academies Press. https://pubmed.ncbi.nlm.nih.gov/25077248/

Isidore, C. (2020). Boeing's 737 Max debacle could be the most expensive corporate blunder ever. Retrieved from https://edition.cnn.com/2020/11/17/business/boeing-737-max-grounding-cost/index.html

Issenberg, S. B., McGaghie, W. C., Petrusa, E. R., Lee Gordon, D., & Scalese, R. J. (2005). Features and uses of high-fidelity medical simulations that lead to effective learning: a BEME systematic review. *Medical Teacher, 27*(1), 10–28. https://doi.org/10.1080/01421590500046924

Jason, Z. (2017). *Bored out of their minds*. Harvard Ed. Retrieved from https://www.gse.harvard.edu/news/ed/17/01/bored-out-their-minds

Kearsley, G., & Shneiderman, B. (1998). Engagement theory: A framework for technology-based teaching and learning. *Educational Technology, 38*, 20–23.

Krietoff, N., & David, G. (2020). In Reversal Boeing recommends 737 max simulator training for pilots. New York Times.

Larsen, T., Jackson, N. J., & Napolitano, J. (2020). A comparison of simulation-based education and problem-based learning in pre-clinical medical undergraduates. *Med Ed Publish*. https://doi.org/10.15694/mep.2020.000172.1

Levin, B. (2012). Why don't we learn at education conferences? *Phi Delta Kappan, 94*(3), 74–75. https://doi.org/10.1177/003172171209400319

Makransky, G., Thisgaard, M. W., & Gadegaard, H. (2016). Virtual simulations as preparation for lab exercises: Assessing learning of key laboratory skills in microbiology and improvement of essential non-cognitive skills. *PLoS One, 11*(6), e0155895. https://doi.org/10.1371/journal.pone.0155895

McGaghie, W. C., Issenberg, S. B., Petrusa, E. R., & Scalese, R. J. (2010). A critical review of simulation-based medical education research: 2003-2009. *Medical Education, 44*(1), 50–63. https://doi.org/10.1111/j.1365-2923.2009.03547.x

Meyers, L., Mahoney, B., Schaffernocker, T., Way, D., Winfield, S., Uribe, A., . . . Lipps, J. (2020). The effect of supplemental high Fidelity simulation training in medical students. *BMC Medical Education, 20*(1), 421. https://doi.org/10.1186/s12909-020-02322-y

Michael, J., Cliff, W., McFarland, J., Modell, H., & Wright, A. (2017). What are the Core concepts of physiology? In *The Core concepts of physiology* (pp. 27–36). Springer.

Motola, I., Devine, L. A., Chung, H. S., Sullivan, J. E., & Issenberg, S. B. (2013). Simulation in healthcare education: a best evidence practical guide. AMEE guide no. 82. *Medical Teacher, 35*(10), e1511–e1530. https://doi.org/10.3109/0142159X.2013.818632

Murphy, B. (2020). USMLE step 1 moves to pass-fail: Answers to 7 key questions. *American Medical Association*. https://www.ama-assn.org/print/pdf/node/48816

Nguyen, K., Ben Khallouq, B., Schuster, A., Beevers, C., Dil, N., Kay, D., . . . Harris, D. M. (2017). Developing a tool for observing group critical thinking skills in first-year medical students: a pilot study using physiology-based, high-fidelity patient simulations. *Advances in Physiology Education, 41*(4), 604–611. https://doi.org/10.1152/advan.00126.2017

Ortiz, G. A. (1994). Effectiveness of PC-based flight simulation. *The International Journal of Aviation Psychology, 4*, 285–291.

Parsons, D. (2019). Full flight simulators incorporate VR for next generation of pilots. *Aviation Today*. Retrieved from https://www.aviationtoday.com/2019/08/01/training-brain-mind/#:~:text=High%2Dfidelity%20full%20flight%20sims,cost%20up%20to%20%241%20million.

Petrizzo, M. C., Barila Labaraco, M. L., Jongco, A., Cassara, M., Agnlim, J., & Stern, J. (2020). Preclinical medical students' perceptions of the utility of high-Fidelity simulation to learn the mechanisms and presentations of hypersensitivity reactions. *The Journal of Allergy and Clinical Immunology, 145*(2).

planecrashinfo.com. Retrieved from planecrashinfo.com

Reinhardt, A., Mullins, I. L., De Blieck, C., & Schultz, P. (2012). IV insertion simulation: Confidence, skill, and performance. *Clinical Simulation in Nursing, 8*(5), 157–167.

Roediger, H. L., 3rd, & Karpicke, J. D. (2006). The power of testing memory: Basic research and implications for educational practice. *Perspectives on Psychological Science, 1*(3), 181–210. https://doi.org/10.1111/j.1745-6916.2006.00012.x

Salas, E., Wilson, K. A., Burke, C. S., & Priest, H. A. (2005). Using simulation-based training to improve patient safety: What does it take? *Joint Commission Journal on Quality and Patient Safety, 31*(7), 363–371. https://doi.org/10.1016/s1553-7250(05)31049-x

SimBio. Retrieved from https://simbio.com/promos/dna-explored

Stark, E. A. (1989). *Aviation psychology*. Gower.

Swamy, M., Bloomfield, T. C., Thomas, R. H., Singh, H., & Searle, R. F. (2013). Role of SimMan in teaching clinical skills to preclinical medical students. *BMC Medical Education, 13*, 20. https://doi.org/10.1186/1472-6920-13-20

Tchaptchet, A., Postnova, S., Finke, C., Schneider, H., & Huber, M. T. (2013). Modeling neuronal activity in relation to experimental voltage–/patch-clamp recordings. *Brain Research, 1536*, 159–167. https://doi.org/10.1016/j.brainres.2013.06.029

TeamSTEPPS. TeamSTEPPS. Retrieved from https://www.ahrq.gov/teamstepps/index.html

Weiner, E. L., Kanki, B. G., & Helmreich, R. L. (1993). *Cockpit resource management*. Academic.

Chapter 17
The Future with Extended Reality, Three-Dimensional, and Advanced Imaging for Molecules, Microscopy, and Anatomy

John Barrow

Abstract Modern visualization methods such as virtual microscopy, three-dimensional anatomy, and augmented or virtual reality technologies have transformed opportunities available to educators. The use of this technology has become ever more widespread, but there is not an abundance of evidence that visualizations improve test performance in students but instead enhance student engagement in their learning. These forms of visualization technologies and methods are well received by students through improved engagement with learning materials. This engagement comes from students spending more time concentrating on their work as visualizations allow them to interact with the subject matter through active learning. These benefits may enhance student understanding, but it is a challenge to measure this in genuine and authentic classroom scenarios. Visualization technologies do have the ability to reduce risk when working with hazardous subject matter, allow teaching using rare resources and materials, demonstrate three-dimensional relationships, and lower the cognitive load for students who struggle with dynamic biological processes. However, the use of these technologies is expensive and often requires specialist equipment and training to implement effectively, so they need far greater investment and pedagogic study to truly realize their potential. There is also a requirement to provide much greater access to the hardware and software required so that all students can benefit from these modern teaching methods as they are so reliant on technology for them to work at their best, especially in an ever-increasing online world and teaching environment.

J. Barrow (✉)
Institute of Education in Healthcare and Medical Sciences, University of Aberdeen, Aberdeen, UK
e-mail: j.barrow@abdn.ac.uk

H. J. Witchel, M. W. Lee (eds.), *Technologies in Biomedical and Life Sciences Education*, Methods in Physiology, https://doi.org/10.1007/978-3-030-95633-2_17

17.1 The Drivers for Visualization Pedagogies in the Biosciences

17.1.1 Biology Is Complex and Challenging to Learn

For a student to understand and learn biological processes, most curricula will focus on a sequential method, starting with basic concepts and processes, then building upon these with more complex functionality and mechanistic descriptions, and then building on these by explaining the impact these processes have on health and disease. For example, if you view the image in Fig. 17.1, this shows a positron emission tomography (PET) scan, the diagnostic technique that requires the loading of a radiotracer dye into a patient which then allows visualization of tumor sites because of their increased metabolic activity (Herst & Berridge, 2013). There are multiple processes at play in this image, whether that be the physics of positron emission, the metastasis of tumors, or the cellular and molecular processes that give rise to cancer. So, for a student to understand these processes, it would require an understanding of molecular processes at play plus physiological processes of the human body and potentially may link to an understanding of types of treatments that could be used and why they can influence a tumor. As such, understanding biology can be thought

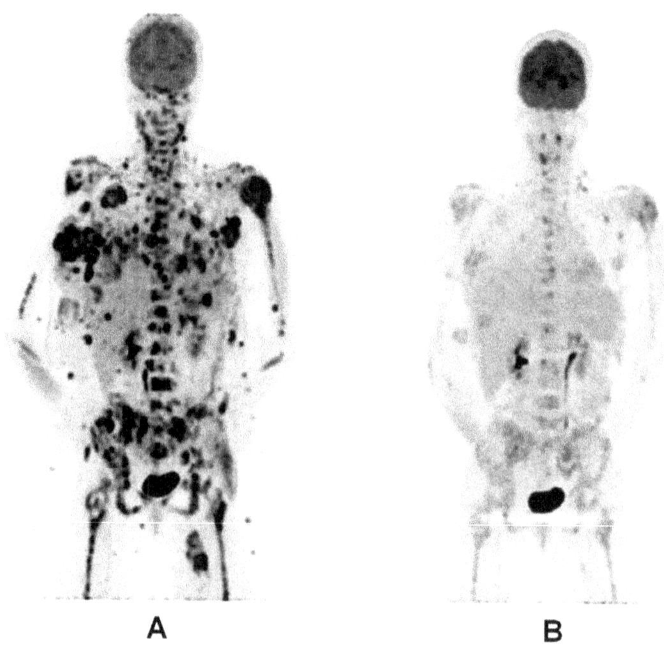

Fig. 17.1 FDG-PET scan images of a patient with metastatic melanoma before (**a**) and after (**b**) treatment with a cancer treatment drug. With kind permission from Peter MacCallum Cancer Centre in Melbourne. See Herst and Berridge (2013)

of as a multilayered process, but these layers of understanding are heavily intertwined and reliant on one another. So, for a holistic understanding, students must comprehend each of the various levels and how they relate to one another in the process. Johnstone describes these layers as macro-, sub-micro-, and symbolic levels of thinking. The need to switch between all three of them to understand a process or concept is critical to a full understanding, with many scientific subjects requiring multiple levels of understanding to allow the learner to fully appreciate and understand processes or concepts being taught (Johnstone, 1991; Johnstone et al., 1994).

This is particularly true in the bioscience disciplines where students may have to assimilate information from the macro-scale of a whole body through to a molecular process so that they can understand consequences of opposing influences in a given system, such as those in cancer metastasis. Coupled with this theoretical and conceptual view of the subject, there also needs to be a practical component to most bioscience teaching as disciplines are rooted in experimentation and the scientific method.

There is a broad consensus among bioscience educators in both the UK (from subject benchmarking) and the USA (from accreditation of programs such as medical school curricula) that bioscience teaching must include a concept-driven ability to relate form and function (LCME, 2019; QAA, 2015). Whether a student is learning about the macro- or micro-scale, the concept of form and function holds. For example, anatomical structures will have a certain shape and structure because of physical and mechanical functions they must carry out within the human body. Likewise, a protein or molecule might have a certain shape so that it can perform its function through interactions with other proteins or molecules at cellular and sub-cellular scales.

In the biosciences, complex concepts are often resistant to immediate understanding, which may elicit misconceptions by the learner. Helping students to master these threshold concepts in both physiology and biomolecular education is key to students' understanding. This idea that students can go through a period of conceptual change and emerge with a correct understanding of a process, function, or notion is critical if students are to gain a deeper understanding of how that process or function is important to a biological system. One example is that of natural selection, where Tibell and Harms describe three main concepts a student must understand to gain a holistic view of the subject: (1) to understand randomness and the variation in biology that leads to evolutionary processes, (2) to envision timeframes and scales of time, and (3) to understand spatial differences in biological organization (see Table 17.1; Tibell & Harms, 2017). This example shows how a whole variety of different ideas intertwine from basic biological concepts of mutations in DNA to population effects over years or millennia. The ability to cross the threshold for a concept can be transformational for students. Crossing a threshold will allow them to progress in their learning, being able to understand further concepts or biological processes that would have otherwise remained elusive or impossible to build into a holistic understanding of how the organism functions. As such, if a student is unable to master a specific threshold concept, they may then struggle to ever understand the entire process or set of processes at play because they are so heavily interwoven.

Table 17.1 Interrelations between the three principles and the key concepts of natural selection and the interweaving threshold concepts. Reprinted with permission (see Tibell & Harms, 2017)

Threshold concept	Principle(s)	Key concepts relating to natural selection
Randomness	Variation	Origin of variation—random mutations and related gene processes
		Random deaths, failure to reproduce, and other stochastic events affect fitness of individuals
Probability	Variation	Differential fitness—different traits give different probabilities of an individual surviving and reproducing
	Inheritance and selection	Differential survival/reproduction—the probability of an individual to reproduce depends on probabilities of various conditions (e.g., survival to a fertile age)
Temporal scale	Variation	Variation arises from chains of events that occur at different timescales, some over minutes to years (e.g., individual variation) and others over millions of years (e.g., population-level variations)
		Inheritance and reproduction happen on generation by generation timescale
	Selection	Selection pressures must be relatively unchanged over several generations for population effects to occur
	Inheritance and selection	Population changes and speciation occur over many generations that can range from hours to millions of years)
Spatial scale	Variation	Variation arises from chains of events at different organizational levels, from nanoscale mutations to cell or whole organisms' changes and to population-level variation on a global scale
	Inheritance	Reproduction happens at the organism level so relatively small scale
	Selection	Population changes and speciation have geographical distributions ranging from millimeters to global/continental scales

They may also struggle to understand other concepts as many science curricula are taught iteratively with foundational knowledge underpinning further understanding.

At a macro-level with subjects such as physiology or anatomy, understanding of form and function is often far more straightforward because they can be directly visualized. At the macroscale, students can see with their own eyes many processes in action, and, as such, these processes are less challenging or perceived to be less challenging, as the idea can be applied to something visible. As mentioned previously, the use of practical classes will often underpin theoretical knowledge with scientific experimentation because physical interaction with a subject is highly beneficial to students' understanding of concepts. It has been shown that practical-based teaching in subjects like physiology can clarify threshold concepts alongside the teaching of key laboratory skills and problem-solving (Horrigan, 2018). This ability for students to visualize processes in action, usually through the work completed in laboratory settings, provides them with an opportunity to transpose physiological function onto themselves and their bodies that in turn allows them to make links between form and functionality. At a cellular and subcellular level, concepts of form and function become increasingly abstract because most of these processes, and their effects on biological mechanisms, remain unseen. However, this does not mean that threshold concepts cannot be elucidated in the molecular bioscience disciplines, and in fact, a series of threshold concepts have been listed for biochemistry by Loertscher and colleagues (Table 17.2; Loertscher et al., 2014). This wide-ranging study identified multiple concepts that are important in understanding biochemistry at a fundamental level, with five principal areas: (1) steady state, (2) dynamic biochemical pathways and their regulation, (3) physical interactions at a molecular level, (4) macromolecular structure formation, and (5) free energy to drive chemical change. As with the previous example of evolution, this set of concepts is heavily interconnected with authors describing them as a web of interconnected concepts, so again, being able to understand one process or set of processes allows students to understand others if they can link and make connections, often through the ability to imagine or visualize processes at play.

17.1.2 Visualization in Bioscience Teaching

Visualization and the ability to link processes in a variety of forms (e.g., numerical data, graphical representations, visualizations of processes, and diagrammatic representations) are key to a student understanding a biological process, but there is limited evidence that visualization aids a truer understanding of a process or concept. Studies cited in 17.1.1 are some of the areas where research has been conducted, but a direct connection between visualization and enhancing academic performance is rare and often anecdotal. A recent meta-analysis conducted using various studies of anatomy education identified a varied efficacy for the use of three-dimensional anatomical teaching and a better understanding of anatomy in a clinical setting. Twelve out of the 21 articles used in the meta-analysis were showing a positive effect of visualizations over traditional teaching methods, but nine showed no learner gain, and three showed a subjective opinion from participants on improving their anatomical knowledge (Triepels et al., 2020). Like anatomy teaching, there have also been numerous studies focusing on chemistry education, which aim to help students understand the three-dimensional interactions of molecules, as textbooks and still images are often limited in how well they can communicate in three dimensions. These include chemical interactions of nanostructures such as carbon nanotubes (Fig. 17.2; Ferrell et al., 2019), specific techniques such as infrared spectroscopy (Fig. 17.3; Dunnagan et al., 2020), and enzyme catalysis using molecular dynamics (Bennie et al., 2019). All three of these studies highlight that many students find the use of three-dimensional visualization a positive experience and that it is often perceived to improve their understanding of concepts being taught. Further to these studies, Fig. 17.4 highlights the use of interactive virtual reality animations for exploring enzyme structure and function. Other studies highlight how the ability to visualize a concept can develop skills but not necessarily attainment, such as

Table 17.2 Biochemistry threshold concepts and associated knowledge statements. Reprinted with permission (see Loertscher et al., 2014)

Name	Knowledge statement (s)	Biochemical ideas that are unlocked once this concept is understood	Connections that were invisible before deep understanding of the concept
Steady state	Living organisms constitute open systems, which constantly exchange matter and energy with their surroundings, yet net concentrations remain relatively constant over time. This dynamic yet outwardly stable condition is referred to as a *steady state*	Steady state is an emergent process that results from regulation of numerous biological reactions	Once the condition of steady state is recognized, the purpose of complex regulatory systems in maintaining steady state and their connections to each other become apparent
	"Steady" is not synonymous with chemically "stable." Concentrations are determined by kinetic, rather than thermodynamic, factors. Hence, biological systems do not exist in a state of chemical equilibrium	Steady state is a metastable condition that can be maintained only because of constant input of energy from the environment	Once the metastable nature of steady state is recognized, the importance of multitiered energy storage systems (starch, glycogen, triglycerides, etc.) becomes apparent
	If an organism reaches chemical equilibrium, its life ceases. Consequently, organisms have evolved extensive regulatory systems for maintaining steady-state conditions	Steady state defines conditions of life under which chemical reactions take place in cells and organisms. Therefore, an understanding of steady state is necessary in order to correctly contextualize all of biochemistry	
Biochemical pathway dynamics and regulation	Reactions and interactions in biological systems are dynamic and reversible	Chemical drivers result in bulk (emergent) properties observed in biological systems	Once these concepts are understood, predictions can be made about (1) how biochemical pathways are likely to respond to changes in environmental conditions and (2) cause and effect of fluctuations in biochemical pathways

(continued)

Table 17.2 (continued)

Name	Knowledge statement (s)	Biochemical ideas that are unlocked once this concept is understood	Connections that were invisible before deep understanding of the concept
	Directionality of processes depends on the free energy and relative concentrations of reactants and products available	Enzyme-mediated regulatory mechanisms allow pathways to be sensitive and responsive to needs of the organism	
	Observable flux is the net result of forward and reverse processes	Enzymes act as gatekeepers rather than drivers of chemical change	
	Enzymes control rates of forward and reverse reactions		
	Enzyme activity is highly regulated		
The physical basis of interactions	Interactions occur because of electrostatic properties of molecules. These properties can involve full, partial, and/or momentary charges	Once this concept is understood, similarities between different types of interactions become clear. Although interactions are given different names, they are all based on the same electrostatic principles	A core biochemical principle is that structure governs function. Correct understanding of noncovalent interactions is essential in integrating structure and function
Thermodynamics of macromolecular structure formation	Interactions in biological systems almost always take place in aqueous solution	Protein folding, the assembly of lipids into micelles and bilayers, the association of polypeptide subunits to form oligomeric proteins, base-pairing of DNA and RNA molecules, and all other biological interactions are driven by a common set of thermodynamic forces	When entropic and enthalpic forces that drive processes like protein folding and binding are understood, predictions can be made about conditions under which these events will occur and what effect perturbations like mutations will have
	Bulk interactions in an aqueous system have an entropic component	The aqueous environment of the cell plays an active and essential role in biochemical structure formation	
	Enthalpic and entropic contributions are responsible for biological structure		

(continued)

Table 17.2 (continued)

Name	Knowledge statement (s)	Biochemical ideas that are unlocked once this concept is understood	Connections that were invisible before deep understanding of the concept
Free energy	The tendency toward equilibrium drives biological processes	Biological systems use favorable processes to drive less favorable processes, which allows for maintenance of steady state	Once this concept is understood, the relationship among free energy, equilibrium, and steady state becomes apparent
	Differences in free energy drive chemical transformations underlying biological function		
	By providing a direct link between a thermodynamically favorable reaction and a thermodynamically unfavorable one, enzymes enable biological systems to drive a normally unfavorable reaction by coupling it to one with a large and favorable free-energy change		
	Enzymes affect reaction rate, yet do not affect equilibrium position		

critical thinking (Shatri & Buza, 2017), which can support active learning within visual skills (Barmaki et al., 2019; Wilson et al., 2018). There is also evidence that visualization can be useful for motivating students to learn especially with more three-dimensional visualization methods like augmented and virtual realities (Khan et al., 2019).

At the macro-level, three-dimensional visualizations are especially important for clinically related disciplines such as anatomy where an understanding of the structure of the human body, and how anatomical structures interplay with each other, is crucial to understanding surgical approaches. If one looks for the number of publications related to augmented and virtual realities linked to an educational context, the number of publications has risen dramatically from the first publication in 1991 and especially sharply in the past 5 years (see Fig. 17.5).

As mentioned previously, there is little evidence that visualization increases knowledge, but there is evidence these technologies can increase motivation in

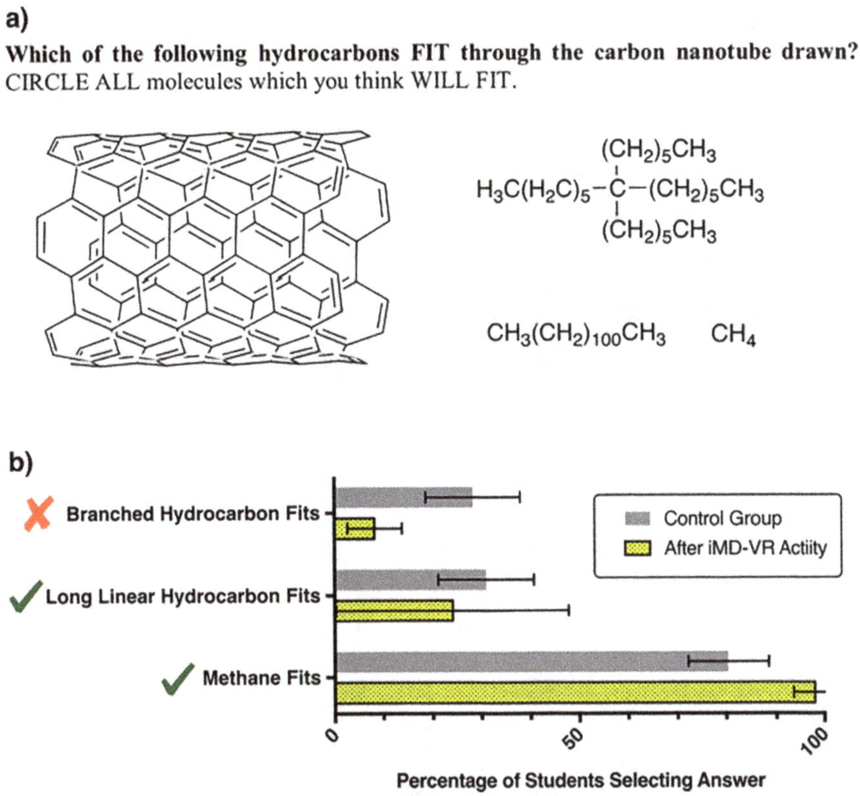

Fig. 17.2 Student understanding relative to control group following a virtual reality experience on molecular interactions of nanotubes. (**a**) Shows the question students were asked to complete and (**b**) shows student answers (iMD-VR group, $n = 70$; control group, $n = 85$). Reprinted with permission (see Ferrell et al., 2019)

students because they are seen by students as more engaging and provide a more modern and technologically driven approach to teaching than other more traditional didactic teaching methodologies. Highly dynamic visualization methods that can be employed also hold a certain mystique when used by students in their learning and could be considered as "edutainment." These kinds of visualizations may add both to engagement [see Chap. 4 in this volume] and to the novelty effect for learning. Tsay and colleagues have identified this as an effect that can enhance student engagement through gamified experiences (Tsay et al., 2020). Therefore, the creation of an entertaining product for students to consume can certainly be a driver to greater participation and a willingness to learn, which then may indirectly impact accomplishment and understanding.

In a more clinical setting, there have been some successes with the use of virtual and augmented realities as they provide the opportunity for students to experiment and to fail, which can aid information retention and allow students to comprehend

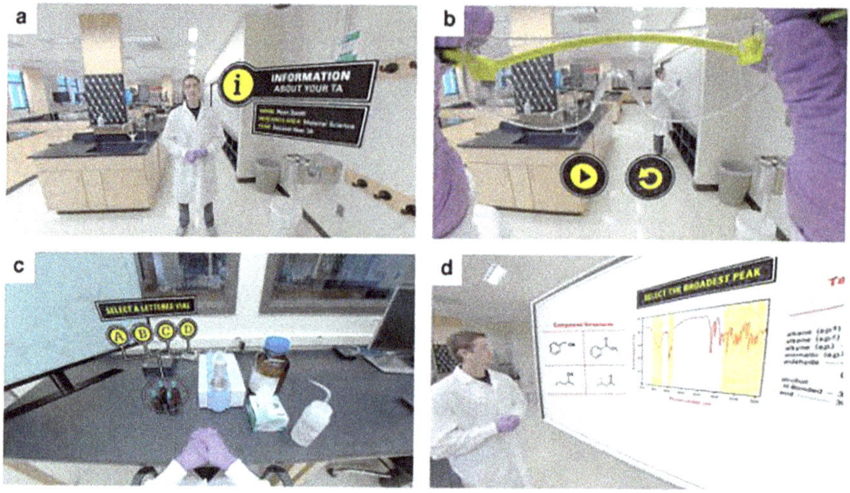

Fig. 17.3 Screenshots from a virtual reality infrared spectroscopy laboratory experience. (**a**) Welcome from teaching assistant, (**b**) first-person view of putting safety glasses on, (**c**) bench view of the experiment with various sample vials labeled, and (**d**) data analysis of experimental results. Reprinted with permission (see Dunnagan et al., 2020)

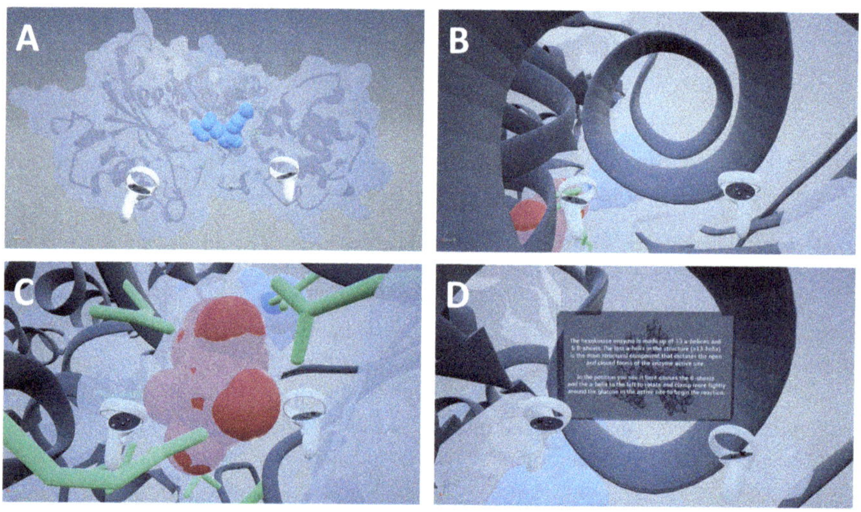

Fig. 17.4 Virtual reality representations of enzyme catalysis. (**a**) View of the enzyme hexokinase showing surface rendering and ribbon structure inside, with glucose and ATP substrates bound at the active site, (**b**) a view inside the enzyme structure moving through an alpha helix, (**c**) the active site showing glucose in the foreground (red) and ATP in the background (blue), and (**d**) an information panel to provide text information about enzyme function and other aspects of catalysis. All images show virtual reality controllers which are used to navigate around the three-dimensional space. Images courtesy of Ben Powell and Andrew Conley, STFC Daresbury Laboratory

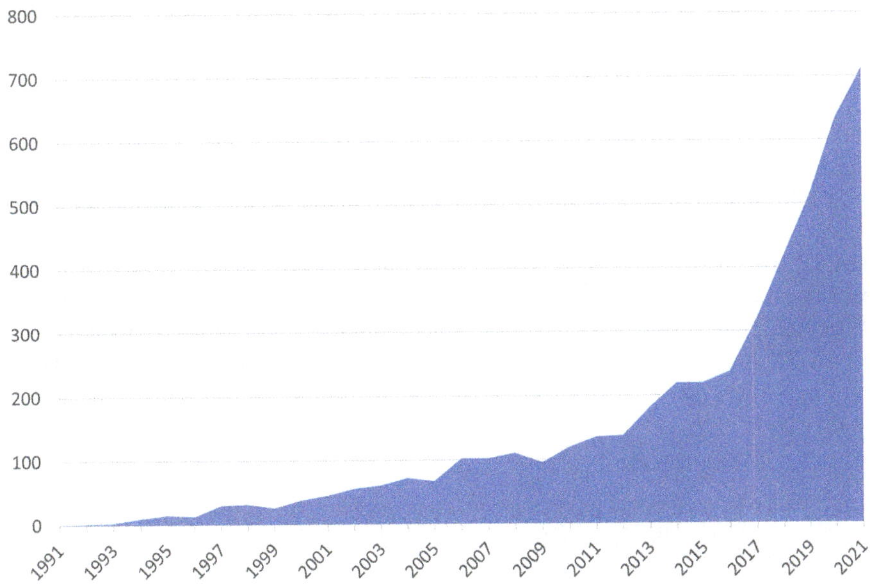

Fig. 17.5 Numbers of publications by year for "augmented reality" and "virtual reality" publications linked to education in PubMed (https://pubmed.ncbi.nlm.nih.gov). Search conducted on April 13, 2021

threshold concepts that will improve their clinical skills, all without posing an extra risk to patients (Ayoub & Pulijala, 2019; Cao & Cerfolio, 2019). This combination of technology enhancement, digital accessibility, and increased requirement for simulation and testing of procedures has in many ways been the catalyst for the increase in the use of visualization methods within the biosciences. As such, these will be the focus of the next section, which identifies some of the areas in which visualization methods are used and some of the approaches that have been taken to using visualization in education.

17.2 Approaches to Visualization

17.2.1 Extended Reality

Extended reality (XR) is a term that defines a suite of related technologies that include augmented reality (AR), virtual reality (VR), and mixed reality (MR). These technology platforms provide the user with a virtual representation of an object or environment and as such can be utilized in an educational setting to provide visual representations of, in theory, any laboratory object, organ, tissue, cell type, or molecule or indeed anything related to teaching including the academic member of staff themselves. An overview of existing technology and commercial products is shown in Table 17.3. This table shows existing headset technologies, but AR can also be used with a smartphone or tablet device and either Web-based applications or custom-designed applications for the operating system of the device.

Figure 17.6 shows a hypothetical scenario and comparison of AR, VR, and MR using a digitized anatomical visualization of the lungs; VR produces an image of the entire imaginary world, whereas AR overlays the synthetic object as a foreground onto a background made from the camera image that is in front of the viewer's headset or device. Thus, in AR, most elements of 3-D motion (head shifts, moving around) require less processor power to generate images being viewed because background rotations that the viewer sees are real rotations that the headset's camera is capturing in real time. That said, the creation of digital three-dimensional imagery on AR devices, which are often smartphones or tablets, can be a demanding task as they have far less processing and graphic rendering capabilities compared to a computer.

Augmented Reality AR can be described as incorporating or overlaying digital objects into the real-world environment and will usually use a mobile device such as a smartphone or tablet. Requirements for this technology for the user are relatively few, with even low-end smartphones being capable of rendering and recreating digital objects, so the technology at the point of use is minimal. Because there is not a requirement for high-end technology (although the better processing and camera functionality a device has, the better it may function) and because the technology usually works on portable devices, it means it is particularly suited to the classroom environment. Once installed on the device, the AR software will capture live images from the camera and using a variety of tracking technologies, usually based on physical markers or trackers to allow the camera to lock on, will position a scaled digital model of the object or process into the video images (Fig. 17.7; Barrow et al., 2019). The model will scale if the user moves toward or away from the marker position and will rotate relative to the device if the user moves around the position where the model is displayed in digital form. Often the digital model will be interactive or animated to show a process in action compared to a static process that would be shown in a textbook or other learning materials. The uncanny way that animations appear on top of the real world the user occupies can have a transformative effect on students as they become more engaged with the activity in question and are also able to see processes in action in a more authentic way. Two examples of recent AR technologies in surgical and medical settings are Augmedics and EchoPixel, respectively. These two companies aim to enhance surgical accuracy and planning of medical procedures through three-dimensional reconstructions of patient data (e.g., scanning imagery). This AR-based approach has recently been applied to deliver surgical implants to a patient undergoing spinal cord surgery (Business Wire, 2020).

Virtual Reality VR is a more immersive technology where the user must wear a headset that allows them to view and interact with a 360° digital "world" that can then provide visual and audio representations of a process or object. This digital environment can aim to deliver a real-world experience for the user or be wholly

Table 17.3 Various existing XR technologies

Technology	Product		Relative Price	Comments
AR/MR	Google Glass		+++	Restricted to enterprise users and very expensive at over $1500
	Epson Moverio		++	Large range of price points with varied features from basic functionality to industrial applications
	Toshiba dynaEdge		+++	Expensive, but carry a mini battery powered PC on the glasses frame and built for enterprise applications
	ZapBox		+	Very inexpensive, has controllers and utlisies Zappar freeware via smartphone
	Magic Leap One		+++	Very expensive and only available for developers so not intended for consumer use
	Microsoft Holo-Lens		+++	Very expensive and limited to enterprise applications, but excellent compatability with the Microsoft environment
VR	Google Cardboard		+	Inexpensive and easy to use but uses smartphone device and as such has limited graphic and processing capability
	HTC Vive		++	Tethered headset requiring powerful PC, has controllers to aid functionality and requires base stations to aid tracking
	Oculus Quest		++	Standalone headset at reasonable price and with a high resolution screen and controllers to enhance interactions
	Oculus Rift		++	Tethered headset, which requires a powerful PC, has controllers to aid interactivity and requires room sensors to aid tracking
	PlayStation VR		++	Requires a PlayStation and is tethered to the console, has developer functionality

(continued)

Table 17.3 (continued)

All commercial products shown in this table are headsets that the user would wear. Augmented reality (AR) and mixed reality (MR) headsets/glasses have standard lenses or make use of smartphone technology to overlay digital imagery on the real-world environment, but the user views the world through the lens of the glasses or the lens of the smartphone camera. Virtual reality (VR) headsets create an entire virtual environment with some VR headsets offering AR functionality as they have outward facing digital cameras that can recreate the environment on the screen the user is looking into. VR headsets are also often tethered, meaning they require a high-powered PC or console to function

Price ranges are correct as of February 2022

+, \$1–200; ++, \$201–1000; +++, >\$1000

AR augmented reality, *MR* mixed reality, *VR* virtual reality

unrealistic. For example, VR has been used to create training scenarios to help autistic spectrum children safely learn to navigate traffic when crossing the street (Dixon et al., 2020) as well as more abstract environments such as an experience that attempts to show effects of becoming partially sighted and then blind (Curzon Artificial Eye, 2016). As with AR, the digital environment can also have interactive components that the user can interact with through the headset or using handsets that can detect movement and can have haptic feedback to enhance the interactivity even further. In medical or science settings, VR has been used to allow clinicians to understand patient scan data better and be able to explore this in detail, such as in ImmersiveTouch. VR has also been taken one step further to allow robotic surgical procedures linked to VR—at the moment, this is experimental, but one company Vicarious Surgical is aiming to commercialize this approach to allow clinicians to perform surgery either remotely or with much greater accuracy using robotics. One of the major differences between AR and VR is that VR users have a single-user experience because the technology usually cannot track another human user and reproduce that person accurately in the virtual world. By contrast, in AR, other people are "part of the background" and appear automatically in the user's viewer. As such, while VR may produce a more immersive and sensory experience compared with using AR, this inability to see other users in VR can be a real issue as it can stifle interactivity in the classroom; however, for single-user activities, the use of VR can be of real benefit to allow full immersion and concentration in the virtual experience.

Mixed Reality (MR) As the name would suggest, this is a mix of technologies and is somewhere in between a VR and AR experience that uses similar headset technology to that of VR. MR creates a virtual digital world that is overlaid onto the real environment the user is in, allowing the user to interact with their actual and their virtual environment simultaneously. As with VR, the user can also interact with the digital environment through the headset or the use of handsets. This is a recent technology and as such has cost implications and less development of applications or use cases but is seen as a technology that could bring the best of both AR and VR technologies together. This would allow greater interactivity for group work but

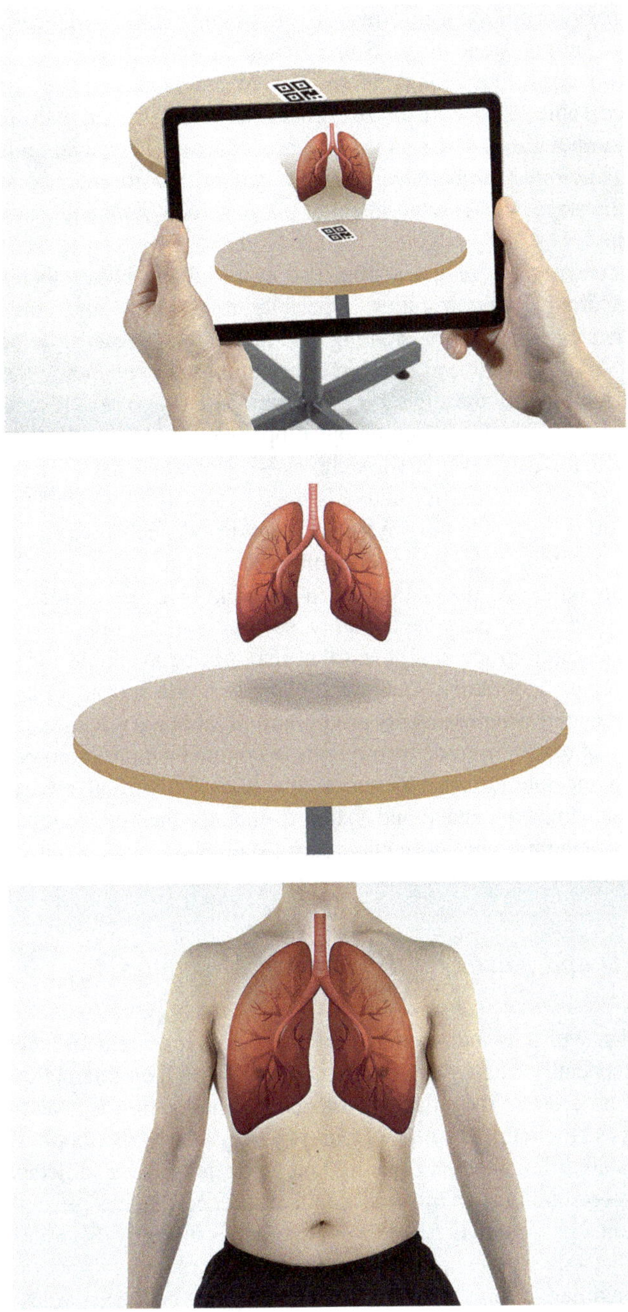

Fig. 17.6 Comparison of augmented, virtual, and mixed realities. Top panel, augmented reality showing a user viewing a digitally represented set of lungs and the device rendering them above the position of the AR marker on the table. Middle panel, a fully digitized virtual reality environment showing the table and lungs above. Bottom panel, a digital rendering of the lungs overlaid on a human torso using mixed reality glasses. Images courtesy of Kalyan Veera and Wai-lum Sung, Medical Illustration, University of Aberdeen

could also provide additional information and augmented experiences for users in the classroom at the same time. Some examples of MR uses are in anatomy education with apps like HoloAnatomy where students can use the Microsoft HoloLens (see Table 17.3) to visualize digital anatomical data on real-life specimens or models. Another use of MR is in surgical procedures, where, again with the use of HoloLens technology, surgeons can overlay patient data to enhance surgical outcomes—at this stage, this is experimental, but one such company working toward this is Proprio.

From the perspective of a teaching use case, all the above technologies are valuable and provide opportunities, especially for training and simulating real-world experiences (for VR) or creating interactive and engaging experiences for students (for AR), and MR can develop opportunities for enhancing the real-world environment with additional digital information. The usage of XR technologies in teaching is mainly limited to adoption by companies allied to medical or bioscience education ("XR for Business, 2018 Archives—Raconteur," n.d.). The potential for these technologies to be transformative is huge as it will allow students to venture into virtual worlds that could not physically be achieved, such as exploring inside the human body or within cells. Extended reality technologies also provide opportunities to develop skills and understanding in practical bioscience areas using virtual worlds that would not be possible in reality, such as working in laboratories that use expensive or research-grade equipment that may not be available for students in a physical university laboratory. One such example is the scenario played out in a study by Makransky where students used a virtual laboratory with equipment from manufacturers of research-grade instruments to conduct a microbiology experiment. Coupled with the laboratory experience, the use of a virtual environment also allowed students to have context and a story to enhance their educational experience compared to a standard laboratory classroom (Makransky et al., 2016).

17.2.2 3-D and Advanced Imaging of Molecules

The molecular world is indiscernible to the naked eye and as such is highly conceptual, especially when that world of molecules and their interactions are linked to molecular processes. This makes it especially challenging to teach concepts in subjects such as biochemistry and molecular biology as most textbook diagrams and visualizations are often abstract or simplified versions of those molecules. A major challenge is representing such molecules so that students can recognize the molecule and its functional properties in such a way that each is instantly differentiated yet not so overcomplicated as to be unrecognizable; for example, it is impossible to differentiate many similarly represented chemical interactions between circles and spheres of assorted colors as is often seen in molecular structures. Furthermore, when textbooks do use chemical or molecular structures, these will often be static images rather than the dynamic molecules that they are, thereby exacerbating their abstract nature. As such, the lack of movement or change shown by molecular structures is

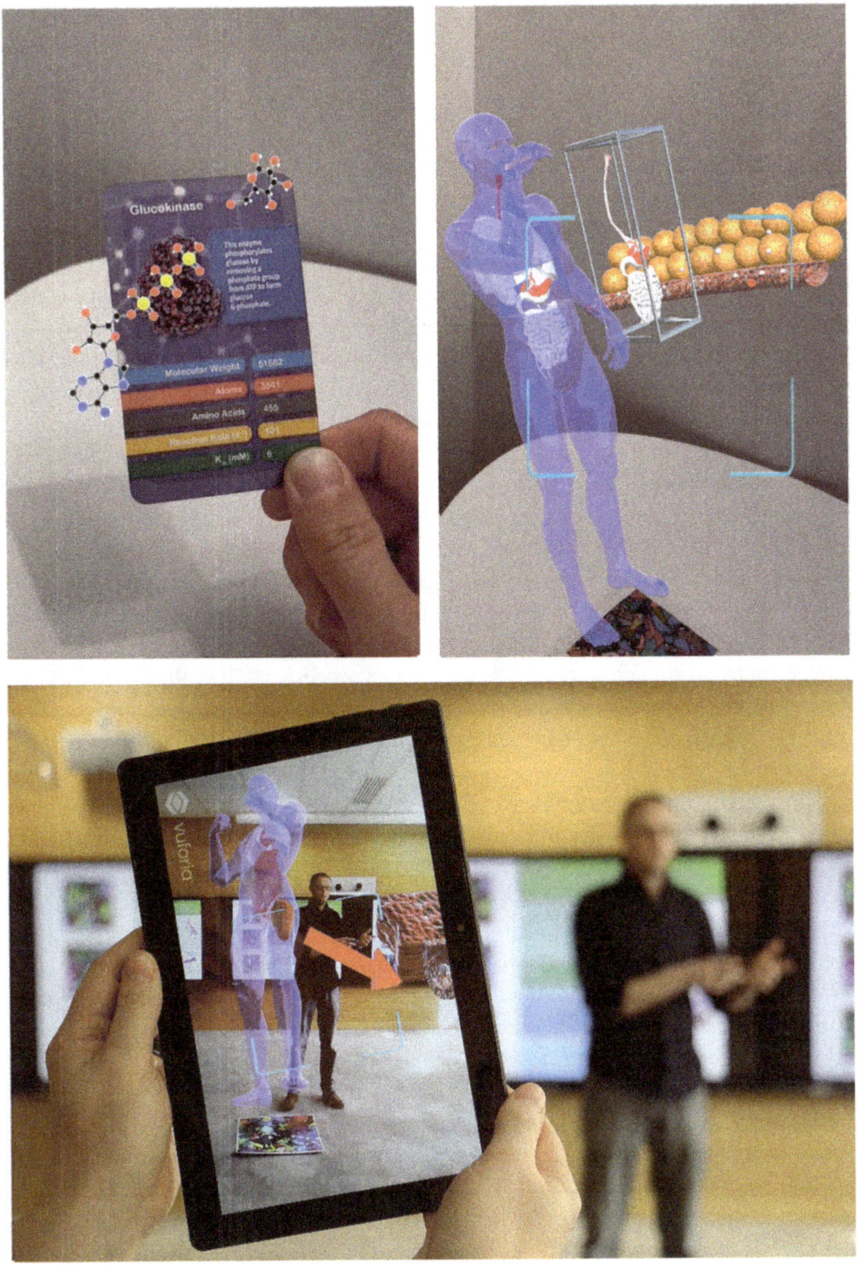

Fig. 17.7 Screenshots of augmented reality examples. Top-left panel, Glycolysis App demonstrating animated reactions of glycolysis (in this example showing the first reaction of glycolysis via the glucokinase enzyme) when triggered by viewing an AR trump card marker; top-right panel, Body App animation of insulin signaling when viewed using an AR marker. Bottom panel, a larger AR marker is used to create a larger Body App animation to show another part of the insulin signaling process with life-size animations. All animations shown utilize a game-engine design approach for their creation and are running through an Android tablet or smartphone device. From Barrow et al. (2019)

not conveyed as well as if a student could see the process in action. However, if a learner can convert two-dimensional imagery from texts and screens into a three-dimensional construct and is able to mentally rotate and manipulate this, they can begin to understand structure and function relationships. This ability then also provides them with a deeper understanding of molecular processes in the same concrete and material way that students can understand gross anatomy. The major benefit comes from helping many students who are unable to visualize static imagery as moving processes in their minds. These individuals will have a large cognitive load imposed on them if they are not able to create these mental constructs (Gonzales et al., 2020; Peterson et al., 2020). Extended reality can alleviate this cognitive overload by presenting the imagery in motion.

One excellent repository of three-dimensional molecular structures is the Protein Data Bank (PDB; Berman et al., 2000). The PDB also has a rich heritage in creating educational content for students and a lay audience through initiatives such as PDB-101, which aims to enhance the understanding of the molecular world and provides direct links to the molecular structure data should the user wish to probe further into visualization methodologies.

Combining data from repositories such as PDB with visualization platforms like XR could allow educators to develop highly motivating visualizations that allow students to see molecular processes in action or understand the interplay between chemical bonds and molecular interactions, which in turn will allow students to gain a better appreciation and understanding of these processes. The other added advantage of using actual molecular structure data is that educators are then teaching with authentic visualizations based on real scientific data rather than the often unrealistic imagery used in texts.

Using three-dimensional real-world data can also provide a learning opportunity that is especially useful for bioscience students as it allows them to understand different approaches that can be taken to visualize protein structure. If we take an enzyme as an example (see Fig. 17.4 and Bennie et al., 2019), the active site of this enzyme will interact with a substrate molecule and in turn allow catalysis of the reaction to occur. The interaction of the substrate and the enzyme will usually be through precise molecular interactions (e.g., hydrogen bonding, hydrophobic interactions, or ionic bonding). There are, therefore, multiple aspects that contribute to the process of a simple substrate-enzyme interaction, all of which can be visualized in a variety of ways. The protein and substrate structures can be shown as space-filled molecules and ball and stick structures, or the surface could be rendered. Protein structures can also be shown as ribbon diagrams which accentuate secondary structures of alpha-helices and beta-sheet conformations. All these structural representations can then be colored in a variety of ways to highlight hydrophobicity, chemical structure, or charge properties among others. Therefore, simultaneously the use of visualization can not only teach students about how molecules function but can also provide them with an opportunity to gain practical skills in molecular visualization by using research standard ways of visualizing molecular structures.

The use of molecular visualizations has not been extensively studied as a tool for educating students. However, the process of creating visualizations and molecular structures can also provide an opportunity for students to understand advanced research methods such as X-ray crystallography, nuclear magnetic resonance (NMR), and cryo-electron microscopy (cryo-EM), which are the most used methods for elucidating molecular structures within the PDB database. This, therefore, provides teachers with an added learning opportunity alongside the teaching of biological processes being discussed in the classroom.

17.2.3 *Microscopy*

Using a microscope correctly is a skill set in itself, and indeed many university curricula will have microscopy practicals that aim to train students in the use of microscopes. Among European medical students, the Trans-European Pedagogic Research Group found that medical students had a favorable view of histology (mode of 3/4 on a scale of 1, extremely favorable, to 11, extremely unfavorable; Moxham et al., 2017); nevertheless, the time spent on histology in the early years of medical teaching in Canadian (Rockarts et al., 2020) and European medical education is diminishing (Leveritt et al., 2016). Given these time pressures, there are several examples of virtual microscopes that allow students to access microscopy data from both light and electron microscopes ("Histology Guide," n.d.; "The Open Science Laboratory: Histology & histopathology," n.d.; Schrader et al., 2006). These examples will often use high-resolution microscopy imagery that can then be manipulated through a Web browser, an applet, or a mobile app (see Fig. 17.8 and Lee et al., 2018). The Open Science Laboratory provides some functionality and mimics the usage of a real microscope with focus, stage, and lighting controls as well as the ability to change magnification through the selection of various "objectives." There are also examples where experimental data can be utilized in a similar way, for example, with the GenePaint database, where annotated in situ hybridization images on sectioned mouse embryos are presented (Visel et al., 2004).

In a variety of subjects, it has been shown that the use of virtual microscopes does not always enhance student understanding and technical proficiency, but instead, the use of virtual microscopes when used to support physical microscopy does provide some additional benefits, even if only perceived. For example, Selvig et al. (2015) state that the substitution of videos, virtual methods, and online resources instead of classroom study will usually negatively impact student success. Upon comparing virtual and physical microscopy, Becerra and colleagues showed that student's perception of their learning was more positive when using an individual virtual microscope when assessed across a variety of parameters such as ease of learning (positive scores of 73% using light microscopy vs. 94% when using virtual microscopy), finding histological structures (positive scores of 69% using light microscopy vs. 93% when using virtual microscopy), and comfort (positive scores of 73% using light microscopy vs. 85% when using virtual microscopy; Becerra et al., 2018). This is in contrast with previous publications from this same research group when using one "digital microscope" for the entire classroom, 74% stated they

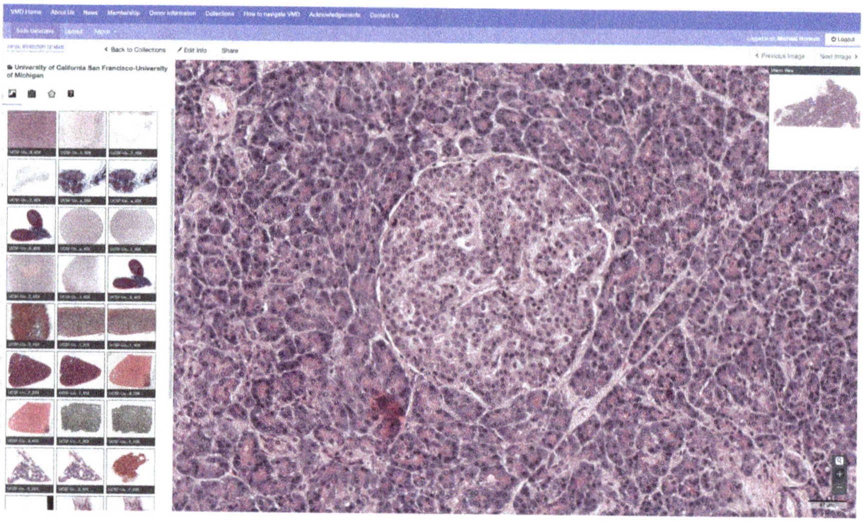

Fig. 17.8 The Virtual Microscope. A screenshot of a VMD virtual tissue slide from the UCSF-University of Michigan collection (Slide #290; Pancreas, human, H&E stain). The BiolucidaVR viewer allows VMD users to examine the virtual image at different magnifications. The insert in the right-upper corner is a navigational thumbprint of the entire slide with a box indicating the magnified area. Images courtesy of Michael Hortsch at the University of Michigan

had been motivated by the light microscope compared to 51% using the virtual microscope (Becerra et al., 2015). Using virtual microscopy alongside physical microscopy can enhance engagement and student satisfaction, so in this instance, it can act to support microscopy classes (Herodotou et al., 2020). As you can see from these three studies, there is a mixture of evidence that is not always clear as to whether virtual microscopy is beneficial to the bioscience educational repertoire, but given the decline in histology education in biomedical curricula, these additional resources must be seen as a positive addition to students' education.

The main advantage of using virtual microscopy is this kind of microscopy only requires access to a computer and does not require physical, and often expensive, microscopes for students. It also has the added advantage of providing the ability for group work as using a traditional microscope is usually a solitary affair where only one student can see a specimen at a time. The ability for students to gain knowledge of microscope specimen preparation is lost through the digital microscope, but this information could be introduced elsewhere if it were required to make sure students understood the basis of good sample preparation before microscopy. Therefore, the use of digitized content can allow students to work together, and this collective experience can potentially offer more engagement and ability to create expansive class sessions.

17.2.4 Anatomy

Traditionally, the study of anatomy required cadaveric specimens and was solely taught using didactic learning approaches through static imagery and instructional dissections. Anatomy is the linchpin of medical disciplines as an understanding of structural arrangements of the human body is crucial for understanding function and the basis of clinical reasoning, especially for invasive techniques and methods such as surgical procedures. With the increasing numbers of medical students, fewer and fewer donations of bodies, and limited budgets, there are definite drivers toward digital pedagogies in anatomy teaching. It has been shown that digital three-dimensional imagery is just as effective, if not better, than physical dissections for teaching anatomy, with one study showing that using digital dissection tables and CT scans improved general anatomy scores by 11.7% compared to similar cohorts of students taught by traditional anatomy teaching and non-digitized radiography images (Fig. 17.9; Paech et al., 2017). There are multiple approaches to digital anatomy visualization including anatomy tables and XR applications. All these approaches make use of three-dimensional anatomical structures that are created digitally with some of them using actual anatomical specimens and others that use artificially created digital imagery. For those methods that use clinical imagery from patients or cadaveric material, it provides the opportunity to teach specific pathologies alongside general anatomy.

One form of technology that can be delivered using this three-dimensional anatomy imagery is the digital anatomy table such as Anatomage. Baratz and colleagues identified an interesting finding that students learning anatomy with the digital table performed significantly better when tested using the digitally created examination (77% Anatomage test score vs. 63% cadaveric test score) compared to students using cadaveric dissection who performed significantly better when tested using a cadaveric specimen examination (74% cadaveric test score vs. 65% Anatomage test score; Baratz et al., 2019). Three-dimensional imagery can also be

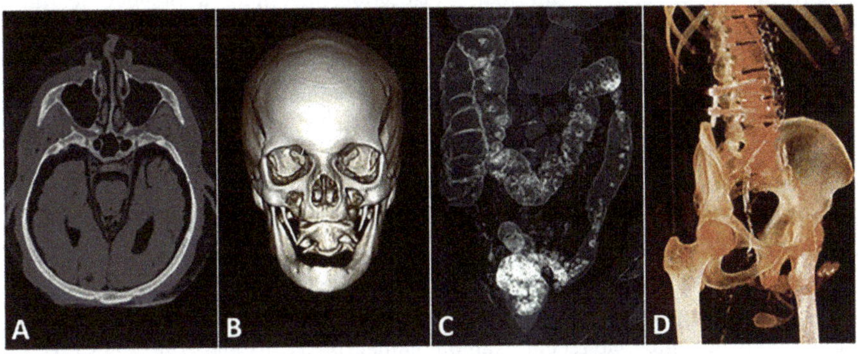

Fig. 17.9 Cadaver-specific CT scans on the virtual dissection table. Cadaver CT scans via an institution-owned CT scanner visualized using AnatomyMap (**a**, **b**) or the virtual dissection Anatomage Table (**c**, **d**). Reprinted with permission (see Paech et al., 2017)

used with XR technologies to increase student engagement and provide the visual benefits of using these enhanced technologies (Bork et al., 2019; Goff et al., 2018; Pei & Wu, 2019).

17.2.5 Distance and Blended Learning

Distance learning structures and processes will usually involve students not attending teaching in person and instead they will consume educational content away from the institutions that deliver the teaching [see Chap. 13 on distance learning in this volume]. Currently, most of this distance learning content will be delivered online via virtual learning environments in an asynchronous manner with supplementary activity, such as online support tutorials delivered synchronously. If a program of study has a practical component, this will usually be delivered in an intensive teaching block to provide practical teaching as efficiently and effectively as possible, especially if students are required to travel for practical sessions because they are in physical laboratories. There is literature for online learning being as good if not better than face-to-face teaching in the biosciences (Brockman et al., 2020); however, for practical teaching that often includes many central learning objectives and threshold concepts for bioscience students [See Chap. 15 on practicals and simulations in this volume] there is far less literature suggesting distance learning versus in-person teaching is better or worse, although there is some evidence that students who have different learning styles will benefit more from either online or in-person learning (Winschel et al., 2015).

Students using on-campus facilities will often have the required hardware and software as well as the equipment (if required), but at home or away from campus, many students do not have resources to supply their own technology, especially if specialist hardware (e.g., VR/AR equipment) is an integral part of a lesson. With the current Covid-19 pandemic, the ability for students to attend in-person classes has become a real challenge with many institutions adopting a blended learning approach that delivers a mixture of classes based on in-person sessions as well as online sessions but also allows for some students, who are not able to attend personally due to restrictions, to move to fully online or shift back into face-to-face teaching rapidly with short notice [see Chap. 15 on teaching during the pandemic in this volume]. This novel and rapidly evolving situation potentially offers an opportunity for technology to be used to deliver a comparable experience for students who cannot attend in person when they are assigned to the same classes as students who can attend in person.

There have been some attempts to mitigate the challenges of using AR and VR applications alongside videoconferencing facilities often used for current teaching during the pandemic. One recent example is that of Reeves and colleagues (2021), who have created an AR application that visualizes protein structures. When tested, those students who used the AR visualization prior to being tested had significantly higher test scores, showing that the enhanced experience of protein visualization aided attainment. This enhanced attainment was also maintained for up to 4 weeks after the AR experience (Reeves et al., 2021). Others have used VR to enhance their teaching, specifically in the field of bioethics, where authors suggest the ethical decision-making process is helped by the use of virtual collaboration and group work (Harfouche & Nakhle, 2020). Byrnes et al. (2021) argue that the current pandemic will enable the "virtualization" of teaching, specifically anatomy teaching, by allowing dissemination of the teaching more creatively using social media and videoconferencing to change how information is distributed.

Many students, even though they may be considered digital natives, will often only have access to a mobile device with some having laptops or computers, but in many cases, this hardware does not have the computing power of university systems. Students may also struggle to have suitable IT infrastructure, such as the Internet or remote access to institutional networks, and, in turn, may then not be able to participate in classes if visualizations would need high bandwidth requirements that students cannot provide themselves. This highlights the real need for digital equity to be considered with the use of technology in a distance learning environment where digital content forms part of the teaching [see Chap. 3 on inclusivity in this volume]. Even basic digital content, such as virtual meetings, recorded classes, and files that must be downloaded, can pose problems for students, so digital visualization content could be a real barrier to inclusion and providing an equitable education for all students on a program. This is supported by a report from the Metropolitan Educational Research Consortium in Virginia, USA, where access to the Internet and computer hardware can vary dramatically dependent on geographical location, socioeconomic status, and race (Becker et al., 2020).

17.3 What Is the Future for Visualization?

Samples or specimens that could potentially be seen in reality, or could be physically handled and manipulated, can now be reliably recreated as digital objects. This transformational change in sample preservation and recording can be a real benefit to educators who may have limited or no access to certain specimens, for example, cadaveric material which may be in limited supply or precious samples such as histological samples from biopsies that are required for diagnostic or research purposes. This situation is especially true of samples that show rare or specific clinical attributes, which would be especially suitable as teaching materials.

For human aspects of teaching in the biosciences, three-dimensional visualization of living or cadaveric specimens is not just valuable for saving money and for increasing accessibility to the traditional teaching of healthy anatomy, but it can supplement a traditional course of in-person anatomy teaching with reconstructions of rare pathological specimens. A further advantage of creating digital model visualizations is that clinical and real-world data can be incorporated into the visualization such as other histological data from samples that link to the classroom specimens or more information and learning materials linked to specimens being dissected (Hanna et al., 2018). Therefore, this allows the incorporation of far more than the visual representation of structures and components by adding more layers of detail or information to a digital object. For example, a specific cadaver or patient that has a unique disease or clinical presentation would allow various clinical visualization tools to overlay with the digital model visualization to provide distinct organ system, anatomical, and other data. This data could be vital information to allow a greater understanding than the model would have provided alone. This also has advantages for these precious cadaveric material samples as a digital model of an anatomical specimen can be repeatedly virtually dissected to show various tissue types and structures. The head and neck are particularly suitable, as specimens are delicate and cadaveric specimens are not always perfectly preserved. For example, dentistry has been supplemented using AR and VR both for education and maxillofacial surgery (Joda et al., 2019). This has also been done with veterinary samples, such as using augmented reality to teach neuroanatomy in the dog (Christ et al., 2018). Digital dissections have the added benefit that they can be carried out as many times as required to explain and understand structures, while the dissection of a physical sample can only be carried out once and following dissection cannot be repeated.

This visualization approach of creating digital models also extends beyond gross anatomy and can be used for visualizing high-resolution imagery from microscopy data (Farahani et al., 2016). Histological specimens will often be included in anatomy teaching as they are still one of the main ways that disease is diagnosed in tissues. The creation of a digital histology specimen again provides an ability to distribute information for teaching and allows students to work with rare or valuable samples that they would otherwise not be able to use. Digital histology can also create three-dimensional images that can be useful for cellular architecture arrangement of cells in tissues, which is something that requires the use of visualization methods (e.g., z-stack imaging) to create the three-dimensional rendering of the microscopy data.

All previous examples are excellent ways to visualize from the microscopic scale through to whole-body imagery. The standout advantage the digitization of models provides is the dissemination of information widely and precisely. In many ways, this is like the dissemination of information from Andreas Vesalius and his book *De humani corporis fabrica* (On the Fabric of the Human Body) that was only made possible due to the printing press, which served two purposes—firstly, to massively increase the number of copies, legally or otherwise, that could be made and, secondly, to increase the accuracy of the publication so every copy is identical, which would not always be the case for handwritten manuscripts (Lanska & Lanska, 2013).

In the same way that Vesalius published to a diverse and widespread audience in the 1500s, the digitization of imagery and models provides opportunities for specific groups to access the material, but it also provides the opportunity to create new ways of learning for geographically diverse or disparate learning communities.

Could this technology usher in a new way to teach and learn? One possibility is the learning of skills for health professionals. For example, when training undergraduate medical students to use ultrasound, when students were provided with a VR-enhanced study as part of a practical workshop (compared to analogous study with a digital atlas), they were faster and scored higher on a series of timed ultrasound tasks (Cohen's $d = 1.03$, Hu et al., 2020). In a small study in Cleveland, second-year medical students were introduced to neurosurgery in a yearlong elective based on virtual reality were they gained positive scores for competency and understanding (Atli et al., 2020). Many of the platforms that are used to create visualizations, especially those linked to creating three-dimensional content and XR, use game engine software to create and develop digital objects and virtual spaces. The same technology and processes that perhaps would only be used to create individual structures and models digitally could be employed to create virtual worlds and environments within which teaching can occur. This technology is already being seen in the conference and event sector with virtual conferences being run online for many varied industries.

When the Covid-19 pandemic spread throughout all areas of the world, it was thought that conferences and events might have to be curtailed and potentially shift the way that researchers meet and exchange ideas, which some feel was long overdue (Roos et al., 2020). There are advantages to hosting conferences online such as reduced fees as there would be no requirement for conference catering, hosting conference delegates and speakers, and hiring of large-scale venues (Jobs, n. d.), but the downside would be a lack of traditional human engagement between delegates and ability for nuanced discussions that rely on nonverbal cues. Digital conferences would also limit the ability to develop relationships that go beyond simply the exchange of information and ideas [see Chap. 15 on teaching during the pandemic in this volume].

Perhaps the use of this technology could see the dawn of the first fully virtual online university that can provide a mixture of real-time learning within a virtual campus. This could allow all students to experience classes as a mix of fully online content and virtual blended learning with synchronous live classes that utilize models and three-dimensional animations to teach bioscience principles. Already many universities employ virtual environments to welcome and recruit prospective students (Larmore et al., 2005), and there have also been VR-based campus evacuation drills and campus safety tours using Oculus Rift technology, such as those at Bowie State University in Maryland (Fig. 17.10; Sharma et al., 2015). The use of virtual environments in this way does not get away from the need for hands-on practical teaching, which is almost impossible to recreate in a virtual world, but it could provide opportunities for the delivery of practical theories without risk and allow students to fail or make mistakes and then repeat processes for an infinite number of times if required.

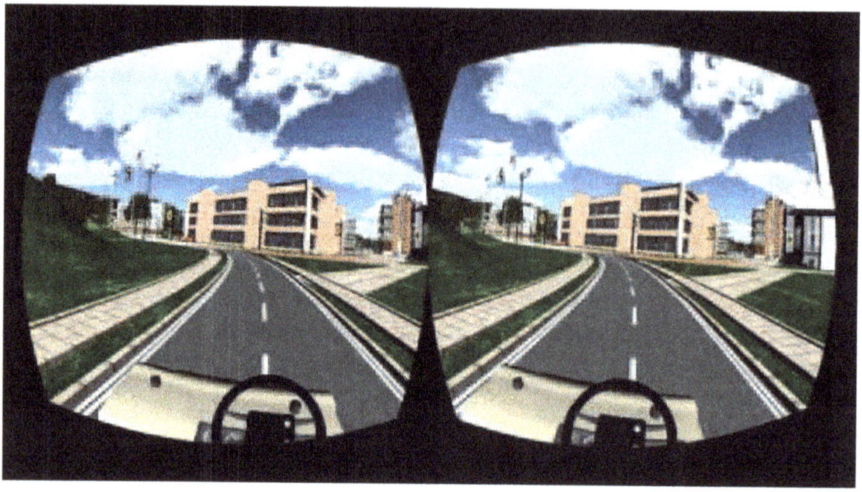

Fig. 17.10 Virtual campus tour using virtual reality. Reprinted with permission (see Sharma et al., 2015)

17.3.1 Digital Limitations

All the above is more than possible theoretically, but even with the right amount of funding, technical expertise to deliver this kind of approach, and students who have the correct hardware and software to access it, there is one major hurdle that at present cannot be overcome—that of broadband infrastructure (Ragnedda & Ragnedda, 2020). The UK has seen an expansion of the broadband network and the introduction of schemes to increase capacity and access to higher-quality broadband in various locales, but there are still many areas of the UK that would not have the bandwidth to cope with high-resolution graphics and large-scale three-dimensional reconstructions of biological models. Students will often cite hardware or software failure for not being able to complete simple coursework activities, or they have a lack of appropriate space to complete assignments safely and within the right environment to work effectively, and this is without further drives toward digitization of educational content. This lack of digital equity within student populations has been brought to the fore with the increased use of existing technologies and digital processes as part of our everyday lives but also in educational settings, so to increase this demand for high-powered computing power and faster Internet speeds would only serve to widen this digital inequality (Gerli et al., 2020).

17.4 Supporting Novel Visualization Teaching Methods

Visualization has been used for decades in areas such as biochemistry for protein structure and function and more recently in areas such as anatomy using three-dimensional digitization of specimens. In these contexts, molecular subjects have used visualization in their research, and tools used for this are now extensively utilized in teaching to create molecular models or dynamic processes, thereby making the abstract more tangible. This contrasts with macro-level subjects like anatomy where the visualization process is often used to aid teaching in place of physical specimens, so these two juxtaposed areas have very differing drivers for the use of visualization techniques—on the one hand, there are techniques to bring an invisible subject to life, and on the other hand, there are techniques to aid teaching when specimens or materials are challenging to work with or obtain. In much of the literature, there may be slight gains in attainment, but a large amount of the modern visualization tools available are acting as an incentive to engage with the subject—there is nothing inherently wrong with this as it will drive student buy-in, but whether it pays dividends for student understanding is questionable. Some of the lack of pedagogic research in this area is also due to the ethics of carrying out a study where individuals would be separated into groups that are utilizing and potentially benefitting from technological advancement and those that are not. That said, the inclusion of digital visualizations as a tool to increase participation and student satisfaction could indirectly contribute to greater retention of ideas and therefore potentially a better understanding of the subject matter, all of which can be measured through student course feedback surveys and focus groups.

While many will promote the idea of a research-intensive university carrying out excellent teaching, the disparity in funding between those discipline-specific research activities and funding that is provided for teaching research is vast. The teaching of various bioscience disciplines requires subject expertise to do it well, but to enact transformational change into degree programs and shift the way we teach these subjects (especially if large-scale investments are required in infrastructure and technology as would be required for the visualization methods cited in this chapter), significant investment in education and novel pedagogies is needed. Currently, income from teaching is over half of all income for UK higher education institutions with 47.3% coming from student fees and 7.8% coming from government grants ("HESA Finance record 2017/18," 2019). When one looks at the amount of money spent on teaching and research activities, it is broadly the same percentage as the income for teaching alone at 53.7% of total operating expenditure but is not broken down further than this by HESA. This implies that the amount of money universities take in from teaching income is spent on both teaching and research, and while this research activity will involve some teaching advancement, most research activity is within subject-specific research and not designed to enhance and develop innovative ideas in teaching. Some disciplines, such as medical education using patient simulation, can be expensive, with Pottle et al. highlighting that replacement of this mode of teaching with virtual simulation can massively reduce the cost per student (Pottle, 2019).

There is funding available for the development of teaching through small pockets of funding often within institutions or from external sources such as the Quality Assurance Agency (QAA) Enhancement Themes (www.enhancementthemes.ac.uk) as one example. However, while this money is welcome and will support some teaching development, the amount of funding is often at most in the thousands of pounds. Further funding opportunities can also be sought for teaching through sources such as scientific outreach funding, but this is often linked to specific research projects, and if it is not, then it will come from charities or learned societies—again, this funding is important, but the amount of funding one can request is very limited and will not be able to be used for direct enhancement of teaching activity. This funding will instead be used to support impact for larger research projects or for engaging younger audiences or the general public in science subjects at schools and within the community. This kind of activity is crucial for the development of young people and educating the public about the importance of universities and their role in civic society, but for the true development of transformational pedagogies, the funding must be for sustained periods to allow serious development of ideas and an understanding of what is effective in a testable way—this cannot be done with the resources that are currently available.

Overall, there are two key things to consider if we want to look forward toward a future of visualization technology in bioscience teaching: (1) reliance on technology and digital visualization, in particular, is something that should form part of a suite of teaching methods and be complementary to them, and (2) evidence-based pedagogic research must be supported through greater levels of funding comparable to scientific research monies. This will ultimately pave the way to firmly embed research-led education into higher education institutes and realize the potential of many aspects of digital media and resources to provide high-quality student experiences, especially as we emerge from the Covid-19 pandemic.

17.5 Future Research and a Commitment to Educational Change

17.5.1 Teaching Infrastructure to Support Educational Innovation

As discussed above, the development of novel digital pedagogies such as XR are on the point of becoming a reality. It is currently not something that is used in widespread educational settings, but with the correct pedagogic experiments and development, this technology could bring a new era of possibilities. Universities are at the forefront of research and development trends so it should be the same for teaching, and indeed many have created education-centered environments that allow collaborative projects to create novel visualization approaches. Liverpool John Moores University has created a physical space that allows educational developments linked to visualization and technological innovations to be tested through behavioral and physiological approaches, and also brings the talent of staff and students together to create new ways of working and educating students (*Live Lab: Liverpool Immersive Experience Lab*, 2019). Another example is the University of Utah, which has The Animation Lab, where they aim to "build visual hypotheses" to aid understanding molecular processes for both research and educational purposes (Nayak et al., 2020).

If more universities were able to fund these collaborative environments, it would allow for the development of a wider set of technologies that could enhance educational theory and practice, so for visualization methods in the bioscience teaching space, it could allow the impact of new teaching methods to be tested extensively. The creation of teaching labs would require significant expenditure by institutions, but this would pay dividends by allowing pedagogic research staff physical spaces to test their ideas and pilot new initiatives without this being something a live class would have to participate in, which could be considered as high risk for students involved, especially if there is no gain in student understanding or obvious benefit for students to take part. Many universities have already invested in infrastructure to deliver teaching, especially for laboratory-based teaching in the sciences, as well as infrastructure to support education such as library and computing facilities, but the creation of learning labs is something few universities have invested in.

For bioscience education at university, whether that is for molecular visualization or simulation of a physiological process, the ability for a learner to utilize XR technology would allow them to master and understand rare situations. For scientists, this has always been why experimentation is so important, but teaching students about processes in a kinesthetic way through physical interactions with molecules and abstract concepts has been unattainable until now. One of the reasons that simulation and visualization have been pioneered in industries such as aviation or healthcare environments is that the expense can be justified. A pilot landing a plane with a damaged engine or a surgeon operating and finding an anomaly may only occur once in a career if at all, so simulation training can aid this by allowing repetition of processes and protocols which would not be possible in reality. Repeating this approach for educational material within universities would allow students to learn how to fail and give them a risk-free way of doing so by developing their knowledge base and skills, especially those linked to more abstract concepts such as molecular or cell biology processes. It would also allow the integration of research ideas and visualizations into the curriculum directly to allow students to see actual structures and processes involved in a concept without the need for simplified and static figures one finds in more traditional learning materials like textbooks or lecture slides.

17.5.2 *Future Research Suggestions: Digital Twinning of Research and Teaching*

As highlighted previously, there is limited evidence that XR technology and other visualization methodologies enhance learner achievement, but they can increase learner engagement (Brown et al., 2020), which must be seen as a good thing and will inspire and influence students to participate for longer and want to take their studies further if they are motivated to do so. If learners can be engaged in their learning through embodied experiences, this merges the two processes of learning through physical training and the incorporation of theoretical concepts being delivered as an authentic training experience. This process of embodied training can work for subjects as abstract as mathematics (Burte et al., 2017) but is also relevant for the biosciences where abstract concepts are taught routinely (Borghi et al., 2017). It could be viewed as "making the invisible visible" by placing the learner within a cell, molecule, or process but in an authentic way using authentic research data and information.

Imagine a future where a classroom could be created virtually with virtual representations of real biological structures, be they on a human, cellular or molecular scale, with real-life data to show concepts and ideas based on patient information or experimental information from research laboratories. In this future, the melding of real-world experimental data with digital models of real-world objects and processes is equivalent to that of digital twinning which currently is utilized for mainly industrial processes and systems (Fig. 17.11).

A digital twin is a digital representation of a physical asset, process, or system that uses real-world data to test predictions or highlight efficiencies in processes in the digital environment that can then be transferred to the real world. Two examples of digital twinning technology are the energy sector and manufacturing processes where digital twinning has been used to highlight efficiencies in designs all based on real-world and real-time data. This allows designs to be translated from the digital environment into real-world enhancements of physical assets and infrastructure used in processes and systems (Gardner et al., 2020).

Digital twinning is not currently used in education settings, but it could be converted to use in this context using the same processes and digital twinning software approach coupled with XR technologies to aid visualizations. In the classroom, it could be used to recreate anatomical data in virtual reality and then overlay real-time clinical data from recordings taken in the clinic—one example may be a three-dimensional heart rendered virtually using VR and then use cardiograph data to show blood flow through the heart (Collins et al., 2019) and how this changes if there were a blood vessel obstruction or damage to the heart tissue which would be seen in the cardiograph data while manipulating or virtually handling the digital heart structure. The same could be carried out at the microscopic scale with virtual models of cells and their organelles that highlight cell functionality, such as metabolic processes where experimental data on metabolic changes (Depaoli et al., 2018; Hou et al., 2018) could be used to show animations of how cells dynamically adapt their energy utilization.

Some further usage cases could include the following:

Molecular visualization—using three-dimensional molecular structures from real experimental structural data (e.g., the Protein Data Bank), proteins or biological molecules could be visualized using XR techniques. Changes to these proteins could then be made such as mutating amino acids, to see how the functionality is affected by structure, but to keep it authentic, this could be carried out using experimental data that is built into the model of animations and structures that have been created virtually. For example, digital models and structures could be used to visualize

Fig. 17.11 Digital twin of visualization data from human and protein samples. Artistic renderings of real-time data presented and displayed within the classroom using augmented or virtual reality technologies that combine clinical data from patients (top panel) or molecular data from experiments in the lab (bottom panel). Images courtesy of Joe Spencer, PixelMax (www.pixelmax.com)

a drug action on its molecular target, the influence of point mutations on protein function and disease processes, and the development of resistance in microorganisms via changes in gene or protein sequences. Furthermore, visualizations could also be used to teach protein-protein interactions, enzyme kinetics, gene expression processes, DNA replication, and almost any other dynamic process that occurs at a molecular level where experimental and structural biology data can be combined in the digital twin.

Cell functionality—the same molecular structures mentioned above could be used to make larger-scale models of cellular components, or indeed whole cells, to show processes that occur in cells to allow them to function. This could be membrane function to highlight the importance of gradients in biological processes (e.g., action potentials in neurons or mitochondrial bioenergetics) or could show receptor signaling cascades and other signaling processes. Further to molecular structures, cellular structures could be created from experimental three-dimensional microscopy data, and cell functionality could be incorporated from signaling and metabolic pathways as well as other system biology data and modeling information to build digital twins of particular cell types.

Tissue functionality—visualizations have recently been produced to highlight the morphology of tissues such as tumors (Waclaw et al., 2015) and displayed in virtual environments using XR technology. Linking back to Sect. 17.1.1 at the start of this chapter, being able to switch back and forth from gross level visualizations to the microscopic may yield a deeper understanding of the molecular basis of diseases like cancer. Varying parameters in growth conditions and mutations of genes will affect growth properties of tumors and as such can be used to experiment and test varied protocols in the digital twin of the tumor. This process could also be applied to developmental biology in model organisms such as nematode worms where well-characterized developmental processes at a cellular level have been imaged and three-dimensional and dynamic animations have been made along with predictive models for cell movement (Sarma et al., 2018), all of which could be visualized using XR technologies and manipulated digitally to assess what would happen if different parameters are changed during the developmental process. One final area where this could work well is in connectome visualizations of neuronal connections in brain tissue as real-world brain scan data, from CT or functional MRI scans, could be linked with connectome data to understand differences in brain function and neuroscience concepts such as network dynamics linked to brain function (Larson et al., 2018; Ozdemir et al., 2020), all within the digital twin model of the brain.

Further to the above examples and pertinent to the final example of brain functionality is that this same idea of digital twinning could be used to analyze neuronal and emotional responses to classroom activities being carried out through the recording of data from wearable sensors that could track in real-time student responses to classroom activities. The digital twinning approach could be used to incorporate the data from wearable sensors, which are sometimes used in consumer advertising and marketing research, to decide how emotionally stimulated and engaged students were in the teaching event. Measurements could include heart rate and the galvanic skin response (electrodermal activity), both of which highlight states of emotional arousal, or measurements of eye movement and tracking as well as gaze analysis to understand which regions of visual stimuli garner more interest for the user (Bell et al., 2018). In this regard, the way they learn within the classroom could be empirically measured to create a digitally twinned classroom environment that can then form part of a teaching laboratory to inform scientific teaching in the biosciences.

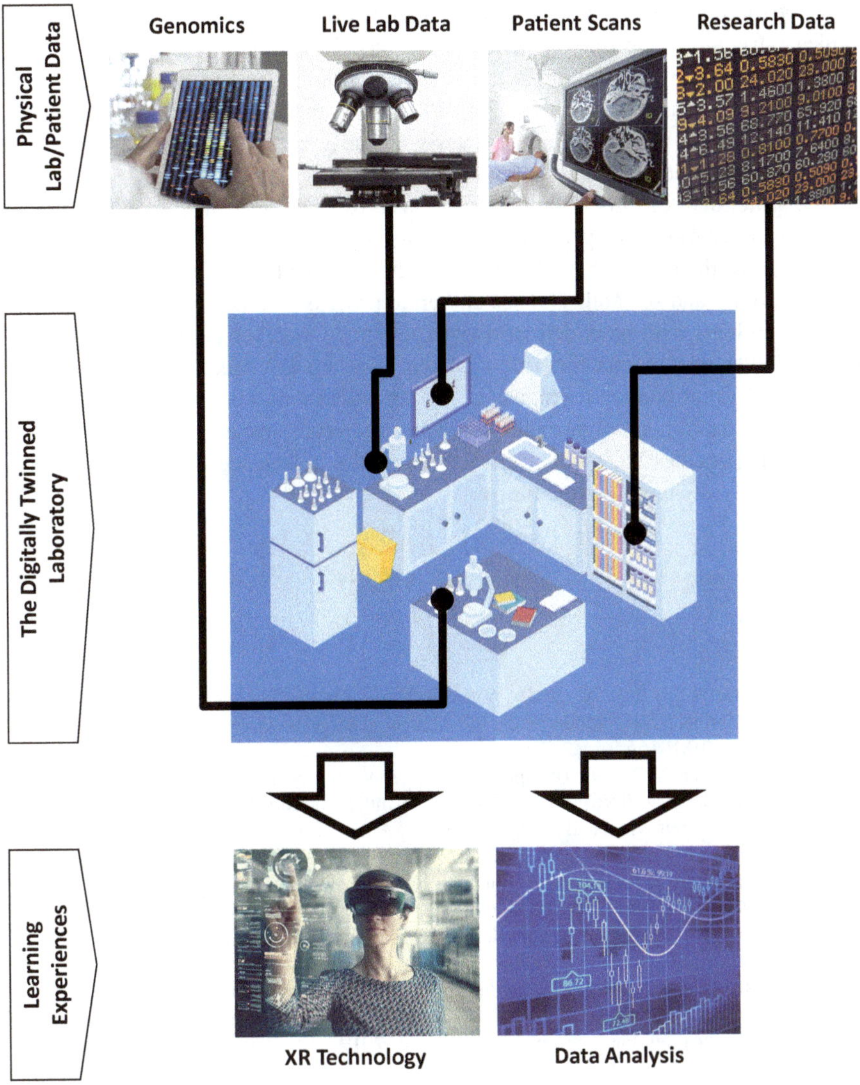

Fig. 17.12 Digital twinning of real-time laboratory data with teaching in the virtual laboratory space. All images are stock images from Microsoft Office 365 except the central image which was created from Vecteezy.com

This ability to digitally twin visualizations and scientific data will allow students to truly experience a novel way to learn (see Fig. 17.12). It will also provide them with opportunities to fail and experiment in a safe environment. However, it does not come without its challenges, such as the requirements for equipment or computing power if high-end visualizations and large datasets are needed, which is especially true if visualizations are based on three-dimensional data from molecular or anatomical structural information. There may also be problems relating to digital inclusion and equity for students from all backgrounds and locations in terms of access to devices and technology. All of that said, the future for XR technologies coupled with real-world biological data is a reality and will hopefully become a viable option for educational institutions.

The potential for this technology is vast, especially with computing power becoming ever more inexpensive and cross-platform working becoming easier. It is the hope of the author that advantages of digital visualization to educate students through doing, and the ability to see authentic processes in action, will converge to provide students with an excellent user experience wherever they are in the world and whatever their socioeconomic status is. We are not there yet, but there are aspects of current pandemic-imposed learning styles that will continue expanding, through digitization of high-quality learning materials, especially for the biosciences. This provides huge opportunity to teach the biosciences in a novel way for a wide variety of student populations.

References

Atli, K., Selman, W., & Ray, A. (2020). A comprehensive multicomponent neurosurgical course with use of virtual reality: Modernizing the medical classroom. *Journal of Surgical Education*. https://doi.org/10.1016/j.jsurg.2020.11.003

Ayoub, A., & Pulijala, Y. (2019). The application of virtual reality and augmented reality in Oral & Maxillofacial Surgery. *BMC Oral Health, 19*, 238. https://doi.org/10.1186/s12903-019-0937-8

Baratz, G., Wilson-Delfosse, A. L., Singelyn, B. M., Allan, K. C., Rieth, G. E., Ratnaparkhi, R., Jenks, B. P., Carlton, C., Freeman, B. K., & Wish-Baratz, S. (2019). Evaluating the Anatomage table compared to cadaveric dissection as a learning modality for gross anatomy. *Medical Science Educator, 29*, 499–506. https://doi.org/10.1007/s40670-019-00719-z

Barmaki, R., Yu, K., Pearlman, R., Shingles, R., Bork, F., Osgood, G. M., & Navab, N. (2019). Enhancement of anatomical education using augmented reality: An empirical study of body painting. *Anatomical Sciences Education, 12*(6), 599–609. https://doi.org/10.1002/ase.1858

Barrow, J., Forker, C., Sands, A., O'Hare, D., & Hurst, W. (2019). Augmented reality for enhancing life science education. Paper presented at VISUAL 2019 - The fourth international conference on applications and Systems of Visual Paradigms, Rome, Italy.

Becerra, D. G., Grob, M., Assadi, J. L., Astorga, C., Tricio, J., Melelli, R., Silva, C., & Sabag, N. (2018). Academic achievement and perception of two teaching methods in histology: Light and digital microscopy. Pilot study. *International Journal of Morphology, 36*(3), 811–816. https://doi.org/10.4067/S0717-95022018000300811

Becerra, D., Grob, M., Rodríguez, A., Barker, M. J., Consiglieri, L., Ferri, G., & Sabag, N. (2015). Academic achievement and perception of two teaching methods in histology: Light microscopy and digital systems. *International Journal of Morphology, 33*(3), 811–816. https://doi.org/10.4067/S0717-95022015000300001

Becker, J., Washington, J., Naff, D., Woodard, A., & Rhodes, J. (2020). *Digital equity in the time of COVID-19: The access issue*. Metropolitan Educational Research Consortium.

Bell, L., Vogt, J., Willemse, C., Routledge, T., Butler, L. T., & Sakaki, M. (2018). Beyond self-report: A review of physiological and neuroscientific methods to investigate consumer behavior. *Frontiers in Psychology, 9*, 1655. https://doi.org/10.3389/fpsyg.2018.01655

Bennie, S. J., Ranaghan, K. E., Deeks, H., Goldsmith, H. E., O'Connor, M. B., Mulholland, A. J., & Glowacki, D. R. (2019). Teaching enzyme catalysis using interactive molecular dynamics in virtual reality. *Journal of Chemical Education, 96*(11), 2488–2496. https://doi.org/10.1021/acs.jchemed.9b00181

Berman, H. M., Westbrook, J., Feng, Z., Gilliland, G., Bhat, T. N., Weissig, H., Shindyalov, I. N., & Bourne, P. E. (2000). The Protein Data Bank. *Nucleic Acids Research, 28*(1), 235–242. https://doi.org/10.1093/nar/28.1.235

Borghi, A. M., Binkofski, F., Castelfranchi, C., Cimatti, F., Scorolli, C., & Tummolini, L. (2017). The challenge of abstract concepts. *Psychological Bulletin, 143*(3), 263–292. https://doi.org/10.1037/bul0000089

Bork, F., Stratmann, L., Enssle, S., Eck, U., Navab, N., Waschke, J., & Kugelmann, D. (2019). The benefits of an augmented reality magic Mirror system for integrated radiology teaching in gross anatomy. *Anatomical Sciences Education, 12*(6), 585–598. https://doi.org/10.1002/ase.1864

Brockman, R. M., Taylor, J. M., Segars, L. W., Selke, V., & Taylor, T. (2020). Student perceptions of online and inperson microbiology laboratory experiences in undergraduate medical education. *Medical Education Online, 25*(1), 1710324. https://doi.org/10.1080/10872981.2019.1710324

Brown, C. E., Whaley, B., & Hyslop, R. M. (2020). Visualizing molecular structures and shapes: A comparison of virtual reality, computer simulation, and traditional modeling. *Chemistry Teacher International.* https://doi.org/10.1515/cti-2019-0009

Burte, H., Gardony, A. L., Hutton, A., & Taylor, H. A. (2017). Think3d!: Improving mathematics learning through embodied spatial training. *Cognitive Research: Principles and Implications, 2*(1), 13. https://doi.org/10.1186/s41235-017-0052-9

Business Wired. (2020). Dr. frank Phillips is first in the world to use augmented reality surgical guidance in minimally invasive spine surgery [WWW document]. https://www.businesswire.com/news/home/20200617005352/en/Dr.-Frank-Phillips-World-Augmented-Reality-Surgical. Accessed 17 May 2021.

Byrnes, K. G., Kiely, P. A., Dunne, C. P., McDermott, K. W., & Coffey, J. C. (2021). Communication, collaboration and contagion: "Virtualisation" of anatomy during COVID-19. *Clinical Anatomy, 34*(1), 82–89. https://doi.org/10.1002/ca.23649

Cao, C., & Cerfolio, R. J. (2019). Virtual or augmented reality to enhance surgical education and surgical planning. *Thoracic Surgery Clinics, 29*(3), 329–337. https://doi.org/10.1016/j.thorsurg.2019.03.010

Christ, R., Guevar, J., Poyade, M., & Rea, P. M. (2018). Proof of concept of a workflow methodology for the creation of basic canine head anatomy veterinary education tool using augmented reality. *PLoS One, 13*(4), e0195866. https://doi.org/10.1371/journal.pone.0195866

Collins, R. T., Laughlin, M. E., Lang, S. M., Bolin, E. H., Daily, J. A., Jensen, H. A., & Jensen, M. O. (2019). Real-time transthoracic vector flow imaging of the heart in pediatric patients. *Progress in Pediatric Cardiology, 53*, 28–36. https://doi.org/10.1016/j.ppedcard.2019.02.003

Curzon Artificial Eye. (2016). Notes on blindness: Into darkness VR experience [WWW document]. Curzon Artificial Eye. https://www.curzonartificialeye.com/notes-on-blindness/articles/notes-on-blindness-into-darkness-vr-experience/. Accessed 26 May 2021.

Depaoli, M. R., Karsten, F., Madreiter-Sokolowski, C. T., Klec, C., Gottschalk, B., Bischof, H., Eroglu, E., Waldeck-Weiermair, M., Simmen, T., Graier, W. F., & Malli, R. (2018). Real-time imaging of mitochondrial ATP dynamics reveals the metabolic setting of single cells. *Cell Reports, 25*(2), 501–512.e3. https://doi.org/10.1016/j.celrep.2018.09.027

Dixon, D. R., Miyake, C. J., Nohelty, K., Novack, M. N., & Granpeesheh, D. (2020). Evaluation of an immersive virtual reality safety training used to teach pedestrian skills to children with autism Spectrum disorder. *Behavior Analysis in Practice, 13*, 631–640. https://doi.org/10.1007/s40617-019-00401-1

Dunnagan, C. L., Dannenberg, D. A., Cuales, M. P., Earnest, A. D., Gurnsey, R. M., & Gallardo-Williams, M. T. (2020). Production and evaluation of a realistic immersive virtual reality organic chemistry laboratory experience: Infrared spectroscopy. *Journal of Chemical Education, 97*(1), 258–262. https://doi.org/10.1021/acs.jchemed.9b00705

Farahani, N., Post, R., Duboy, J., Ahmed, I., Kolowitz, B., Krinchai, T., Monaco, S., Fine, J., Hartman, D., & Pantanowitz, L. (2016). Exploring virtual reality technology and the oculus rift for the examination of digital pathology slides. *Journal of Pathology Informatics, 7*, 22. https://doi.org/10.4103/2153-3539.181766

Ferrell, J. B., Campbell, J. P., McCarthy, D. R., McKay, K. T., Hensinger, M., Srinivasan, R., Zhao, X., Wurthmann, A., Li, J., & Schneebeli, S. T. (2019). Chemical exploration with virtual reality in organic teaching laboratories. *Journal of Chemical Education, 96*(9), 1961–1966. https://doi.org/10.1021/acs.jchemed.9b00036

Gardner, P., Dal Borgo, M., Ruffini, V., Hughes, A. J., Zhu, Y., & Wagg, D. J. (2020). Towards the development of an operational digital twin. *Vibration, 3*(3), 235–265. https://doi.org/10.3390/vibration3030018

Gerli, P., Matteucci, N., & Whalley, J. (2020). Infrastructure provision on the margins: An assessment of broadband delivery UK. *International Journal of Public Administration, 43*(6), 540–551. https://doi.org/10.1080/01900692.2019.1638932

Goff, E. E., Reindl, K. M., Johnson, C., McClean, P., Offerdahl, E. G., Schroeder, N. L., & White, A. R. (2018). Investigation of a stand-alone online learning module for cellular respiration instruction. *Journal of Microbiology & Biology Education, 19*(2), article 1460. https://doi.org/10.1128/2Fjmbe.v19i2.1460

Gonzales, R. A., Ferns, G., Vorstenbosch, M. A. T. M., & Smith, C. F. (2020). Does spatial awareness training affect anatomy learning in medical students? *Anatomical Sciences Education, 13*(6), 707–720. https://doi.org/10.1002/ase.1949

Hanna, M. G., Ahmed, I., Nine, J., Prajapati, S., & Pantanowitz, L. (2018). Augmented reality technology using Microsoft HoloLens in anatomic pathology. *Archives of Pathology and Laboratory Medicine, 142*(5), 638–644. https://doi.org/10.5858/arpa.2017-0189-OA

Harfouche, A. L., & Nakhle, F. (2020). Creating bioethics distance learning through virtual reality. *Trends in Biotechnology., 38*(11), 1187–1192. https://doi.org/10.1016/j.tibtech.2020.05.005

Herodotou, C., Muirhead, D. K., Aristeidou, M., Hole, M. J., Kelley, S., Scanlon, E., & Duffy, M. (2020). Blended and online learning: A comparative study of virtual microscopy in higher education. *Interactive Learning Environments, 28*(6), 713–728. https://doi.org/10.1080/10494820.2018.1552874

Herst, P., & Berridge, M. (2013). Cell hierarchy, metabolic flexibility and systems approaches to cancer treatment. *Current Pharmaceutical Biotechnology, 14*(3), 289–299. https://doi.org/10.2174/13892010113140030005

HESA Finance record 2017/18 [WWW Document] 2017/18 [WWW Document]. (2019). https://www.hesa.ac.uk/collection/c17031. Accessed 15 April 2021.

Histology Guide [WWW Document]. (n.d.). http://histologyguide.org/index.html. Accessed 18 Nov 2020.

Horrigan, L. A. (2018). Tackling the threshold concepts in physiology: What is the role of the laboratory class? *Advances in Physiology Education, 42*(3), 507–515. https://doi.org/10.1152/advan.00123.2017

Hou, J., Williams, J., Botvinick, E. L., Potma, E. O., & Tromberg, B. J. (2018). Visualization of breast cancer metabolism using multimodal nonlinear optical microscopy of cellular lipids and redox state. *Cancer Research, 78*(10), 2503–2512. https://doi.org/10.1158/0008-5472.CAN-17-2618

Hu, K.-C., Salcedo, D., Kang, Y.-N., Lin, C.-W., Hsu, C.-W., Cheng, C.-Y., Suk, F.-M., & Huang, W.-C. (2020). Impact of virtual reality anatomy training on ultrasound competency development: A randomized controlled trial. *PLoS One, 15*(11), e0242731. https://doi.org/10.1371/journal.pone.0242731

Jobs, S. (n.d.) Steve Jobs on the importance of face-to-face meetings (even in the age of iPhones) | by SpeedNetworking.com | medium [WWW document]. https://medium.com/@shannonkelly_

80469/steve-jobs-on-the-importance-of-face-to-face-meetings-even-in-the-age-of-iphones-a5a4b83621a6. Accessed 23 Nov 2020.

Joda, T., Gallucci, G. O., Wismeijer, D., & Zitzmann, N. U. (2019). Augmented and virtual reality in dental medicine: A systematic review. *Computers in Biology and Medicine, 108*, 930–100. https://doi.org/10.1016/j.compbiomed.2019.03.012

Johnstone, A. H. (1991). Why is science difficult to learn? Things are seldom what they seem. *Journal of Computer Assisted Learning, 7*(2), 75–83. https://doi.org/10.1111/j.1365-2729.1991.tb00230.x

Johnstone, A. H., Sleet, R. J., & Vianna, J. F. (1994). An information processing model of learning: Its application to an undergraduate laboratory course in chemistry. *Studies in Higher Education, 19*(1), 77–87. https://doi.org/10.1080/03075079412331382163

Khan, T., Johnston, K., & Ophoff, J. (2019). The impact of an augmented reality application on learning motivation of students. *Advances in Human-Computer Interaction, 2019*. https://doi.org/10.1155/2019/7208494

Lanska, D. J., & Lanska, J. R. (2013). Medieval and renaissance anatomists: The printing and unauthorized copying of illustrations, and the dissemination of ideas. In *Progress in brain research* (pp. 33–74). Elsevier B.V.. https://doi.org/10.1016/B978-0-444-62730-8.00002-5

Larmore, R., Knaus, M., Dascalu, S., & Harris, F. C. (2005). Virtual environment for on-campus orientation. In *Proceedings - 2005 International Symposium on Collaborative Technologies and Systems* (pp. 259–266). IEEE. https://doi.org/10.1109/ISCST.2005.1553321

Larson, S. D., Gleeson, P., & Brown, A. E. X. (2018). Connectome to behaviour: Modeling Caenorhabditis elegans at cellular resolution. *Philosophical Transactions of the Royal Society B: Biological Sciences, 373*, 20170366. https://doi.org/10.1098/rstb.2017.0366

LCME. (2019). Functions and structure of a medical school standards for accreditation of medical education programs leading to the MD degree.

Lee, L. M. J., Goldman, H. M., & Hortsch, M. (2018). The virtual microscopy database-sharing digital microscope images for research and education. *Anatomical Sciences Education, 11*(5), 510–515. https://doi.org/10.1002/ase.1774

Leveritt, S., McKnight, G., Edwards, K., Pratten, M., & Merrick, D. (2016). What anatomy is clinically useful and when should we be teaching it? *Anatomical Sciences Education, 9*(5), 468–475. https://doi.org/10.1002/ase.1596

Live Lab: Liverpool Immersive Experience Lab. (2019). [WWW Document]. https://www.ljmu.ac.uk/~/media/files/ljmu/about-us/faculties-and-schools/fet/live-lab.pdf?la=en. Accessed 23 May 2021.

Loertscher, J., Green, D., Lewis, J. E., Lin, S., & Minderhout, V. (2014). Identification of threshold concepts for biochemistry. *CBE Life Sciences Education, 13*(3), 516–528. https://doi.org/10.1187/cbe.14-04-0066

Makransky, G., Thisgaard, M. W., & Gadegaard, H. (2016). Virtual simulations as preparation for lab exercises: Assessing learning of key laboratory skills in microbiology and improvement of essential non-cognitive skills. *PLoS One, 11*, e0155895. https://doi.org/10.1371/journal.pone.0155895

Moxham, B. J., Emmanouil-Nikoloussi, E., Brenner, E., Plaisant, O., Brichova, H., Kucera, T., Pais, D., Stabile, I., Borg, J., Scholz, M., Paulsen, F., Luis Bueno-López, J., Arráez-Aybar, L. A., de Caro, R., Arsic, S., Lignier, B., & Chirculescu, A. (2017). The attitudes of medical students in Europe toward the clinical importance of histology. *Clinical Anatomy, 30*(5), 635–643. https://doi.org/10.1002/ca.22889

Nayak, S., Liu, H., Hsu, G. I., & Iwasa, J. H. (2020). Using 3D animation to visualize hypotheses. *Trends in Biochemical Sciences, 45*(7), 633–634. https://doi.org/10.1016/j.tibs.2020.04.009

Ozdemir, R. A., Tadayon, E., Boucher, P., Momi, D., Karakhanyan, K. A., Fox, M. D., Halko, M. A., Pascual-Leone, A., Shafi, M. M., & Santarnecchi, E. (2020). Individualized perturbation of the human connectome reveals reproducible biomarkers of network dynamics relevant to cognition. *Proceedings of the National Academy of Sciences of the United States of America, 117*(14), 8115–8125. https://doi.org/10.1073/pnas.1911240117

Paech, D., Giesel, F. L., Unterhinninghofen, R., Schlemmer, H. P., Kuner, T., & Doll, S. (2017). Cadaver-specific CT scans visualized at the dissection table combined with virtual dissection tables improve learning performance in general gross anatomy. *European Radiology, 27*, 2153–2160. https://doi.org/10.1007/s00330-016-4554-5

Pei, L., & Wu, H. (2019). Does online learning work better than offline learning in undergraduate medical education? A systematic review and meta-analysis. *Medical Education Online, 24*, 1. https://doi.org/10.1080/10872981.2019.1666538

Peterson, C. N., Tavana, S. Z., Akinleye, O. P., Johnson, W. H., & Berkmen, M. B. (2020). An idea to explore: Use of augmented reality for teaching three-dimensional biomolecular structures. *Biochemistry and Molecular Biology Education, 48*(3), 276–282. https://doi.org/10.1002/bmb.21341

Pottle, J. (2019). Virtual reality and the transformation of medical education. *Future Healthcare Journal, 6*(3), 181–185. https://doi.org/10.7861/fhj.2019-0036

QAA. (2015). Subject benchmark statement UK quality code for higher education part A: Setting and maintaining academic standards Biosciences [WWW Document]. https://www.qaa.ac.uk/docs/qaa/subject-benchmark-statements/subject-benchmark-statement-biosciences.pdf. Accessed 27 May 2021.

Ragnedda, M., & Ragnedda, M. (2020). Connecting the digital underclass. In *Enhancing digital equity* (pp. 85–104). Springer International Publishing. https://doi.org/10.1007/978-3-030-49079-9_5

Reeves, L., Bolton, E., Bulpitt, M., Scott, A., Tomey, I., Gates, M., & Baldock, R. A. (2021). Use of augmented reality (AR) to aid bioscience education and enrich student experience. *Research in Learning Technology, 29*. https://doi.org/10.25304/rlt.v29.2572

Rockarts, J., Brewer-Deluce, D., Shali, A., Mohialdin, V., & Wainman, B. (2020). National Survey on Canadian undergraduate medical programs: The decline of the anatomical sciences in Canadian medical education. *Anatomical Sciences Education, 13*(3), 381–389. https://doi.org/10.1002/ase.1960

Roos, G., Oláh, J., Ingle, R., Kobayashi, R., & Feldt, M. (2020). Online conferences – Towards a new (virtual) reality. *Computational and Theoretical Chemistry*. Elsevier B.V.*, 1189*, 112975. https://doi.org/10.1016/j.comptc.2020.112975

Sarma, G. P., Lee, C. W., Portegys, T., Ghayoomie, V., Jacobs, T., Alicea, B., Cantarelli, M., Currie, M., Gerkin, R. C., Gingell, S., Gleeson, P., Gordon, R., Hasani, R. M., Idili, G., Khayrulin, S., Lung, D., Palyanov, A., Watts, M., & Larson, S. D. (2018). OpenWorm: Overview and recent advances in integrative biological simulation of *Caenorhabditis elegans*. *Philosophical Transactions of the Royal Society B: Biological Sciences, 373*, 20170382. https://doi.org/10.1098/rstb.2017.0382

Schrader, T., Niepage, S., Leuthold, T., Saeger, K., Schluns, K., Hufnagl, P., Kayser, K., & Dietel, M. (2006). The diagnostic path, a useful visualisation tool in virtual microscopy. *Diagnostic Pathology, 1*, 40. https://doi.org/10.1186/1746-1596-1-40

Selvig, D., Holaday, L. W., Purkiss, J., & Hortsch, M. (2015). Correlating students' educational background, study habits, and resource usage with learning success in medical histology. *Anatomical Sciences Education, 8*(1), 1–11. https://doi.org/10.1002/ase.1449

Sharma, S., Rajeev, S. P., & Devearux, P. (2015). An immersive collaborative virtual environment of a university campus for performing virtual campus evacuation drills and tours for campus safety. In *2015 International Conference on Collaboration Technologies and Systems, CTS 2015* (pp. 84–89). Institute of Electrical and Electronics Engineers Inc.. https://doi.org/10.1109/CTS.2015.7210404

Shatri, K., & Buza, K. (2017). The use of visualization in teaching and learning process for developing critical thinking of students. *European Journal of Social Sciences Education and Research, 4*(1), 71–74. https://doi.org/10.26417/ejser.v9i1.p71-74

The Open Science Laboratory: Histology & histopathology. (n.d.). [WWW document]. https://learn5.open.ac.uk/course/format/sciencelab/section.php?name=histology_microscope. Accessed 18 Nov 2020.

Tibell, L. A. E., & Harms, U. (2017). Biological principles and threshold concepts for understanding natural selection: Implications for developing visualizations as a pedagogic tool. *Science and Education, 26*, 953–973. https://doi.org/10.1007/s11191-017-9935-x

Triepels, C. P. R., Smeets, C. F. A., Notten, K. J. B., Kruitwagen, R. F. P. M., Futterer, J. J., Vergeldt, T. F. M., & van Kuijk, S. M. J. (2020). Does three-dimensional anatomy improve student understanding? *Clinical Anatomy, 33*(1), 25–33. https://doi.org/10.1002/ca.23405

Tsay, C. H., Kofinas, A. K., Trivedi, S. K., & Yang, Y. (2020). Overcoming the novelty effect in online gamified learning systems: An empirical evaluation of student engagement and performance. *Journal of Computer Assisted Learning, 36*(2), 128–146. https://doi.org/10.1111/jcal.12385

Visel, A., Thaller, C., & Eichele, G., (2004). GenePaint.org: An atlas of gene expression patterns in the mouse embryo. *Nucleic Acids Research 32*, D552–D556. https://doi.org/10.1093/nar/gkh029

Waclaw, B., Bozic, I., Pittman, M. E., Hruban, R. H., Vogelstein, B., & Nowak, M. A. (2015). A spatial model predicts that dispersal and cell turnover limit intratumour heterogeneity. *Nature, 525*, 261–264. https://doi.org/10.1038/nature14971

Wilson, A. B., Miller, C. H., Klein, B. A., Taylor, M. A., Goodwin, M., Boyle, E. K., Brown, K., Hoppe, C., & Lazarus, M. (2018). A meta-analysis of anatomy laboratory pedagogies. *Clinical Anatomy, 31*(1), 122–133. https://doi.org/10.1002/ca.22934

Winschel, G. A., Everett, R. K., Coppola, B. P., Shultz, G., & v, Lonn, S. (2015). Using jigsaw-style spectroscopy problem-solving to elucidate molecular structure through online cooperative learning. *Journal of Chemical Education, 92*(7), 1188–1193. https://doi.org/10.1021/acs.jchemed.5b00114

XR for Business. (2018). Archives - Raconteur [WWW Document]. https://www.raconteur.net/report/xr-business-2018/. Accessed 17 Nov 2020.

Chapter 18
The Future of Biomedical and Life Science Education: Evidence-Based Future Directions

Kristina Lisk, Maria Mylopoulos, and Nicole N. Woods

Abstract In this chapter, we outline principles and practices for designing future education in the biomedical and life sciences that aim to support the development of adaptive expertise. We draw on examples from training in health professions and literature from cognitive psychology and learning sciences to make the case for a new way forward, emphasizing the development of the learner as a future expert. This goal of developing adaptive experts concurrently promotes both applying learners' past knowledge to solve known problems and preparing learners to create new knowledge and solutions when faced with novelty, complexity, and ambiguity. We propose that when instruction is designed to support the ongoing development of adaptive expertise, learners benefit longer term, both in terms of retention and application/transfer of knowledge. We argue that rather than an exclusively process-based focus on the learner experience and the modality by which instruction

K. Lisk (✉)
Faculty of Health Sciences and Wellness, Humber College Institute of Technology and Advanced Learning, Toronto, ON, Canada

The Wilson Centre, University Health Network, Toronto, ON, Canada

Division of Anatomy, Department of Surgery, Temerty Faculty of Medicine, University of Toronto, Toronto, ON, Canada
e-mail: kristina.lisk@utoronto.ca

M. Mylopoulos
The Wilson Centre, University Health Network, Toronto, ON, Canada

Department of Paediatrics, Temerty Faculty of Medicine, University of Toronto, Toronto, ON, Canada

Health Professions Education Research, Institute of Health Policy Management and Evaluation, Dalla Lana School of Public Health, University of Toronto, Toronto, ON, Canada
e-mail: mylopoulos@utoronto.ca

N. N. Woods
The Wilson Centre, University Health Network, Toronto, ON, Canada

The Institute for Education Research (TIER), University Health Network, Toronto, ON, Canada

Department of Family and Community Medicine, University of Toronto, Toronto, ON, Canada
e-mail: nikki.woods@utoronto.ca

H. J. Witchel, M. W. Lee (eds.), *Technologies in Biomedical and Life Sciences Education*, Methods in Physiology, https://doi.org/10.1007/978-3-030-95633-2_18

is delivered, educators should approach curriculum design with a lens that values outcomes of learning (e.g., clinical competencies) and underlying mechanisms that must be invoked to achieve these desired outcomes. As an example, when the biomedical sciences are explicitly integrated with clinical instruction, learners develop better conceptual knowledge, which will be retained long term and utilized when learning novel but related concepts or when solving future problems. When the purpose of education is redirected to focus on desired outcomes of learner, educators can shift their attention to incorporating evidence-informed processes into their curricula that can have profound effects on student learning.

18.1 Introduction

Much like their counterparts in the natural sciences, social sciences, arts, and humanities, teachers in biomedical and life sciences are faced with the challenge of providing high-quality postsecondary education for society's future experts. Despite the enormity of the task, few teachers at the world's leading universities have advanced training in education or pedagogy. Nonetheless, many teachers and entire programs, committed to nurturing future generations, are eager to adopt new education technologies and approaches. In fact, the future of education in the biomedical and life sciences depends both on universities creating the opportunities and on faculty improving teaching and assessment approaches. With supportive administration and leadership, education programs in the biomedical and life sciences can provide ample opportunity for education innovation. Curriculum designers and educators can create, implement, and evaluate new ways of teaching and learning that can stand as examples for education in other disciplines.

Ideally, curricular innovation in these programs should be informed by formal requirements (accreditation bodies, institutional mandates, etc.), understanding of the learner experience, education science, and a clear picture of the intended outcome of instruction. Yet, history tells us that these core components rarely converge in the creation of new education programming (Whitehead et al., 2013; Whitehead, 2017). Instead, there is a tendency for curriculum designers and teachers to focus on formal expectations (e.g., accreditation standards), experience of the individual learners (e.g., learner satisfaction data), and pedagogical tools most popular at any given time (e.g., virtual learning and competency-based education). In creating a new program or when modifying an existing one, it is common for educators to forgo a careful examination of intended outcomes and the wealth of guidance available from the learning sciences. Undue emphasis on the learner and medium is a heuristic that often leads to change for the sake of change and ultimately no impact on the development of future experts. With the best of intentions, educators might find themselves on a never-ending cycle of reform in search of the latest and greatest initiatives (Whitehead, 2017). Creating a curriculum that is evidence based, meets needs of learners, and supports the development of experts society needs requires a significant shift in ways in which we design curriculum. In this chapter, we outline some principles and practices for the future of education in the biomedical and life sciences. We will draw on examples from training in health professions and as well the literature from cognitive psychology, education, and cognitive science, to make the case for a new way forward. Rather than exclusive focus on the learner experience and medium of instruction, we will challenge educators to approach curriculum design with a lens that values outcomes of learning and mechanisms that must be invoked to achieve desired outcomes.

18.2 Integration of Biomedical Sciences in North American Medical Education: A Case Study on the "Curricular Carousel"

Health profession education typically includes training in both the clinical skills necessary for the profession and a solid foundation in the basic biomedical sciences. Traditionally, the basic biomedical sciences considered core to human health and disease include subject areas, such as clinical chemistry, human biology, and pharmacology (Quality Assurance Agency for Higher Education, 2019). An understanding of this content is usually viewed as foundational to modern medicine and significantly contributes to the foundational training of health professionals. Despite the recognized importance of basic sciences, historically, the basic and clinical sciences were separated structurally and functionally within the medical school, basic science courses taught by basic science instructors and clinical courses by clinical instructors (Koens et al., 2006). Irrespective of consistent acknowledgment of the value of basic science education in accreditation standards etc., the two forms of training were often seen as entirely separate worlds (Patel et al., 1988), with clinical sciences being the more relevant and important to eventual medical expertise (Koens et al., 2005; Patel & Groen, 1986).

This segregation between the basic and clinical sciences has been a long-standing issue within health profession education, and a recurring theme within curricular reform efforts across health professions is the call for better integration of the basic and clinical sciences (Finnerty et al., 2010; Association of Faculties of Medicine of Canada, 2010). Physicians, basic science teachers, and students have continuously lamented the need to shift to a more integrated model. However, despite the recognized importance of integrated knowledge and ongoing reform efforts, basic science instruction continues to be largely compartmentalized from clinical training (Benner et al., 2010; Irby et al., 2010). As Whitehead (2017) has argued, reform efforts have been historically focused on localized learning outcomes (e.g., performance on assessments) and shifts in learning modalities (e.g., problem-based learning), rather than the broader question of what integration offers to future health professionals. As a result, through each iteration of curriculum renewal, ongoing challenges related to integration have simply been positioned as new and minor adjustments to the existing curriculum are often considered acceptable solutions which essentially has led to "reform without change" (Ludmerer, 1999; Whitehead et al., 2013). This cyclical process of tinkering with the curriculum and focusing on the experience of the learner is akin to ponies on a carousel, where recurring themes of curricular reform are ponies on the merry-go-round, perpetually coming around and around, over and over again (Whitehead, 2017). A closer look at how biomedical and life science curriculum within health professions is traditionally designed may provide insight as to why current integration reform efforts have effectively led to little change. Broadly, across health professions, the development of programs and curriculum is guided by conceptual frameworks which outline essential competencies and standards that learners are to achieve in order to enter practice safely, competently, and ethically (College of Nurses of Ontario, 2018; Frank et al., 2015). Further, entry-to-practice competencies set by each respective college (e.g., nursing, physicians, and pharmacists) are used as a benchmark by these governing bodies to evaluate and approve educational programs while also functioning to guide the assessment of learners' competence as they proceed through their training programs. Program planners and curriculum designers are also tasked with ensuring that new and existing programs meet external accreditation requirements, and thus, these standards also play an integral role in curriculum development, revision, and renewal (Postsecondary Education Quality Assessment Board, 2020). Together, these standards ultimately become building blocks which guide the development of program learning outcomes and course-level learning objectives. Often, it then becomes the responsibility of curriculum committees and educators to design learning activities/curricula that will facilitate the achievement of these learning outcomes, including the integration of biomedical and clinical sciences within their specific program structure (Koens et al., 2006). As such, the focus of curriculum design becomes centered around the curriculum in effort to meet formal learning requirements.

18.3 Lessons Learned and a Way Forward

Our mapping of changes in health profession education landscape regarding basic science integration provides many valuable lessons on how to approach education in the biomedical and life sciences moving forward. Most notably, reform efforts have consistently continued to emphasize what has been termed "unproductive success" for learners (Kapur, 2016). Unproductive success is the use of instruction and assessments that reward performance in the short term, rather than long-term learning. For example, repetition and rote memorization are highly effective for success on immediate assessments that focus on retention. This might appear to be resulting in student success, but learning science has repeatedly shown that these types of short-term outcomes do not transfer to long-term learning (Dochy et al., 2003; Derry et al., 2006; Hmelo, 1998). Thus, they are understood to be unproductive. As an example, Dean and Kuhn (2006) demonstrate that with respect to learning about the scientific inquiry process, both direct instruction and direct instruction followed by problem-solving produce a significant level of correct performance immediately after learning. However, over the longer term, direct instruction was neither necessary nor sufficient for long-term learning. Instruction that focuses solely on clinical science (rather than integrated basic science) can also be understood as a form of instruction that results in unproductive success.

One way to shift curricula away from unproductive success requires that rather than focusing on curricular outcomes or learner experiences, instead educators focus on the development of the learner as a future expert. We argue that a powerful framework for this is adaptive expertise. Adaptive expertise is a framework of expert development and performance that emphasizes complementary routine and nonroutine problem-solving capabilities (Mylopoulos et al., 2018). Adaptive experts are able to both efficiently apply their past knowledge to effectively solve known problems and create new solutions when faced with novelty, complexity, and ambiguity in their work (Schwartz et al., 2005). While the former capability can be the product of instruction and assessment that emphasizes retention and replication, the latter is understood to require instruction and assessment that support the development of conceptual knowledge and "preparation for future learning," essentially decentering the curriculum away from acquisition of knowledge toward the acquisition and generation of new knowledge to support later problem-solving.

18.4 Cognition Matters: Supporting the Development of Conceptual Knowledge at the Session Level

When the purpose of health profession education is redirected to focus on preparing learners to become future adaptive experts rather than the learner experience or medium through which the curriculum is delivered, educators can shift their attention to evidence-based processes that will directly impact outcomes in student learning. Central to this shift is a focus on student cognition (Kulasegaram et al., 2013). Importantly, benefits of this focus are not limited to classroom-based instruction but can also be applied to the design of workplace learning which relies on apprenticeship. For example, a recent study demonstrated benefits of understanding how different patterns of learner participation in the workplace can result in different trajectories of expert development, concluding that more purposeful consideration of specific cognitive affordances of workplace learning experiences is needed to optimize learner outcomes (Brydges et al., 2020). Specific to the biomedical and life sciences, we suggest that educators and curriculum designers need to carefully consider not only *what* content is essential to be included within a curriculum but also *how* this content is presented to learners. It is suggested that we need to move beyond the historical conversation of how much basic science is required or the insufficiency of basic science in health profession training and turn our attention to reform efforts that will optimize educational opportunities created for learners (Whitehead, 2013).

A well-studied instructional approach that has been shown to maximize the impact of basic science knowledge is integrated instruction (Woods, 2007). This type of integration occurs at the individual session level and can be contrasted with integration efforts that occur within the curriculum, such as horizontal and vertical integration or contextualized teaching (Kulasegaram et al., 2013). Key to this instructional approach is the explicit and mechanistic use of basic science (Kulasegaram et al., 2015). For example, in one of the foundational studies by Woods et al. (2006), medical students were taught signs and symptoms of neurological and rheumatic diseases with either accompanying causal biomedical information or associated epidemiology data. Immediately following this instruction and also 1 week later, students in both groups completed a diagnostic accuracy test and a test on basic recall of signs and symptoms associated with each disease. Results of this study showed no difference on diagnostic performance between the two groups immediately after learning. However, after the 1-week time delay, students who were taught underlying causal mechanisms demonstrated superior diagnostic accuracy compared to those who were taught signs and symptoms of each disease along with the epidemiological information. At first glance, it seems plausible that learning causal biomedical explanations enabled learners to simply better remember clinical features associated with each disease, but performance on the basic recall test was the same for both groups. Taken together, findings of this study suggest that explicitly exposing learners to underlying causal mechanisms that govern why specific clinical features are associated with each disease helps learners to develop a coherent mental representation which they can rely on over time (Woods et al., 2006). In this way, integration is understood as a cognitive activity that occurs within the learner's mind rather than a manifestation of the curriculum (Kulasegaram et al., 2013).

Further research in clinical reasoning has examined the effect of presentation and sequence of causal information with clinical instruction during individual teaching sessions. This experimental work has importantly shown that simply creating proximity between basic and clinical science instructions does not guarantee cognitive integration (Baghdady et al., 2013; Kulasegaram et al., 2015; Lisk et al., 2017). Thus, simply rearranging modules or courses within a program structure, such as having students in the anatomy lab in the morning, followed by clinical skill instruction in the afternoon, may not sufficiently support learners in developing rich conceptual knowledge. Again, the key strategy is to create learning opportunities that will explicitly provide or make evident, causal links between different learning domains. As an example, while students work through a dissection of the coronary vessels of the heart in the anatomy lab, they can also be simultaneously guided through a discussion on why specific clinical symptoms of heart failure present as a result of occlusion in these arteries. The facilitation of this discussion could be supported in a variety of ways, including making use of technology, such as an e-module that provides students with specific prompts and feedback. Likewise, during clinical skill instruction on the precordial exam, an emphasis should be placed on why normal and abnormal heart sounds present. These two simple examples demonstrate how the overt use of causal biomedical mechanisms can be used to

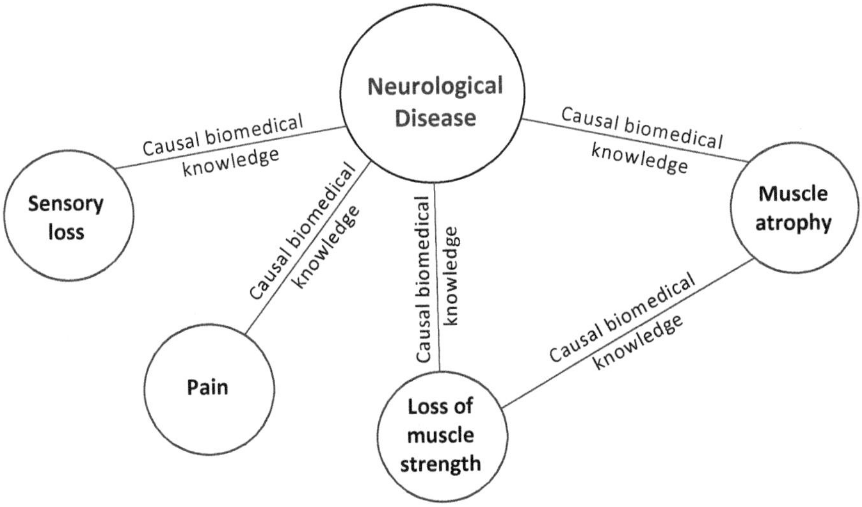

Fig. 18.1 Semantic network example of a neurological disease

facilitate cognitive integration and the development of learners' conceptual knowledge.

Experimental studies described thus far, along with others from across health disciplines, show that one way to maximize learning of the biomedical and life sciences is to use them as a tool to help learners more effectively understand and organize clinical concepts (Baghdady et al., 2013; Chaudhary et al., 2019; Goldszmidt et al., 2012; Lisk et al., 2016; Woods et al., 2006). More specifically, in the context of teaching and learning diagnostic reasoning, the explicit use of causal biomedical mechanisms during instruction can support learners' conceptual coherence of disease categories. To understand the concept of conceptual coherence, it is helpful to think about the role of biomedical and clinical knowledge in the mental semantic network of learners. A mental semantic network is a knowledge representation that consists of concepts (nodes) and links between concepts (lines) that exist in a person's memory (Murphy & Medin, 1985). As shown in Fig. 18.1, for the mental representation of a disease category, the nodes of the semantic network represent clinical features associated with the disease, and the connectors represent causal biomedical knowledge. Thus, for a categorization task such as diagnosis, it is obvious that causal biomedical knowledge plays a critical role in understanding why specific signs and symptoms are associated with a particular disease (Woods et al., 2005). Further, these critical connections between clinical features help create coherence for learners and thereby support the development of conceptual knowledge. So far, we have highlighted the value of conceptual knowledge through experimental studies which have shown that integrated basic science instruction compared to basic or clinical science instruction alone results in superior diagnostic performance (Baghdady et al., 2013; Kulasegaram et al., 2015). However, conceptual knowledge has also been shown to enhance problem-solving of complex cases and improve procedural skills training in novice learners (Woods et al., 2007; Cheung et al., 2018).

18.5 Preparation for Future Learning

For teachers outside of health professions, it is important to note that beyond clinical reasoning and procedural skill development, integrated basic science instruction has also been shown to support learners' later acquisition of new knowledge (Mylopoulos et al., 2016; Mylopoulos & Woods, 2014). This concept is known as preparation for future learning (PFL), and it is generally understood as "the ability to learn new information from available resources, relate new learning to past experiences and demonstrate innovation and flexibility in problem solving" (Bransford & Schwartz, 1999; Mylopoulos & Woods, 2014). It can be argued that a key goal or outcome of training in the biomedical and life sciences is to provide learners with tools that they will need to solve future problems since it is unlikely that formal educational experiences will expose learners to all scenarios they may encounter in practice. In the context of health profession education, it has been shown that integrated basic science instruction is an educational approach that supports PFL. As an example, when medical students were taught neurological conditions using either clinically focused instruction or basic science instruction, it was found that students who were taught the underlying anatomical pathology associated with each condition were able to capitalize on this knowledge when completing a future learning task. It is suggested that by learning underlying causal mechanisms during the initial learning phase, students developed conceptual knowledge that was then applied to learning the novel, related content (Mylopoulos & Woods, 2014).

18.6 Productive Failure

Similarly, the instructional approach known as productive failure, which initially engages learners with problem-solving activities beyond their initial capabilities, followed by direct instruction, also has been found to support the development of conceptual knowledge and future successful problem-solving (Kapur, 2008; Kapur & Bielaczyc, 2012). While this type of instruction minimizes learners' success in the short term, it is understood to be productive since the struggle experienced during the initial problem-solving phase supports learners in achieving success in the longer term (Kapur, 2016; Schwartz & Bransford, 1998). Examples of learning interventions which support productive failure include guided discovery learning experiences, such as invention activities or the use of contrasting cases, which are then followed by direct instruction (Schwartz et al., 2011; Schwartz & Martin, 2004). As an example, in one study (Steenhof et al., 2019) that involved teaching pharmacy students about creatinine clearance, those assigned to the productive failure group were first told about the problem of estimating creatinine clearance and then were asked to invent their own formula, using data provided, that would best serve as a solution to this problem. In comparison, students assigned to the direct instruction group were told about the same problem of estimating creatinine clearance; however, they were given a formula that served as a potential solution. Following this first phase, all students were then provided with direct instruction along with an opportunity to practice calculating creatinine clearance using a standard formula. After this consolidation phase, all students completed a series of questions which assessed knowledge acquisition, application, and PFL. No differences were observed between groups on content knowledge or the ability to apply learned content; however, students in the productive failure group performed significantly better on the PFL assessment (Steenhof et al., 2019). Essentially the "failure activity" was productive as students who completed this activity were better able to solve novel but related problems. These findings contribute to the growing body of evidence which demonstrates a clear benefit of providing learners with an opportunity to struggle and problem-solve prior to receiving direct instruction (Kapur, 2016; Steenhof et al., 2019). In this way, challenging learners with ill-structured problems before providing structured teaching creates an opportunity for learners to activate their relevant prior knowledge, which may, in turn, help them to recognize both inconsistencies and limits of their prior knowledge related to the particular concept understudy (DeCaro & Rittle-Johnson, 2012; Schwartz et al., 2011). In addition, it is suggested that when direct instruction follows a failed problem-solving activity, learners are able to better attend to and encode the novel concept as they compare and contrast their incorrect answers with correct solutions (Kapur, 2014; Schwartz et al., 2011). These cognitive processes all support the development of conceptual knowledge which contributes to the development of adaptive expertise. As showcased here, designing learning experiences and assessments that support PFL is another way that curriculum designers and educators can support and prepare learners to be future adaptive experts that society needs.

Importantly, these types of learning experiences can be implemented using various pedagogical tools, including synchronous and asynchronous virtual platforms. They may also vary in complexity and do not necessarily need to be time intensive. In one recent study, we utilized a popular digital learning platform to maximize benefits of productive failure. The platform is designed to allow learners to choose the level of difficulty of test questions used for self-directed learning (easy, medium, or hard). While the natural tendency of most learners left to their own devices is to master easier questions before proceeding to more difficult questions, participants in our study were randomized to either "high struggle" (hard questions followed by easy questions) or "low struggle" (easy questions followed by more difficult) instructional conditions. Our data shows that, compared to the low struggle condition, learners required to complete difficult questions first showed superior performance on a preparation for future learning assessment (Steenhof, Mylopoulos, & Woods, in progress). Thus, we were able to successfully utilize an existing learning technology to optimize productive struggle with minimal technical modifications. This study is one demonstration that existing digital platforms can be easily adapted to engage cognitive mechanisms of learning.

18.7 Self-Explanation

A number of studies have shown that specific learning techniques, including self-explanation and retrieval practice, foster and support the development of learners' conceptual knowledge (Baghdady et al., 2014; Chamberland et al., 2011, 2013, 2015a, b; Larsen et al., 2013b; Versteeg et al., 2019). As an example, Chamberland and colleagues have shown that when medical students generate self-explanations aloud when problem-solving through unfamiliar clinical cases, they develop superior diagnostic performance over time, in comparison with students who do not self-explain during initial clinical problem-solving (Chamberland et al., 2011). Interestingly, the added benefit of generating inferences to oneself was not observed when reasoning through familiar clinical problems. Further investigation of students' self-explanations revealed that students generated more self-explanation segments along with an increased number of biomedical inferences when reasoning through less familiar clinical cases. Thus, when students were less familiar with a clinical context, they were more likely to actively retrieve their basic science knowledge to try to make sense of presenting clinical signs and symptoms (Chamberland et al., 2011). Findings of this study, along with others, demonstrate that knowledge elaboration fosters deep learning and that this learning technique holds the potential to support cognitive integration. It is suggested that as learners generate inferences to themselves, they are able to integrate new knowledge with their prior knowledge and that through the process of overtly trying to make sense of learning material, they may identify misconceptions in their current understanding and as a result restructure their mental model (Chi, 2000).

It is also important to acknowledge the evidence that suggests that there are certain conditions when self-explanation may not be the optimal learning technique to support the development of conceptual knowledge. Two recent studies have shown that prompting learners to elaborate on underlying causal mechanisms while learning disease categories did not improve diagnostic accuracy compared to not self-explaining at all (Lisk et al., 2017; Peixoto et al., 2017). Findings of these studies seem almost counterintuitive to what we have just described as beneficial. However, a closer look at learning materials used in these experimental studies sheds light on additional parameters that should be carefully considered when designing and implementing learning strategies to support the development of conceptual knowledge at the session, course, or program level. First, the learning strategy should foster holistic processing. This means that prompting learners to only elaborate on part of a concept (or disease) likely will not foster a holistic understanding and as a result the potential benefit of the technique may be lost. Second, not all integrated learning materials will require additional learning techniques to enhance integrated understanding. Foundational work by Chi (1994) has shown that self-explanation is most beneficial when learners work on nonroutine, complex problems. Further, research in higher education and cognitive psychology have also reported that the use of certain instructional strategies results in an expertise reversal effect, where learners with high prior domain/procedural knowledge are negatively impacted by employing the same strategies that benefit learners with low prior domain knowledge and vice versa (Leppink et al., 2012; van Gog & Rummel, 2010). As an example, Rey and Fischer (2013) report that when teaching statistical data analysis in a computer-based learning environment, novices who received explanations as part of their instruction performed better on a transfer task in comparison with novices who did not receive explanations. However, learners with advanced prior knowledge, who were classified as experts in this study, who received explanations during instruction performed worse on the transfer task compared to experts who did not receive explanations. This suggests that learners' prior knowledge and experience need to be carefully considered when implementing additional learning techniques to support instruction.

18.8 Retrieval Practice

As previously discussed in Chap. 5 (adaptive learning) and Chap. 6 (flipped classroom) in this volume, retrieval practice is another robust strategy that has been shown to enhance higher-order learning over the long term, including critical evaluation, clinical skills, and diagnostic reasoning (Agarwal, 2019; Baghdady et al., 2014; Dobson et al., 2018; Kromann et al., 2009; Larsen et al., 2013a). For example, in one study that examined the impact of retrieval practice on diagnostic accuracy, undergraduate dental students were taught radiographic features and the underlying pathophysiology of intrabony abnormalities followed by either a comprehension test on pathophysiology mechanisms or additional study materials which highlighted these mechanisms. More specifically, the comprehension test consisted of simple multiple-choice questions, and corrective feedback was provided sequentially after each question was completed. On both immediate and delayed testing (1 week), students who completed the comprehension test had superior diagnostic accuracy compared to those who were given additional study materials (Baghdady et al., 2014). This observed phenomenon is known as the testing effect, and findings of this study example highlight the value and potential of using learning-oriented assessments to support deep and meaningful learning (Roediger & Karpicke, 2006). Decades of research unequivocally show that practice tests and other techniques that support active retrieval, including flash cards and practice problems, are more beneficial than other non-testing learning conditions (Adesope et al., 2017). However, we emphasize that in order to support higher-order learning and transfer of knowledge, retrieval techniques need to be carefully designed so that they emphasize connections between learning materials and underlying concepts. It would be futile to simply add in a myriad of fact-based tests or activities that focus only on the recall of isolated facts if the goal is to enhance learners' conceptual knowledge and their subsequent development of adaptive expertise.

18.9 Context Matters

Teaching and learning in the biomedical and life sciences can occur in complex learning environments, including large lecture halls, clinics, laboratories, and simulation centers. Considering the context in which teaching and learning occur is also especially relevant right now (year 2021) given that many educational programs have had to adapt and shift their curricula to be facilitated online due to COVID-19. Thus, while, up to this point, we have presented much evidence on specific cognitive mechanisms that are known to support the development of adaptive expertise, it is also important to acknowledge and consider challenges and opportunities that different contextual factors have on learning. These include, but are not limited to, the impact of individual differences and lived experiences of both learners and teachers and the interaction between the two, along with a multitude of environmental factors that can impact learning.

18.10 Cohort Effects

As curriculum designers and educators develop instruction to prepare future experts, attention should be given to individual differences of learners in which the curriculum is being developed for. Within a single cohort of learners, context- and content-specific knowledge can vary significantly based on students' past educational and lived experiences. This can have a significant impact on learning. As mentioned briefly before, it has been shown that certain instructional strategies that are found to benefit novice learners may negatively impact learners with more advanced knowledge (Kalyuga & Renkl, 2010). For example, in statistics education, learners with low prior knowledge were found to benefit from instruction that made use of worked examples, whereas performance of those with high prior knowledge was negatively impacted when employing the same approach (Leppink et al., 2012). This highlights an expertise reversal effect which is postulated to occur due to the overlap of learners' existing knowledge structures and the instructional guidance provided via the learning strategy (e.g., worked examples; Kalyuga et al., 2012). Thus, an awareness of learners' current knowledge is critical when designing instruction and selecting specific learning strategies to support desired outcomes.

As another example, a recent study which utilized a virtual platform to deliver integrated basic science instruction on musculoskeletal pathologies found that allied health students who completed holistic self-explanation tasks during learning developed a more coherent understanding of each pathology in comparison with students who reviewed complete explanations of underlying mechanisms of same pathologies

(Lisk, Silvestri, & Woods, forthcoming). Further analysis of self-explanation responses also revealed that students did very well at connecting their basic science knowledge with clinical features of each disease. In this context, it was required that all learners had a basic understanding of relevant anatomy before completing study materials. Thus, in this instance, it makes sense that learners would benefit more from an instructional strategy that supports them in actively integrating new knowledge with their prior knowledge rather than providing them with an additional task that does not capitalize on their current abilities. Considering the level of prior knowledge of learners can be especially important when developing or modifying curricula for entry-level courses. For instance, a first-year nursing or medical class likely includes students with educational backgrounds outside of the biomedical and life sciences along with mature students who bring a wealth of life experience. Thus, when it is suspected that a learner's current knowledge may change their learning trajectory of specific content or concepts, one possible way to tailor the curricula to individual learners is to design instruction for online, asynchronous delivery. In doing so, a learning management system can be optimized to incorporate real-time feedback of a learner's current expertise and as result adapt the sequence and learning strategies offered to each individual.

Our position on the importance of ensuring that learners meaningfully engage in challenging learning opportunities at every level of development (i.e., K-12, higher education, and novice to expert) is supported by theories of expert development (Bereiter & Scardamalia, 1993; Bransford et al., 2000). These theories suggest that waiting until learners begin practice to introduce them to more complex problem-solving requiring "innovation" makes it less likely that they will achieve "true" expertise. In particular, research in adaptive expertise suggests that there are two distinct trajectories of expert development, one toward "true" or adaptive expertise and one toward "swelling the ranks of the mediocre" (Bereiter & Scardamalia, 1993; Bransford et al., 2000). Thus, adaptive expertise in any domain is thought to be the product of a learned skill set that develops throughout training as learners balance opportunities to learn and apply all forms of knowledge with opportunities to cope with uncertainty and novelty through the construction of new knowledge.

18.11 Motivational Factors

There is also evidence from the motivation science literature which shows that some individuals will thrive and respond positively to evidence-based interventions, whereas others may not benefit at all (Schwartz et al., 2016). Several factors may contribute to this including a student's perceived value of a particular learning task along with their attitudes toward learning. For example, a study by Selvig et al. (2015) reports that learner success in a medical histology course was positively associated with students' perception that histology was relevant to their future career. Similarly, an undergraduate life science student may put minimal effort into engaging with a biochemistry lab assignment that was designed to support deep learning if they feel it has little relevancy to their future aspirations. Thus, even though the student completes the assignment and demonstrates the acquisition of basic knowledge (e.g., recall of procedural steps), it is possible that they have developed little conceptual understanding that can be transferred to a future problem-solving task. It is also suggested that a student's motivation to learn may also be impacted by their previous learning experiences. As an example, some students have a less than optimal learning experience when they are first introduced to a particular discipline, and then they carry these negative perceptions with them as they continue along their academic journey. As a teacher of a first-year undergraduate anatomy course, I have heard first-hand from several students who are having difficulty that they have "never been good" at memorizing facts or that they previously struggled with similar content in high school. Thus, despite designing instruction that emphasizes mechanisms over rote memorization, some learners may need additional encouragement or motivation to seek alternative study strategies (see examples in Dunlosky et al., 2013) in order to recognize their potential to do well in a particular course. This is also an example of a student-teacher interaction that can positively impact learning. Teachers can play a very important role in helping learners to recognize that their intelligence is not fixed and that both effort and strategic selection of evidence-based learning strategies can help to facilitate learning (Park et al., 2016). In addition, research has also shown that when teachers emphasize mastery and conceptual understanding over performance (e.g., posting or rewarding highest grades), a student's motivation and learning behavior can be positively impacted (Young, 1997). Considering that these contextual factors can greatly influence student learning, educators and curriculum designers should be encouraged to reflect on their current instructional and assessment practices to consider how they can better support all learners in their classroom.

18.12 Environmental Factors

The context of where learning is situated also can have important implications on learning and the development of adaptive expertise. This can be especially relevant for biomedical and life science courses that typically involve hands-on learning and technical skills training. While instructional strategies that support the development of conceptual knowledge can easily be applied in traditional learning settings (e.g., small or large classrooms), laboratories and authentic learning environments may also afford learners with the unique opportunity to embrace struggle and provide exposure to meaningful variation (Mylopoulos et al., 2018). Further, these types of learning environments are also easily conducive to active learning activities which can promote knowledge synthesis and analysis, along with the generation of new knowledge, in response to novel and complex learning tasks assigned (Mylopoulos et al., 2018). Within the learning environment itself, the educator also plays a critical role in fostering the development of key capabilities of adaptive expertise, including preparation for future learning (Auerbach et al., 2020). This can involve creating an environment that purposely supports failure and curiosity while also ensuring that curricular activities are guided appropriately to support deep learning. As discussed in Chap. 6 in this volume and eloquently stated by Kapur (2016), there is a large design space that exists between the two extremes of heavily guided instruction and unguided problem-solving, and it is suggested that both educators and curriculum designers should capitalize on this space to design instruction that supports optimal learning.

18.13 Looking to the Future: Research and Reform

Research in higher education, professional education, and science education across the continuum demonstrates that carefully constructed educational experiences can lay the foundation for future learning and future expertise. In this chapter, we have provided just a snapshot of the potential for harnessing formal education to amplify basic cognitive mechanisms to support the development of adaptive expertise. Drawing on lessons from cognitive psychology and the learning sciences, educators in the biomedical and life sciences can use their time in the classroom and during assessments to foster cognitive integration of clinical concepts and underlying basic science mechanisms. There are many ways in which teachers can provide space for struggle (and failure) that will ultimately support longer-term success outside of the classroom. Ideally, any emphasis on learner cognition should be in service of the broader goal of preparing students to continue to learn long after the course is over. This opens doors for a variety of educational innovations that, if designed well, transform classroom instruction into basic building blocks for expert development. However, in acknowledging the value of taking a cognition-focused approach to fostering expertise, it is important to also note that neither education nor expertise and cognition exist in a vacuum. Rather, any attempt at creating customized educational experiences in the biomedical and life sciences must account for the meaningful interaction between cognition and educational context. We have described environmental factors that might impact implementation and uptake; however, there remain several key points for consideration before putting these cognitive mechanisms and classroom innovations into practice.

First, curriculum and assessment methods must be aligned. When modifying or designing new curricula, it is imperative that learning objectives, instructional approaches, and assessment practices reinforce one another. By ensuring that this alignment exists, it is more likely that the curriculum will better support learners in achieving intended outcomes (Kulasegaram & Rangachari, 2018; Squires, 2012). As discussed earlier, it is also important to consider how assessments can be used to support learning and the achievement of desired outcomes, rather than only as measures of success of learning. Decades of research have shown that formative assessments, or assessments *for* learning, can result in significant learning gains, especially in low-achieving students (Black & Wiliam, 1998; see Chap. 9 on feedback in this volume). The feedback provided through formative assessments can also be used by learners to identify and address gaps in their existing knowledge while simultaneously supporting the creation of personalized learning plans that can be utilized by each individual student to guide their future learning. For example, a program could choose to implement a programmatic assessment model to guide assessment practices across a curriculum (Schuwirth & Van der Vleuten, 2011) that would include frequent formative assessment and meaningful feedback with the aim of a more holistic view of student performance (Heeneman et al., 2015; Swan Sein et al., 2020). Importantly, the ongoing development of adaptive expertise can also be fostered through the implementation of assessments that challenge learners to innovate or generate new knowledge through novel problem-solving activities or transfer tasks (Mylopoulos & Woods, 2009). Thus, the inclusion and alignment of such assessments within and across the curriculum will also help to ensure that intended outcomes of the learner are maximally achieved.

Second, it is also important to acknowledge that change or reform to any type of curriculum will inevitably come with a cost and will require a commitment from all those involved, including students, faculty, and administration. Financial support and dedicated time for faculty development are two critical factors necessary for successful change to occur. As highlighted in a recent article by Hendrick and Heal (2020), most educators likely require specific training that demonstrates how evidence-based learning approaches that support deep learning can be applied into teaching practice. Further, ongoing professional development is also necessary to address perpetuating educational myths (e.g., the supposed efficacy of using learning styles for teaching) that are widespread across teaching professions and consequently impacting the delivery of day-to-day instruction (de Bruin, 2020; Howard-Jones et al., 2020). As suggested here, transformative change to any curriculum will require an investment beyond simply changing the modality by which instruction is delivered.

Third, the success of curricular reform and innovation in education depends on supportive academic structures that value high-quality teaching and create space for faculty to incorporate education science in their instruction and assessments. Faculty development and program evaluation are particularly important considerations. Well-resourced faculty development helps foster faculty buy-in as well as provide training and resources to support the implementation of new curricular activities. Similarly, program evaluation that considers the alignment between learning processes and intended program outcomes is vital for ensuring a cycle of continuous quality improvement (Kulasegaram et al., 2018). For example, a well-designed program evaluation could reveal potential challenges students might be facing with intentional experiences of productive struggle and frequent formative assessments, leading to changes in how a program might provide learner support and orientation. Similar to the academic structures needed to support faculty, systems must be in place to ensure the learning environment offers suitable supports for learner wellness and students in difficulty. Building a curriculum that pushes learners to work at the edge of their abilities and experience intentional struggle (and possibly failure) requires creating space for students to struggle safely, to develop new study habits, and to reorient their learning objectives (Paul et al., 2009; Valentine & Bolyard, 2018).

Strategies, mechanisms, and approaches outlined in this chapter are a starting point for educators in the biomedical and life sciences; they are intended to introduce evidence-based educational innovation. Our focus has been primarily on evidence from the cognitive and learning sciences, with recognition that education is a rich interdisciplinary field. Educators will find additional relevant perspectives on education in life sciences in other dominant scientific areas within education research. Perhaps most notably, these include sociocultural theories of learning that further explore learning in context through participation and critical theory that emphasizes revealing and disrupting oppressive power structures in society. These other literatures may provide an especially useful lens for future research examining the impact of technology on education in the biomedical and life sciences. While we have addressed many fundamental questions linked to learning and assessment, it is important to acknowledge that learning technologies will continue to play an integral role in transforming how education can be accessed and delivered. Many universities strive to keep up with the rapid pace of technological advances in medicine, science, and education. Today's educators have ample opportunity to explore and refine the use of simulation, augmented reality, virtual reality, and artificial intelligence in everyday teaching. However, the introduction of technology within a curriculum will not necessarily translate to more effective teaching and learning (Grainger et al., 2021). Future research must build on cognitive mechanisms outlined here and expand on sociocultural and critical dimensions of biomedical and life science education. The adoption or inclusion of technology will unquestionably create a rich context for deep exploration of the relationship between learners and their context and the unique ways in which power might operate in technology-enriched spaces (Selwyn & Facer, 2014).

References

Adesope, O. O., Trevisan, D. A., & Sundararajan, N. (2017). Rethinking the use of tests: A meta-analysis of practice testing. *Review of Educational Research, 87*(3), 659–701.

Agarwal, P. K. (2019). Retrieval practice & Bloom's taxonomy: Do students need fact knowledge before higher order learning? *Journal of Educational Psychology, 111*(2), 189–209.

Association of Faculties of Medicine of Canada. (2010). *The future of medical education in Canada (FMEC): A collective vision for MD education.* https://afmc.ca/sites/default/files/pdf/2010-FMEC-MD_EN.pdf. Accessed 3 Nov 2020.

Auerbach, L., Santen, S. A., Cutrer, W. B., Daniel, M., Wilson-Delfosse, A. L., & Roberts, N. K. (2020). The educators' experience: Learning environments that support the master adaptive learning. *Medical Teacher, 42*(11), 1270–1274.

Baghdady, M., Carnahan, H., Lam, E. W. N., & Woods, N. N. (2013). Integration of basic sciences and clinical sciences in oral radiology education for dental students. *Journal of Dental Education, 77*(6), 757–763.

Baghdady, M., Carnahan, H., Lam, E. W. N., & Woods, N. N. (2014). Test-enhanced learning and its effect on comprehension and diagnostic accuracy. *Medical Education, 48*(2), 181–188.

Benner, P., Sutphen, M., Leonard, V., & Day, L. (2010). *Educating nurses: A call for radical transformation*. The Carnegie Foundation for the Advancement of Teaching.

Bereiter, C., & Scardamalia, M. (1993). *Surpassing ourselves: An inquiry into the nature and implications of expertise*. Open Court.

Black, P., & Wiliam, D. (1998). Inside the black box: Raising standards through classroom assessment. *Phi Delta Kappan, 80*(20), 139–148.

Bransford, J. D., Brown, A. L., & Cocking, R. R. (2000). *How people learn: Brain, mind, experience, and school*. National Academy Press.

Bransford, J. D., & Schwartz, D. L. (1999). Rethinking transfer: A simple proposal with multiple implications. *Review of Research in Education, 24*(1), 61–100.

Brydges, R., Tran, J., Goffi, A., Lee, C., Miller, D., & Mylopoulos, M. (2020). Resident learning trajectories in the workplace: A self-regulated learning analysis. *Medical Education, 54*(12), 1120–1128.

Chamberland, M., Mamede, S., St-Onge, C., Rivard, M.-A., Setrakian, J., Lévesque, A., Lanthier, L., Schmidt, H. G., & Rikers, R. M. J. P. (2013). Students' self-explanations while solving unfamiliar cases: The role of biomedical knowledge. *Medical Education, 47*(11), 1109–1116.

Chamberland, M., Mamede, S., St-Onge, C., Setrakian, J., Bergeron, L., & Schmidt, H. (2015a). Self- explanation in learning clinical reasoning: The added value of examples and prompts. *Medical Education, 49*(2), 193–202.

Chamberland, M., Mamede, S., St-Onge, C., Setrakian, J., & Schmidt, H. G. (2015b). Does medical students' diagnostic performance improve by observing examples of self-explanation provided by peers or experts. *Advances in Health Sciences Education, 20*(4), 981–993.

Chamberland, M., St-Onge, C., Setrakian, J., Lantheir, L., Bergeron, L., Bourget, A., Mamed, S., Schmidt, H., & Rikers, R. (2011). The influence of medical students' self-explanations on diagnostic performance. *Medical Education, 45*(7), 688–695.

Chaudhary, Z. K., Mylopoulos, M., Barnett, R., Sockalingam, S., Hawkins, M., O'Brien, J. D., & Woods, N. N. (2019). Reconsidering basic: Integrating social and behavioral sciences to support learning. *Academic Medicine, 94*(11S), S73–S78.

Cheung, J. J. H., Kulasegaram, K. M., Woods, N. N., Moulton, C.-A., Ringsted, C. V., & Brydges, R. (2018). Knowing how and knowing why: Testing the effect of instruction designed for cognitive integration on procedural skills transfer. *Advances in Health Sciences Education, 23*(1), 61–74.

Chi, M. T., Leeuw, N. D., Chiu, M.-H., & LaVancher, C. (1994). Eliciting self-explanations improves understanding. *Cognitive Science, 18*(3), 439–477.

Chi, M. T. H. (2000). Self-explaining expository texts: The dual process of generating inferences and repairing mental models. In R. Glaser (Ed.), *Advances in instructional psychology* (pp. 161–238). Lawrence Erlbaum Associates.

College of Nurses of Ontario. (2018). *Entry-to-practice competencies for registered nurses*. https://www.cno.org/globalassets/docs/reg/41037-entry-to-practice-competencies-2020.pdf. Accessed 27 Nov 2020.

de Bruin, A. B. H. (2020). Debunking myths in medical education: The science of refutation. *Medical Education, 54*(1), 6–8.

Dean, D., & Kuhn, D. (2006). Direction instruction vs. discovery: The long view. *Science Education, 91*, 384–397.

DeCaro, M. S., & Rittle-Johnson, B. (2012). Exploring mathematics problems prepares children to learn from instruction. *Journal of Experimental Child Psychology, 113*(4), 552–568.

Derry, S. J., Hmelo-Silver, C. E., Nagarajan, A., Chemobilsky, E., & Beitzel, B. (2006). Cognitive transfer revisited: Can we exploit new media to solve old problems on a large scale? *Journal of Educational Computing Research, 35*(2), 145–162.

Dobson, J., Linderholm, T., & Perez, J. (2018). Retrieval practice enhances the ability to evaluate complex physiology information. *Medical Education, 52*(5), 513–525.

Dochy, F., Segers, M., Van den Bossche, P., & Gijebs, D. (2003). Effects of problem-based learning: A meta-analysis. *Learning and Instruction, 13*(5), 533–568.

Dunlosky, J., Rawson, K. A., Marsh, E. J., Nathan, M. J., & Willingham, D. T. (2013). Improving students' learning with effective learning techniques: Promising directions from cognitive and educational psychology. *Psychological Science in the Public Interest, 14*(1), 4–58.

Finnerty, E. P., Chauvin, S., Bonaminio, G., Andrews, M., Carroll, R. G., & Pangaro, L. N. (2010). Flexner revisited: The role and value of the basic sciences in medical education. *Academic Medicine, 85*(2), 349–355.

Frank, J. R., Snell, L., & Sherbino, J. eds. (2015). *CanMEDS 2015 physician competency framework.* Royal College of Physicians and Surgeons of Canada. http://canmeds.royalcollege.ca/uploads/en/framework/CanMEDS%202015%20Framework_EN_Reduced.pdf. Accessed 27 Nov 2020.

Goldszmidt, M., Minda, J. P., Devantier, S. L., Skye, A. L., & Woods, N. N. (2012). Expanding the basic science debate: The role of physics knowledge in interpreting clinical findings. *Advances in Health Sciences Education, 17*(4), 547–555.

Grainger, R., Liu, Q., & Geertshuis, S. (2021). Learning technologies: A medium for the transformation of medical education? *Medical Education, 55*(1), 23–29.

Heeneman, S., Oudkerk, P. A., Schuwirth, L. W. T., van der Vleuten, C. P. M., & Driessen, E. W. (2015). The impact of programmatic assessment on student learning: Theory versus practice. *Medical Education, 49*(5), 487–498.

Hendrick, C., & Heal, J. (2020). *Just because they are engaged, it doesn't mean that they are learning*. Retrieved March 1, 2021, from https://impact.chartered.college/article/just-because-theyre-engaged-doesnt-mean-learning/

Hmelo, C. E. (1998). Problem-based learning: Effects on the early acquisition of cognitive skill in medicine. *Journal of the Learning Sciences, 7*(2), 173–236.

Howard-Jones, P. A., Jay, T., & Galeano, L. (2020). Professional development on the science of learning and teachers' performative thinking – A pilot study. *Mind, Brain, and Education, 14*(3), 267–278.

Irby, D. M., Cooke, M., & O'Brien, B. C. (2010). Calls for reform of medical education by the Carnegie Foundation for the Advancement of Teaching: 1910 and 2010. *Academic Medicine, 85*(2), 220–227.

Kalyuga, S., & Renkl, A. (2010). Expertise reversal effect and its instructional implications: Introduction to the special issue. *Instructional Science, 38*(3), 209–215.

Kalyuga, S., Rikers, R., & Paas, F. (2012). Educational implications of expertise reversal effects in learning and performance of complex cognitive and sensorimotor skills. *Educational Psychology Review, 24*(2), 313–337.

Kapur, M. (2008). Productive failure. *Cognition and Instruction, 26*(3), 379–424.

Kapur, M. (2014). Productive failure in learning math. *Cognitive Science, 38*(5), 1008–1022.

Kapur, M. (2016). Examining productive failure, productive success, unproductive failure, and unproductive success in learning. *Educational Psychologist, 51*(2), 289–299.

Kapur, M., & Bielaczyc, K. (2012). Designing for productive failure. *Journal of the Learning Sciences, 21*(1), 45–83.

Koens, F., Custers, E. J. F. M., & ten Cate, O. T. J. (2006). Clinical and basic science teachers' opinions about the required depth of biomedical knowledge for medical students. *Medical Teacher, 28*(3), 234–238.

Koens, F., Rademakers, J. J. D. J. M., & ten Cate, O. T. J. (2005). Validation of core medical knowledge by postgraduates and specialists. *Medical Education, 39*(9), 911–917.

Kromann, C. B., Jensen, M. L., & Ringsted, C. (2009). The effect of testing on skills learning. *Medical Education, 43*(1), 21–27.

Kulasegaram, K., Myolopoulos, M., Tonin, P., Bernstein, S., Bryden, P., Law, M., Lazor, J., Pittini, R., Sockalingam, S., Tait, G. R., & Houston, P. (2018). The alignment imperative in curriculum renewal. *Medical Teacher, 40*(5), 443–448.

Kulasegaram, K. M., Manzone, J. C., Ku, C., Skye, A., Wadey, V., & Woods, N. N. (2015). Cause and effect: Testing a mechanism for cognitive integration of basic science. *Academic Medicine, 90*, S63–S69.

Kulasegaram, K. M., Martimianakis, M. A., Mylopoulos, M., Whitehead, C. R., & Woods, N. N. (2013). Cognition before curriculum: Rethinking the integration of basic science and clinical learning. *Academic Medicine, 88*(10), 1578–1585.

Kulasegaram, K. M., & Rangachari, P. K. (2018). Beyond "formative": Assessment to enrich student learning. *Advances in Physiology Education, 42*(1), 5–14.

Larsen, D. P., Butler, A. C., Lawson, A. L., & Roediger, H. L., III. (2013a). The importance of seeing the patient: Test-enhanced learning with standardized patients and written tests improves clinical application of knowledge. *Advances in Health Sciences Education, 18*(3), 409–425.

Larsen, D. P., Butler, A. C., & Roediger, H. L., III. (2013b). Comparative effects of test-enhanced learning and self-explanation on long-term retention. *Medical Education, 47*(7), 674–682.

Leppink, J., Broers, N. J., Imbos, T., van der Vleuten, C. P. M., & Berger, M. P. F. (2012). Prior knowledge moderates instructional effects on conceptual understanding of statistics. *Educational Research and Evaluation, 18*(1), 37–51.

Lisk, K., Agur, A. M. R., & Woods, N. N. (2016). Exploring cognitive integration of basic science and its effect on diagnostic reasoning in novices. *Perspectives on Medical Education, 5*(3), 147–153.

Lisk, K., Agur, A. M. R., & Woods, N. N. (2017). Examining the effect of self-explanation on cognitive integration of basic and clinical sciences in novices. *Advances in Health Sciences Education, 22*(5), 1071–1083.

Ludmerer, K. M. (1999). *Time to Heal: American medical education from the turn of the century to the era of managed care*. Oxford University Press.

Murphy, G. L., & Medin, D. L. (1985). The role of theories in conceptual coherence. *Psychology Review, 92*(3), 289–314.

Mylopoulos, M., Brydges, R., Woods, N. N., Manzone, J., & Schwartz, D. L. (2016). Preparation for future learning: A missing competency in health professions education? *Medical Education., 50*(1), 115–123.

Mylopoulos, M., Kulasegaram, M., & Woods, N. N. (2018). Developing the experts we need: Fostering adaptive expertise through education. *Journal of Evaluation in Clinical Practice., 24*(3), 674–677.

Mylopoulos, M., & Woods, N. N. (2009). Having our cake and eating it too. Seeking the best of worlds in expertise research. *Medical Education, 43*(5), 406–413.

Mylopoulos, M., & Woods, N. N. (2014). Preparing medical students for future learning using basic science instruction. *Medical Education, 48*(7), 667–673.

Park, D., Gunderson, E. A., Tsukayama, E., Levine, S. C., & Beilock, S. L. (2016). Young children's motivational frameworks and math achievement: Relation to teacher-reported instructional practices, but not teach theory of intelligence. *Journal of Education Psychology, 108*(3), 200–313.

Patel, V. L., & Groen, G. J. (1986). Knowledge based solution strategies in medical reasoning. *Cognitive Science., 10*(1), 91–116.

Patel, V. L., Groen, G. J., & Scott, H. M. (1988). Biomedical knowledge in explanations of clinical problems by medical students. *Medical Education, 22*(5), 398–406.

Paul, G., Hinman, G., Dottl, S., & Passon, J. (2009). Academic development: A survey of academic difficulties experienced by medical students and support services provided. *Teaching and Learning in Medicine, 21*(3), 254–260.

Peixoto, J. M., Mamede, S., de Faria, R. M. D., Moura, A. S., Santos, S. M. E., & Schmidt, H. G. (2017). The effect of self-explanation of pathophysiological mechanisms of diseases on medical students' diagnostic performance. *Advances in Health Sciences Education, 22*(5), 1183–1197.

Postsecondary Education Quality Assessment Board. (2020). *Manual for public organizations (including ontario colleges)*. http://www.peqab.ca/Publications/Handbooks%20Guidelines/MANUAL_COLLEGES_PUBLICS%20November2020.pdf. Accessed 1 Oct 2020.

Rey, G. D., & Fischer, A. (2013). The expertise reversal effect concerning instructional explanations. *Instructional Science, 41*, 407–429.

Roediger, H. L., & Karpicke, J. D. (2006). The power of testing memory: Basic research and implications for educational practice. *Perspectives on Psychological Science, 1*(3), 181–276.

Schuwirth, L. W. T., & Van der Vleuten, C. P. M. (2011). Programmatic assessment: From assessment of learning to assessment for learning. *Medical Teacher, 33*(6), 478–485.

Schwartz, D., & Bransford, J. D. (1998). A time for telling. *Cognition and Instruction, 16*(4), 475–5223.

Schwartz, D. L., Bransford, J. D., & Sears, D. (2005). Efficiency and innovation in transfer. In J. P. Mestre (Ed.), *Transfer of learning from a modern multidisciplinary perspective* (pp. 1–52). Information Age Publishing.

Schwartz, D. L., Chase, C. C., Oppezzo, M. A., & Chin, D. B. (2011). Practicing versus inventing with contrasting cases: The effects of telling first on learning and transfer. *Journal of Educational Psychology, 103*(4), 759–775.

Schwartz, D. L., Cheng, K. M., Salehi, S., & Wieman, C. (2016). The half empty questions for socio-cognitive interventions. *Journal of Educational Psychology, 108*(3), 397–404.

Schwartz, D. L., & Martin, T. (2004). Inventing to prepare for future learning: The hidden efficiency of encouraging original student production in statistics instruction. *Cognition and Instruction, 22*(2), 129–184.

Selvig, D., Holaday, L. W., Purkiss, J., & Hortch, M. (2015). Correlating students' educational background, study habits, and resource usage with learning success in medical histology. *Anatomical Sciences Education, 8*(1), 1–11.

Selwyn, N., & Facer, K. (2014). The sociology of education and digital technology: Past, present and future. *Oxford Review of Education, 40*(4), 482–496.

Squires, D. (2012). Curriculum alignment research suggests that alignment can improve student achievement. *The Clearing House: A Journal of Educational Strategies, Issues and Ideas, 85*(4), 129–135.

Steenhof, N., Van Woods, N. N., Gerven, P. W. M., & Mylopoulos, M. (2019). Productive failure as an instructional approach to promote future learning. *Advances in Health Sciences Education, 24*(4), 739–749.

Swan Sein, A., Rashid, H., Meka, J., Amiel, J., & Pluta, W. (2020). Twelve tips for embedding assessment for and as learning practices in a programmatic assessment system. *Medical Teacher, 43*(3), 300–306.

The Quality Assurance Agency for Higher Education. (2019). *Subject benchmark statement for biomedical sciences.* https://www.qaa.ac.uk/docs/qaa/subject-benchmark-statements/subject-benchmark-statement-biomedical-sciences.pdf?sfvrsn=2bf2c881_12. Accessed 3 Nov 2020.

Valentine, K. D., & Bolyard, J. J. (2018). *Creating a classroom culture that supports productive struggle: Pre-service teachers' reflections on teaching mathematics.*

van Gog, T., & Rummel, N. (2010). Example-based learning: Integrating cognitive and social cognitive research perspectives. *Educational Psychology Review, 22*(2), 155–174.

Versteeg, M., van Blankenstein, F. M., Putter, H., & Steendijk, P. (2019). Peer instruction improves comprehension and transfer of physiological concepts: A randomized comparison with self-explanation. *Advances in Health Sciences Education, 24*(1), 151–165.

Whitehead, C. R. (2013). Scientist or science-stuffed? Discourses of science in north American medical education. *Medical Education, 47*(1), 26–32.

Whitehead, C. R. (2017). Getting off the carousel: de-centring the curriculum in medical education. *Perspectives in Medical Education, 6*(5), 283–285.

Whitehead, C. R., Hodges, B. D., & Austin, Z. (2013). Captive on a carousel: Discourses of new in medical education 1910-2010. *Advances in Health Sciences Education, 18*(4), 755–768.

Woods, N. N. (2007). Science is fundamental: The role of biomedical knowledge in clinical reasoning. *Medical Education, 41*(12), 1173–1177.

Woods, N. N., Brooks, L. R., & Norman, G. R. (2005). The value of basic science in clinical diagnosis: Creating coherence among signs and symptoms. *Medical Education, 39*(1), 107–112.

Woods, N. N., Brooks, L. R., & Norman, G. R. (2007). The role of biomedical knowledge in diagnosis of difficult clinical cases. *Advances in Health Sciences Education, 12*(4), 417–426.

Woods, N. N., Neville, A. J., Levinson, A. J., Howey, E. H., Oczkowski, W. J., & Norman, G. R. (2006). The value of basic science in clinical diagnosis. *Academic Medicine., 81*(10S), S124–S127.

Young, A. J. (1997). I think, therefore I'm motivated: The relations among cognitive strategy use, motivational orientation and classroom perceptions over time. *Learning and Individual Differences, 9*(3), 249–283.

www.ingramcontent.com/pod-product-compliance
Lightning Source LLC
Chambersburg PA
CBHW060744150926
51501CB00003B/85
* 9 7 8 3 0 3 0 9 5 6 3 2 5 *